T0325804

Random Processes for Engineers

This engaging introduction to random processes provides students with the critical tools needed to design and evaluate engineering systems that must operate reliably in uncertain environments.

A brief review of probability theory and real analysis of deterministic functions sets the stage for understanding random processes, while the underlying measure theoretic notions are explained in an intuitive, straightforward style. Students will learn to manage the complexity of randomness through the use of simple classes of random processes, statistical means and correlations, asymptotic analysis, sampling, and effective algorithms. Key topics covered include:

- Calculus of random processes in linear systems
- Kalman and Wiener filtering
- Hidden Markov models for statistical inference
- The estimation maximization (EM) algorithm
- An introduction to martingales and concentration inequalities

Understanding of the key concepts is reinforced through more than 100 worked examples and 300 thoroughly tested homework problems (half of which are solved in detail at the end of the book, with the remaining solutions available online for instructors at *www.cambridge.org/hajek*).

Bruce Hajek has been an avid student, instructor, and user of probability theory for his entire career. He is the Mary Lou and Leonard C. Hoeft Chair of Engineering, Center for Advanced Study Professor of Electrical and Computer Engineering, and Professor in the Coordinated Science Laboratory at the University of Illinois. Among his many awards, he is a member of the US National Academy of Engineering and a recipient of the IEEE Koji Kobayashi Computers and Communications Award. He is co-author, with E. Wong, of the more advanced book, *Stochastic Processes in Engineering Systems*, 2nd edn, 1985.

"A comprehensive exposition of random processes ... Abstract concepts are nicely explained through many examples ... The book will be very helpful for beginning graduate students who want a firm foundational understanding of random processes. It will also serve as a nice reference for the advanced reader."

Anima Anandkumar, The University of California, Irvine

"This is a fantastic book from one of the eminent experts in the field and is the standard text for the graduate class I teach. The material covered is perfect for a first-year graduate class in Probability and Stochastic Processes."

Sanjay Shakkottai, The University of Texas at Austin

"This is an excellent introductory book on random processes and basic estimation theory from the foremost expert and is suitable for advanced undergraduate students and/or first-year graduate students who are interested in stochastic analysis. It covers an extensive set of topics that are very much applicable to a wide range of engineering fields."

Richard La, University of Maryland

"Bruce Hajek has created a textbook for engineering students with interest in control, signal processing, communications, machine learning, amongst other disciplines in electrical engineering and computer science. Anyone who knows Bruce Hajek knows that he cares deeply about the foundations of probability and statistics, and he is equally engaged in applications. Bruce is a dedicated teacher and author in the spirit of Prof. Joe Doob, formally at the statistics department at the University of Illinois.

I was fortunate to have a mature draft of his book when I introduced a stochastic processes course to my department in the spring of 2014. The book provides an entirely accessible introduction to the foundations of stochastic processes. I was surprised to find that the students in my course enjoyed Hajek's introduction to measure theory, and (at least by the end of the course) could appreciate the value of the abstract concepts introduced at the start of the text.

It includes applications of this general theory to many topics that are of tremendous interest to students and practitioners, such as nonlinear filtering, statistical methods such as the EM-algorithm, and stability theory for Markov processes. Because the book establishes strong foundations, in a course it is not difficult to substitute other applications, such as Monte-Carlo methods or reinforcement learning. Graduate students will be thrilled to learn these exciting techniques from an accessible source."

Sean Meyn, University of Florida

Random Processes for Engineers

BRUCE HAJEK

University of Illinois

CAMBRIDGE
UNIVERSITY PRESS

University Printing House, Cambridge CB2 8BS, United Kingdom

One Liberty Plaza, 20th Floor, New York, NY 10006, USA

477 Williamstown Road, Port Melbourne, VIC 3207, Australia

314-321, 3rd Floor, Plot 3, Splendor Forum, Jasola District Centre, New Delhi - 110025, India

103 Penang Road, #05-06/07, Visioncrest Commercial, Singapore 238467

Cambridge University Press is part of the University of Cambridge.

It furthers the University's mission by disseminating knowledge in the pursuit of education, learning and research at the highest international levels of excellence.

www.cambridge.org
Information on this title: www.cambridge.org/9781107100121

First published 2015

A catalogue record for this publication is available from the British Library

Library of Congress Cataloging in Publication data
Hajek, Bruce.
Random processes for engineers / Bruce Hajek, Illinois.
Includes bibliographical references and index.
ISBN 978-1-107-10012-1 (Hardback)
1. Engineering mathematics–Study and teaching (Graduate) 2. Stochastic processes–Study and teaching (Graduate) 3. Numbers, Random–Study and teaching (Graduate) I. Title.
TA340.H325 2015
519.2′3–dc23 2014035900

ISBN 978-1-107-10012-1 Hardback

Additional resources for this publication at www.cambridge.org/hajek

To Beth, for her loving support.

Contents

Preface

From an applications viewpoint, the main reason to study the subject of this book is to help deal with the complexity of describing random, time-varying functions. A random variable can be interpreted as the result of a single measurement. The distribution of a single random variable is fairly simple to describe. It is completely specified by the cumulative distribution function $F(x)$, a function of one variable. It is relatively easy to approximately represent a cumulative distribution function on a computer. The joint distribution of several random variables is much more complex, for in general it is described by a joint cumulative probability distribution function, $F(x_1, x_2, \ldots, x_n)$, which is much more complicated than n functions of one variable. A random process, for example a model of time-varying fading in a communication channel, involves many, possibly infinitely many (one for each time instant t within an observation interval) random variables. Woe the complexity!

This book helps prepare the reader to understand and use the following methods for dealing with the complexity of random processes:

- Work with moments, such as means and covariances.
- Use extensively processes with special properties. Most notably, Gaussian processes are characterized entirely by means and covariances, Markov processes are characterized by one-step transition probabilities or transition rates, and initial distributions. Independent increment processes are characterized by the distributions of single increments.
- Appeal to models or approximations based on limit theorems for reduced complexity descriptions, especially in connection with averages of independent, identically distributed random variables. The law of large numbers tells us, in a certain sense, that a probability distribution can be characterized by its mean alone. The central limit theorem similarly tells us that a probability distribution can be characterized by its mean and variance. These limit theorems are analogous to, and in fact examples of, perhaps the most powerful tool ever discovered for dealing with the complexity of functions: Taylor's theorem, in which a function in a small interval can be approximated using its value and a small number of derivatives at a single point.
- Diagonalize. A change of coordinates reduces an arbitrary n-dimensional Gaussian vector into a Gaussian vector with n independent coordinates. In the new coordinates the joint probability distribution is the product of n one-dimensional

distributions, representing a great reduction of complexity. Similarly, a random process on an interval of time is diagonalized by the Karhunen–Loève representation. A periodic random process is diagonalized by a Fourier series representation. Stationary random processes are diagonalized by Fourier transforms.

- Sample. A narrowband continuous time random process can be exactly represented by its samples taken with sampling rate twice the highest frequency of the random process. The samples offer a reduced complexity representation of the original process.
- Work with baseband equivalent. The range of frequencies in a typical wireless transmission is much smaller than the center frequency, or carrier frequency, of the transmission. The signal could be represented directly by sampling at twice the largest frequency component. However, the sampling frequency, and hence the complexity, can be dramatically reduced by sampling a baseband equivalent random process.

This book was written for the first semester graduate course on random processes offered by the Department of Electrical and Computer Engineering at the University of Illinois at Urbana-Champaign. Students in the class are assumed to have had a previous course in probability, which is briefly reviewed in the first chapter. Students are also expected to have some familiarity with real analysis and elementary linear algebra, such as the notions of limits, definitions of derivatives, Riemann integration, and diagonalization of symmetric matrices. These topics are reviewed in the appendix. Finally, students are expected to have some familiarity with transform methods and complex analysis, though the concepts used are reviewed in the relevant chapters.

Each chapter represents roughly two weeks of lectures, and includes homework problems. Solutions to the even numbered problems without stars can be found at the end of the book. Students are encouraged to first read a chapter, then try doing the even numbered problems before looking at the solutions. Problems with stars, for the most part, investigate additional theoretical issues, and solutions are not provided.

Hopefully some students reading this book will find the problems useful for understanding the diverse technical literature on systems engineering, ranging from control systems, signal and image processing, communication theory, and analysis of a variety of networks and algorithms. Hopefully some students will go on to design systems, and define and analyze stochastic models. Hopefully others will be motivated to continue study in probability theory, going on to learn measure theory and its applications to probability and analysis in general.

A brief comment is in order on the level of rigor and generality at which this book is written. Engineers and scientists have great intuition and ingenuity, and routinely use methods that are not typically taught in undergraduate mathematics courses. For example, engineers generally have good experience and intuition about transforms, such as Fourier transforms, Fourier series, and z-transforms, and some associated methods of complex analysis. In addition, they routinely use generalized functions, in particular the delta function is frequently used. The use of these concepts in this book leverages on this knowledge, and it is consistent with mathematical definitions, but full mathematical justification is not given in every instance. The mathematical background required for a full mathematically rigorous treatment of the material in this book is roughly at the level of a second year graduate course in measure theoretic probability, pursued after a course on measure theory.

The author gratefully acknowledges the many students and faculty members, including Todd Coleman, Christoforos Hadjicostis, Jonathan Ligo, Andrew Singer, R. Srikant, and Venu Veeravalli who gave many helpful comments and suggestions.

Organization

The first four chapters of the book are used heavily in the remaining chapters, so most readers should cover those chapters before moving on.

Chapter 1 is meant primarily as a review of concepts found in a typical first course on probability theory, with an emphasis on axioms and the definition of expectation. Readers desiring a more extensive review of basic probability are referred to the author's notes for ECE 313 at the University of Illinois.

Chapter 2 focuses on various ways in which a sequence of random variables can converge, and the basic limit theorems of probability: law of large numbers, central limit theorem, and the asymptotic behavior of large deviations.

Chapter 3 focuses on minimum mean square error estimation and the orthogonality principle. Kalman filtering is explained from the geometric standpoint based on innovations sequences.

Chapter 4 introduces the notion of a random process, and briefly covers several key examples and classes of random processes. Markov processes and martingales are introduced in this chapter, but are covered in greater depth in later chapters.

After Chapter 4 is covered, the following four topics can be covered independently of each other.

Chapter 5 describes the use of Markov processes for modeling and statistical inference. Applications include natural language processing.

Chapter 6 describes the use of Markov processes for modeling and analysis of dynamical systems. Applications include modeling of queueing systems.

Chapters 7–9 develop calculus for random processes based on mean square convergence, moving to linear filtering, orthogonal expansions, and ending with causal and noncausal Wiener filtering.

Chapter 10 explores martingales with respect to filtrations, with emphasis on elementary concentration inequalities, and on the optional sampling theorem.

In recent one-semester course offerings, the author covered Chapters 1–5, Sections 6.1–6.8, Chapter 7, Sections 8.1–8.4, and Section 9.1. Time did not permit covering the Foster–Lyapunov stability criteria, noncausal Wiener filtering, and the chapter on martingales.

A number of background topics are covered in the appendix, including basic notation.

1 A selective review of basic probability

This chapter reviews many of the main concepts in a first level course on probability theory, with more emphasis on axioms and the definition of expectation than is typical of a first course.

1.1 The axioms of probability theory

Random processes are widely used to model systems in engineering and scientific applications. This book adopts the most widely used framework of probability and random processes, namely the one based on Kolmogorov's axioms of probability. The idea is to assume a mathematically solid definition of the model. This structure encourages a modeler to have a consistent, if not accurate, model.

A *probability space* is a triplet $(\Omega, \mathcal{F}, \mathcal{P})$. The first component, Ω, is a nonempty set. Each element ω of Ω is called an *outcome* and Ω is called the *sample space*. The second component, $\mathcal{F}$, is a set of subsets of Ω called *events*. The set of events $\mathcal{F}$ is assumed to be a σ-algebra, meaning it satisfies the following axioms (see Appendix 11.1 for set notation):

A.1 $\Omega \in \mathcal{F}$,

A.2 If $A \in \mathcal{F}$ then $A^c \in \mathcal{F}$,

A.3 If $A, B \in \mathcal{F}$ then $A \cup B \in \mathcal{F}$. Also, if $A_1, A_2, \ldots$ is a sequence of elements in $\mathcal{F}$ then $\bigcup_{i=1}^{\infty} A_i \in \mathcal{F}$.

If $A,\ B \in \mathcal{F}$, then $AB \in \mathcal{F}$ by A.2, A.3 and the fact $AB = (A^c \cup B^c)^c$. By the same reasoning, if $A_1, A_2, \ldots$ is a sequence of elements in a σ-algebra $\mathcal{F}$, then $\bigcap_{i=1}^{\infty} A_i \in \mathcal{F}$.

Events A_i, $i \in I$, indexed by a set I are called *mutually exclusive* if the intersection $A_iA_j = \emptyset$ for all $i, j \in I$ with $i \neq j$. The final component, P, of the triplet $(\Omega, \mathcal{F}, P)$ is a probability measure on $\mathcal{F}$ satisfying the following axioms:

P.1 $P(A) \geq 0$ for all $A \in \mathcal{F}$,

P.2 If $A, B \in \mathcal{F}$ and A and B are mutually exclusive, $P(A \cup B) = P(A) + P(B)$. Also, if $A_1, A_2, \ldots$ is a sequence of mutually exclusive events in $\mathcal{F}$ then $P\left(\bigcup_{i=1}^{\infty} A_i\right) = \sum_{i=1}^{\infty} P(A_i)$.

P.3 $P(\Omega) = 1$.

The axioms imply a host of properties including the following. For any subsets A, B, C of $\mathcal{F}$:

- If $A \subset B$ then $P(A) \leq P(B)$,
- $P(A \cup B) = P(A) + P(B) - P(AB)$,
- $P(A \cup B \cup C) = P(A) + P(B) + P(C) - P(AB) - P(AC) - P(BC) + P(ABC)$,
- $P(A) + P(A^c) = 1$,
- $P(\emptyset) = 0$.

Example 1.1 (Toss of a fair coin) Using "H" for "heads" and "T" for "tails," the toss of a fair coin is modeled by

$$\Omega = \{H, T\} \quad \mathcal{F} = \{\{H\}, \{T\}, \{H, T\}, \emptyset\}$$

$$P\{H\} = P\{T\} = \frac{1}{2} \quad P\{H, T\} = 1 \quad P(\emptyset) = 0.$$

Note that, for brevity, we omitted the parentheses and wrote $P\{H\}$ instead of $P(\{H\})$.

Example 1.2 (Standard unit-interval probability space) Take $\Omega = \{\omega : 0 \leq \omega \leq 1\}$. Imagine an experiment in which the outcome ω is drawn from Ω with no preference towards any subset. In particular, we want the set of events $\mathcal{F}$ to include intervals, and the probability of an interval $[a, b]$ with $0 \leq a \leq b \leq 1$ to be given by:

$$P(\ [a, b]\ ) = b - a. \tag{1.1}$$

Taking $a = b$, we see that $\mathcal{F}$ contains singleton sets $\{a\}$, and these sets have probability zero. Since $\mathcal{F}$ is to be a σ-algebra, it must also contain all the open intervals (a, b) in Ω, and for such an open interval, $P(\ (a, b)\ ) = b - a$. Any open subset of Ω is the union of a finite or countably infinite set of open intervals, so that $\mathcal{F}$ should contain all open and all closed subsets of Ω. Thus, $\mathcal{F}$ must contain any set that is the intersection of countably many open sets, the union of countably many such sets, and so on. The specification of the probability function P must be extended from intervals to all of $\mathcal{F}$. It is not a priori clear how large $\mathcal{F}$ can be. It is tempting to take $\mathcal{F}$ to be the set of all subsets of Ω. However, that idea doesn't work – see Problem 1.37 showing that the length of all subsets of $\mathbb{R}$ can't be defined in a consistent way. The problem is resolved by taking $\mathcal{F}$ to be the smallest σ-algebra containing all the subintervals of Ω, or equivalently, containing all the open subsets of Ω. This σ-algebra is called the Borel σ-algebra for $[0, 1]$, and the sets in it are called Borel sets. While not every subset of Ω is a Borel subset, any set we are likely to encounter in applications is a Borel set. The existence of the Borel σ-algebra is discussed in Problem 1.38. Furthermore, extension theorems of measure theory[1] imply that P can be extended from its definition (1.1) for interval sets to all Borel sets.

---

[1] See, for example, (Royden 1968) or (Varadhan 2001). The σ-algebra $\mathcal{F}$ can be extended somewhat further by requiring the following completeness property: if $B \subset A \in \mathcal{F}$ with $P(A) = 0$, then $B \in \mathcal{F}$ (and also $P(B) = 0$).

The smallest σ-algebra, $\mathcal{B}$, containing the open subsets of $\mathbb{R}$ is called the Borel σ-algebra for $\mathbb{R}$, and the sets in it are called *Borel subsets of* $\mathbb{R}$. Similarly, the Borel σ-algebra $\mathcal{B}^n$ of subsets of $\mathbb{R}^n$ is the smallest σ-algebra containing all sets of the form $[a_1, b_1] \times [a_2, b_2] \times \cdots \times [a_n, b_n]$. Sets in $\mathcal{B}^n$ are called *Borel subsets of* $\mathbb{R}^n$. The class of Borel sets includes not only rectangle sets and countable unions of rectangle sets, but all open sets and all closed sets. Virtually any subset of $\mathbb{R}^n$ arising in applications is a Borel set.

Example 1.3 (Repeated binary trials) Suppose we would like to represent an infinite sequence of binary observations, where each observation is a zero or one with equal probability. For example, the experiment could consist of repeatedly flipping a fair coin, and recording a one each time it shows heads and a zero each time it shows tails. Then an outcome ω would be an infinite sequence, $\omega = (\omega_1, \omega_2, \cdots)$, such that for each $i \geq 1$, $\omega_i \in \{0, 1\}$. Let Ω be the set of all such ωs. The set of events can be taken to be large enough so that any set that can be defined in terms of only finitely many of the observations is an event. In particular, for any binary sequence $(b_1, \cdots, b_n)$ of some finite length n, the set $\{\omega \in \Omega : \omega_i = b_i \text{ for } 1 \leq i \leq n\}$ should be in $\mathcal{F}$, and the probability of such a set is taken to be 2^{-n}.

There are also events that don't depend on a fixed, finite number of observations. For example, let F be the event that an even number of observations is needed until a one is observed. Show that F is an event and then find its probability.

Solution

For $k \geq 1$, let E_k be the event that the first one occurs on the kth observation. So $E_k = \{\omega : \omega_1 = \omega_2 = \cdots = \omega_{k-1} = 0 \text{ and } \omega_k = 1\}$. Then E_k depends on only a finite number of observations, so it is an event, and $P\{E_k\} = 2^{-k}$. Observe that $F = E_2 \cup E_4 \cup E_6 \cup \ldots$, so F is an event by Axiom A.3. Also, the events $E_2, E_4, \ldots$ are mutually exclusive, so by the full version of Axiom P.2:

$$P(F) = P(E_2) + P(E_4) + \cdots = \frac{1}{4}\left(1 + \left(\frac{1}{4}\right) + \left(\frac{1}{4}\right)^2 + \cdots\right) = \frac{\frac{1}{4}}{1 - \frac{1}{4}} = \frac{1}{3}.$$

The following lemma gives a continuity property of probability measures which is analogous to continuity of functions on $\mathbb{R}^n$, reviewed in Appendix 11.3. If $B_1, B_2, \ldots$ is a sequence of events such that $B_1 \subset B_2 \subset B_3 \subset \ldots$, then we can think that B_j converges to the set $\cup_{i=1}^{\infty} B_i$ as $j \to \infty$. The lemma states that in this case, $P(B_j)$ converges to the probability of the limit set as $j \to \infty$.

Lemma 1.1 *(Continuity of probability) Suppose $B_1, B_2, \ldots$ is a sequence of events.*

(a) *If $B_1 \subset B_2 \subset \cdots$ then $\lim_{j\to\infty} P(B_j) = P\left(\bigcup_{i=1}^{\infty} B_i\right)$.*

(b) *If $B_1 \supset B_2 \supset \cdots$ then $\lim_{j\to\infty} P(B_j) = P\left(\bigcap_{i=1}^{\infty} B_i\right)$.*

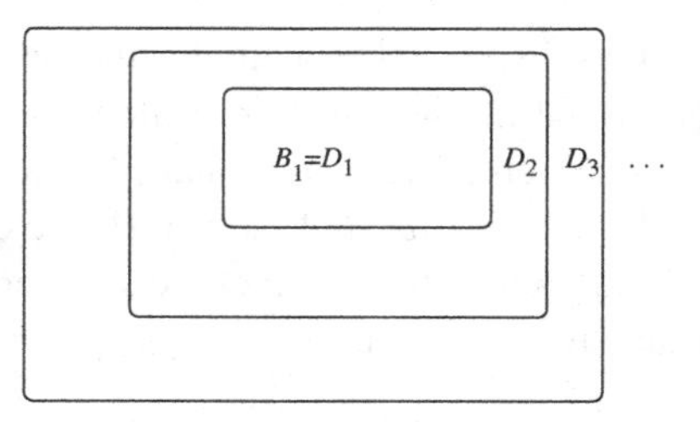

Figure 1.1 A sequence of nested sets.

Proof Suppose $B_1 \subset B_2 \subset \ldots$. Let $D_1 = B_1$, $D_2 = B_2 - B_1$, and, in general, let $D_i = B_i - B_{i-1}$ for $i \geq 2$, as shown in Figure 1.1. Then $P(B_j) = \sum_{i=1}^{j} P(D_i)$ for each $j \geq 1$, so

$$
\begin{aligned}
\lim_{j\to\infty} P(B_j) &= \lim_{j\to\infty} \sum_{i=1}^{j} P(D_i) \\
&\overset{(a)}{=} \sum_{i=1}^{\infty} P(D_i) \\
&\overset{(b)}{=} P\left(\bigcup_{i=1}^{\infty} D_i\right) = P\left(\bigcup_{i=1}^{\infty} B_i\right)
\end{aligned}
$$

where (a) is true by the definition of the sum of an infinite series, and (b) is true by axiom $P.2$. This proves Lemma 1.1(a). Lemma 1.1(b) can be proved similarly, or can be derived by applying Lemma 1.1(a) to the sets B_j^c. □

Example 1.4 (Selection of a point in a square) Take Ω to be the square region in the plane,

$$
\Omega = \{(x, y) : x, y \in [0, 1]\}.
$$

Let $\mathcal{F}$ be the Borel σ-algebra for Ω, which is the smallest σ-algebra containing all the rectangular subsets of Ω that are aligned with the axes. Take P so that for any rectangle R,

$$
P(R) = \text{area of } R.
$$

(It can be shown that $\mathcal{F}$ and P exist.) Let T be the triangular region $T = \{(x, y) : 0 \leq y \leq x \leq 1\}$. Since T is not rectangular, it is not immediately clear that $T \in \mathcal{F}$, nor is it clear what $P(T)$ is. That is where the axioms come in. For $n \geq 1$, let T_n denote the region shown in Figure 1.2. Since T_n can be written as a union of finitely many mutually exclusive rectangles, it follows that $T_n \in \mathcal{F}$ and it is easily seen that $P(T_n) = \frac{1+2+\cdots+n}{n^2} = \frac{n+1}{2n}$. Since $T_1 \supset T_2 \supset T_4 \supset T_8 \cdots$ and $\cap_j T_{2^j} = T$, it follows that $T \in \mathcal{F}$ and $P(T) = \lim_{n\to\infty} P(T_n) = \frac{1}{2}$.

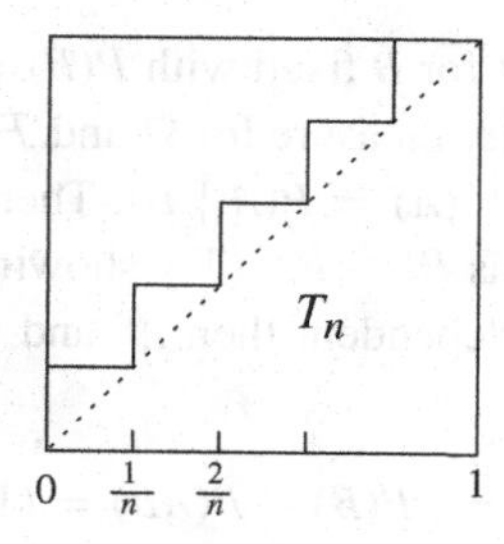

Figure 1.2 Approximation of a triangular region.

The reader is encouraged to show that if C is the diameter one disk inscribed within Ω then $P(C) = $ (area of C) $= \frac{\pi}{4}$.

1.2 Independence and conditional probability

Events A_1 and A_2 are defined to be *independent* if $P(A_1A_2) = P(A_1)P(A_2)$. More generally, events $A_1, A_2, \ldots, A_k$ are defined to be independent if

$$P(A_{i_1}A_{i_2}\ldots A_{i_j}) = P(A_{i_1})P(A_{i_2})\ldots P(A_{i_j})$$

whenever j and $i_1, i_2, \ldots, i_j$ are integers with $j \geq 1$ and $1 \leq i_1 < i_2 < \cdots < i_j \leq k$. For example, events A_1, A_2, A_3 are independent if the following four conditions hold:

$$P(A_1A_2) = P(A_1)P(A_2),$$
$$P(A_1A_3) = P(A_1)P(A_3),$$
$$P(A_2A_3) = P(A_2)P(A_3),$$
$$P(A_1A_2A_3) = P(A_1)P(A_2)P(A_3).$$

A weaker condition is sometimes useful: Events $A_1, \ldots, A_k$ are defined to be *pairwise independent* if A_i is independent of A_j whenever $1 \leq i < j \leq k$. Independence of k events requires that $2^k - k - 1$ equations hold: one for each subset of $\{1, 2, \ldots, k\}$ of size at least two. Pairwise independence only requires that $\binom{k}{2} = \frac{k(k-1)}{2}$ equations hold.

If A and B are events and $P(B) \neq 0$, then the *conditional probability* of A given B is defined by

$$P(A \mid B) = \frac{P(AB)}{P(B)}.$$

It is not defined if $P(B) = 0$, which has the following meaning. If you were to write a computer routine to compute $P(A \mid B)$ and the inputs are $P(AB) = 0$ and $P(B) = 0$, your routine shouldn't simply return the value 0. Rather, your routine should generate an error message such as "input error – conditioning on event of probability zero." Such an error message would help you or others find errors in larger computer programs which use the routine.

As a function of A for B fixed with $P(B) \neq 0$, the conditional probability of A given B is itself a probability measure for Ω and $\mathcal{F}$. More explicitly, fix B with $P(B) \neq 0$. For each event A define $P'(A) = P(A \mid B)$. Then $(\Omega, \mathcal{F}, P')$ is a probability space, because P' satisfies the axioms $P1 - P3$. (Try showing that.)

If A and B are independent then A^c and B are independent. Indeed, if A and B are independent then

$$P(A^cB) = P(B) - P(AB) = (1 - P(A))P(B) = P(A^c)P(B).$$

Similarly, if A, B, and C are independent events then AB is independent of C. More generally, suppose $E_1, E_2, \ldots, E_n$ are independent events, suppose $n = n_1 + \cdots + n_k$ with $n_i \geq 1$ for each i, and suppose F_1 is defined by Boolean operations (intersections, complements, and unions) of the first n_1 events $E_1, \ldots, E_{n_1}$, F_2 is defined by Boolean operations on the next n_2 events, $E_{n_1+1}, \ldots, E_{n_1+n_2}$, and so on. Then $F_1, \ldots, F_k$ are independent.

Events $E_1, \ldots, E_k$ are said to form a *partition* of Ω if the events are mutually exclusive and $\Omega = E_1 \cup \ldots \cup E_k$. Of course for a partition, $P(E_1) + \ldots + P(E_k) = 1$. More generally, for any event A, the *law of total probability* holds because A is the union of the mutually exclusive sets $AE_1, AE_2, \ldots, AE_k$:

$$P(A) = P(AE_1) + \ldots + P(AE_k).$$

If $P(E_i) \neq 0$ for each i, this can be written as

$$P(A) = P(A \mid E_1)P(E_1) + \ldots + P(A \mid E_k)P(E_k).$$

Figure 1.3 illustrates the condition of the law of total probability.

Judicious use of the definition of conditional probability and the law of total probability leads to *Bayes' formula* for $P(E_i \mid A)$ (if $P(A) \neq 0$) in simple form

$$P(E_i \mid A) = \frac{P(AE_i)}{P(A)} = \frac{P(A \mid E_i)P(E_i)}{P(A)},$$

or in expanded form:

$$P(E_i \mid A) = \frac{P(A \mid E_i)P(E_i)}{P(A \mid E_1)P(E_1) + \ldots + P(A \mid E_k)P(E_k)}.$$

The remainder of this section gives the Borel–Cantelli lemma. It is a simple result based on continuity of probability and independence of events, but it is not typically

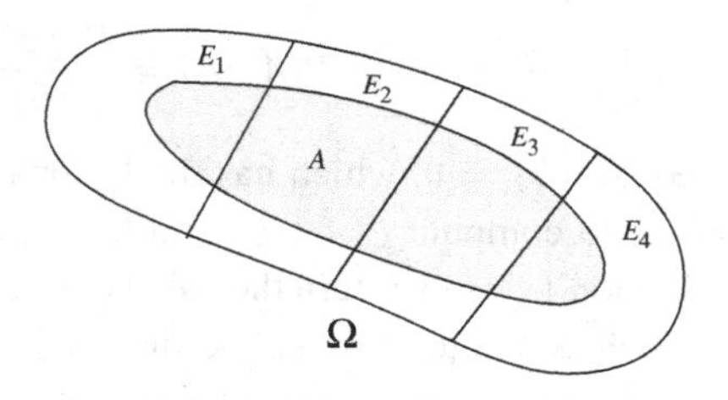

Figure 1.3 Partitioning a set A using a partition of Ω.

encountered in a first course on probability. Let $(A_n : n \geq 0)$ be a sequence of events for a probability space $(\Omega, \mathcal{F}, P)$.

Definition 1.2 The event $\{A_n$ infinitely often$\}$ is the set of $\omega \in \Omega$ such that $\omega \in A_n$ for infinitely many values of n.

Another way to describe $\{A_n$ infinitely often$\}$ is that it is the set of ω such that for any k, there is an $n \geq k$ such that $\omega \in A_n$. Therefore,

$$\{A_n \text{ infinitely often}\} = \cap_{k\geq 1} \left(\cup_{n\geq k} A_n\right).$$

For each k, the set $\cup_{n\geq k} A_n$ is a countable union of events, so it is an event, and $\{A_n$ infinitely often$\}$ is an intersection of countably many such events, so that $\{A_n$ infinitely often$\}$ is also an event.

Lemma 1.3 *(Borel–Cantelli lemma) Let $(A_n : n \geq 1)$ be a sequence of events and let $p_n = P(A_n)$.*

(a) If $\sum_{n=1}^{\infty} p_n < \infty$, then $P\{A_n$ infinitely often$\} = 0$.
(b) If $\sum_{n=1}^{\infty} p_n = \infty$ and $A_1, A_2, \ldots$ are mutually independent, then $P\{A_n$ infinitely often$\} = 1$.

Proof (a) Since $\{A_n$ infinitely often$\}$ is the intersection of the monotonically non-increasing sequence of events $\cup_{n\geq k} A_n$, it follows from the continuity of probability for monotone sequences of events (Lemma 1.1) that $P\{A_n$ infinitely often$\} = \lim_{k\to\infty} P(\cup_{n\geq k} A_n)$. Lemma 1.1, the fact that the probability of a union of events is less than or equal to the sum of the probabilities of the events, and the definition of the sum of a sequence of numbers, yield that for any $k \geq 1$,

$$P(\cup_{n\geq k} A_n) = \lim_{m\to\infty} P(\cup_{n=k}^{m} A_n) \leq \lim_{m\to\infty} \sum_{n=k}^{m} p_n = \sum_{n=k}^{\infty} p_n.$$

Therefore, $P\{A_n$ infinitely often$\} \leq \lim_{k\to\infty} \sum_{n=k}^{\infty} p_n$. If $\sum_{n=1}^{\infty} p_n < \infty$, then $\lim_{k\to\infty} \sum_{n=k}^{\infty} p_n = 0$, which implies part (a) of the lemma.
(b) Suppose that $\sum_{n=1}^{\infty} p_n = +\infty$ and that the events $A_1, A_2, \ldots$ are mutually independent. For any $k \geq 1$, using the fact $1 - u \leq \exp(-u)$ for all u,

$$\begin{aligned} &P(\cup_{n\geq k} A_n) \\ &= \lim_{m\to\infty} P(\cup_{n=k}^{m} A_n) = \lim_{m\to\infty} 1 - \prod_{n=k}^{m} (1 - p_n) \\ &\geq \lim_{m\to\infty} 1 - \exp\left(-\sum_{n=k}^{m} p_n\right) = 1 - \exp\left(-\sum_{n=k}^{\infty} p_n\right) = 1 - \exp(-\infty) = 1. \end{aligned}$$

Therefore, $P\{A_n$ infinitely often$\} = \lim_{k\to\infty} P(\cup_{n\geq k} A_n) = 1$. □

Example 1.5 Consider independent coin tosses using biased coins, such that $P(A_n) = p_n = \frac{1}{n}$, where A_n is the event of getting heads on the nth toss. Since $\sum_{n=1}^{\infty} \frac{1}{n} = +\infty$, the part of the Borel–Cantelli lemma for independent events implies that $P\{A_n \text{ infinitely often}\} = 1$.

Example 1.6 Let $(\Omega, \mathcal{F}, P)$ be the standard unit-interval probability space defined in Example 1.2, and let $A_n = [0, \frac{1}{n})$. Then $p_n = \frac{1}{n}$ and $A_{n+1} \subset A_n$ for $n \geq 1$. The events are not independent, because for $m < n$, $P(A_m A_n) = P(A_n) = \frac{1}{n} \neq P(A_m)P(A_n)$. Of course $0 \in A_n$ for all n. But for any $\omega \in (0, 1]$, $\omega \notin A_n$ for $n > \frac{1}{\omega}$. Therefore, $\{A_n \text{ infinitely often}\} = \{0\}$. The single point set $\{0\}$ has probability zero, so $P\{A_n \text{ infinitely often}\} = 0$. This conclusion holds even though $\sum_{n=1}^{\infty} p_n = +\infty$, illustrating the need for the independence assumption in Lemma 1.3(b).

1.3 Random variables and their distribution

Let a probability space $(\Omega, \mathcal{F}, P)$ be given. By definition, a random variable is a function X from Ω to the real line $\mathbb{R}$ that is $\mathcal{F}$ measurable, meaning that for any number c,

$$\{\omega : X(\omega) \leq c\} \in \mathcal{F}. \tag{1.2}$$

If Ω is finite or countably infinite, then $\mathcal{F}$ can be the set of all subsets of Ω, in which case any real-valued function on Ω is a random variable.

If $(\Omega, \mathcal{F}, P)$ is the standard unit-interval probability space described in Example 1.2, then the random variables on $(\Omega, \mathcal{F}, P)$ are called the Borel measurable functions on Ω. Since the Borel σ-algebra contains all subsets of $[0, 1]$ that come up in applications, for practical purposes we can think of any function on $[0, 1]$ as being a random variable. For example, any piecewise continuous or piecewise monotone function on $[0, 1]$ is a random variable for the standard unit-interval probability space.

The cumulative distribution function (CDF) of a random variable X is denoted by F_X. It is the function, with domain the real line $\mathbb{R}$, defined by

$$\begin{aligned} F_X(c) &= P\{\omega : X(\omega) \leq c\} \\ &= P\{X \leq c\} \text{ (for short).} \end{aligned}$$

If X denotes the outcome of the roll of a fair die ("die" is singular of "dice") and if Y is uniformly distributed on the interval $[0, 1]$, then F_X and F_Y are shown in Figure 1.4.

The CDF of a random variable X determines $P\{X \leq c\}$ for any real number c. But what about $P\{X < c\}$ and $P\{X = c\}$? Let $c_1, c_2, \ldots$ be a monotone nondecreasing sequence that converges to c from the left. This means $c_i \leq c_j < c$ for $i < j$ and $\lim_{j \to \infty} c_j = c$. Then the events $\{X \leq c_j\}$ are nested: $\{X \leq c_i\} \subset \{X \leq c_j\}$ for $i < j$, and the union of all such events is the event $\{X < c\}$. Thus, by Lemma 1.1

$$P\{X < c\} = \lim_{i \to \infty} P\{X \leq c_i\} = \lim_{i \to \infty} F_X(c_i) = F_X(c-).$$

Therefore, $P\{X = c\} = F_X(c) - F_X(c-) = \Delta F_X(c)$, where $\Delta F_X(c)$ is defined to be the size of the jump of F at c. For example, if X has the CDF shown in Figure 1.5 then $P\{X = 0\} = \frac{1}{2}$. The collection of all events A such that $P\{X \in A\}$ is determined by F_X

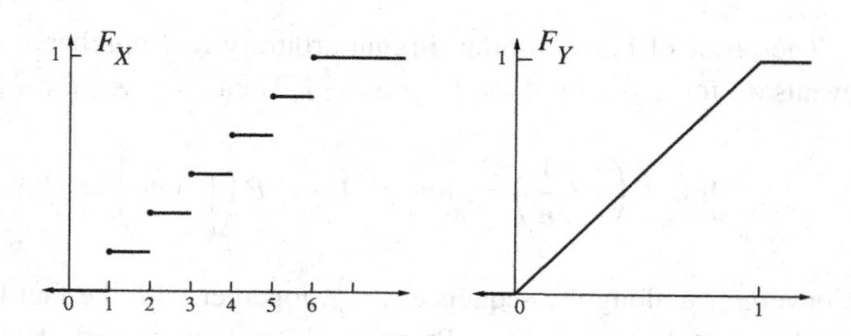

Figure 1.4 Examples of CDFs.

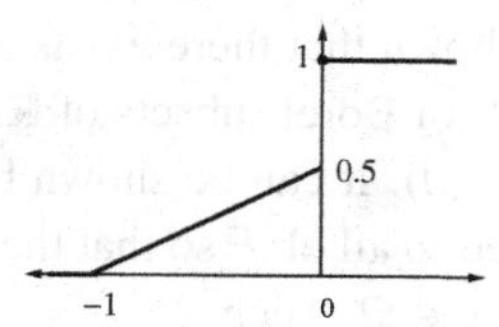

Figure 1.5 An example of a CDF.

is a σ-algebra containing the intervals, and thus this collection contains all Borel sets. That is, $P\{X \in A\}$ is determined by F_X for any Borel set A.

Proposition 1.4 *A function F is the CDF of some random variable if and only if it has the following three properties:*

F.1 F is nondecreasing,
F.2 $\lim_{x\to+\infty} F(x) = 1$ *and* $\lim_{x\to-\infty} F(x) = 0$,
F.3 F is right continuous.

Proof The "only if" part is proved first. Suppose that F is the CDF of some random variable X. If $x < y$, $F(y) = P\{X \leq y\} = P\{X \leq x\} + P\{x < X \leq y\} \geq P\{X \leq x\} = F(x)$ so that F.1 is true. Consider the events $B_n = \{X \leq n\}$. Then $B_n \subset B_m$ for $n \leq m$. Thus, by Lemma 1.1,

$$\lim_{n\to\infty} F(n) = \lim_{n\to\infty} P(B_n) = P\left(\bigcup_{n=1}^{\infty} B_n\right) = P(\Omega) = 1.$$

This and the fact F is nondecreasing imply the following. Given any $\epsilon > 0$, there exists N_ϵ so large that $F(x) \geq 1 - \epsilon$ for all $x \geq N_\epsilon$. That is, $F(x) \to 1$ as $x \to +\infty$. Similarly,

$$\lim_{n\to-\infty} F(n) = \lim_{n\to\infty} P(B_{-n}) = P\left(\bigcap_{n=1}^{\infty} B_{-n}\right) = P(\emptyset) = 0.$$

so that $F(x) \to 0$ as $x \to -\infty$. Property F.2 is proved.

The proof of F.3 is similar. Fix an arbitrary real number x. Define the sequence of events A_n for $n \geq 1$ by $A_n = \{X \leq x + \frac{1}{n}\}$. Then $A_n \subset A_m$ for $n \geq m$ so

$$\lim_{n\to\infty} F\left(x + \frac{1}{n}\right) = \lim_{n\to\infty} P(A_n) = P\left(\bigcap_{k=1}^{\infty} A_k\right) = P\{X \leq x\} = F_X(x).$$

Convergence along the sequence $x + \frac{1}{n}$, together with the fact that F is nondecreasing, implies that $F(x+) = F(x)$. Property F.3 is thus proved. The proof of the "only if" portion of Proposition 1.4 is complete.

To prove the "if" part of Proposition 1.4, let F be a function satisfying properties F.1–F.3. It must be shown that there exists a random variable with CDF F. Let $\Omega = \mathbb{R}$ and let $\mathcal{F}$ be the set $\mathcal{B}$ of Borel subsets of $\mathbb{R}$. Define $\tilde{P}$ on intervals of the form $(a, b]$ by $\tilde{P}((a, b]) = F(b) - F(a)$. It can be shown by an extension theorem of measure theory that $\tilde{P}$ can be extended to all of $\mathcal{F}$ so that the axioms of probability are satisfied. Finally, let $\tilde{X}(\omega) = \omega$ for all $\omega \in \Omega$. Then

$$\tilde{P}(\tilde{X} \in (a, b]) = \tilde{P}((a, b]) = F(b) - F(a).$$

Therefore, $\tilde{X}$ has CDF F. So F is a CDF, as was to be proved. □

The vast majority of random variables described in applications are one of two types, to be described next. A random variable X is a discrete random variable if there is a finite or countably infinite set of values $\{x_i : i \in I\}$ such that $P\{X \in \{x_i : i \in I\}\} = 1$. The probability mass function (pmf) of a discrete random variable X, denoted $p_X(x)$, is defined by $p_X(x) = P\{X = x\}$. Typically the pmf of a discrete random variable is much more useful than the CDF. However, the pmf and CDF of a discrete random variable are related by $p_X(x) = \Delta F_X(x)$ and conversely,

$$F_X(x) = \sum_{y: y \leq x} p_X(y), \tag{1.3}$$

where the sum in (1.3) is taken only over y such that $p_X(y) \neq 0$. If X is a discrete random variable with only finitely many mass points in any finite interval, then F_X is a piecewise constant function.

A random variable X is a *continuous* random variable if the CDF is the integral of a function:

$$F_X(x) = \int_{-\infty}^{x} f_X(y)dy.$$

The function f_X is called the *probability density function* (pdf). If the pdf f_X is continuous at a point x, then the value $f_X(x)$ has the following nice interpretation:

$$\begin{aligned} f_X(x) &= \lim_{\varepsilon\to 0} \frac{1}{\varepsilon} \int_{x}^{x+\varepsilon} f_X(y)dy \\ &= \lim_{\varepsilon\to 0} \frac{1}{\varepsilon} P\{x \leq X \leq x + \varepsilon\}. \end{aligned}$$

If A is any Borel subset of $\mathbb{R}$, then

$$P\{X \in A\} = \int_A f_X(x)dx. \tag{1.4}$$

The integral in (1.4) can be understood as a Riemann integral if A is a finite union of intervals and f is piecewise continuous or monotone. In general, f_X is required to be Borel measurable and the integral is defined by Lebesgue integration.[2]

Any random variable X on an arbitrary probability space has a CDF F_X. As noted in the proof of Proposition 1.4 there exists a probability measure P_X (called $\tilde{P}$ in the proof) on the Borel subsets of $\mathbb{R}$ such that for any interval $(a, b]$,

$$P_X((a, b]) = P\{X \in (a, b]\}.$$

We define the *probability distribution* of X to be the probability measure P_X. The distribution P_X is determined uniquely by the CDF F_X. The distribution is also determined by the pdf f_X if X is continuous type, or the pmf p_X if X is discrete type. In common usage, the response to the question "What is the distribution of X?" is answered by giving one or more of F_X, f_X, or p_X, or possibly a transform of one of these, whichever is most convenient.

1.4 Functions of a random variable

Recall that a random variable X on a probability space $(\Omega, \mathcal{F}, P)$ is a function mapping Ω to the real line $\mathbb{R}$, satisfying the condition $\{\omega : X(\omega) \leq a\} \in \mathcal{F}$ for all $a \in \mathbb{R}$. Suppose g is a function mapping $\mathbb{R}$ to $\mathbb{R}$ that is not too bizarre. Specifically, suppose for any constant c that $\{x : g(x) \leq c\}$ is a Borel subset of $\mathbb{R}$. Let $Y(\omega) = g(X(\omega))$. Then Y maps Ω to $\mathbb{R}$ and Y is a random variable, see Figure 1.6. We write $Y = g(X)$.

Often we'd like to compute the distribution of Y from knowledge of g and the distribution of X. When X is a continuous random variable with known distribution, the following three step procedure works well:

(1) Examine the ranges of possible values of X and Y. Sketch the function g.

(2) Find the CDF of Y, using $F_Y(c) = P\{Y \leq c\} = P\{g(X) \leq c\}$. The idea is to express the event $\{g(X) \leq c\}$ as $\{X \in A\}$ for some set A depending on c.

(3) If F_Y has a piecewise continuous derivative, and if the pdf f_Y is desired, differentiate F_Y.

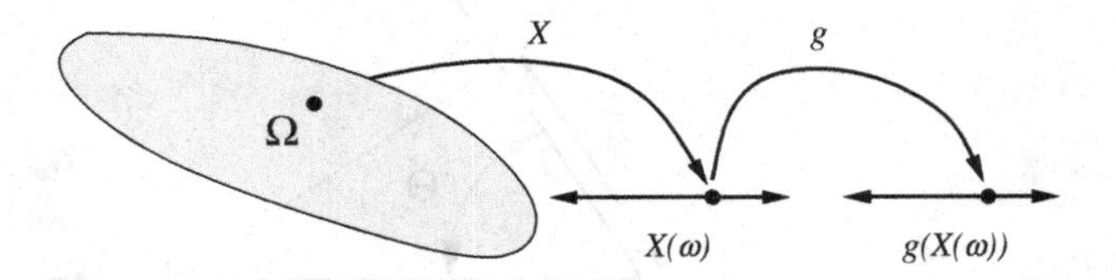

Figure 1.6 A function of a random variable as a composition of mappings.

[2] Lebesgue integration is defined in Sections 1.5 and 11.5.

If instead X is a discrete random variable then step 1 should be followed. After that the pmf of Y can be found from the pmf of X using

$$p_Y(y) = P\{g(X) = y\} = \sum_{x:g(x)=y} p_X(x).$$

Example 1.7 Suppose X is an $N(\mu = 2, \sigma^2 = 3)$ random variable (see Section 1.6 for the definition) and $Y = X^2$. Let us describe the density of Y. Note that $Y = g(X)$ where $g(x) = x^2$. The support of the distribution of X is the whole real line, and the range of g over this support is $\mathbb{R}_+$. Next we find the CDF, F_Y. Since $P\{Y \geq 0\} = 1$, $F_Y(c) = 0$ for $c < 0$. For $c \geq 0$,

$$\begin{aligned} F_Y(c) &= P\{X^2 \leq c\} = P\{-\sqrt{c} \leq X \leq \sqrt{c}\} \\ &= P\left\{\frac{-\sqrt{c}-2}{\sqrt{3}} \leq \frac{X-2}{\sqrt{3}} \leq \frac{\sqrt{c}-2}{\sqrt{3}}\right\} \\ &= \Phi\left(\frac{\sqrt{c}-2}{\sqrt{3}}\right) - \Phi\left(\frac{-\sqrt{c}-2}{\sqrt{3}}\right). \end{aligned}$$

Differentiate with respect to c, using the chain rule and $\Phi'(s) = \frac{1}{\sqrt{2\pi}}\exp(-\frac{s^2}{2})$, to obtain

$$f_Y(c) = \begin{cases} \frac{1}{\sqrt{24\pi c}}\left\{\exp\left(-\left[\frac{\sqrt{c}-2}{\sqrt{6}}\right]^2\right) + \exp\left(-\left[\frac{-\sqrt{c}-2}{\sqrt{6}}\right]^2\right)\right\} & \text{if } c \geq 0 \\ 0 & \text{if } c < 0 \end{cases}.$$

Example 1.8 Suppose a vehicle is traveling in a straight line at speed a, and that a random direction is selected, subtending an angle Θ from the direction of travel which is uniformly distributed over the interval $[0, \pi]$, see Figure 1.7. Then the effective speed of the vehicle in the random direction is $B = a\cos(\Theta)$.

Let us find the pdf of B.

The range of $a\cos(\theta)$, as θ ranges over $[0, \pi]$, is the interval $[-a, a]$. Therefore, $F_B(c) = 0$ for $c \leq -a$ and $F_B(c) = 1$ for $c \geq a$. Let $-a < c < a$. Then, because cos is monotone nonincreasing on the interval $[0, \pi]$,

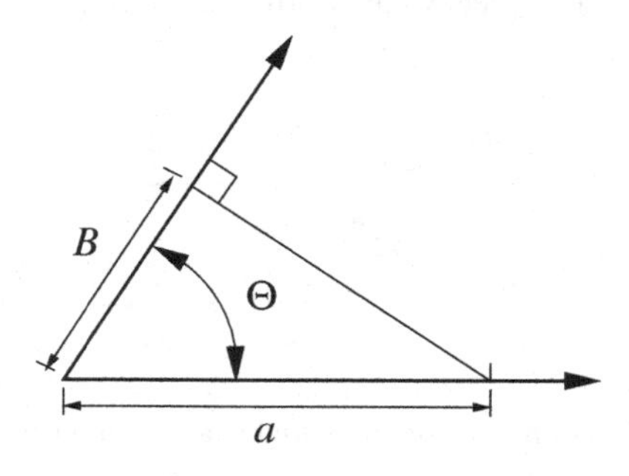

Figure 1.7 Direction of travel and a random direction.

$$\begin{aligned} F_B(c) &= P\{a\cos(\Theta) \le c\} = P\left\{\cos(\Theta) \le \frac{c}{a}\right\} \\ &= P\left\{\Theta \ge \cos^{-1}\left(\frac{c}{a}\right)\right\} \\ &= 1 - \frac{\cos^{-1}\left(\frac{c}{a}\right)}{\pi}. \end{aligned}$$

Therefore, because $\cos^{-1}(y)$ has derivative, $-(1-y^2)^{-\frac{1}{2}}$,

$$f_B(c) = \begin{cases} \frac{1}{\pi\sqrt{a^2-c^2}} & |c| < a \\ 0 & |c| > a \end{cases}.$$

A sketch of the density is given in Figure 1.8.

Example 1.9 Suppose $Y = \tan(\Theta)$, as illustrated in Figure 1.9, where Θ is uniformly distributed over the interval $\left(-\frac{\pi}{2}, \frac{\pi}{2}\right)$. Let us find the pdf of Y. The function $\tan(\theta)$ increases from $-\infty$ to ∞ as θ ranges over the interval $\left(-\frac{\pi}{2}, \frac{\pi}{2}\right)$.

For any real c,

$$\begin{aligned} F_Y(c) &= P\{Y \le c\} \\ &= P\{\tan(\Theta) \le c\} \\ &= P\{\Theta \le \tan^{-1}(c)\} = \frac{\tan^{-1}(c) + \frac{\pi}{2}}{\pi}. \end{aligned}$$

Differentiating the CDF with respect to c yields that Y has the Cauchy pdf:

$$f_Y(c) = \frac{1}{\pi(1+c^2)} \qquad -\infty < c < \infty.$$

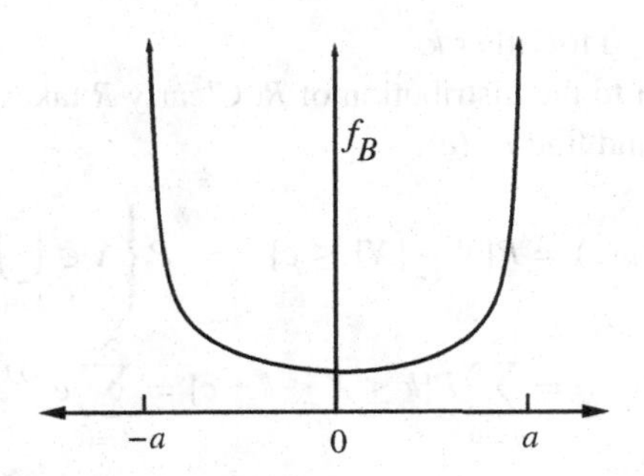

Figure 1.8 The pdf of the effective speed in a uniformly distributed direction in two dimensions.

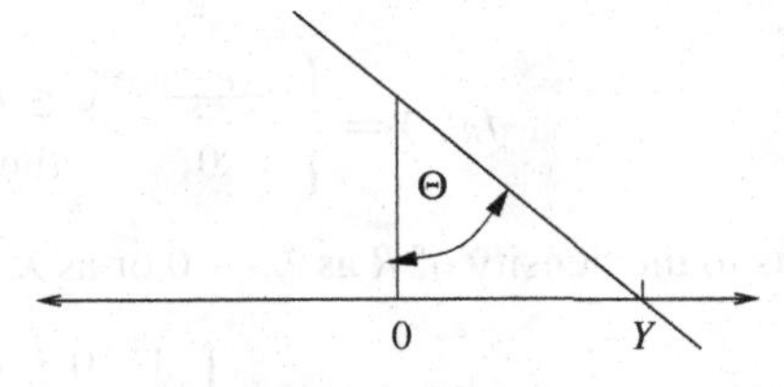

Figure 1.9 A horizontal line, a fixed point at unit distance, and a line through the point with random direction.

Example 1.10 Given an angle θ expressed in radians, let ($\theta \bmod 2\pi$) denote the equivalent angle in the interval $[0, 2\pi]$. Thus, ($\theta \bmod 2\pi$) is equal to $\theta + 2\pi n$, where the integer n is such that $0 \le \theta + 2\pi n < 2\pi$.

Let Θ be uniformly distributed over $[0, 2\pi]$, let h be a constant, and let

$$\tilde{\Theta} = (\Theta + h \bmod 2\pi).$$

Let us find the distribution of $\tilde{\Theta}$.

Clearly $\tilde{\Theta}$ takes values in the interval $[0, 2\pi]$, so fix c with $0 \le c < 2\pi$ and seek to find $P\{\tilde{\Theta} \le c\}$. Let A denote the interval $[h, h+2\pi]$. Thus, $\Theta + h$ is uniformly distributed over A. Let $B = \bigcup_n [2\pi n, 2\pi n + c]$. Thus $\tilde{\Theta} \le c$ if and only if $\Theta + h \in B$. Therefore,

$$P\{\tilde{\Theta} \le c\} = \int_{A \bigcap B} \frac{1}{2\pi} d\theta.$$

By sketching the set B, it is easy to see that $A \bigcap B$ is either a single interval of length c, or the union of two intervals with lengths adding to c. Therefore, $P\{\tilde{\Theta} \le c\} = \frac{c}{2\pi}$, so that $\tilde{\Theta}$ is itself uniformly distributed over $[0, 2\pi]$.

Example 1.11 Let X be an exponentially distributed random variable with parameter λ. Let $Y = \lfloor X \rfloor$, which is the integer part of X, and let $R = X - \lfloor X \rfloor$, which is the remainder. We shall describe the distributions of Y and R.

Clearly Y is a discrete random variable with possible values $0, 1, 2, \ldots$, so it is sufficient to find the pmf of Y. For integers $k \ge 0$,

$$p_Y(k) = P\{k \le X < k+1\} = \int_k^{k+1} \lambda e^{-\lambda x} dx = e^{-\lambda k}(1 - e^{-\lambda})$$

and $p_Y(k) = 0$ for other k.

Turn next to the distribution of R. Clearly R takes values in the interval $[0, 1]$. So let $0 < c < 1$ and find $F_R(c)$:

$$\begin{aligned} F_R(c) = P\{X - \lfloor X \rfloor \le c\} &= P\left\{X \in \bigcup_{k=0}^{\infty} [k, k+c]\right\} \\ &= \sum_{k=0}^{\infty} P\{k \le X \le k+c\} = \sum_{k=0}^{\infty} e^{-\lambda k}(1 - e^{-\lambda c}) = \frac{1 - e^{-\lambda c}}{1 - e^{-\lambda}}, \end{aligned}$$

where we used the fact $1 + \alpha + \alpha^2 + \cdots = \frac{1}{1-\alpha}$ for $|\alpha| < 1$. Differentiating F_R yields the pmf:

$$f_R(c) = \begin{cases} \frac{\lambda e^{-\lambda c}}{1 - e^{-\lambda}} & 0 \le c \le 1 \\ 0 & \text{otherwise} \end{cases}.$$

What happens to the density of R as $\lambda \to 0$ or as $\lambda \to \infty$? By l'Hôpital's rule,

$$\lim_{\lambda \to 0} f_R(c) = \begin{cases} 1 & 0 \le c \le 1 \\ 0 & \text{otherwise} \end{cases}.$$

That is, in the limit as $\lambda \to 0$, the density of X becomes more and more evenly spread out, and R becomes uniformly distributed over the interval $[0, 1]$. If λ is very large then the factor $1 - e^{-\lambda}$ is nearly one , and the density of R is nearly the same as the exponential density with parameter λ.

An important step in many computer simulations of random systems is to generate a random variable with a specified CDF, by applying a function to a random variable that is uniformly distributed on the interval $[0, 1]$. Let F be a function satisfying the three properties required of a CDF, and let U be uniformly distributed over the interval $[0, 1]$. The problem is to find a function g so that F is the CDF of $g(U)$. An appropriate function g is given by the inverse function of F. Although F may not be strictly increasing, a suitable version of F^{-1} always exists, defined for $0 < u < 1$ by

$$F^{-1}(u) = \min\{x : F(x) \geq u\}. \tag{1.5}$$

If the graphs of F and F^{-1} are closed up by adding vertical lines at jump points, then the graphs are reflections of each other about the $x = y$ line, as illustrated in Figure 1.10. It is not hard to check that for any real x_o and u_o with $0 < u_o < 1$,

$$F^{-1}(u_o) \leq x_o \text{ if and only if } u_o \leq F(x_o).$$

Thus, if $X = F^{-1}(U)$ then

$$F_X(x) = P\{F^{-1}(U) \leq x\} = P\{U \leq F(x)\} \quad = \quad F(x),$$

so that indeed F is the CDF of X.

Example 1.12 Suppose $F(x) = 1 - e^{-x}$ for $x \geq 0$ and $F(x) = 0$ for $x < 0$. Since F is continuously increasing in this case, we can identify its inverse by solving for x as a function of u so that $F(x) = u$. That is, for $0 < u < 1$, we'd like $1 - e^{-x} = u$ which is equivalent to $e^{-x} = 1 - u$, or $x = -\ln(1 - u)$. Thus, $F^{-1}(u) = -\ln(1 - u)$. So we can take $g(u) = -\ln(1 - u)$ for $0 < u < 1$. That is, if U is uniformly distributed on the interval $[0, 1]$, then the CDF of $-\ln(1 - U)$ is F. The choice of g is not unique in

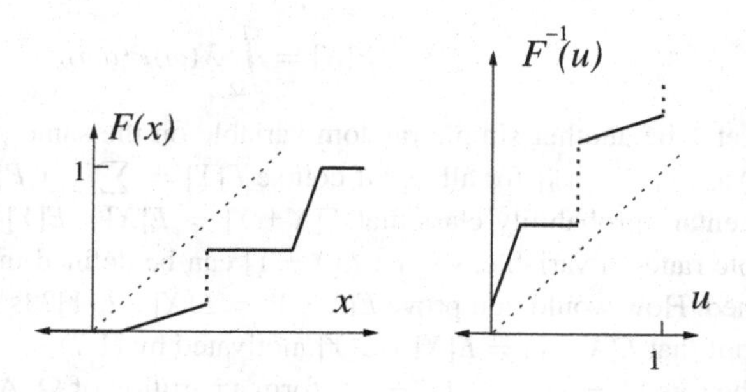

Figure 1.10 A CDF and its inverse.

general. For example, $1-U$ has the same distribution as U, so the CDF of $-\ln(U)$ is also F. To double check the answer, note that if $x \geq 0$, then

$$P\{-\ln(1-U) \leq x\} = P\{\ln(1-U) \geq -x\}$$
$$= P\{1-U \geq e^{-x}\} = P\{U \leq 1-e^{-x}\} = F(x).$$

Example 1.13 Suppose F is the CDF for the experiment of rolling a fair die, shown on the left half of Figure 1.4. One way to generate a random variable with CDF F is to actually roll a die. To simulate that on a computer, we'd seek a function g so that $g(U)$ has the same CDF. Using $g = F^{-1}$ and using (1.5) or the graphical method illustrated in Figure 1.10 to find F^{-1}, we get that for $0 < u < 1$, $g(u) = i$ for $\frac{i-1}{6} < u \leq \frac{i}{6}$ for $1 \leq i \leq 6$. To double check the answer, note that if $1 \leq i \leq 6$, then

$$P\{g(U) = i\} = P\left\{\frac{i-1}{6} < U \leq \frac{i}{6}\right\} = \frac{1}{6}$$

so that $g(U)$ has the correct pmf, and hence the correct CDF.

1.5 Expectation of a random variable

The *expectation*, alternatively called the *mean*, of a random variable X can be defined in several different ways. Before giving a general definition, we shall consider a straightforward case. A random variable X is called simple if there is a finite set $\{x_1, \ldots, x_m\}$ such that $X(\omega) \in \{x_1, \ldots, x_m\}$ for all ω. The expectation of such a random variable is defined by

$$E[X] = \sum_{i=1}^{m} x_i P\{X = x_i\}. \tag{1.6}$$

The definition (1.6) clearly shows that $E[X]$ for a simple random variable X depends only on the pmf of X.

Like all random variables, X is a function on a probability space $(\Omega, \mathcal{F}, P)$. Figure 1.11 illustrates that the sum defining $E[X]$ in (1.6) can be viewed as an integral over Ω. This suggests writing

$$E[X] = \int_{\Omega} X(\omega) P(d\omega). \tag{1.7}$$

Let Y be another simple random variable on the same probability space as X, with $Y(\omega) \in \{y_1, \ldots, y_n\}$ for all ω. Of course $E[Y] = \sum_{i=1}^{n} y_i P\{Y = y_i\}$. One learns in any elementary probability class that $E[X+Y] = E[X] + E[Y]$. Note that $X+Y$ is again a simple random variable, so that $E[X+Y]$ can be defined in the same way as $E[X]$ was defined. How would you prove $E[X+Y] = E[X] + E[Y]$? Is (1.6) helpful? We shall give a proof that $E[X+Y] = E[X] + E[Y]$ motivated by (1.7).

The sets $\{X = x_1\}, \ldots, \{X = x_m\}$ form a partition of Ω. A refinement of this partition consists of another partition $C_1, \ldots, C_{m'}$ such that X is constant over each C_j. If we let

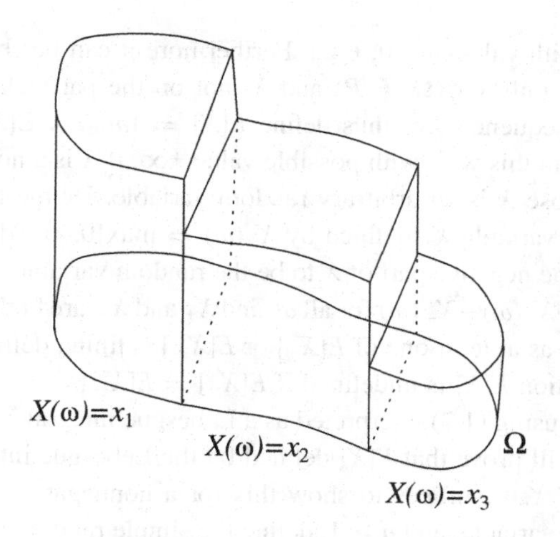

Figure 1.11 A simple random variable with three possible values.

x'_j denote the value of X on C_j, then clearly

$$E[X] = \sum_j x'_j P(C_j).$$

Now, it is possible to select the partition $C_1, \ldots, C_{m'}$ so that both X and Y are constant over each C_j. For example, each C_j could have the form $\{X = x_i\} \cap \{Y = y_k\}$ for some i, k. Let x'_j denote the value of X and y'_j denote the value of Y on C_j. Then $x'_j + y'_j$ is the value of $X + Y$ on C_j. Therefore,

$$E[X+Y] = \sum_j (x'_j + y'_j)P(C_j) = \sum_j x'_j P(C_j) + \sum_j y'_j P(C_j) = E[X] + E[Y].$$

While the expression (1.7) is rather indicative, it would be overly restrictive to interpret it as a Riemann integral over Ω. For example, if X is a random variable for the standard unit-interval probability space defined in Example 1.2, then it is tempting to define $E[X]$ by Riemann integration (see the Appendix):

$$E[X] = \int_0^1 X(\omega)d\omega. \tag{1.8}$$

However, suppose X is the simple random variable such that $X(w) = 1$ for rational values of ω and $X(\omega) = 0$ otherwise. Since the set of rational numbers in Ω is countably infinite, such X satisfies $P\{X = 0\} = 1$. Clearly we'd like $E[X] = 0$, but the Riemann integral (1.8) is not convergent for this choice of X.

The expression (1.7) can be used to define $E[X]$ in great generality if it is interpreted as a Lebesgue integral, defined as follows: suppose X is an arbitrary nonnegative random variable. Then there exists a sequence of simple random variables $X_1, X_2, \ldots$ such that for every $\omega \in \Omega$, $X_1(\omega) \leq X_2(\omega) \leq \ldots$ and $X_n(\omega) \to X(\omega)$ as $n \to \infty$. Then $E[X_n]$ is well defined for each n and is nondecreasing in n, so the limit of $E[X_n]$ as $n \to \infty$

exists with values in $[0, +\infty]$. Furthermore it can be shown that the value of the limit depends only on $(\Omega, \mathcal{F}, P)$ and X, not on the particular choice of the approximating simple sequence. We thus define $E[X] = \lim_{n\to\infty} E[X_n]$. Thus, $E[X]$ is always well defined in this way, with possible value $+\infty$, if X is a nonnegative random variable.

Suppose X is an arbitrary random variable. Define the positive part of X to be the random variable X_+ defined by $X_+(\omega) = \max\{0, X(\omega)\}$ for each value of ω. Similarly define the negative part of X to be the random variable $X_-(\omega) = \max\{0, -X(\omega)\}$. Then $X(\omega) = X_+(\omega) - X_-(\omega)$ for all ω, and X_+ and X_- are both nonnegative random variables. As long as at least one of $E[X_+]$ or $E[X_-]$ is finite, define $E[X] = E[X_+] - E[X_-]$. The expectation $E[X]$ is undefined if $E[X_+] = E[X_-] = +\infty$. This completes the definition of $E[X]$ using (1.7) interpreted as a Lebesgue integral.

We will prove that $E[X]$ defined by the Lebesgue integral (1.7) depends only on the CDF of X. It suffices to show this for a nonnegative random variable X. For such a random variable, and $n \geq 1$, define the simple random variable X_n by

$$X_n(\omega) = \begin{cases} k2^{-n} & \text{if} \quad k2^{-n} \leq X(\omega) < (k+1)2^{-n}, \quad k = 0, 1, \ldots, 2^{2n} - 1 \\ 0 & \text{else} \end{cases}.$$

Then

$$E[X_n] = \sum_{k=0}^{2^{2n}-1} k2^{-n}(F_X((k+1)2^{-n}) - F_X(k2^{-n})),$$

so that $E[X_n]$ is determined by the CDF F_X for each n. Furthermore, the X_ns are nondecreasing in n and converge to X. Thus, $E[X] = \lim_{n\to\infty} E[X_n]$, and therefore the limit $E[X]$ is determined by F_X.

In Section 1.3 we defined the probability distribution P_X of a random variable such that the canonical random variable $\tilde{X}(\omega) = \omega$ on $(\mathbb{R}, \mathcal{B}, P_X)$ has the same CDF as X. Therefore $E[X] = E[\tilde{X}]$, or

$$E[X] = \int_{-\infty}^{\infty} x P_X(dx) \qquad \text{(Lebesgue)}. \tag{1.9}$$

By definition, the integral (1.9) is the Lebesgue–Stieltjes integral of x with respect to F_X, so that

$$E[X] = \int_{-\infty}^{\infty} x dF_X(x) \qquad \text{(Lebesgue–Stieltjes)}. \tag{1.10}$$

Expectation has the following properties. Let X, Y be random variables and c be a constant.

E.1 (Linearity) $E[cX] = cE[X]$. If $E[X]$, $E[Y]$, and $E[X] + E[Y]$ are well defined, then $E[X+Y]$ is well defined and $E[X+Y] = E[X] + E[Y]$.

E.2 (Preservation of order) If $P\{X \geq Y\} = 1$ and $E[Y] > -\infty$, then $E[X]$ is well defined and $E[X] \geq E[Y]$.

E.3 If X has pdf f_X then

$$E[X] = \int_{-\infty}^{\infty} x f_X(x) dx \qquad \text{(Lebesgue)}.$$

E.4 If X has pmf p_X then

$$E[X] = \sum_{x>0} x p_X(x) + \sum_{x<0} x p_X(x).$$

E.5 (Law of the unconscious statistician (LOTUS)) If g is Borel measurable,

$$E[g(X)] = \int_\Omega g(X(\omega))P(d\omega) \qquad \text{(Lebesgue)}$$

$$= \int_{-\infty}^{\infty} g(x)dF_X(x) \qquad \text{(Lebesgue–Stieltjes)},$$

and when X is a continuous type random variable

$$E[g(X)] = \int_{-\infty}^{\infty} g(x)f_X(x)dx \qquad \text{(Lebesgue)}.$$

E.6 (Integration by parts formula)

$$E[X] = \int_0^{\infty} (1 - F_X(x))dx - \int_{-\infty}^{0} F_X(x)dx, \tag{1.11}$$

which is well defined whenever at least one of the two integrals in (1.11) is finite. There is a simple graphical interpretation of (1.11). Namely, $E[X]$ is equal to the area of the region between the horizontal line $\{y = 1\}$ and the graph of F_X and contained in $\{x \geq 0\}$, minus the area of the region bounded by the x axis and the graph of F_X and contained in $\{x \leq 0\}$, as long as at least one of these regions has finite area, see Figure 1.12.

Properties E.1 and E.2 are true for simple random variables and they carry over to general random variables in the limit defining the Lebesgue integral (1.7). Properties E.3 and E.4 follow from the equivalent definition (1.9) and properties of Lebesgue–Stieltjes integrals. Property E.5 can be proved by approximating g by piecewise constant functions. Property E.6 can be proved by integration by parts applied to (1.10). Alternatively, since $F_X^{-1}(U)$ has the same distribution as X, if U is uniformly distributed on the interval $[0, 1]$, the law of the unconscious statistician yields that $E[X] = \int_0^1 F_X^{-1}(u)du$, and this integral can also be interpreted as the difference of the areas of the same two regions.

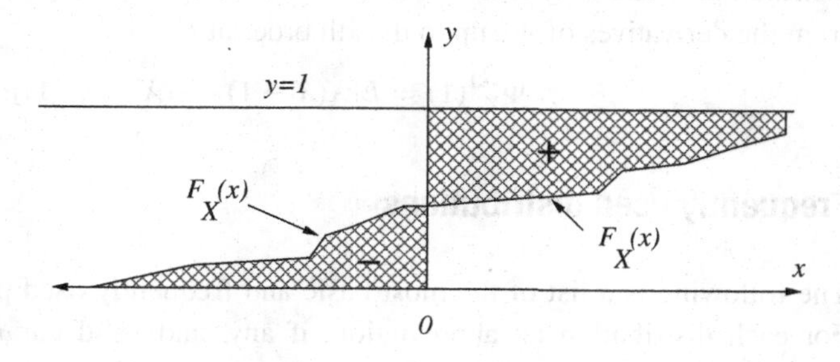

Figure 1.12 $E[X]$ is the difference of two areas.

The variance of a random variable X with $E[X]$ finite is defined by $\mathrm{Var}(X) = E[(X - E[X])^2]$. By the linearity of expectation, if $E[X]$ is finite, the variance of X satisfies the useful relation: $\mathrm{Var}(X) = E[X^2 - 2XE[X] + E[X]^2] = E[X^2] - E[X]^2$.

The following two inequalities are simple and fundamental. The *Markov inequality* states that if Y is a nonnegative random variable, then for $c > 0$,

$$P\{Y \geq c\} \leq \frac{E[Y]}{c}.$$

To prove Markov's inequality, note that $I_{\{Y \geq c\}} \leq \frac{Y}{c}$, and take expectations on each side. The *Chebychev inequality* states that if X is a random variable with finite mean μ and variance σ^2, then for any $d > 0$,

$$P\{|X - \mu| \geq d\} \leq \frac{\sigma^2}{d^2}.$$

The Chebychev inequality follows by applying the Markov inequality with $Y = |X - \mu|^2$ and $c = d^2$.

The characteristic function Φ_X of a random variable X is defined by

$$\Phi_X(u) = E[e^{juX}]$$

for real values of u, where $j = \sqrt{-1}$. For example, if X has pdf f, then

$$\Phi_X(u) = \int_{-\infty}^{\infty} \exp(jux) f_X(x) dx,$$

which is 2π times the inverse Fourier transform of f_X.

Two random variables have the same probability distribution if and only if they have the same characteristic function. If $E[X^k]$ exists and is finite for an integer $k \geq 1$, then the derivatives of Φ_X up to order k exist and are continuous, and

$$\Phi_X^{(k)}(0) = j^k E[X^k].$$

For a nonnegative integer-valued random variable X it is often more convenient to work with the z transform of the pmf, defined by

$$\Psi_X(z) = E[z^X] = \sum_{k=0}^{\infty} z^k p_X(k)$$

for real or complex z with $|z| \leq 1$. Two such random variables have the same probability distribution if and only if their z transforms are equal. If $E[X^k]$ is finite it can be found from the derivatives of Ψ_X up to the kth order at $z = 1$,

$$\Psi_X^{(k)}(1) = E[X(X-1)\cdots(X-k+1)].$$

1.6 Frequently used distributions

The following is a list of the most basic and frequently used probability distributions. For each distribution an abbreviation, if any, and valid parameter values are given, followed by either the CDF, pdf, or pmf, then the mean, variance, a typical example, and significance of the distribution.

The constants p, λ, μ, σ, a, b, and α are real-valued, and n and i are integer-valued, except n can be noninteger-valued in the case of the gamma distribution.

Bernoulli:
$Be(p)$, $0 \le p \le 1$

$$\text{pmf: } p(i) = \begin{cases} p & i = 1 \\ 1-p & i = 0 \\ 0 & \text{else} \end{cases}$$

z-transform: $1 - p + pz$

mean: p variance: $p(1-p)$

Example: Number of heads appearing in one flip of a coin. The coin is called fair if $p = \frac{1}{2}$ and biased otherwise.

Binomial:
$Bi(n,p)$, $n \ge 1$, $0 \le p \le 1$

$$\text{pmf:} p(i) = \binom{n}{i} p^i (1-p)^{n-i} \qquad 0 \le i \le n$$

z-transform: $(1 - p + pz)^n$

mean: np variance: $np(1-p)$

Example: Number of heads appearing in n independent flips of a coin.

Poisson:

$$\text{pmf: } p(i) = \frac{\lambda^i e^{-\lambda}}{i!} \qquad i \ge 0$$

z-transform: $\exp(\lambda(z-1))$

mean: λ variance: λ

Example: Number of phone calls placed during a ten second interval in a large city.

Significance: The Poisson pmf is the limit of the binomial pmf as $n \to +\infty$ and $p \to 0$ in such a way that $np \to \lambda$.

Geometric:
$Geo(p)$, $0 < p \le 1$

$$\text{pmf: } p(i) = (1-p)^{i-1} p \qquad i \ge 1$$

z-transform: $\dfrac{pz}{1 - z + pz}$

mean: $\dfrac{1}{p}$ variance: $\dfrac{1-p}{p^2}$

Example: Number of independent flips of a coin until heads first appears.

Significant property: If X has the geometric distribution, $P\{X > i\} = (1-p)^i$ for integers $i \geq 1$. So X has the *memoryless property*:

$$P(X > i+j \mid X > i) = P\{X > j\} \text{ for } i,j \geq 1.$$

Any positive integer-valued random variable with this property has a geometric distribution.

Gaussian
(also called Normal): $N(\mu, \sigma^2)$, $\mu \in \mathbb{R}$, $\sigma \geq 0$

$$\text{pdf (if } \sigma^2 > 0\text{): } f(x) = \frac{1}{\sqrt{2\pi\sigma^2}} \exp\left(-\frac{(x-\mu)^2}{2\sigma^2}\right)$$

$$\text{pmf (if } \sigma^2 = 0\text{): } p(x) = \begin{cases} 1 & x = \mu \\ 0 & \text{else} \end{cases}$$

$$\text{characteristic function: } \exp\left(ju\mu - \frac{u^2\sigma^2}{2}\right)$$

$$\text{mean: } \mu \qquad \text{variance: } \sigma^2$$

Example: Instantaneous voltage difference (due to thermal noise) measured across a resistor held at a fixed temperature.

Notation: The character Φ is often used to denote the CDF of an $N(0,1)$ random variable,[3] and Q is often used for the complementary CDF:

$$Q(c) = 1 - \Phi(c) = \int_c^\infty \frac{1}{\sqrt{2\pi}} e^{-\frac{x^2}{2}} dx.$$

Significant property (Central limit theorem): If $X_1, X_2, \ldots$ are independent and identically distributed with mean μ and nonzero variance σ^2, then for any constant c,

$$\lim_{n\to\infty} P\left\{\frac{X_1 + \cdots + X_n - n\mu}{\sqrt{n\sigma^2}} \leq c\right\} = \Phi(c).$$

Exponential:
$\text{Exp}(\lambda)$, $\lambda > 0$

$$\text{pdf: } f(x) = \lambda e^{-\lambda x} \qquad x \geq 0$$

$$\text{characteristic function: } \frac{\lambda}{\lambda - ju}$$

$$\text{mean: } \frac{1}{\lambda} \qquad \text{variance: } \frac{1}{\lambda^2}$$

[3] As noted earlier, Φ is also used to denote characteristic functions. The meaning should be clear from the context.

Example: Time elapsed between noon sharp and the first telephone call placed in a large city, on a given day.

Significance: If X has the Exp(λ) distribution, $P\{X \geq t\} = e^{-\lambda t}$ for $t \geq 0$. So X has the memoryless property:

$$P\{X \geq s+t \mid X \geq s\} = P\{X \geq t\} \qquad s, t \geq 0.$$

Any nonnegative random variable with this property is exponentially distributed.

Uniform:
$U(a, b) \quad -\infty < a < b < \infty$

$$\text{pdf: } f(x) = \begin{cases} \frac{1}{b-a} & a \leq x \leq b \\ 0 & \text{else} \end{cases}$$

$$\text{characteristic function: } \frac{e^{jub} - e^{jua}}{ju(b-a)}$$

$$\text{mean: } \frac{a+b}{2} \qquad \text{variance: } \frac{(b-a)^2}{12}$$

Example: The phase difference between two independent oscillators operating at the same frequency may be modeled as uniformly distributed over $[0, 2\pi]$.

Significance: Uniform is uniform.

Gamma(n, α):
$n, \alpha > 0$ (n real valued)

$$\text{pdf: } f(x) = \frac{\alpha^n x^{n-1} e^{-\alpha x}}{\Gamma(n)} \qquad x \geq 0$$

$$\text{where } \Gamma(n) = \int_0^\infty s^{n-1} e^{-s} ds$$

$$\text{characteristic function: } \left(\frac{\alpha}{\alpha - ju}\right)^n$$

$$\text{mean: } \frac{n}{\alpha} \qquad \text{variance: } \frac{n}{\alpha^2}$$

Significance: If n is a positive integer then $\Gamma(n) = (n-1)!$ and a Gamma(n, α) random variable has the same distribution as the sum of n independent, Exp(α) distributed random variables.

Rayleigh(σ^2):

$$\text{pdf: } f(r) = \frac{r}{\sigma^2}\exp\left(-\frac{r^2}{2\sigma^2}\right) \qquad r > 0$$

$$\text{CDF} : 1 - \exp\left(-\frac{r^2}{2\sigma^2}\right)$$

$$\text{mean: } \sigma\sqrt{\frac{\pi}{2}} \qquad \text{variance: } \sigma^2\left(2 - \frac{\pi}{2}\right)$$

Example: Instantaneous value of the envelope of a mean zero, narrow band noise signal.

Significance: If X and Y are independent, $N(0, \sigma^2)$ random variables, $(X^2 + Y^2)^{\frac{1}{2}}$ has the Rayleigh(σ^2) distribution. Also notable is the simple form of the CDF.

1.7 Failure rate functions

Eventually a system or a component of a particular system will fail. Let T be a random variable that denotes the lifetime of this item. Suppose T is a positive random variable with pdf f_T. The *failure rate function*, $h = (h(t) : t \geq 0)$, of T (and of the item itself) is defined by the following limit:

$$h(t) \stackrel{\triangle}{=} \lim_{\epsilon \to 0} \frac{P(t < T \leq t+\epsilon | T > t)}{\epsilon}.$$

That is, given the item is still working after t time units, the probability the item fails within the next ϵ time units is $h(t)\epsilon + o(\epsilon)$.

The failure rate function is determined by the distribution of T as follows:

$$\begin{aligned} h(t) &= \lim_{\epsilon \to 0} \frac{P\{t < T \leq t+\epsilon\}}{P\{T > t\}\epsilon} \\ &= \lim_{\epsilon \to 0} \frac{F_T(t+\epsilon) - F_T(t)}{(1 - F_T(t))\epsilon} \\ &= \frac{f_T(t)}{1 - F_T(t)}, \end{aligned} \tag{1.12}$$

because the pdf f_T is the derivative of the CDF F_T.

Conversely, a nonnegative function $h = (h(t) : t \geq 0)$ with $\int_0^\infty h(t)dt = \infty$ determines a probability distribution with failure rate function h as follows. The CDF is given by

$$F(t) = 1 - e^{-\int_0^t h(s)ds}. \tag{1.13}$$

It is easy to check that F given by (1.13) has failure rate function h. To derive (1.13), and hence show it gives the unique distribution with failure rate function h, start with $F'/(1-F) = h$. Equivalently, $(\ln(1-F))' = -h$ or $\ln(1-F) = \ln(1-F(0)) - \int_0^t h(s)ds$, which is equivalent to (1.13).

Example 1.14 (a) Find the failure rate function for an exponentially distributed random variable with parameter λ. (b) Find the distribution with the linear failure rate function $h(t) = \frac{t}{\sigma^2}$ for $t \geq 0$. (c) Find the failure rate function of $T = \min\{T_1, T_2\}$, where T_1 and T_2 are independent random variables such that T_1 has failure rate function h_1 and T_2 has failure rate function h_2.

Solution

(a) If T has the exponential distribution with parameter λ, then for $t \geq 0$, $f_T(t) = \lambda e^{-\lambda t}$ and $1 - F_T(t) = e^{-\lambda t}$, so by (1.12), $h(t) = \lambda$ for all $t \geq 0$. That is, the exponential distribution with parameter λ has constant failure rate λ. The constant failure rate property is connected with the memoryless property of the exponential distribution; the memoryless property implies that $P(t < T \leq T + \epsilon | T > t) = P\{T > \epsilon\}$,, which in view of the definition of h shows that h is constant.

(b) If $h(t) = \frac{t}{\sigma^2}$ for $t \geq 0$, then by (1.13), $F_T(t) = 1 - e^{-\frac{t^2}{2\sigma^2}}$. The corresponding pdf is given by

$$f_T(t) = \begin{cases} \frac{t}{\sigma^2} e^{-\frac{t^2}{2\sigma^2}} & t \geq 0 \\ 0 & \text{else} \end{cases}.$$

This is the pdf of the Rayleigh distribution with parameter σ^2.

(c) By the independence and (1.12) applied to T_1 and T_2,

$$\begin{aligned} P\{T > t\} &= P\{T_1 > t \text{ and } T_2 > t\} = P\{T_1 > t\}P\{T_2 > t\} \\ &= e^{\int_0^t -h_1(s)ds} e^{\int_0^t -h_2(s)ds} = e^{-\int_0^t h(s)ds}, \end{aligned}$$

where $h = h_1 + h_2$. Therefore, the failure rate function for the minimum of two independent random variables is the sum of their failure rate functions. This makes intuitive sense; if there is a system that fails when either of one of two components fails, then the rate of system failure is the sum of the rates of component failure.

1.8 Jointly distributed random variables

Let $X_1, X_2, \ldots, X_m$ be random variables on a single probability space $(\Omega, \mathcal{F}, P)$. The *joint cumulative distribution function* (CDF) is the function on $\mathbb{R}^m$ defined by

$$F_{X_1 X_2 \cdots X_m}(x_1, \ldots, x_m) = P\{X_1 \leq x_1, X_2 \leq x_2, \ldots, X_m \leq x_m\}.$$

The CDF determines the probabilities of all events concerning $X_1, \ldots, X_m$. For example, if R is the rectangular region $(a, b] \times (a', b']$ in the plane, then

$$P\{(X_1, X_2) \in R\} = F_{X_1X_2}(b, b') - F_{X_1X_2}(a, b') - F_{X_1X_2}(b, a') + F_{X_1X_2}(a, a').$$

We write $+\infty$ as an argument of F_X in place of x_i to denote the limit as $x_i \to +\infty$. By the countable additivity axiom of probability,

$$F_{X_1X_2}(x_1, +\infty) = \lim_{x_2 \to \infty} F_{X_1X_2}(x_1, x_2) = F_{X_1}(x_1).$$

The random variables are jointly continuous if there exists a function $f_{X_1X_2\cdots X_m}$, called the *joint probability density function* (pdf), such that

$$F_{X_1X_2\cdots X_m}(x_1, \ldots, x_m) = \int_{-\infty}^{x_1} \cdots \int_{-\infty}^{x_m} f_{X_1X_2\cdots X_m}(u_1, \ldots, u_m) du_m \cdots du_1.$$

Note that if X_1 and X_2 are jointly continuous, then

$$\begin{aligned} F_{X_1}(x_1) &= F_{X_1X_2}(x_1, +\infty) \\ &= \int_{-\infty}^{x_1} \left[\int_{-\infty}^{\infty} f_{X_1X_2}(u_1, u_2) du_2 \right] du_1, \end{aligned}$$

so that X_1 has pdf given by

$$f_{X_1}(u_1) = \int_{-\infty}^{\infty} f_{X_1X_2}(u_1, u_2) du_2.$$

The pdfs f_{X_1} and f_{X_2} are called the marginal pdfs for the joint pdf f_{X_1,X_2}.

If $X_1, X_2, \ldots, X_m$ are each discrete random variables, then they have a joint pmf $p_{X_1X_2\cdots X_m}$ defined by

$$p_{X_1X_2\cdots X_m}(u_1, u_2, \ldots, u_m) = P\{X_1 = u_1, X_2 = u_2, \ldots, X_m = u_m\}.$$

The sum of the probability masses is one, and for any subset A of $\mathbb{R}^m$

$$P\{(X_1, \ldots, X_m) \in A\} = \sum_{(u_1, \ldots, u_m) \in A} p_X(u_1, u_2, \ldots, u_m).$$

The joint pmf of subsets of $X_1, \ldots X_m$ can be obtained by summing out the other coordinates of the joint pmf. For example,

$$p_{X_1}(u_1) = \sum_{u_2} p_{X_1X_2}(u_1, u_2).$$

The joint characteristic function of $X_1, \ldots, X_m$ is the function on $\mathbb{R}^m$ defined by

$$\Phi_{X_1X_2\cdots X_m}(u_1, u_2, \ldots, u_m) = E[e^{j(X_1u_1 + X_2u_x + \cdots + X_mu_m)}].$$

Random variables $X_1, \ldots, X_m$ are defined to be *independent* if for any Borel subsets $A_1, \ldots, A_m$ of $\mathbb{R}$, the events $\{X_1 \in A_1\}, \ldots, \{X_m \in A_m\}$ are independent. The random variables are independent if and only if the joint CDF factors.

$$F_{X_1X_2\cdots X_m}(x_1, \ldots, x_m) = F_{X_1}(x_1) \cdots F_{X_m}(x_m).$$

If the random variables are jointly continuous, independence is equivalent to the condition that the joint pdf factors. If the random variables are discrete, independence is equivalent to the condition that the joint pmf factors. Similarly, the random variables are independent if and only if the joint characteristic function factors.

1.9 Conditional densities

Suppose that X and Y have a joint pdf f_{XY}. Recall that the pdf f_Y, the second marginal density of f_{XY}, is given by

$$f_Y(y) = \int_{-\infty}^{\infty} f_{XY}(x, y)dx.$$

The conditional pdf of X given Y, denoted by $f_{X|Y}(x \mid y)$, is undefined if $f_Y(y) = 0$. It is defined for y such that $f_Y(y) > 0$ by

$$f_{X|Y}(x \mid y) = \frac{f_{XY}(x, y)}{f_Y(y)} \qquad -\infty < x < +\infty.$$

If y is fixed and $f_Y(y) > 0$, then as a function of x, $f_{X|Y}(x \mid y)$ is itself a pdf.

The expectation of the conditional pdf is called the conditional expectation (or conditional mean) of X given $Y = y$, written as

$$E[X \mid Y = y] = \int_{-\infty}^{\infty} x f_{X|Y}(x \mid y)dx.$$

If the deterministic function $E[X \mid Y = y]$ is applied to the random variable Y, the result is a random variable denoted by $E[X \mid Y]$.

Note that conditional pdf and conditional expectation were so far defined in case X and Y have a joint pdf. If instead, X and Y are both discrete random variables, the conditional pmf $p_{X|Y}$ and the conditional expectation $E[X \mid Y = y]$ can be defined in a similar way. More general notions of conditional expectation are considered later.

1.10 Correlation and covariance

Let X and Y be random variables on the same probability space with finite second moments. Three important related quantities are:

the correlation: $E[XY]$,

the covariance: $\text{Cov}(X, Y) = E[(X - E[X])(Y - E[Y])]$,

the correlation coefficient: $\rho_{XY} = \dfrac{\text{Cov}(X, Y)}{\sqrt{\text{Var}(X)\text{Var}(Y)}}$.

A fundamental inequality is *Schwarz's inequality:*

$$| E[XY] | \leq \sqrt{E[X^2]E[Y^2]}. \tag{1.14}$$

Furthermore, if $E[Y^2] \neq 0$, equality holds if and only if $P(X = cY) = 1$ for some constant c. Schwarz's inequality (1.14) is equivalent to the L^2 *triangle inequality* for random variables:

$$E[(X + Y)^2]^{\frac{1}{2}} \leq E[X^2]^{\frac{1}{2}} + E[Y^2]^{\frac{1}{2}}. \tag{1.15}$$

Schwarz's inequality can be proved as follows. If $P\{Y = 0\} = 1$ the inequality is trivial, so suppose $E[Y^2] > 0$. By the inequality $(a+b)^2 \leq 2a^2 + 2b^2$ it follows that $E[(X - \lambda Y)^2] < \infty$ for any constant λ. Take $\lambda = E[XY]/E[Y^2]$ and note that

$$\begin{aligned} 0 \leq E[(X - \lambda Y)^2] &= E[X^2] - 2\lambda E[XY] + \lambda^2 E[Y^2] \\ &= E[X^2] - \frac{E[XY]^2}{E[Y^2]}, \end{aligned}$$

which is clearly equivalent to the Schwarz inequality. If $P(X = cY) = 1$ for some c then equality holds in (1.14), and conversely, if equality holds in (1.14) then $P(X = cY) = 1$ for $c = \lambda$.

Application of Schwarz's inequality to $X - E[X]$ and $Y - E[Y]$ in place of X and Y yields that

$$\mid \mathrm{Cov}(X, Y) \mid \leq \sqrt{\mathrm{Var}(X)\mathrm{Var}(Y)}.$$

Furthermore, if $\mathrm{Var}(Y) \neq 0$ then equality holds if and only if $X = aY + b$ for some constants a and b. Consequently, if $\mathrm{Var}(X)$ and $\mathrm{Var}(Y)$ are not zero, so that the correlation coefficient ρ_{XY} is well defined, then $\mid \rho_{XY} \mid \leq 1$ with equality if and only if $X = aY + b$ for some constants a, b.

The following alternative expressions for $\mathrm{Cov}(X, Y)$ are often useful in calculations:

$$\mathrm{Cov}(X, Y) \;=\; E[X(Y - E[Y])] = E[(X - E[X])Y] = E[XY] - E[X]E[Y].$$

In particular, if either X or Y has mean zero then $E[XY] = \mathrm{Cov}(X, Y)$.

Random variables X and Y are called orthogonal if $E[XY] = 0$ and are called uncorrelated if $\mathrm{Cov}(X, Y) = 0$. If X and Y are independent then they are uncorrelated. The converse is far from true. Independence requires a large number of equations to be true, namely $F_{XY}(x, y) = F_X(x)F_Y(y)$ for every real value of x and y. The condition of being uncorrelated involves only a single equation to hold.

Covariance generalizes variance, in that $\mathrm{Var}(X) = \mathrm{Cov}(X, X)$. Covariance is linear in each of its two arguments:

$$\begin{aligned} \mathrm{Cov}(X + Y, U + V) &= \mathrm{Cov}(X, U) + \mathrm{Cov}(X, V) + \mathrm{Cov}(Y, U) + \mathrm{Cov}(Y, V) \\ \mathrm{Cov}(aX + b, cY + d) &= ac\mathrm{Cov}(X, Y). \end{aligned}$$

for constants a, b, c, d. For example, consider the sum $S_m = X_1 + \ldots + X_m$, such that $X_1, \ldots, X_m$ are (pairwise) uncorrelated with $E[X_i] = \mu$ and $\mathrm{Var}(X_i) = \sigma^2$ for $1 \leq i \leq m$. Then $E[S_m] = m\mu$ and

$$\begin{aligned} \mathrm{Var}(S_m) &= \mathrm{Cov}(S_m, S_m) \\ &= \sum_i \mathrm{Var}(X_i) + \sum_{i,j:i\neq j} \mathrm{Cov}(X_i, X_j) \\ &= m\sigma^2. \end{aligned}$$

Therefore, $\frac{S_m - m\mu}{\sqrt{m\sigma^2}}$ has mean zero and variance one.

1.11 Transformation of random vectors

A random vector X of dimension m has the form

$$X = \begin{pmatrix} X_1 \\ X_2 \\ \vdots \\ X_m \end{pmatrix},$$

where $X_1, \ldots, X_m$ are random variables. The joint distribution of $X_1, \ldots, X_m$ can be considered to be the distribution of the vector X. For example, if $X_1, \ldots, X_m$ are jointly continuous, the joint pdf $f_{X_1 X_2 \cdots X_m}(x_1, \ldots, x_m)$ can as well be written as $f_X(x)$, and be thought of as the pdf of the random vector X.

Let X be a continuous type random vector on $\mathbb{R}^m$. Let g be a one-to-one mapping from $\mathbb{R}^m$ to $\mathbb{R}^m$. Think of g as mapping x-space (here x is lower case, representing a coordinate value) into y-space. As x varies over $\mathbb{R}^m$, y varies over the range of g. All the while, $y = g(x)$ or, equivalently, $x = g^{-1}(y)$.

Suppose that the Jacobian matrix of derivatives $\frac{\partial y}{\partial x}(x)$ is continuous in x and nonsingular for all x. By the inverse function theorem of vector calculus, it follows that the Jacobian matrix of the inverse mapping (from y to x) exists and satisfies $\frac{\partial x}{\partial y}(y) = \left(\frac{\partial y}{\partial x}(x)\right)^{-1}$. Use $\mid K \mid$ for a square matrix K to denote $|\det(K)|$.

Proposition 1.5 *Under the above assumptions, Y is a continuous type random vector and for y in the range of g:*

$$f_Y(y) = \frac{f_X(x)}{\mid \frac{\partial y}{\partial x}(x) \mid} = f_X(x) \left| \frac{\partial x}{\partial y}(y) \right|.$$

Example 1.15 Let U, V have the joint pdf:

$$f_{UV}(u, v) = \begin{cases} u + v & 0 \le u, v \le 1 \\ 0 & \text{else} \end{cases}$$

and let $X = U^2$ and $Y = U(1 + V)$. Let us find the pdf f_{XY}. The vector (U, V) in the $u - v$ plane is transformed into the vector (X, Y) in the $x - y$ plane under a mapping g that maps u, v to $x = u^2$ and $y = u(1 + v)$. The image in the $x - y$ plane of the square $[0, 1]^2$ in the $u - v$ plane is the set A given by

$$A = \{(x, y) : 0 \le x \le 1, \text{ and } \sqrt{x} \le y \le 2\sqrt{x}\}.$$

The mapping from the square is one to one, for if $(x, y) \in A$ then (u, v) can be recovered by $u = \sqrt{x}$ and $v = \frac{y}{\sqrt{x}} - 1$, see Figure 1.13. The Jacobian determinant is

$$\begin{vmatrix} \frac{\partial x}{\partial u} & \frac{\partial x}{\partial v} \\ \frac{\partial y}{\partial u} & \frac{\partial y}{\partial v} \end{vmatrix} = \begin{vmatrix} 2u & 0 \\ 1 + v & u \end{vmatrix} = 2u^2.$$

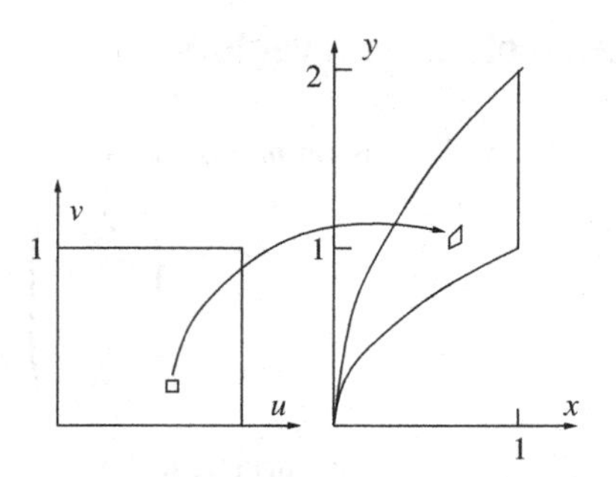

Figure 1.13 Transformation from the $u - v$ plane to the $x - y$ plane.

Therefore, using the transformation formula and expressing u and v in terms of x and y yields

$$f_{XY}(x,y) = \begin{cases} \frac{\sqrt{x}+(\frac{y}{\sqrt{x}}-1)}{2x} & \text{if } (x,y) \in A \\ 0 & \text{else} \end{cases}.$$

Example 1.16 Let U and V be independent continuous type random variables. Let $X = U + V$ and $Y = V$. Let us find the joint density of X, Y and the marginal density of X. The mapping

$$g : \begin{pmatrix} u & v \end{pmatrix} \to \begin{pmatrix} u & v \end{pmatrix} = \begin{pmatrix} u+v & v \end{pmatrix}$$

is invertible, with inverse given by $u = x - y$ and $v = y$. The absolute value of the Jacobian determinant is given by

$$\left| \begin{array}{cc} \frac{\partial x}{\partial u} & \frac{\partial x}{\partial v} \\ \frac{\partial y}{\partial u} & \frac{\partial y}{\partial v} \end{array} \right| = \left| \begin{array}{cc} 1 & 1 \\ 0 & 1 \end{array} \right| = 1.$$

Therefore

$$f_{XY}(x,y) = f_{UV}(u,v) = f_U(x-y)f_V(y).$$

The marginal density of X is given by

$$f_X(x) = \int_{-\infty}^{\infty} f_{XY}(x,y)dy = \int_{-\infty}^{\infty} f_U(x-y)f_V(y)dy.$$

That is $f_X = f_U * f_V$.

Example 1.17 Let X_1 and X_2 be independent $N(0,\sigma^2)$ random variables, and let $X = (X_1, X_2)^T$ denote the two-dimensional random vector with coordinates X_1 and X_2. Any point of $x \in \mathbb{R}^2$ can be represented in polar coordinates by the vector $(r,\theta)^T$ such that $r = \|x\| = (x_1^2 + x_2^2)^{\frac{1}{2}}$ and $\theta = \tan^{-1}\left(\frac{x_2}{x_1}\right)$ with values $r \geq 0$ and $0 \leq \theta < 2\pi$. The inverse of this mapping is given by

$$x_1 = r\cos(\theta)$$
$$x_2 = r\sin(\theta).$$

We endeavor to find the pdf of the random vector $(R,\Theta)^T$, the polar coordinates of X. The pdf of X is given by

$$f_X(x) = f_{X_1}(x_1)f_{X_2}(x_2) = \frac{1}{2\pi\sigma^2}e^{-\frac{r^2}{2\sigma^2}}.$$

The range of the mapping is the set $r > 0$ and $0 < \theta \leq 2\pi$. On the range,

$$\left|\frac{\partial x}{\partial\begin{pmatrix} r \\ \theta \end{pmatrix}}\right| = \left|\begin{matrix} \frac{\partial x_1}{\partial r} & \frac{\partial x_1}{\partial \theta} \\ \frac{\partial x_2}{\partial r} & \frac{\partial x_2}{\partial \theta} \end{matrix}\right| = \left|\begin{matrix} \cos(\theta) & -r\sin(\theta) \\ \sin(\theta) & r\cos(\theta) \end{matrix}\right| = r.$$

Therefore for $(r,\theta)^T$ in the range of the mapping,

$$f_{R,\Theta}(r,\theta) = f_X(x)\left|\frac{\partial x}{\partial\begin{pmatrix} r \\ \theta \end{pmatrix}}\right| = \frac{r}{2\pi\sigma^2}e^{-\frac{r^2}{2\sigma^2}}.$$

Of course $f_{R,\Theta}(r,\theta) = 0$ off the range of the mapping. The joint density factors into a function of r and a function of θ, so R and Θ are independent. Moreover, R has the Rayleigh density with parameter σ^2, and Θ is uniformly distributed on $[0, 2\pi]$.

Problems

1.1 Simple events A register contains eight random binary digits which are mutually independent. Each digit is a zero or a one with equal probability.

(a) Describe an appropriate probability space $(\Omega, \mathcal{F}, P)$ corresponding to looking at the contents of the register.

(b) Express each of the following four events explicitly as subsets of Ω, and find their probabilities:

E_1 = "No two neighboring digits are the same,"

E_2 = "Some cyclic shift of the register contents is equal to 01100110,"

E_3 = "The register contains exactly four zeros,"

E_4 = "There is a run of at least six consecutive ones."

(c) Find $P(E_1|E_3)$ and $P(E_2|E_3)$.

1.2 A ballot problem Suppose there is an election with two candidates and six ballots turned in, such that four of the ballots are for the winning candidate and two of the ballots are for the other candidate. The ballots are opened and counted one at a time, in random order, with all orders equally likely. Find the probability that from the time the first ballot is counted until all the ballots are counted, the winning candidate has the majority of the ballots counted. ("Majority" means there are strictly more votes for the winning candidate than for the other candidate.)

1.3 Ordering of three random variables Suppose X, Y, and U are mutually independent, such that X and Y are each exponentially distributed with some common parameter $\lambda > 0$, and U is uniformly distributed on the interval $[0, 1]$. Express $P\{X < U < Y\}$ in terms of λ. Simplify your answer.

1.4 Independent vs. mutually exclusive (a) Suppose that an event E is independent of itself. Show that either $P(E) = 0$ or $P(E) = 1$.
(b) Events A and B have probabilities $P(A) = 0.3$ and $P(B) = 0.4$. What is $P(A \cup B)$ if A and B are independent? What is $P(A \cup B)$ if A and B are mutually exclusive?
(c) Now suppose that $P(A) = 0.6$ and $P(B) = 0.8$. In this case, could the events A and B be independent? Could they be mutually exclusive?

1.5 Congestion at output ports Consider a packet switch with some number of input ports and eight output ports. Suppose four packets simultaneously arrive on different input ports, and each is routed toward an output port. Assume the choices of output ports are mutually independent, and for each packet, each output port has equal probability.
(a) Specify a probability space $(\Omega, \mathcal{F}, P)$ to describe this situation.
(b) Let X_i denote the number of packets routed to output port i for $1 \leq i \leq 8$. Describe the joint pmf of $X_1, \ldots, X_8$.
(c) Find $\mathrm{Cov}(X_1, X_2)$.
(d) Find $P\{X_i \leq 1 \text{ for all } i\}$.
(e) Find $P\{X_i \leq 2 \text{ for all } i\}$.

1.6 Frantic search At the end of each day Professor Plum puts her glasses in her drawer with probability .90, leaves them on the table with probability .06, leaves them in her briefcase with probability 0.03, and she actually leaves them at the office with probability 0.01. The next morning she has no recollection of where she left the glasses. She looks for them, but each time she looks in a place the glasses are actually located, she misses finding them with probability 0.1, whether or not she already looked in the same place. (After all, she doesn't have her glasses on and she is in a hurry.)
(a) Given that Professor Plum didn't find the glasses in her drawer after looking one time, what is the conditional probability the glasses are on the table?
(b) Given that she didn't find the glasses after looking for them in the drawer and on the table once each, what is the conditional probability they are in the briefcase?
(c) Given that she failed to find the glasses after looking in the drawer twice, on the table twice, and in the briefcase once, what is the conditional probability she left the glasses at the office?

1.7 Conditional probability of failed device given failed attempts A particular webserver may be working or not working. If the webserver is not working, any attempt to access it fails. Even if the webserver is working, an attempt to access it can fail due to network congestion beyond the control of the webserver. Suppose that the a-priori probability that the server is working is 0.8. Suppose that if the server is working, then each access attempt is successful with probability 0.9, independently of other access attempts. Find the following quantities.
(a) $P($ first access attempt fails$)$,
(b) $P($server is working $|$ first access attempt fails$)$,

(c) P(second access attempt fails | first access attempt fails),
(d) P(server is working | first and second access attempts fail).

1.8 Conditional probabilities – basic computations of iterative decoding
(a) Suppose $B_1, \ldots, B_n, Y_1, \ldots, Y_n$ are discrete random variables with joint pmf

$$p(b_1, \ldots, b_n, y_1, \ldots, y_n) = \begin{cases} 2^{-n} \prod_{i=1}^{n} q_i(y_i|b_i) & \text{if } b_i \in \{0,1\} \text{ for } 1 \le i \le n \\ 0 & \text{else} \end{cases},$$

where $q_i(y_i|b_i)$ as a function of y_i is a pmf for $b_i \in \{0,1\}$. Finally, let $B = B_1 \oplus \cdots \oplus B_n$ represent the modulo two sum of $B_1, \cdots, B_n$. Thus, the ordinary sum of the $n+1$ random variables $B_1, \ldots, B_n, B$ is even. Express $P(B = 1|Y_1 = y_1, \cdots, Y_n = y_n)$ in terms of the y_i and the functions q_i. Simplify your answer.
(b) Suppose B and $Z_1, \ldots, Z_k$ are discrete random variables with joint pmf

$$p(b, z_1, \ldots, z_k) = \begin{cases} \frac{1}{2} \prod_{j=1}^{k} r_j(z_j|b) & \text{if } b \in \{0,1\} \\ 0 & \text{else} \end{cases},$$

where $r_j(z_j|b)$ as a function of z_j is a pmf for $b \in \{0,1\}$ fixed.
Express $P(B = 1|Z_1 = z_1, \ldots, Z_k = z_k)$ in terms of the z_j and the functions r_j.

1.9 Conditional lifetimes; memoryless property of the geometric distribution (a) Let X represent the lifetime, rounded up to an integer number of years, of a certain car battery. Suppose that the pmf of X is given by $p_X(k) = 0.2$ if $3 \le k \le 7$ and $p_X(k) = 0$ otherwise.
(i) Find the probability, $P\{X > 3\}$, that a three year old battery is still working.
(ii) Given that the battery is still working after five years, what is the conditional probability that the battery will still be working three years later? (i.e. what is $P(X > 8|X > 5)$?).
(b) A certain basketball player shoots the ball repeatedly from half court during practice. Each shot is a success with probability p and a miss with probability $1-p$, independently of the outcomes of previous shots. Let Y denote the number of shots required for the first success.
(i) Express the probability that she needs more than three shots for a success, $P\{Y > 3\}$, in terms of p.
(ii) Given that she already missed the first five shots, what is the conditional probability that she will need more than three additional shots for a success? (i.e. what is $P(Y > 8|Y > 5)$?).
(iii) What type of probability distribution does Y have?

1.10 Blue corners Suppose each corner of a cube is colored blue, independently of the other corners, with some probability p. Let B denote the event that at least one face of the cube has all four corners colored blue.
(a) Find the conditional probability of B given that exactly five corners of the cube are colored blue.
(b) Find $P(B)$, the unconditional probability of B.

1.11 Distribution of the flow capacity of a network A communication network is shown. The link capacities in megabits per second (Mbps) are given by $C_1 = C_3 = 5$, $C_2 = C_5 = 10$ and $C_4 = 8$, and are the same in each direction. Information flow from the source to the destination can be split among multiple paths. For example, if all links

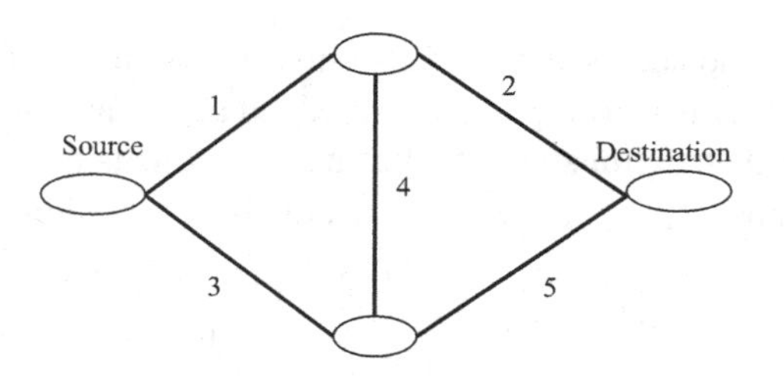

are working, then the maximum communication rate is 10 Mbps: 5 Mbps can be routed over links 1 and 2, and 5 Mbps can be routed over links 3 and 5. Let F_i be the event that link i fails. Suppose that F_1, F_2, F_3, F_4, and F_5 are independent and $P(F_i) = 0.2$ for each i. Let X be defined as the maximum rate (in Mbits per second) at which data can be sent from the source node to the destination node. Find the pmf p_X.

1.12 Recognizing cumulative distribution functions Which of the following are valid CDFs? For each that is not valid, state at least one reason why. For each that is valid, find $P\{X^2 > 5\}$.

$$F_1(x) = \begin{cases} \frac{e^{-x^2}}{4} & x < 0 \\ 1 - \frac{e^{-x^2}}{4} & x \geq 0 \end{cases}, \quad F_2(x) = \begin{cases} 0 & x < 0 \\ 0.5 + e^{-x} & 0 \leq x < 3 \\ 1 & x \geq 3 \end{cases},$$

$$F_3(x) = \begin{cases} 0 & x \leq 0 \\ 0.5 + \frac{x}{20} & 0 < x \leq 10 \\ 1 & x \geq 10 \end{cases}.$$

1.13 A CDF of mixed type Let X have the CDF shown.

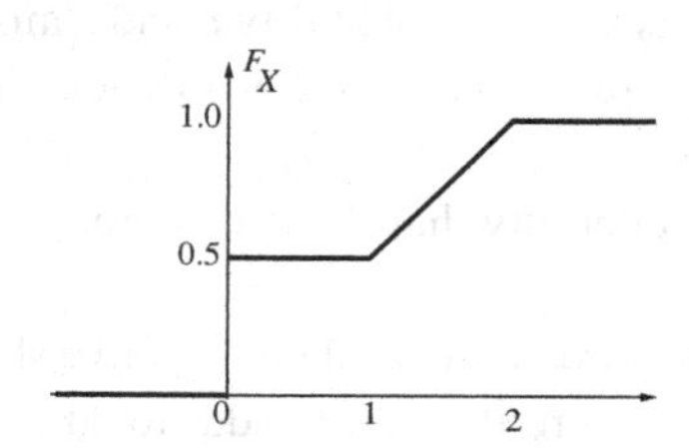

(a) Find $P\{X \leq 0.8\}$.
(b) Find $E[X]$.
(c) Find $\mathrm{Var}(X)$.

1.14 CDF and characteristic function of a mixed type random variable Let $X = (U - 0.5)_+$, where U is uniformly distributed over the interval $[0, 1]$. That is, $X = U - 0.5$ if $U - 0.5 \geq 0$, and $X = 0$ if $U - 0.5 < 0$.
(a) Find and carefully sketch the CDF F_X. In particular, what is $F_X(0)$?
(b) Find the characteristic function $\Phi_X(u)$ for real values of u.

1.15 Poisson and geometric random variables with conditioning Let Y be a Poisson random variable with mean $\mu > 0$ and let Z be a geometrically distributed random variable with parameter p with $0 < p < 1$. Assume Y and Z are independent.
(a) Find $P\{Y < Z\}$. Express your answer as a simple function of μ and p.

(b) Find $P(Y < Z|Z = i)$ for $i \geq 1$. (Hint: This is a conditional probability for events.)
(c) Find $P(Y = i|Y < Z)$ for $i \geq 0$. Express your answer as a simple function of p, μ and i. (Hint: This is a conditional probability for events.)
(d) Find $E[Y|Y < Z]$, which is the expected value computed according to the conditional distribution found in part (c). Express your answer as a simple function of μ and p.

1.16 Conditional expectation for uniform density over a triangular region Let (X, Y) be uniformly distributed over the triangle with coordinates $(0, 0)$, $(1, 0)$, and $(2, 1)$.
(a) What is the value of the joint pdf inside the triangle?
(b) Find the marginal density of X, $f_X(x)$. Be sure to specify your answer for all real values of x.
(c) Find the conditional density function $f_{Y|X}(y|x)$. Be sure to specify which values of x the conditional density is well defined for, and for such x specify the conditional density for all y. Also, for such x briefly describe the conditional density of y in words.
(d) Find the conditional expectation $E[Y|X = x]$. Be sure to specify which values of x this conditional expectation is well defined for.

1.17 Transformation of a random variable Let X be exponentially distributed with mean λ^{-1}. Find and carefully sketch the distribution functions for the random variables $Y = \exp(X)$ and $Z = \min(X, 3)$.

1.18 Density of a function of a random variable Suppose X is a random variable with probability density function

$$f_X(x) = \begin{cases} 2x & 0 \leq x \leq 1 \\ 0 & \text{else} \end{cases}$$

(a) Find $P(X \geq 0.4|X \leq 0.8)$.
(b) Find the density function of Y defined by $Y = -\log(X)$.

1.19 Moments and densities of functions of a random variable Suppose the length L and width W of a rectangle are independent and each uniformly distributed over the interval [0, 1]. Let $C = 2L + 2W$ (the length of the perimeter) and $A = LW$ (the area). Find the means, variances, and probability densities of C and A.

1.20 Functions of independent exponential random variables Let X_1 and X_2 be independent random variables, with X_i being exponentially distributed with parameter λ_i. (a) Find the pdf of $Z = \min\{X_1, X_2\}$. (b) Find the pdf of $R = \frac{X_1}{X_2}$.

1.21 Using the Gaussian Q function Express each of the given probabilities in terms of the standard Gaussian complementary CDF Q.
(a) $P\{X \geq 16\}$, where X has the $N(10, 9)$ distribution.
(b) $P\{X^2 \geq 16\}$, where X has the $N(10, 9)$ distribution.
(c) $P\{|X - 2Y| > 1\}$, where X and Y are independent, $N(0, 1)$ random variables. (Hint: Linear combinations of independent Gaussian random variables are Gaussian.)

1.22 Gaussians and the Q function Let X and Y be independent, $N(0, 1)$ random variables.
(a) Find $\text{Cov}(3X + 2Y, X + 5Y + 10)$.
(b) Express $P\{X + 4Y \geq 2\}$ in terms of the Q function.
(c) Express $P\{(X - Y)^2 > 9\}$ in terms of the Q function.

1.23 Correlation of histogram values Suppose that n fair dice are independently rolled. Let

$$X_i = \begin{cases} 1 & \text{if a 1 shows on the } i\text{th roll} \\ 0 & \text{else} \end{cases}, \quad Y_i = \begin{cases} 1 & \text{if a 2 shows on the } i\text{th roll} \\ 0 & \text{else} \end{cases}.$$

Let X denote the sum of the X_is, which is simply the number of 1s rolled. Let Y denote the sum of the Y_is, which is simply the number of 2s rolled. Note that if a histogram is made recording the number of occurrences of each of the six numbers, then X and Y are the heights of the first two entries in the histogram.
(a) Find $E[X_1]$ and $\text{Var}(X_1)$.
(b) Find $E[X]$ and $\text{Var}(X)$.
(c) Find $\text{Cov}(X_i, Y_j)$ if $1 \le i,j \le n$ (Hint: Does it make a difference if $i = j$?)
(d) Find $\text{Cov}(X, Y)$ and the correlation coefficient $\rho(X, Y)$.
(e) Find $E[Y|X = x]$ for any integer x with $0 \le x \le n$. Note that your answer should depend on x and n, but otherwise your answer is deterministic.

1.24 Working with a joint density Suppose X and Y have joint density function $f_{X,Y}(x, y) = c(1 + xy)$ if $2 \le x \le 3$ and $1 \le y \le 2$, and $f_{X,Y}(x, y) = 0$ otherwise. (a) Find c. (b) Find f_X and f_Y. (c) Find $f_{X|Y}$.

1.25 A function of jointly distributed random variables Suppose (U, V) is uniformly distributed over the square with corners (0,0), (1,0), (1,1), and (0,1), and let $X = UV$. Find the CDF and pdf of X.

1.26 Density of a difference Let X and Y be independent, exponentially distributed random variables with parameter λ, such that $\lambda > 0$. Find the pdf of $Z = |X - Y|$.

1.27 Working with a two-dimensional density Let the random variables X and Y be jointly uniformly distributed over the region shown.

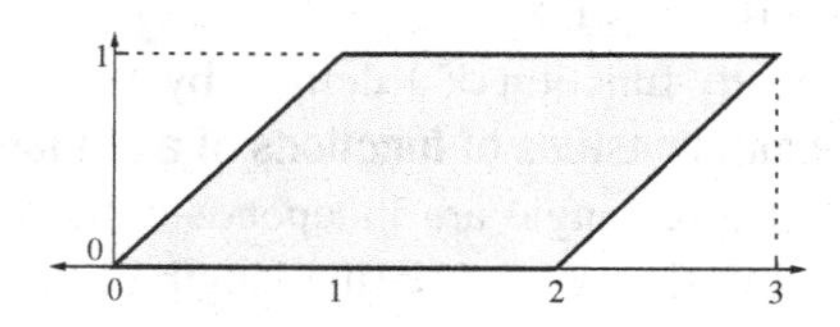

(a) Determine the value of $f_{X,Y}$ on the region shown.
(b) Find f_X, the marginal pdf of X.
(c) Find the mean and variance of X.
(d) Find the conditional pdf of Y given that $X = x$, for $0 \le x \le 1$.
(e) Find the conditional pdf of Y given that $X = x$, for $1 \le x \le 2$.
(f) Find and sketch $E[Y|X = x]$ as a function of x. Be sure to specify which range of x this conditional expectation is well defined for.

1.28 Some characteristic functions Find the mean and variance of random variables with the following characteristic functions: (a) $\Phi(u) = \exp(-5u^2 + 2ju)$; (b) $\Phi(u) = (e^{ju} - 1)/ju$; and (c) $\Phi(u) = \exp(\lambda(e^{ju} - 1))$.

1.29 Uniform density over a union of two square regions Let X and Y be jointly uniformly distributed on the region $\{0 \le u \le 1, 0 \le v \le 1\} \cup \{-1 \le u < 0, -1 \le v < 0\}$.
(a) Determine the value of f_{XY} on the region shown.

(b) Find f_X, the marginal pdf of X.
(c) Find the conditional pdf of Y given that $X = a$, for $0 < a \leq 1$.
(d) Find the conditional pdf of Y given that $X = a$, for $-1 \leq a < 0$.
(e) Find $E[Y|X = a]$ for $|a| \leq 1$.
(f) What is the correlation coefficient of X and Y?
(g) Are X and Y independent?
(h) What is the pdf of $Z = X + Y$?

1.30 A transformation of jointly continuous random variables Suppose (U, V) has joint pdf

$$f_{U,V}(u, v) = \begin{cases} 9u^2v^2 & \text{if } 0 \leq u \leq 1 \,\&\, 0 \leq v \leq 1 \\ 0 & \text{else} \end{cases}.$$

Let $X = 3U$ and $Y = UV$.
(a) Find the joint pdf of X and Y, being sure to specify where the joint pdf is zero.
(b) Using the joint pdf of X and Y, find the conditional pdf, $f_{Y|X}(y|x)$, of Y given X. (Be sure to indicate which values of x the conditional pdf is well defined for, and for each such x specify the conditional pdf for all real values of y.)

1.31 Transformation of densities Let U and V have the joint pdf:

$$f_{UV}(u, v) = \begin{cases} c(u - v)^2 & 0 \leq u, v \leq 1 \\ 0 & \text{else} \end{cases}$$

for some constant c.
(a) Find the constant c.
(b) Suppose $X = U^2$ and $Y = U^2V^2$. Describe the joint pdf $f_{X,Y}(x, y)$ of X and Y. Be sure to indicate where the joint pdf is zero.

1.32 Opening a bicycle combination lock A certain bicycle combination lock has 10^4 possible combinations, ranging from 0000 to 9999. Suppose the combination required to open the lock takes any one of the possible values with equal probability. Suppose it takes two seconds to try opening the lock with a given combination. Find the mean and standard deviation of the amount of time, each to within a minute, of how long it would take to open the lock by cycling through the combinations without repetition. (Hint: You can approximate the random amount of time required by a continuous type random variable.)

1.33 Transformation of joint densities Assume X and Y are independent, each with the exponential pdf with parameter $\lambda > 0$. Let $W = X - Y$ and $Z = X^2 + X - Y$. Find the joint pdf of (W, Z). Be sure to specify its support (i.e. where it is not zero).

1.34 Computing some covariances Suppose X, Y, and Z are random variables, each with mean zero and variance 20, such that $\text{Cov}(X, Y) = \text{Cov}(X, Z) = 10$ and $\text{Cov}(Y, Z) = 5$.
(a) Find $\text{Cov}(X + Y, X - Y)$.
(b) Find Cov(3X+Z,3X+Y). (c) Find $E[(X + Y)^2]$.

1.35 Conditional densities and expectations Suppose that random variables X and Y have the joint pdf:

$$f_{XY}(u,v) = \begin{cases} 4u^2, & 0 < v < u < 1 \\ 0, & \text{elsewhere} \end{cases}.$$

(a) Find $E[XY]$.
(b) Find $f_Y(v)$. Be sure to specify it for all values of v.
(c) Find $f_{X|Y}(u|v)$. Be sure to specify where it is undefined, and where it is zero.
(d) Find $E[X^2|Y=v]$ for $0 < v < 1$.

1.36 Jointly distributed variables Let U and V be independent random variables, such that U is uniformly distributed over the interval $[0, 1]$, and V has the exponential probability density function.
(a) Calculate $E[\frac{V^2}{1+U}]$.
(b) Calculate $P\{U \leq V\}$.
(c) Find the joint probability density function of Y and Z, where $Y = U^2$ and $Z = UV$.

1.37* Why not every set has a length Suppose a length (actually, "one-dimensional volume" would be a better name) of any subset $A \subset \mathbb{R}$ could be defined, so that the following axioms are satisfied:

L0: $0 \leq \text{length}(A) \leq \infty$ for any $A \subset \mathbb{R}$,

L1: $\text{Length}([a,b]) = b - a$ for $a < b$,

L2: $\text{Length}(A) = \text{length}(A+y)$, for any $A \subset \mathbb{R}$ and $y \in R$, where $A+y$ represents the translation of A by y, defined by $A + y = \{x + y : x \in A\}$,

L3: If $A = \cup_{i=1}^{\infty} B_i$ such that $B_1, B_2, \ldots$ are disjoint, then $\text{length}(A) = \sum_{i=1}^{\infty} \text{length}(B_i)$.

The purpose of this problem is to show that the above supposition leads to a contradiction. Let $\mathbb{Q}$ denote the set of rational numbers, $\mathbb{Q} = \{p/q : p, q \in \mathbb{Z}, q \neq 0\}$.
(a) Show that the set of rational numbers can be expressed as $\mathbb{Q} = \{q_1, q_2, \ldots\}$, which means that $\mathbb{Q}$ is *countably* infinite. Say that $x, y \in \mathbb{R}$ are equivalent, and write $x \sim y$, if $x - y \in \mathbb{Q}$.
(b) Show that $\sim$ is an *equivalence relation*, meaning it is reflexive ($a \sim a$ for all $a \in \mathbb{R}$), symmetric ($a \sim b$ implies $b \sim a$), and transitive ($a \sim b$ and $b \sim c$ implies $a \sim c$). For any $x \in \mathbb{R}$, let $Q_x = \mathbb{Q} + x$.
(c) Show that for any $x, y \in \mathbb{R}$, either $Q_x = Q_y$ or $Q_x \cap Q_y = \emptyset$. Sets of the form Q_x are called *equivalence classes* of the equivalence relation $\sim$.
(d) Show that $Q_x \cap [0,1] \neq \emptyset$ for all $x \in R$, or in other words, each equivalence class contains at least one element from the interval $[0, 1]$. Let V be a set obtained by choosing exactly one element in $[0, 1]$ from each equivalence class (by accepting that V is well defined, you'll be accepting what is called the *Axiom of Choice*). So V is a subset of $[0, 1]$. Suppose $q'_1, q'_2, \ldots$ is an enumeration of all the rational numbers in the interval $[-1, 1]$, with no number appearing twice in the list. Let $V_i = V + q'_i$ for $i \geq 1$.
(e) Verify that the sets V_i are disjoint, and $[0,1] \subset \cup_{i=1}^{\infty} V_i \subset [-1, 2]$. Since the V_is are translations of V, they should all have the same length as V. If the length of V is defined to be zero, then $[0, 1]$ would be covered by a countable union of disjoint sets of length zero, so $[0, 1]$ would also have length zero. If the length of V were strictly positive, then

the countable union would have infinite length, and hence the interval $[-1,2]$ would have infinite length. Either way there is a contradiction.

1.38* On sigma-algebras, random variables, and measurable functions Prove the seven statements lettered (a)–(g) in what follows.

Definition. Let Ω be an arbitrary set. A nonempty collection $\mathcal{F}$ of subsets of Ω is defined to be an algebra if: (i) $A^c \in \mathcal{F}$ whenever $A \in F$ and (ii) $A \cup B \in \mathcal{F}$ whenever $A, B \in \mathcal{F}$.

(a) If $\mathcal{F}$ is an algebra then $\emptyset \in \mathcal{F}$, $\Omega \in \mathcal{F}$, and the union or intersection of any finite collection of sets in $\mathcal{F}$ is in $\mathcal{F}$. **Definition.** $\mathcal{F}$ is called a σ-algebra if $\mathcal{F}$ is an algebra such that whenever $A_1, A_2, \ldots$ are each in F, so is the union, $\cup A_i$.

(b) If $\mathcal{F}$ is a σ-algebra and $B_1, B_2, \ldots$ are in F, then so is the intersection, $\cap B_i$.

(c) Let U be an arbitrary nonempty set, and suppose that $\mathcal{F}_u$ is a σ-algebra of subsets of Ω for each $u \in U$. Then the intersection $\cap_{u \in U} \mathcal{F}_u$ is also a σ-algebra.

(d) The collection of all subsets of Ω is a σ-algebra.

(e) If $\mathcal{F}_o$ is any collection of subsets of Ω then there is a smallest σ-algebra containing $\mathcal{F}_o$ (Hint: use (c) and (d).)

Definitions. $\mathcal{B}(R)$ is the smallest σ-algebra of subsets of R which contains all sets of the form $(-\infty, a]$. Sets in $\mathcal{B}(R)$ are called Borel sets. A real-valued random variable on a probability space $(\Omega, \mathcal{F}, P)$ is a real-valued function X on Ω such that $\{\omega : X(\omega) \leq a\} \in \mathcal{F}$ for any $a \in R$.

(f) If X is a random variable on $(\Omega, \mathcal{F}, P)$ and $A \in \mathcal{B}(R)$ then $\{\omega : X(\omega) \in A\} \in \mathcal{F}$. (Hint: Fix a random variable X. Let $\mathcal{D}$ be the collection of all subsets A of $\mathcal{B}(R)$ for which the conclusion is true. It is enough (why?) to show that $\mathcal{D}$ contains all sets of the form $(-\infty, a]$ and that $\mathcal{D}$ is a σ-algebra of subsets of R. You must use the fact that $\mathcal{F}$ is a σ-algebra.)

Remark. By (f), $P\{\omega : X(\omega) \in A\}$, or $P\{X \in A\}$ for short, is well defined for $A \in \mathcal{B}(R)$.

Definition. A function g mapping R to R is called Borel measurable if $\{x : g(x) \in A\} \in \mathcal{B}(R)$ whenever $A \in \mathcal{B}(R)$.

(g) If X is a real-valued random variable on $(\Omega, \mathcal{F}, P)$ and g is a Borel measurable function, then Y defined by $Y = g(X)$ is also a random variable on $(\Omega, \mathcal{F}, P)$.

2 Convergence of a sequence of random variables

Convergence to limits is a central concept in the theory of calculus. Limits are used to define derivatives and integrals. So to study integrals and derivatives of random functions it is natural to begin by examining what it means for a sequence of random variables to converge. Convergence of sequences of random variables is also central to important tools in probability theory, such as the law of large numbers and central limit theorem. See the Appendix for a review of the definition of convergence for a sequence of numbers.

2.1 Four definitions of convergence of random variables

Recall that a random variable X is a function on Ω for some probability space $(\Omega, \mathcal{F}, P)$. A sequence of random variables $(X_n(\omega) : n \geq 1)$ is hence a sequence of functions. There are many possible definitions for convergence of a sequence of random variables. One idea is to require $X_n(\omega)$ to converge for each fixed ω. However, at least intuitively, what happens on an event of probability zero is not important. Thus, we use the following definition.

Definition 2.1 A sequence of random variables $(X_n : n \geq 1)$ *converges almost surely* to a random variable X, if all the random variables are defined on the same probability space, and $P\{\lim_{n\to\infty} X_n = X\} = 1$. Almost sure convergence is denoted by $\lim_{n\to\infty} X_n = X$ *a.s.* or $X_n \stackrel{a.s.}{\to} X$.

Conceptually, to check almost sure convergence, one can first find the set $\{\omega : \lim_{n\to\infty} X_n(\omega) = X(\omega)\}$ and then see if it has probability one.

We shall construct some examples using the standard unit-interval probability space defined in Example 1.2. This particular choice of $(\Omega, \mathcal{F}, P)$ is useful for generating examples, because random variables, being functions on Ω, can be simply specified by their graphs. For example, consider the random variable X pictured in Figure 2.1. The probability mass function for such X is given by $P\{X = 1\} = P\{X = 2\} = \frac{1}{4}$ and $P\{X = 3\} = \frac{1}{2}$. Figure 2.1 is a bit ambiguous, in that it is not clear what the values of X are at the jump points, $\omega = 1/4$ or $\omega = 1/2$. However, each of these points has probability zero, so the distribution of X is the same no matter how X is defined at those points.

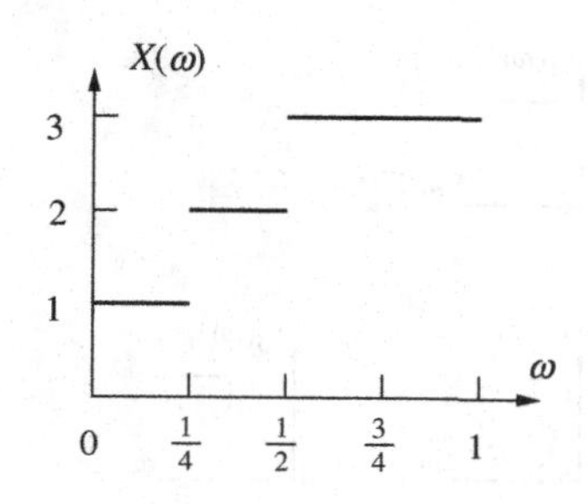

Figure 2.1 A random variable on $(\Omega, \mathcal{F}, P)$.

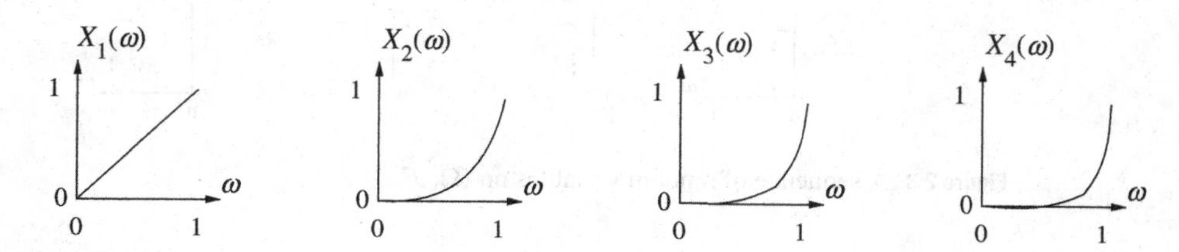

Figure 2.2 $X_n(\omega) = \omega^n$ on the standard unit-interval probability space.

Example 2.1 Let $(X_n : n \geq 1)$ be the sequence of random variables on the standard unit-interval probability space defined by $X_n(\omega) = \omega^n$, illustrated in Figure 2.2.

This sequence converges for all $\omega \in \Omega$, with the limit

$$\lim_{n\to\infty} X_n(\omega) = \begin{cases} 0 & \text{if } 0 \leq \omega < 1 \\ 1 & \text{if } \omega = 1 \end{cases}.$$

The single point set $\{1\}$ has probability zero, so it is also true (and simpler to say) that $(X_n : n \geq 1)$ converges a.s. to zero. In other words, if we let X be the zero random variable, defined by $X(\omega) = 0$ for all ω, then $X_n \stackrel{a.s.}{\to} X$.

Example 2.2 (Moving, shrinking rectangles) Let $(X_n : n \geq 1)$ be the sequence of random variables on the standard unit-interval probability space, as shown in Figure 2.3. The variable X_1 is identically one. The variables X_2 and X_3 are one on intervals of length $\frac{1}{2}$. The variables X_4, X_5, X_6, and X_7 are one on intervals of length $\frac{1}{4}$. In general, each $n \geq 1$ can be written as $n = 2^k + j$ where $k = \lfloor \ln_2 n \rfloor$ and $0 \leq j < 2^k$. The variable X_n is one on the length 2^{-k} interval $(j2^{-k}, (j+1)2^{-k}]$.

To investigate a.s. convergence, fix an arbitrary value for ω. Then for each $k \geq 1$, there is one value of n with $2^k \leq n < 2^{k+1}$ such that $X_n(\omega) = 1$, and $X_n(\omega) = 0$ for all other n. Therefore, $\lim_{n\to\infty} X_n(\omega)$ does not exist. That is, $\{\omega : \lim_{n\to\infty} X_n \text{ exists}\} = \emptyset$, so of course, $P\{\lim_{n\to\infty} X_n \text{ exists}\} = 0$. Thus, X_n does not converge in the a.s. sense.

However, for large n, $P\{X_n = 0\}$ is close to one. This suggests that X_n converges to the zero random variable in some weaker sense.

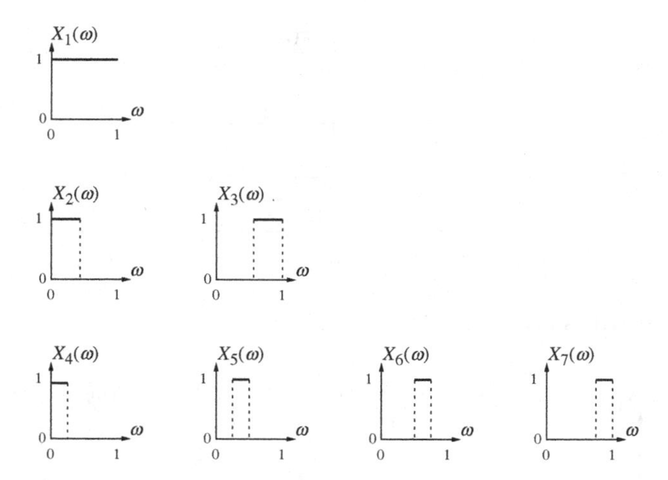

Figure 2.3 A sequence of random variables on $(\Omega, \mathcal{F}, P)$.

Example 2.2 motivates us to consider the following weaker notion of convergence of a sequence of random variables.

Definition 2.2 A sequence of random variables (X_n) converges to a random variable X *in probability* if all the random variables are defined on the same probability space, and for any $\epsilon > 0$, $\lim_{n\to\infty} P\{|X - X_n| \geq \epsilon\} = 0$. Convergence in probability is denoted by $\lim_{n\to\infty} X_n = X p.$, or $X_n \xrightarrow{p.} X$.

Convergence in probability requires that $|X - X_n|$ be small with high probability (to be precise, less than or equal to ϵ with probability that converges to one as $n \to \infty$), but on the small probability event that $|X - X_n|$ is not small, it can be arbitrarily large. For some applications that is unacceptable. Roughly speaking, the next definition of convergence requires that $|X - X_n|$ be small with high probability for large n, and even if it is not small, the average squared value has to be small enough.

Definition 2.3 A sequence of random variables (X_n) converges to a random variable X *in the mean square* sense if all the random variables are defined on the same probability space, $E[X_n^2] < +\infty$ for all n, and $\lim_{n\to\infty} E[(X_n - X)^2] = 0$. Mean square convergence is denoted by $\lim_{n\to\infty} X_n = X\, m.s.$ or $X_n \xrightarrow{m.s.} X$.

Although it is not explicitly stated in the definition of m.s. convergence, the limit random variable must also have a finite second moment:

Proposition 2.4 *If $X_n \xrightarrow{m.s.} X$, then $E[X^2] < +\infty$.*

Proof Suppose $X_n \xrightarrow{m.s.} X$. By definition, $E[X_n^2] < \infty$ for all n. Also, there exists some n_o so $E[(X - X_n)^2] < 1$ for all $n \geq n_o$. The L^2 triangle inequality for random variables, (1.15), yields $E[(X_\infty)^2]^{\frac{1}{2}} \leq E[(X_\infty - X_{n_o})^2]^{\frac{1}{2}} + E[X_{n_o}^2]^{\frac{1}{2}} < +\infty$. □

Example 2.3 (More moving, shrinking rectangles) This example is along the same lines as Example 2.2, using the standard unit-interval probability space. Each random variable of the sequence $(X_n : n \geq 1)$ is defined as indicated in Figure 2.4, where the value $a_n > 0$ is some constant depending on n. The graph of X_n for $n \geq 1$ has height a_n over some subinterval of Ω of length $\frac{1}{n}$. We don't explicitly identify the location of the interval, but we require that for any fixed ω, $X_n(\omega) = a_n$ for infinitely many values of n, and $X_n(\omega) = 0$ for infinitely many values of n. Such a choice of the locations of the intervals is possible because the sum of the lengths of the intervals, $\sum_{n=1}^{\infty} \frac{1}{n}$, is infinite.

Of course $X_n \stackrel{a.s.}{\to} 0$ if the deterministic sequence (a_n) converges to zero. However, if there is a constant $\epsilon > 0$ such that $a_n \geq \epsilon$ for all n (for example if $a_n = 1$ for all n), then $\{\omega : \lim_{n\to\infty} X_n(\omega) \text{ exists}\} = \emptyset$, just as in Example 2.2. The sequence converges to zero in probability for any choice of the constants (a_n), because for any $\epsilon > 0$,

$$P\{|X_n - 0| \geq \epsilon\} \leq P\{X_n \neq 0\} = \frac{1}{n} \to 0.$$

Finally, to investigate mean square convergence, note that $E[|X_n - 0|^2] = \frac{a_n^2}{n}$. Hence, $X_n \stackrel{m.s.}{\to} 0$ if and only if the sequence of constants (a_n) is such that $\lim_{n\to\infty} \frac{a_n^2}{n} = 0$. For example, if $a_n = \ln(n)$ for all n, then $X_n \stackrel{m.s.}{\to} 0$, but if $a_n = \sqrt{n}$, then (X_n) does not converge to zero in the m.s. sense. (Proposition 2.7 below shows that a sequence can have only one limit in the a.s., p., or m.s. senses, so the fact $X_n \stackrel{p.}{\to} 0$, implies that zero is the only possible limit in the m.s. sense. So if $\frac{a_n^2}{n} \not\to 0$, then (X_n) doesn't converge to any random variable in the m.s. sense.)

Example 2.4 (Anchored, shrinking rectangles) Let $(X_n : n \geq 1)$ be a sequence of random variables defined on the standard unit-interval probability space, as indicated in Figure 2.5, where the value $a_n > 0$ is some constant depending on n. That is, $X_n(\omega)$ is equal to a_n if $0 \leq \omega \leq 1/n$, and to zero otherwise. For any nonzero ω in Ω, $X_n(\omega) = 0$ for all n such that $n > 1/\omega$. Therefore, $X_n \stackrel{a.s.}{\to} 0$.

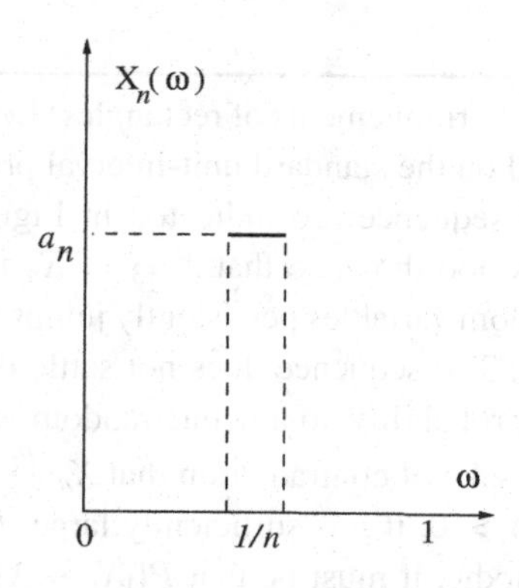

Figure 2.4 A sequence of random variables corresponding to moving, shrinking rectangles.

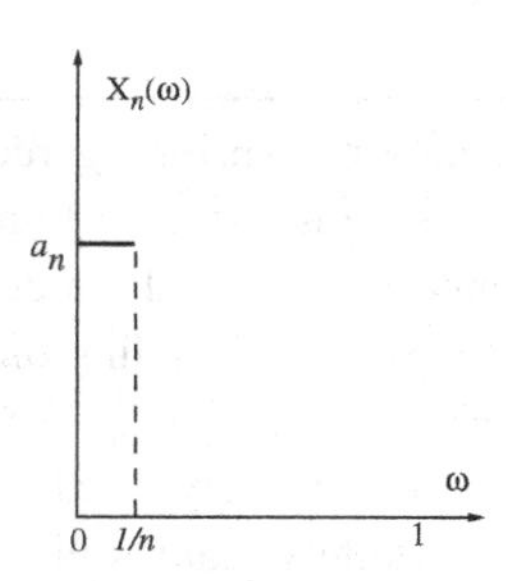

Figure 2.5 A sequence of random variables corresponding to anchored, shrinking rectangles.

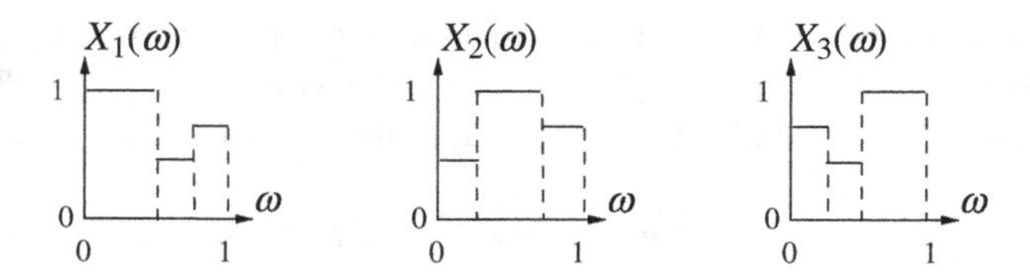

Figure 2.6 A sequence of random variables obtained by rearrangement of rectangles.

Whether the sequence (X_n) converges in p. or m.s. sense for this example is exactly the same as in Example 2.3. That is, for convergence in probability or mean square sense, the locations of the shrinking intervals of support don't matter. So $X_n \xrightarrow{p.} 0$. And $X_n \xrightarrow{m.s.} 0$ if and only if $\frac{a_n^2}{n} \to 0$.

It is shown in Proposition 2.7 below that either a.s. or m.s. convergence implies convergence in probability. Example 2.4 shows that a.s. convergence, like convergence in probability, can allow $|X_n(\omega) - X(\omega)|$ to be extremely large for ω in a small probability set. So neither convergence in probability, nor a.s. convergence, implies m.s. convergence, unless an additional assumption is made to control $|X_n(\omega) - X(\omega)|$ everywhere on Ω.

Example 2.5 (Rearrangements of rectangles) Let $(X_n : n \geq 1)$ be a sequence of random variables defined on the standard unit-interval probability space. The first three random variables in the sequence are indicated in Figure 2.6. Suppose that the sequence is periodic, with period three, so that $X_{n+3} = X_n$ for all $n \geq 1$. Intuitively speaking, the sequence of random variables persistently jumps around. Obviously it does not converge in the a.s. sense. The sequence does not settle down to converge, even in the sense of convergence in probability, to any one random variable. This can be proved as follows. Suppose for the sake of contradiction that $X_n \xrightarrow{p.} X$ for some random variable. Then for any $\epsilon > 0$ and $\delta > 0$, if n is sufficiently large, $P\{|X_n - X| \geq \epsilon\} \leq \delta$. But because the sequence is periodic, it must be that $P\{|X_n - X| \geq \epsilon\} \leq \delta$ for $1 \leq n \leq 3$. Since δ is arbitrary it must be that $P\{|X_n - X| \geq \epsilon\} = 0$ for $1 \leq n \leq 3$. Since ϵ is arbitrary it must be that $P\{X = X_n\} = 1$ for $1 \le n \le 3$. Hence, $P\{X_1 = X_2 = X_3\} = 1$, which is a contradiction. Thus, the sequence does not converge in probability. A similar argument shows it does not converge in the m.s. sense, either.

Even though the sequence fails to converge in a.s., m.s., or p. senses, it can be observed that all of the X_ns have the same probability distribution. The variables are only different in that the places they take their possible values are rearranged.

Example 2.5 suggests that it would be useful to have a notion of convergence that just depends on the distributions of the random variables. One idea for a definition of convergence in distribution is to require that the sequence of CDFs $F_{X_n}(x)$ converge as $n \to \infty$ for all n. The following example shows such a definition could give unexpected results in some cases.

Example 2.6 Let U be uniformly distributed on the interval $[0, 1]$, and for $n \ge 1$, let $X_n = \frac{(-1)^n U}{n}$. Let X denote the random variable such that $X = 0$ for all ω. It is easy to verify that $X_n \stackrel{a.s.}{\to} X$ and $X_n \stackrel{p.}{\to} X$. Does the CDF of X_n converge to the CDF of X? The CDF of X_n is graphed in Figure 2.7. The CDF $F_{X_n}(x)$ converges to 0 for $x < 0$ and to one for $x > 0$. However, $F_{X_n}(0)$ alternates between 0 and 1 and hence does not converge to anything. In particular, it does not converge to $F_X(0)$. Thus, $F_{X_n}(x)$ converges to $F_X(x)$ for all x except $x = 0$.

Recall that the distribution of a random variable X has probability mass Δ at some value x_o, i.e. $P\{X = x_o\} = \Delta > 0$, if and only if the CDF has a jump of size Δ at x_o: $F(x_o) - F(x_o-) = \Delta$. Example 2.6 illustrates the fact that if the limit random variable X has such a point mass, then even if X_n is very close to X, the value $F_{X_n}(x)$ need not converge. To overcome this phenomenon, we adopt a definition of convergence in distribution which requires convergence of the CDFs only at the continuity points of the limit CDF. Continuity points are defined for general functions in Appendix 11.3. Since CDFs are right-continuous and nondecreasing, a point x is a continuity point of a CDF F if and only if there is no jump of F at X: i.e. if $F_X(x) = F_X(x-)$.

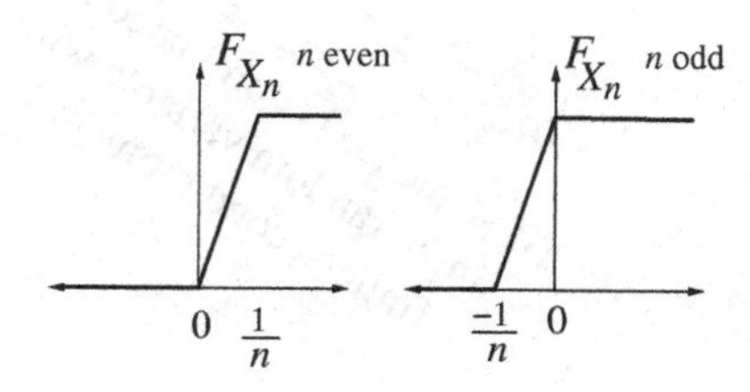

Figure 2.7 CDF of $X_n = \frac{(-1)^n}{n}$.

Definition 2.5 A sequence $(X_n : n \geq 1)$ of random variables *converges in distribution* to a random variable X if

$$\lim_{n\to\infty} F_{X_n}(x) = F_X(x) \text{ at all continuity points } x \text{ of } F_X.$$

Convergence in distribution is denoted by $\lim_{n\to\infty} X_n = X\, d.$ or $X_n \stackrel{d.}{\to} X$.

One way to investigate convergence in distribution is through the use of characteristic functions.

Proposition 2.6 *Let (X_n) be a sequence of random variables and let X be a random variable. Then the following are equivalent:*

(i) $X_n \stackrel{d.}{\to} X$,
(ii) $E[f(X_n)] \to E[f(X)]$ *for any bounded continuous function f,*
(iii) $\Phi_{X_n}(u) \to \Phi_X(u)$ *for each $u \in \mathbb{R}$ (i.e. pointwise convergence of characteristic functions).*

The relationships among the four types of convergence discussed in this section are given in the following proposition, and are pictured in Figure 2.8. The definitions use differing amounts of information about the random variables $(X_n : n \geq 1)$ and X involved. Convergence in the a.s. sense involves joint properties of all the random variables. Convergence in the p. or m.s. sense involves only pairwise joint distributions – namely those of (X_n, X) for all n. Convergence in distribution involves only the individual distributions of the random variables to have a convergence property. Convergence in the a.s., m.s., and p. senses requires the variables to all be defined on the same probability space. For convergence in distribution, the random variables need not be defined on the same probability space.

Proposition 2.7 *(a)* *If $X_n \stackrel{a.s.}{\to} X$ then $X_n \stackrel{p.}{\to} X$.*
(b) *If $X_n \stackrel{m.s.}{\to} X$ then $X_n \stackrel{p.}{\to} X$.*

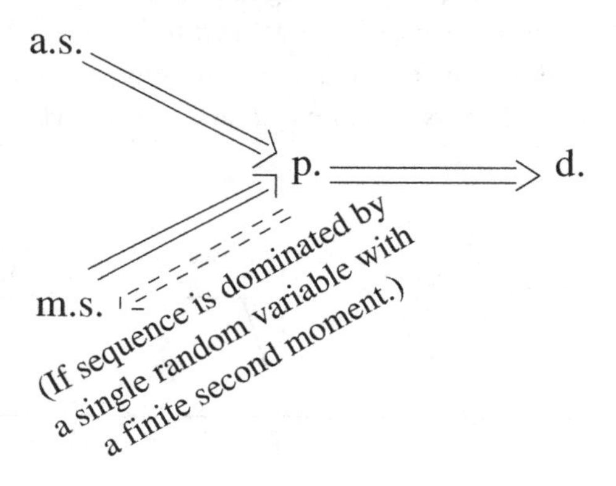

Figure 2.8 Relationships among four types of convergence of random variables.

(c) *If $P\{|X_n| \leq Y\} = 1$ for all n for some fixed random variable Y with $E[Y^2] < \infty$, and if $X_n \xrightarrow{p.} X$, then $X_n \xrightarrow{m.s.} X$.*

(d) *If $X_n \xrightarrow{p.} X$ then $X_n \xrightarrow{d.} X$.*

(e) *Suppose $X_n \to X$ in the p., m.s., or a.s. sense and $X_n \to Y$ in the p., m.s., or a.s. sense. Then $P\{X = Y\} = 1$. That is, if differences on sets of probability zero are ignored, a sequence of random variables can have only one limit (if p., m.s., and/or a.s. senses are used).*

(f) *Suppose $X_n \xrightarrow{d.} X$ and $X_n \xrightarrow{d.} Y$. Then X and Y have the same distribution.*

Proof (a) Suppose $X_n \xrightarrow{a.s.} X$ and let $\epsilon > 0$. Define a sequence of events A_n by

$$A_n = \{\omega :| X_n(\omega) - X(\omega) |< \epsilon\}.$$

We only need to show that $P(A_n) \to 1$. Define B_n by

$$B_n = \{\omega :| X_k(\omega) - X(\omega) |< \epsilon \text{ for all } k \geq n\}.$$

Note that $B_n \subset A_n$ and $B_1 \subset B_2 \subset \ldots$ so $\lim_{n\to\infty} P(B_n) = P(B)$ where $B = \bigcup_{n=1}^{\infty} B_n$. Clearly

$$B \supset \{\omega : \lim_{n\to\infty} X_n(\omega) = X(\omega)\},$$

so $1 = P(B) = lim_{n\to\infty} P(B_n)$. Since $P(A_n)$ is squeezed between $P(B_n)$ and 1, $\lim_{n\to\infty} P(A_n) = 1$, so $X_n \xrightarrow{p.} X$.

(b) Suppose $X_n \xrightarrow{m.s.} X$ and let $\epsilon > 0$. By the Markov inequality applied to $|X - X_n|^2$,

$$P\{| X - X_n |\geq \epsilon\} \leq \frac{E[| X - X_n |^2]}{\epsilon^2}. \tag{2.1}$$

The right side of (2.1), and hence the left side of (2.1), converge to zero as n goes to infinity. Therefore $X_n \xrightarrow{p.} X$ as $n \to \infty$.

(c) Suppose $X_n \xrightarrow{p.} X$. Then for any $\epsilon > 0$,

$$P\{| X |\geq Y + \epsilon\} \leq P\{| X - X_n |\geq \epsilon\} \to 0,$$

so that $P\{| X |\geq Y + \epsilon\} = 0$ for every $\epsilon > 0$. Thus, $P\{| X |\leq Y\} = 1$, so that $P\{| X - X_n |^2 \leq 4Y^2\} = 1$. Therefore, with probability one, for any $\epsilon > 0$,

$$| X - X_n |^2 \leq 4Y^2 I_{\{|X - X_n| \geq \epsilon\}} + \epsilon^2$$

so

$$E[| X - X_n |^2] \leq 4E[Y^2 I_{\{|X - X_n| \geq \epsilon\}}] + \epsilon^2.$$

In the special case that $P\{Y = L\} = 1$ for a constant L, the term $E[Y^2 I_{\{|X - X_n| \geq \epsilon\}}]$ is equal to $L^2 P\{|X - X_n| \geq \epsilon\}$, and by the hypotheses, $P\{|X - X_n| \geq \epsilon\} \to 0$. Even if Y is random, since $E[Y^2] < \infty$ and $P\{|X - X_n| \geq \epsilon\} \to 0$, it still follows that $E[Y^2 I_{\{|X - X_n| \geq \epsilon\}}] \to 0$ as

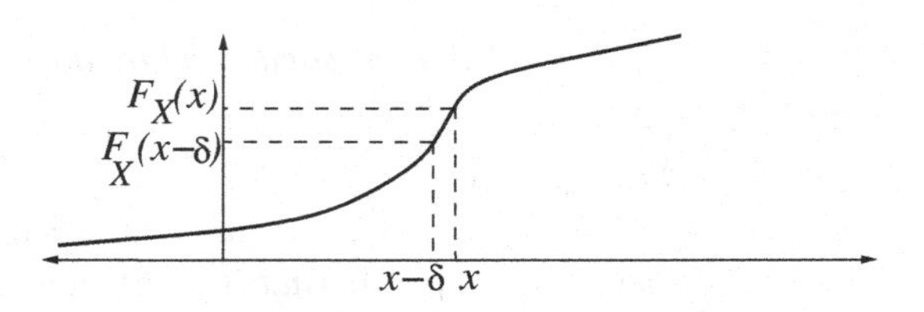

Figure 2.9 A CDF at a continuity point.

$n \to \infty$, by Corollary 11.13 (see Appendix). So, for n large enough, $E[|X-X_n|^2] \le 2\epsilon^2$. Since ϵ was arbitrary, $X_n \stackrel{m.s.}{\to} X$.

(d) Assume $X_n \stackrel{p.}{\to} X$. Select any continuity point x of F_X. It must be proved that $\lim_{n\to\infty} F_{X_n}(x) = F_X(x)$. Let $\epsilon > 0$. Then there exists $\delta > 0$ so that $F_X(x) \le F_X(x-\delta) + \frac{\epsilon}{2}$ (see Figure 2.9). Now

$$\begin{aligned}\{X \le x-\delta\} &= \{X \le x-\delta,\ X_n \le x\} \cup \{X \le x-\delta,\ X_n > x\} \\ &\subset \{X_n \le x\} \cup \{|X-X_n| \ge \delta\}\end{aligned}$$

so

$$F_X(x-\delta) \le F_{X_n}(x) + P\{|\, X_n - X \,| \ge \delta\}.$$

For all n sufficiently large, $P\{|\, X_n - X \,| \ge \delta\} \le \frac{\epsilon}{2}$. This and the choice of δ yield, for all n sufficiently large, $F_X(x) \le F_{X_n}(x) + \epsilon$. Similarly, for all n sufficiently large, $F_X(x) \ge F_{X_N}(x) - \epsilon$. So for all n sufficiently large, $|F_{X_n}(x) - F_X(x)| \le \epsilon$. Since ϵ was arbitrary, $\lim_{n\to\infty} F_{X_n}(x) = F_X(x)$.

(e) By parts (a) and (b), already proved, we can assume that $X_n \stackrel{p.}{\to} X$ and $X_n \stackrel{p.}{\to} Y$. Let $\epsilon > 0$ and $\delta > 0$, and select N so large that $P\{|X_n - X| \ge \epsilon\} \le \delta$ and $P\{|X_n - Y| \ge \epsilon\} \le \delta$ for all $n \ge N$. By the triangle inequality, $|X - Y| \le |X_N - X| + |X_N - Y|$. Thus,
$\{|X-Y| \ge 2\epsilon\} \subset \{|X_N - X| \ge \epsilon\} \cup \{|Y_N - X| \ge \epsilon\}$ so that
$P\{|X-Y| \ge 2\epsilon\} \le P\{|X_N - X| \ge \epsilon\} + P\{|X_N - Y| \ge \epsilon\} \le 2\delta$. We have proved that $P\{|X-Y| \ge 2\epsilon\} \le 2\delta$. Since δ was arbitrary, it must be that $P\{|X-Y| \ge 2\epsilon\} = 0$. Since ϵ was arbitrary, it must be that $P\{|X-Y| = 0\} = 1$.

(f) Suppose $X_n \stackrel{d.}{\to} X$ and $X_n \stackrel{d.}{\to} Y$. Then $F_X(x) = F_Y(y)$ whenever x is a continuity point of both x and y. Since F_X and F_Y are nondecreasing and bounded, they can have only finitely many discontinuities of size greater than $1/n$ for any n, so that the total number of discontinuities is at most countably infinite. Hence, in any nonempty interval, there is a point of continuity of both functions. So for any $x \in \mathbb{R}$, there is a strictly decreasing sequence of numbers converging to x, such that x_n is a point of continuity of both F_X and F_Y. So $F_X(x_n) = F_Y(x_n)$ for all n. Taking the limit as $n \to \infty$ and using the right-continuity of CDFs, we have $F_X(x) = F_Y(x)$. □

Example 2.7 Suppose $P\{X_0 \geq 0\} = 1$ and $X_n = 6 + \sqrt{X_{n-1}}$ for $n \geq 1$. For example, if for some ω it happens that $X_0(\omega) = 12$, then

$$X_1(\omega) = 6 + \sqrt{12} = 9.465\ldots$$
$$X_2(\omega) = 6 + \sqrt{9.46} = 9.076\ldots$$
$$X_3(\omega) = 6 + \sqrt{9.076} = 9.0127\ldots$$

Examining Figure 2.10, it is clear that for any ω with $X_0(\omega) > 0$, the sequence of numbers $X_n(\omega)$ converges to 9. Therefore, $X_n \stackrel{a.s.}{\to} 9$. The rate of convergence can be bounded as follows.

Note that for each $x \geq 0$, $|\ 6 + \sqrt{x} - 9\ | \ \leq\ |\ 6 + \frac{x}{3} - 9\ |$. Therefore,

$$|\ X_n(\omega) - 9\ | \ \leq\ |\ 6 + \frac{X_{n-1}(\omega)}{3} - 9\ | = \frac{1}{3}\ |\ X_{n-1}(\omega) - 9\ |$$

so that by induction on n,

$$|\ X_n(\omega) - 9\ | \leq 3^{-n}\ |\ X_0(\omega) - 9\ |\ . \tag{2.2}$$

Since $X_n \stackrel{a.s.}{\to} 9$ it follows that $X_n \stackrel{p.}{\to} 9$.

Finally, we investigate m.s. convergence under the assumption that $E[X_0^2] < +\infty$. By the inequality $(a+b)^2 \leq 2a^2 + 2b^2$, it follows that

$$E[(X_0 - 9)^2] \leq 2(E[X_0^2] + 81). \tag{2.3}$$

Squaring and taking expectations on each side of (2.2) and using (2.3) thus yields

$$E[|\ X_n - 9\ |^2] \leq 2 \cdot 3^{-2n}\{E[X_0^2] + 81\}.$$

Therefore, $X_n \stackrel{m.s.}{\to} 9$.

Example 2.8 Let $W_0, W_1, \ldots$ be independent, normal random variables with mean 0 and variance 1. Let $X_{-1} = 0$ and

$$X_n = (.9)X_{n-1} + W_n \qquad n \geq 0.$$

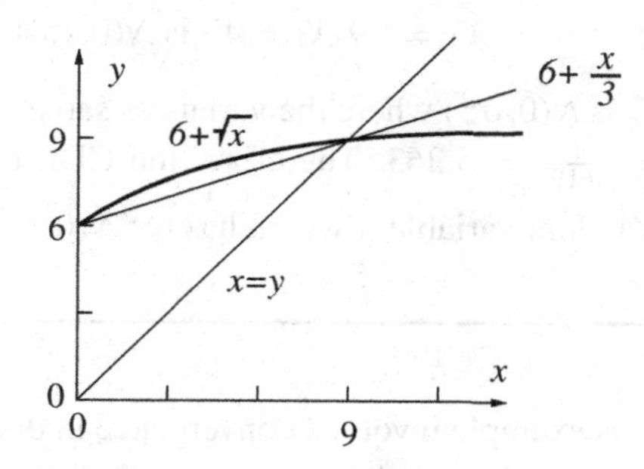

Figure 2.10 Graph of the functions $6 + \sqrt{x}$ and $6 + \frac{x}{3}$.

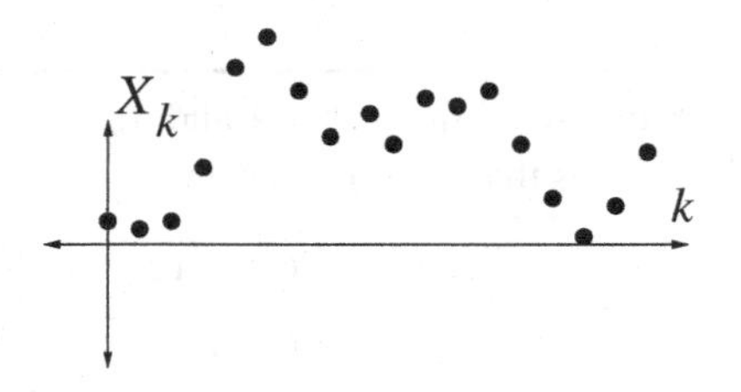

Figure 2.11 A typical sample sequence of X.

In what sense does X_n converge as n goes to infinity? For fixed ω, the sequence of numbers $X_0(\omega), X_1(\omega), \ldots$ might appear as in Figure 2.11.

Intuitively speaking, X_n persistently moves. We claim that X_n does not converge in probability (so also not in the a.s. or m.s. senses). Here is a proof of the claim. Examination of a table for the normal distribution yields that $P\{W_n \geq 2\} = P\{W_n \leq -2\} \geq 0.02$. Then

$$\begin{aligned} P\{| X_n - X_{n-1} |\geq 2\} &\geq P\{X_{n-1} \geq 0, W_n \leq -2\} + P\{X_{n-1} < 0, W_n \geq 2\} \\ &= P\{X_{n-1} \geq 0\}P\{W_n \leq -2\} + P\{X_{n-1} < 0\}P\{W_n \geq 2\} \\ &= P\{W_n \geq 2\} \geq 0.02. \end{aligned}$$

Therefore, for any random variable X,

$$\begin{aligned} &P\{| X_n - X |\geq 1\} + P\{| X_{n-1} - X |\geq 1\} \\ &\quad \geq P\{| X_n - X |\geq 1 \text{ or } | X_{n-1} - X |\geq 1\} \\ &\quad \geq P\{| X_n - X_{n-1} |\geq 2\} \geq 0.02, \end{aligned}$$

so $P\{| X_n - X |\geq 1\}$ does not converge to zero as $n \to \infty$. So X_n does not converge in probability to any random variable X. The claim is proved.

Although X_n does not converge in probability, or in the a.s. or m.s. senses, it nevertheless seems to asymptotically settle into an equilibrium. To probe this point further, let us find the distribution of X_n for each n:

$$\begin{aligned} X_0 &= W_0 \text{ is } N(0, 1), \\ X_1 &= (.9)X_0 + W_1 \text{ is } N(0, 1.81), \\ X_2 &= (.9)X_1 + W_2 \text{ is } N(0, (.81)(1.81 + 1)). \end{aligned}$$

In general, X_n is $N(0, \sigma_n^2)$ where the variances satisfy $\sigma_n^2 = (0.81)\sigma_{n-1}^2 + 1$ so $\sigma_n^2 \to \sigma_\infty^2$ where $\sigma_\infty^2 = \frac{1}{0.19} = 5.263$. Therefore, the CDF of X_n converges everywhere to the CDF of any random variable X which has the $N(0, \sigma_\infty^2)$ distribution. So $X_n \xrightarrow{d.} X$ for any such X.

The previous example involved convergence in distribution of Gaussian random variables. The limit random variable was also Gaussian. In fact, we close this section by showing that limits of Gaussian random variables are always Gaussian. Recall that X

is a Gaussian random variable with mean μ and variance σ^2 if either $\sigma^2 > 0$ and $F_X(c) = \Phi(\frac{c-\mu}{\sigma})$ for all c, where Φ is the CDF of the standard $N(0,1)$ distribution, or $\sigma^2 = 0$, in which case $F_X(c) = I_{\{c \geq \mu\}}$ and $P\{X = \mu\} = 1$.

Proposition 2.8 *Suppose X_n is a Gaussian random variable for each n, and $X_n \to X_\infty$ as $n \to \infty$, in any one of the four senses, a.s., m.s., p., or d. Then X_∞ is also a Gaussian random variable.*

Proof Since convergence in the other senses implies convergence in distribution, we can assume that the sequence converges in distribution. Let μ_n and σ_n^2 denote the mean and variance of X_n. The first step is to show that the sequence σ_n^2 is bounded. Intuitively, if it weren't bounded, the distribution of X_n would get too spread out to converge. Since F_{X_∞} is a valid CDF, there exists a value L so large that $F_{X_\infty}(-L) < \frac{1}{3}$ and $F_{X_\infty}(L) > \frac{2}{3}$. By increasing L if necessary, we can also assume that L and $-L$ are continuity points of F_{X_∞}. So there exists n_o so that, whenever $n \geq n_o$, $F_{X_n}(-L) \leq \frac{1}{3}$ and $F_{X_n}(L) \geq \frac{2}{3}$. Therefore, for $n \geq n_o$, $P\{|X_n| \leq L\} \geq F_{X_n}(\frac{2}{3}) - F_{X_n}(\frac{1}{3}) \geq \frac{1}{3}$. For σ_n^2 fixed, the probability $P\{|X_n| \leq L\}$ is maximized by $\mu_n = 0$, so no matter what the value of μ_n is, $2\Phi(\frac{L}{\sigma_n}) - 1 \geq P\{|X_n| \leq L\}$. Therefore, for $n \geq n_o$, $\Phi(\frac{L}{\sigma_n}) \geq \frac{2}{3}$, or equivalently, $\sigma_n \leq L/\Phi^{-1}(\frac{2}{3})$, where Φ^{-1} is the inverse of Φ. The first $n_o - 1$ terms of the sequence (σ_n^2) are finite. Therefore, the whole sequence (σ_n^2) is bounded.

Constant random variables are considered to be Gaussian random variables – namely degenerate ones with zero variance. So assume without loss of generality that X_∞ is not a constant random variable. Then there exists a value c_o so that $F_{X_\infty}(c_o)$ is strictly between zero and one. Since F_{X_∞} is right-continuous, the function must lie strictly between zero and one over some interval of positive length, with left endpoint c_o. The function can only have countably many points of discontinuity, so it has infinitely many points of continuity such that the function value is strictly between zero and one. Let c_1 and c_2 be two distinct such points, and let p_1 and p_2 denote the values of F_{X_∞} at those two points, and let $b_i = \Phi^{-1}(p_i)$ for $i = 1, 2$. It follows that $\lim_{n\to\infty} \frac{c_i - \mu_n}{\sigma_n} = b_i$ for $i = 1, 2$. The limit of the difference of the sequences is the difference of the limits, so $\lim_{n\to\infty} \frac{c_1 - c_2}{\sigma_n} = b_1 - b_2$. Since $c_1 - c_2 \neq 0$ and the sequence (σ_n) is bounded, it follows that (σ_n) has a finite limit, σ_∞, and therefore also (μ_n) has a finite limit, μ_∞. Therefore, the CDFs F_{X_n} converge pointwise to the CDF for the $N(\mu_\infty, \sigma_\infty^2)$ distribution. Thus, X_∞ has the $N(\mu_\infty, \sigma_\infty^2)$ distribution. □

2.2 Cauchy criteria for convergence of random variables

It is important to be able to show that a limit exists even if the limit value is not known. For example, it is useful to determine if the sum of an infinite series of numbers is convergent without needing to know the value of the sum. One useful result for this purpose is that if $(x_n : n \geq 1)$ is monotone nondecreasing, i.e. $x_1 \leq x_2 \leq \ldots$, and if it satisfies $x_n \leq L$ for all n for some finite constant L, then the sequence is convergent. This result carries over immediately to random variables: if $(X_n : n \geq 1)$ is a sequence of random variables such $P\{X_n \le X_{n+1}\} = 1$ for all n and if there is a random variable Y such that $P\{X_n \le Y\} = 1$ for all n, then (X_n) converges a.s.

For deterministic sequences that are not monotone, the Cauchy criteria give a simple yet general condition that implies convergence to a finite limit. A deterministic sequence $(x_n : n \ge 1)$ is said to be a Cauchy sequence if $\lim_{m,n\to\infty} |x_m - x_n| = 0$. This means that, for any $\epsilon > 0$, there exists N sufficiently large, such that $|x_m - x_n| < \epsilon$ for all $m, n \ge N$. If the sequence (x_n) has a finite limit x_∞, then the triangle inequality for distances between numbers, $|x_m - x_n| \le |x_m - x_\infty| + |x_n - x_\infty|$, implies that the sequence is a Cauchy sequence. More useful is the converse statement, called the Cauchy criteria for convergence, or the completeness property of $\mathbb{R}$: if (x_n) is a Cauchy sequence then (x_n) converges to a finite limit as $n \to \infty$. The following proposition gives similar criteria for convergence of random variables.

Proposition 2.9 *(Cauchy criteria for random variables) Let (X_n) be a sequence of random variables on a probability space $(\Omega, \mathcal{F}, P)$.*

(a) *X_n converges a.s. to some random variable if and only if*

$$P\{\omega : \lim_{m,n\to\infty} |X_m(\omega) - X_n(\omega)| = 0\} = 1.$$

(b) *X_n converges m.s. to some random variable if and only if (X_n) is a* Cauchy sequence *in the m.s. sense, meaning $E[X_n^2] < +\infty$ for all n and*

$$\lim_{m,n\to\infty} E[(X_m - X_n)^2] = 0. \tag{2.4}$$

(c) *X_n converges p. to some random variable if and only if for every $\epsilon > 0$,*

$$\lim_{m,n\to\infty} P\{|X_m - X_n| \ge \epsilon\} = 0. \tag{2.5}$$

Proof (a) For any ω fixed, $(X_n(\omega) : n \ge 1)$ is a sequence of numbers. So by the Cauchy criterion for convergence of a sequence of numbers, the following equality of sets holds:

$$\{\omega : \lim_{n\to\infty} X_n(\omega) \text{ exists and is finite}\} = \{\omega : \lim_{m,n\to\infty} |X_m(\omega) - X_n(\omega)| = 0\}.$$

Thus, the set on the left has probability one (i.e. X converges a.s. to a random variable) if and only if the set on the right has probability one. Part (a) is proved.

(b) First the "only if" part is proved. Suppose $X_n \overset{m.s.}{\to} X_\infty$. By the L^2 triangle inequality for random variables,

$$E[(X_n - X_m)^2]^{\frac{1}{2}} \le E[(X_m - X_\infty)^2]^{\frac{1}{2}} + E[(X_n - X_\infty)^2]^{\frac{1}{2}}. \tag{2.6}$$

Since $X_n \overset{m.s.}{\to} X_\infty$, the right side of (2.6) converges to zero as $m, n \to \infty$, so that (2.4) holds. The "only if" part of (b) is proved.

Moving to the proof of the "if" part, suppose that (2.4) holds. Choose the sequence $k_1 < k_2 < \ldots$ recursively as follows. Let k_1 be so large that $E[(X_n - X_{k_1})^2] \le 1/2$ for all $n \ge k_1$. Given $k_1, \ldots, k_{i-1}$, let k_i be so large that $k_i > k_{i-1}$ and $E[(X_n - X_{k_i})^2] \le 2^{-i}$ for all $n \ge k_i$. It follows from this choice of the k_is that $E[(X_{k_{i+1}} - X_{k_i})^2] \le 2^{-i}$ for all $i \ge 1$. Let $S_n = |X_{k_1}| + \sum_{i=1}^{n-1} |X_{k_{i+1}} - X_{k_i}|$. Note that $|X_{k_i}| \le S_n$ for $1 \le i \le k$ by the triangle inequality for differences of real numbers. By the L^2 triangle inequality for random variables (1.15),

$$E[S_n^2]^{\frac{1}{2}} \leq E[X_{k_1}^2]^{\frac{1}{2}} + \sum_{i=1}^{n-1} E[(X_{k_{i+1}} - X_{k_i})^2]^{\frac{1}{2}} \leq E[X_{k_1}^2]^{\frac{1}{2}} + 1.$$

Since S_n is monotonically increasing, it converges a.s. to a limit S_∞. Note that $|X_{k_i}| \leq S_\infty$ for all $i \geq 1$. By the monotone convergence theorem, $E[S_\infty^2] = \lim_{n\to\infty} E[S_n^2] \leq (E[X_{k_1}^2]^{\frac{1}{2}} + 1)^2$. So, S_∞ is in $L^2(\Omega, \mathcal{F}, P)$. In particular, S_∞ is finite a.s., and for any ω such that $S_\infty(\omega)$ is finite, the sequence of numbers $(X_{k_i}(\omega) : i \geq 1)$ is a Cauchy sequence (see Example 11.3 in the Appendix). By completeness of $\mathbb{R}$, for ω in that set, the limit $X_\infty(\omega)$ exists. Let $X_\infty(\omega) = 0$ on the zero probability event that $(X_{k_i}(\omega) : i \geq 1)$ does not converge. Summarizing, we have $\lim_{i\to\infty} X_{k_i} = X_\infty$ a.s. and $|X_{k_i}| \leq S_\infty$ where $S_\infty \in L^2(\Omega, \mathcal{F}, P)$. It therefore follows from Proposition 2.7(c) that $X_{k_i} \stackrel{m.s.}{\to} X_\infty$.

The final step is to prove that the entire sequence (X_n) converges in the m.s. sense to X_∞. For this purpose, let $\epsilon > 0$. Select i so large that $E[(X_n - X_{k_i})^2] < \epsilon^2$ for all $n \geq k_i$, and $E[(X_{k_i} - X_\infty)^2] \leq \epsilon^2$. Then, by the L^2 triangle inequality, for any $n \geq k_i$,

$$E[(X_n - X_\infty)^2]^{\frac{1}{2}} \leq E(X_n - X_{k_i})^2]^{\frac{1}{2}} + E[(X_{k_i} - X_\infty)^2]^{\frac{1}{2}} \leq 2\epsilon.$$

Since ϵ was arbitrary, $X_n \stackrel{m.s.}{\to} X_\infty$. The proof of (b) is complete.

(c) First the "only if" part is proved. Suppose $X_n \stackrel{p.}{\to} X_\infty$. Then for any $\epsilon > 0$,

$$P\{|X_m - X_n| \geq 2\epsilon\} \leq P\{|X_m - X_\infty| \geq \epsilon\} + P\{|X_m - X_\infty| \geq \epsilon\} \to 0$$

as $m, n \to \infty$, so that (2.5) holds. The "only if" part is proved.

Moving to the proof of the "if" part, suppose (2.5) holds. Select an increasing sequence of integers k_i so that $P\{|X_n - X_m| \geq 2^{-i}\} \leq 2^{-i}$ for all $m, n \geq k_i$. It follows, in particular, that $P\{|X_{k_{i+1}} - X_{k_i}| \geq 2^{-i}\} \leq 2^{-i}$. Since the sum of the probabilities of these events is finite, the probability that infinitely many of the events is true is zero, by the Borel–Cantelli lemma (specifically, Lemma 1.3(a)). Thus, $P\{|X_{k_{i+1}} - X_{k_i}| \leq 2^{-i}$ for all large enough $i\} = 1$. Thus, for all ω in a set with probability one, $(X_{k_i}(\omega) : i \geq 1)$ is a Cauchy sequence of numbers. By completeness of $\mathbb{R}$, for ω in that set, the limit $X_\infty(\omega)$ exists. Let $X_\infty(\omega) = 0$ on the zero probability event that $(X_{k_i}(\omega) : i \geq 1)$ does not converge. Then, $X_{k_i} \stackrel{a.s.}{\to} X_\infty$. It follows that $X_{k_i} \stackrel{p.}{\to} X_\infty$ as well.

The final step is to prove that the entire sequence (X_n) converges in the p. sense to X_∞. For this purpose, let $\epsilon > 0$. Select i so large that $P\{||X_n - X_{k_i}|| \geq \epsilon\} < \epsilon$ for all $n \geq k_i$, and $P\{|X_{k_i} - X_\infty| \geq \epsilon\} < \epsilon$. Then $P\{|X_n - X_\infty| \geq 2\epsilon\} \leq 2\epsilon$ for all $n \geq k_i$. Since ϵ was arbitrary, $X_n \stackrel{p.}{\to} X_\infty$. The proof of (c) is complete. □

The following is a corollary of Proposition 2.9(c) and its proof.

Corollary 2.10 *If $X_n \stackrel{p.}{\to} X_\infty$, then there is a subsequence $(X_{k_i} : i \geq 1)$ such that $\lim_{i\to\infty} X_{k_i} = X_\infty$ a.s.*

Proof By Proposition 2.9(c), the sequence satisfies (2.9). By the proof of Proposition 2.9(c) there is a subsequence (X_{k_i}) that converges a.s. By uniqueness of limits in the p. or a.s. senses, the limit of the subsequence is the same random variable, X_∞ (up to differences on a set of measure zero). □

Proposition 2.9(b), the Cauchy criterion for mean square convergence, is used extensively in these notes. The remainder of this section concerns a more convenient form of the Cauchy criteria for m.s. convergence.

Proposition 2.11 *(Correlation version of the Cauchy criterion for m.s. convergence) Let (X_n) be a sequence of random variables with $E[X_n^2] < +\infty$ for each n. Then there exists a random variable X such that $X_n \stackrel{m.s.}{\to} X$ if and only if the limit $\lim_{m,n\to\infty} E[X_nX_m]$ exists and is finite. Furthermore, if $X_n \stackrel{m.s.}{\to} X$, then $\lim_{m,n\to\infty} E[X_nX_m] = E[X^2]$.*

Proof The "if" part is proved first. Suppose $\lim_{m,n\to\infty} E[X_nX_m] = c$ for a finite constant c. Then

$$\begin{aligned} E\lfloor (X_n - X_m)^2 \rfloor &= E[X_n^2] - 2E[X_nX_m] + E[X_m^2] \\ &\to c - 2c + c = 0 \text{ as } m, n \to \infty. \end{aligned}$$

Thus, X_n is Cauchy in the m.s. sense, so $X_n \stackrel{m.s.}{\to} X$ for some random variable X.

To prove the "only if" part, suppose $X_n \stackrel{m.s.}{\to} X$. Observe next that

$$\begin{aligned} E[X_mX_n] &= E[(X + (X_m - X))(X + (X_n - X))] \\ &= E[X^2 + (X_m - X)X + X(X_n - X) + (X_m - X)(X_n - X)]. \end{aligned}$$

By the Cauchy–Schwarz inequality,

$$\begin{aligned} E[| (X_m - X)X |] &\le E[(X_m - X)^2]^{\frac{1}{2}} E[X^2]^{\frac{1}{2}} \to 0, \\ E[| (X_m - X)(X_n - X) |] &\le E[(X_m - X)^2]^{\frac{1}{2}} E[(X_n - X)^2]^{\frac{1}{2}} \to 0, \end{aligned}$$

and similarly $E[| X(X_n - X) |] \to 0$. Thus $E[X_mX_n] \to E[X^2]$. This establishes both the "only if" part of the proposition and the last statement of the proposition. The proof of the proposition is complete. □

Corollary 2.12 *Suppose $X_n \stackrel{m.s.}{\to} X$ and $Y_n \stackrel{m.s.}{\to} Y$. Then $E[X_nY_n] \to E[XY]$.*

Proof By the inequality $(a+b)^2 \le 2a^2+2b^2$, it follows that $X_n+Y_n \stackrel{m.s.}{\to} X+Y$ as $n \to \infty$. Proposition 2.11 therefore implies that $E[(X_n + Y_n)^2] \to E[(X + Y)^2]$, $E[X_n^2] \to E[X^2]$, and $E[Y_n^2] \to E[Y^2]$. Since $X_nY_n = ((X_n+Y_n)^2 - X_n^2 - Y_n^2)/2$, the corollary follows. □

Corollary 2.13 *Suppose $X_n \stackrel{m.s.}{\to} X$. Then $E[X_n] \to E[X]$.*

Proof Corollary 2.13 follows from Corollary 2.12 by taking $Y_n = 1$ for all n. □

Example 2.9 This example illustrates the use of Proposition 2.11. Let $X_1, X_2, \ldots$ be mean zero random variables such that

$$E[X_iX_j] = \begin{cases} 1 & \text{if } i = j \\ 0 & \text{else} \end{cases}.$$

Does the series $\sum_{k=1}^{\infty} \frac{X_k}{k}$ converge in the mean square sense to a random variable with a finite second moment? Let $Y_n = \sum_{k=1}^{n} \frac{X_k}{k}$. The question is whether Y_n converges in the mean square sense to a random variable with finite second moment. The answer is yes if and only if $\lim_{m,n\to\infty} E[Y_m Y_n]$ exists and is finite. Observe that

$$E[Y_m Y_n] = \sum_{k=1}^{\min(m,n)} \frac{1}{k^2} \to \sum_{k=1}^{\infty} \frac{1}{k^2} \text{ as } m, n \to \infty.$$

This sum is smaller than $1 + \int_1^{\infty} \frac{1}{x^2} dx = 2 < \infty$.[1] Therefore, by Proposition 2.11, the series $\sum_{k=1}^{\infty} \frac{X_k}{k}$ indeed converges in the m.s. sense.

2.3 Limit theorems for sums of independent random variables

Sums of many independent random variables often have distributions that can be characterized by a small number of parameters. For engineering applications, this represents a low complexity method for describing the random variables. An analogous tool is the Taylor series approximation. A continuously differentiable function f can be approximated near zero by the first order Taylor's approximation

$$f(x) \approx f(0) + x f'(0).$$

A second order approximation, in case f is twice continuously differentiable, is

$$f(x) \approx f(0) + x f'(0) + \frac{x^2}{2} f''(0).$$

Bounds on the approximation error are given by Taylor's theorem, found in Appendix 11.4. In essence, Taylor's approximation lets us represent the function by the numbers $f(0)$, $f'(0)$, and $f''(0)$. We shall see that the law of large numbers and central limit theorem can be viewed not just as analogies of the first and second order Taylor's approximations, but actually as consequences of them.

Lemma 2.14 *Let $(z_n : n \geq 1)$ be a sequence of real or complex numbers with limit z. Then $\left(1 + \frac{z_n}{n}\right)^n \to e^z$ as $n \to \infty$.*

Proof The basic idea is to note that $(1+s)^n = \exp(n \ln(1+s))$, and apply a power series expansion of $\ln(1+s)$ about the point $s = 0$. The details are given next. Since the sequence (z_n) converges to a finite limit, $|\frac{z_n}{n}| \leq \frac{1}{2}$ for all sufficiently large n, so it suffices to consider $\ln(1+s)$ for complex s with $|s| \leq \frac{1}{2}$. Note that the kth derivative of $\ln(1+s)$ evaluated at $s = 0$ is $(-1)^{(k-1)}(k-1)!$ for $k \geq 1$. Since the function $\ln(1+s)$ is analytic in an open region containing $|s| \leq \frac{1}{2}$, its power series expansion converges absolutely:

[1] In fact, the sum is equal to $\frac{\pi^2}{6}$, but the technique of comparing the sum to an integral to show the sum is finite is the main point here.

$$\ln(1+s) = \sum_{k=1}^{\infty} \frac{s^k(-1)^{(k+1)}}{k}.$$

Therefore, for $|s| \leq \frac{1}{2}$,

$$|\ln(1+s) - s| = \left| \sum_{k=2}^{\infty} \frac{s^k(-1)^{(k+1)}}{k} \right| \leq |s|^2 \sum_{k=2}^{\infty} \frac{2^{-k}}{k} \leq \frac{|s|^2}{4}.$$

So, for $|s| \leq \frac{1}{2}$, $\ln(1+s) = s + |s|^2 h(s)$, where h is a function such that $|h(s)| \leq \frac{1}{4}$. Thus, for n sufficiently large,

$$\begin{aligned}\left(1+\frac{z_n}{n}\right)^n &= \exp\left(n\ln\left(1+\frac{z_n}{n}\right)\right)\\ &= \exp\left(z_n + \frac{|z_n|^2 h(z_n/n)}{n}\right),\end{aligned}$$

and, by continuity of the exponential function, the conclusion of the lemma follows. □

A sequence of random variables (X_n) is said to be independent and identically distributed (*iid*) if the X_is are mutually independent and identically distributed.

Proposition 2.15 *(Law of large numbers) Suppose that $X_1, X_2, \ldots$ is a sequence of random variables such that each X_i has finite mean m. Let $S_n = X_1 + \cdots + X_n$. Then*

(a) $\frac{S_n}{n} \stackrel{m.s.}{\to} m$ *(hence also* $\frac{S_n}{n} \stackrel{p.}{\to} m$ *and* $\frac{S_n}{n} \stackrel{d.}{\to} m$*) if for some constant c, $\mathrm{Var}(X_i) \leq c$ for all i, and $\mathrm{Cov}(X_i, X_j) = 0$ $i \neq j$ (i.e. if the variances are bounded and the X_is are uncorrelated).*

(b) $\frac{S_n}{n} \stackrel{p.}{\to} m$ *if $X_1, X_2, \ldots$ are iid. (This version is the weak law of large numbers.)*

(c) $\frac{S_n}{n} \stackrel{a.s.}{\to} m$ *if $X_1, X_2, \ldots$ are iid. (This version is the strong law of large numbers.)*

We give a proof of (a) and (b), but prove (c) only under an extra condition. Suppose the conditions of (a) are true. Then

$$\begin{aligned}E\left[\left(\frac{S_n}{n} - m\right)^2\right] &= \mathrm{Var}\left(\frac{S_n}{n}\right) = \frac{1}{n^2}\mathrm{Var}(S_n)\\ &= \frac{1}{n^2}\sum_i\sum_j \mathrm{Cov}(X_i, X_j) = \frac{1}{n^2}\sum_i \mathrm{Var}(X_i) \leq \frac{c}{n}.\end{aligned}$$

Therefore $\frac{S_n}{n} \stackrel{m.s.}{\to} m$.

Turn next to part (b). If in addition to the conditions of (b) it is assumed that $\mathrm{Var}(X_1) < +\infty$, then the conditions of part (a) are true. Since mean square convergence implies convergence in probability, the conclusion of part (b) follows. An extra credit problem shows how to use the same approach to verify (b) even if $\mathrm{Var}(X_1) = +\infty$.

Here a second approach to proving (b) is given. The characteristic function of $\frac{X_i}{n}$ is given by

$$E\left[\exp\left(\frac{juX_i}{n}\right)\right] = E\left[\exp\left(j\left(\frac{u}{n}\right)X_i\right)\right] = \Phi_X\left(\frac{u}{n}\right),$$

where Φ_X denotes the characteristic function of X_1. Since the characteristic function of the sum of independent random variables is the product of the characteristic functions,

$$\Phi_{\frac{S_n}{n}}(u) = \left(\Phi_X\left(\frac{u}{n}\right)\right)^n.$$

Since $E[X_1] = m$ it follows that Φ_X is differentiable with $\Phi_X(0) = 1$, $\Phi'_X(0) = jm$ and Φ' is continuous. By Taylor's theorem (Theorem 11.5) applied separately to the real and imaginary parts of Φ_X, for any u fixed,

$$\Phi_X\left(\frac{u}{n}\right) = 1 + \frac{u}{n}\left(Re(\Phi'_X(u_n)) + jIm(\Phi'_X(v_n))\right),$$

for some u_n and v_n between 0 and $\frac{u}{n}$ for all n. Since $\Phi'(u_n) \to jm$ and $\Phi'(v_n) \to jm$ as $n \to \infty$, it follows that $Re(\Phi'_X(u_n)) + jIm(\Phi'_X(v_n)) \to jm$ as $n \to \infty$. So Lemma 2.14 yields $\Phi_X(\frac{u}{n})^n \to \exp(jum)$ as $n \to \infty$. Note that $\exp(jum)$ is the characteristic function of a random variable equal to m with probability one. Since pointwise convergence of characteristic functions to a valid characteristic function implies convergence in distribution, it follows that $\frac{S_n}{n} \stackrel{d.}{\to} m$. However, convergence in distribution to a constant implies convergence in probability, so (b) is proved.

Part (c) is proved under the additional assumption that $E[X_1^4] < +\infty$. Without loss of generality we assume that $EX_1 = 0$. Consider expanding S_n^4. There are n terms of the form X_i^4 and $3n(n-1)$ terms of the form $X_i^2X_j^2$ with $1 \leq i,j \leq n$ and $i \neq j$. The other terms have the form $X_i^3X_j, X_i^2X_jX_k$ or $X_iX_jX_kX_l$ for distinct i, j, k, l, and these terms have mean zero. Thus,

$$E[S_n^4] = nE[X_1^4] + 3n(n-1)E[X_1^2]^2.$$

Let $Y = \sum_{n=1}^{\infty}(\frac{S_n}{n})^4$. The value of Y is well defined but it is a priori possible that $Y(\omega)$ is infinite for some ω. However, by the monotone convergence theorem, the expectation of the sum of nonnegative random variables is the sum of the expectations, so

$$E[Y] = \sum_{n=1}^{\infty} E\left[\left(\frac{S_n}{n}\right)^4\right] = \sum_{n=1}^{\infty} \frac{nE[X_1^4] + 3n(n-1)E[X_1^2]^2}{n^4} < +\infty,$$

Therefore, $P\{Y < +\infty\} = 1$. However, $\{Y < +\infty\}$ is a subset of the event of convergence $\left\{w : \frac{S_n(w)}{n} \to 0 \text{ as } n \to \infty\right\}$, so the event of convergence also has probability one. Thus, part (c) under the extra fourth moment condition is proved. □

Proposition 2.16 *(Central limit theorem) Suppose that $X_1, X_2, \ldots$ are iid, each with mean μ and variance σ^2. Let $S_n = X_1 + \ldots + X_n$. Then the normalized sum*

$$\frac{S_n - n\mu}{\sqrt{n}}$$

converges in distribution to the $N(0, \sigma^2)$ distribution as $n \to \infty$.

Proof Without loss of generality, assume that $\mu = 0$. Then the characteristic function of the normalized sum $\frac{S_n}{\sqrt{n}}$ is given by $\Phi_X\left(\frac{u}{\sqrt{n}}\right)^n$, where Φ_X denotes the characteristic function of X_1. Since X_1 has mean 0 and finite second moment σ^2, it follows that Φ_X is twice differentiable with $\Phi_X(0) = 1$, $\Phi'_X(0) = 0$, $\Phi''_X(0) = -\sigma^2$, and Φ''_X is continuous. By Taylor's theorem (Theorem 11.5) applied separately to the real and imaginary parts of Φ_X, for any u fixed,

$$\Phi_X\left(\frac{u}{\sqrt{n}}\right) = 1 + \frac{u^2}{2n}\left(Re(\Phi''_X(u_n)) + jIm(\Phi''_X(v_n))\right),$$

for some u_n and v_n between 0 and $\frac{u}{\sqrt{n}}$ for all n. Note that $u_n \to 0$ and $v_n \to 0$ as $n \to \infty$, so $\Phi''(u_n) \to -\sigma^2$ and $\Phi''(v_n) \to -\sigma^2$ as $n \to \infty$. It follows that $\mathrm{Re}(\Phi''_X(u_n)) + j\mathrm{Im}(\Phi''_X(v_n)) \to -\sigma^2$ as $n \to \infty$. Lemma 2.14 yields $\Phi_X\left(\frac{u}{\sqrt{n}}\right)^n \to \exp\left(-\frac{u^2\sigma^2}{2}\right)$ as $n \to \infty$. Since pointwise convergence of characteristic functions to a valid characteristic function implies convergence in distribution, the proposition is proved. □

2.4 Convex functions and Jensen's inequality

Let φ be a function on $\mathbb{R}$ with values in $\mathbb{R} \cup \{+\infty\}$ such that $\varphi(x) < \infty$ for at least one value of x. Then φ is said to be *convex* if for any a, b and λ with $a < b$ and $0 \le \lambda \le 1$

$$\varphi(a\lambda + b(1-\lambda)) \le \lambda\varphi(a) + (1-\lambda)\varphi(b).$$

This means that the graph of φ on any interval $[a, b]$ lies below the line segment equal to φ at the endpoints of the interval.

Proposition 2.17 *Suppose f is a function with domain $\mathbb{R}$. (a) If f is continuously differentiable, f is convex if and only if f' is nondecreasing. (b) If f is twice continuously differentiable, f is convex if and only if $f''(v) \ge 0$ for all v.*

Proof (a) (if) Suppose f is continuously differentiable. Given $s \le t$, define $D_{s,t} = \lambda f(s) + (1-\lambda)f(t) - f(\lambda s + (1-\lambda)t)$. We claim that

$$D_{s,t} = (1-\lambda)\int_s^t \left(f'(x) - f'(\lambda s + (1-\lambda)x)\right) dx. \tag{2.7}$$

To verify (2.7), fix s and note that (2.7) is true if $t = s$, for then both sides are zero, and the derivative with respect to t of each side of (2.7) is the same, equal to $(1-\lambda)\left(f'(t) - f'(\lambda s + (1-\lambda)t)\right)$. If f' is nondecreasing, then the integrand in (2.7) is nonnegative, so $D_{s,t} \ge 0$, so f is convex.

(only if) Turning to the "only if" part of (a), suppose f is convex, and let $s < t$. For any $h > 0$ small enough that $s < s+h < t < t+h$,

$$f(s+h)(t-s+h) \leq (t-s)f(s) + hf(t+h), \tag{2.8}$$

$$f(t)(t-s+h) \leq hf(s) + (t-s)f(t+h), \tag{2.9}$$

by the convexity of f. Combining (2.8) and (2.9) by summing the lefthand sides and righthand sides, rearranging, and multiplying by a positive constant, yields

$$\frac{f(s+h)-f(s)}{h} \leq \frac{f(t+h)-f(t)}{h}. \tag{2.10}$$

Letting $h \to 0$ in (2.10) yields $f'(s) \leq f'(t)$, so f' is nondecreasing. Part (a) is proved.
(b) Suppose f is twice continuously differentiable. Part (b) follows from part (a) and the fact that f' is nondecreasing if and only if $f''(v) \geq 0$ for all v. □

Examples of convex functions include:

$$ax^2 + bx + c \text{ for constants } a, b, c \text{ with } a \geq 0,$$
$$e^{\lambda x} \text{ for } \lambda \text{ constant},$$

$$\varphi(x) = \begin{cases} -\ln x & x > 0 \\ +\infty & x \leq 0, \end{cases} \qquad \psi(x) = \begin{cases} x\ln x & x > 0 \\ 0 & x = 0 \\ +\infty & x < 0. \end{cases}$$

Theorem 2.18 *(Jensen's inequality) Let φ be a convex function and let X be a random variable such that $E[X]$ is finite. Then $E[\varphi(X)] \geq \varphi(E[X])$.*

For example, Jensen's inequality implies that $E[X^2] \geq E[X]^2$, which also follows from the fact $\mathrm{Var}(X) = E[X^2] - E[X]^2$.

Proof Since φ is convex, there is a tangent to the graph of φ at $E[X]$, meaning there is a function L of the form $L(x) = a + bx$ such that $\varphi(x) \geq L(x)$ for all x and $\varphi(E[X]) = L(E[X])$, see Figure 2.12. Therefore $E[\varphi(X)] \geq E[L(X)] = L(E[X]) = \varphi(E[X])$, which establishes the theorem. □

A function φ is *concave* if $-\varphi$ is convex. If φ is concave, $E[\varphi(X)] \leq \varphi(E[X])$.

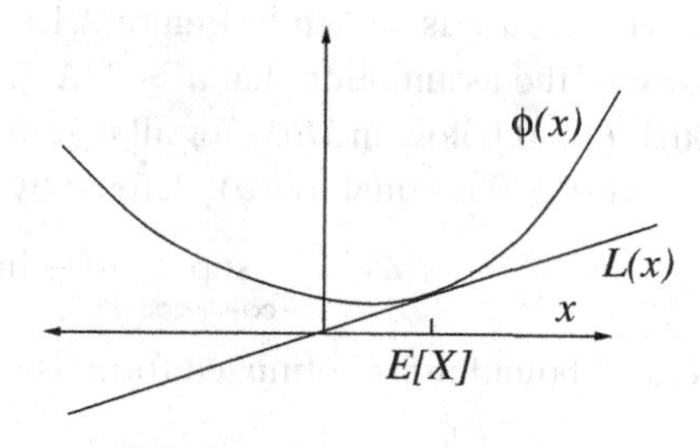

Figure 2.12 A convex function and a tangent linear function.

2.5 Chernoff bound and large deviations theory

Let $X_1, X_2, \ldots$ be an *iid* sequence of random variables with finite mean μ, and let $S_n = X_1 + \ldots + X_n$. The weak law of large numbers implies that for fixed a with $a > \mu$, $P\left\{\frac{S_n}{n} \geq a\right\} \to 0$ as $n \to \infty$. When the X_is have finite variance, the central limit theorem offers a refinement of the law of large numbers, by identifying the limit of $P\left\{\frac{S_n}{n} \geq a_n\right\}$, where (a_n) is a sequence that converges to μ in the particular manner: $a_n = \mu + \frac{c}{\sqrt{n}}$. For fixed c, the limit is not zero. One can think of the central limit theorem, therefore, as concerning "normal" deviations of S_n from its mean. Large deviations theory, by contrast, addresses $P\left\{\frac{S_n}{n} \geq a\right\}$ for a fixed, and in particular it identifies how quickly $P\left\{\frac{S_n}{n} \geq a\right\}$ converges to zero as $n \to \infty$. We shall first describe the Chernoff bound, which is a simple upper bound on $P\left\{\frac{S_n}{n} \geq a\right\}$. Then Cramér's theorem, to the effect that the Chernoff bound is in a certain sense tight, is stated.

The moment generating function of X_1 is defined by $M(\theta) = E[e^{\theta X_1}]$, and $\ln M(\theta)$ is called the *log moment generating function*. Since $e^{\theta X_1}$ is a positive random variable, the expectation, and hence $M(\theta)$ itself, is well-defined for all real values of θ, with possible value $+\infty$. The Chernoff bound is simply given as

$$P\left\{\frac{S_n}{n} \geq a\right\} \leq \exp(-n[\theta a - \ln M(\theta)]) \quad \text{for } \theta \geq 0. \tag{2.11}$$

The bound (2.11), like the Chebychev inequality, is a consequence of Markov's inequality applied to an appropriate function. For $\theta \geq 0$:

$$\begin{aligned} P\left\{\frac{S_n}{n} \geq a\right\} &= P\{e^{\theta(X_1+\cdots+X_n-na)} \geq 1\} \\ &\leq E[e^{\theta(X_1+\cdots+X_n-na)}] \\ &= E[e^{\theta X_1}]^n e^{-n\theta a} = \exp(-n[\theta a - \ln M(\theta)]). \end{aligned}$$

To make the best use of the Chernoff bound we can optimize the bound by selecting the best θ. Thus, we wish to select $\theta \geq 0$ to maximize $a\theta - \ln M(\theta)$.

In general the log moment generating function $\ln M$ is convex. Note that $\ln M(0) = 0$. Let us suppose that $M(\theta)$ is finite for some $\theta > 0$. Then

$$\left.\frac{d \ln M(\theta)}{d\theta}\right|_{\theta=0} = \left.\frac{E[X_1 e^{\theta X_1}]}{E[e^{\theta X_1}]}\right|_{\theta=0} = E[X_1].$$

A sketch of a typical case is shown in Figure 2.13. Figure 2.13 also shows the line of slope a. Because of the assumption that $a > E[X_1]$, the line lies strictly above $\ln M(\theta)$ for small enough θ and below $\ln M(\theta)$ for all $\theta < 0$. Therefore, the maximum value of $\theta a - \ln M(\theta)$ over $\theta \geq 0$ is equal to $l(a)$, defined by

$$l(a) = \sup_{-\infty < \theta < \infty} \theta a - \ln M(\theta). \tag{2.12}$$

Thus, the Chernoff bound in its optimized form, is

$$P\left\{\frac{S_n}{n} \geq a\right\} \leq \exp(-nl(a)) \quad a > E[X_1].$$

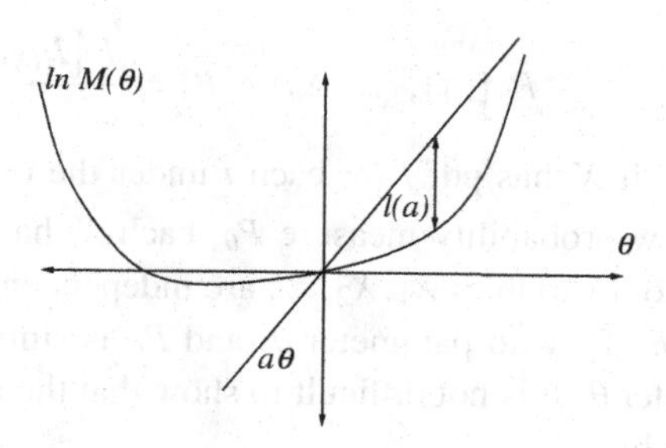

Figure 2.13 A log moment generating function and a line of slope a.

There does not exist such a clean lower bound on the large deviation probability $P\left\{\frac{S_n}{n} \geq a\right\}$, but by the celebrated theorem of Cramér stated next, the Chernoff bound gives the right exponent.

Theorem 2.19 *(Cramér's theorem) Suppose $E[X_1]$ is finite, and that $E[X_1] < a$. Then for $\epsilon > 0$ there exists a number n_ϵ such that*

$$P\left\{\frac{S_n}{n} \geq a\right\} \geq \exp(-n(l(a)+\epsilon)) \tag{2.13}$$

for all $n \geq n_\epsilon$. Combining this bound with the Chernoff inequality yields

$$\lim_{n\to\infty} \frac{1}{n} \ln P\left\{\frac{S_n}{n} \geq a\right\} = -l(a).$$

In particular, if $l(a)$ is finite (equivalently if $P\{X_1 \geq a\} > 0$) then

$$P\left\{\frac{S_n}{n} \geq a\right\} = \exp(-n(l(a)+\epsilon_n)),$$

where (ϵ_n) is a sequence with $\epsilon_n \geq 0$ and $\lim_{n\to\infty} \epsilon_n = 0$.

Similarly, if $a < E[X_1]$ and $l(a)$ is finite, then

$$P\left\{\frac{S_n}{n} \leq a\right\} = \exp(-n(l(a)+\epsilon_n)),$$

where ϵ_n is a sequence with $\epsilon_n \geq 0$ and $\lim_{n\to\infty} \epsilon_n = 0$. Informally, we can write for n large:

$$P\left\{\frac{S_n}{n} \in da\right\} \approx e^{-nl(a)}da.$$

Proof The lower bound (2.13) is proved here under the additional assumption that X_1 is a bounded random variable: $P\{|X_1| \leq C\} = 1$ for some constant C; this assumption can be removed by a truncation argument covered in Problem 2.43. Also, to avoid trivialities, suppose $P\{X_1 > a\} > 0$. The assumption that X_1 is bounded and the monotone convergence theorem imply that the function $M(\theta)$ is finite and infinitely differentiable over $\theta \in \mathbb{R}$. Given $\theta \in \mathbb{R}$, let P_θ denote a new probability measure on the same probability space that $X_1, X_2, \ldots$ are defined on such that for any n and any event of the form $\{(X_1, \ldots, X_n) \in B\}$,

$$P_\theta\{(X_1,\ldots,X_n)\in B\} = \frac{E\left[I_{\{(X_1,\ldots,X_n)\in B\}}e^{\theta S_n}\right]}{M(\theta)^n}.$$

In particular, if X_i has pdf f for each i under the original probability measure P, then under the new probability measure P_θ, each X_i has pdf f_θ defined by $f_\theta(x) = \frac{f(x)e^{\theta x}}{M(\theta)}$, and the random variables $X_1, X_2, \ldots$ are independent under P_θ. The pdf f_θ is called the *tilted version* of f with parameter θ, and P_θ is similarly called the tilted version of P with parameter θ. It is not difficult to show that the mean and variance of the X_is under P_θ are given by:

$$E_\theta[X_1] = \frac{E\left[X_1 e^{\theta X_1}\right]}{M(\theta)} = (\ln M(\theta))',$$
$$\mathrm{Var}_\theta[X_1] = E_\theta[X_1^2] - E_\theta[X_1]^2 = (\ln M(\theta))''.$$

Under the assumptions we have made, X_1 has strictly positive variance under P_θ for all θ, so that $\ln M(\theta)$ is strictly convex.

The assumption $P\{X_1 > a\} > 0$ implies that $(a\theta - \ln M(\theta)) \to -\infty$ as $\theta \to \infty$. Together with the fact that $\ln M(\theta)$ is differentiable and strictly convex, there thus exists a unique value θ^* of θ that maximizes $a\theta - \ln M(\theta)$. So $l(a) = a\theta^* - \ln M(\theta^*)$. Also, the derivative of $a\theta - \ln M(\theta)$ at $\theta = \theta^*$ is zero, so that $E_{\theta^*}[X] = (\ln M(\theta))'\Big|_{\theta=\theta^*} = a$. Observe that for any b with $b > a$,

$$\begin{aligned}
P\left\{\frac{S_n}{n} \geq a\right\} &= \int_{\{\omega: na \leq S_n\}} 1\, dP \\
&= \int_{\{\omega: na \leq S_n\}} M(\theta^*)^n e^{-\theta^* S_n} \frac{e^{\theta^* S_n} dP}{M(\theta^*)^n} \\
&= M(\theta^*)^n \int_{\{\omega: na \leq S_n\}} e^{-\theta^* S_n} dP_{\theta^*} \\
&\geq M(\theta^*)^n \int_{\{\omega: na \leq S_n \leq nb\}} e^{-\theta^* S_n} dP_{\theta^*} \\
&\geq M(\theta^*)^n e^{-\theta^* nb} P_{\theta^*}\{na \leq S_n \leq nb\}.
\end{aligned}$$

Now $M(\theta^*)^n e^{-\theta^* nb} = \exp(-n\{l(a) + \theta^*(b-a)\})$, and by the central limit theorem, $P_{\theta^*}\{na \leq S_n \leq nb\} \to \frac{1}{2}$ as $n \to \infty$ so $P_{\theta^*}\{na \leq S_n \leq nb\} \geq 1/3$ for n large enough. Therefore, for n large enough,

$$P\left\{\frac{S_n}{n} \geq a\right\} \geq \exp\left(-n\left(l(a) + \theta^*(b-a) + \frac{\ln 3}{n}\right)\right).$$

Taking b close enough to a, implies (2.13) for large enough n. □

Example 2.10 Let $X_1, X_2, \ldots$ be independent and exponentially distributed with parameter $\lambda = 1$. Then (see Figure 2.14):

$$\ln M(\theta) = \ln \int_0^\infty e^{\theta x} e^{-x} dx = \begin{cases} -\ln(1-\theta) & \theta < 1 \\ +\infty & \theta \geq 1 \end{cases}.$$

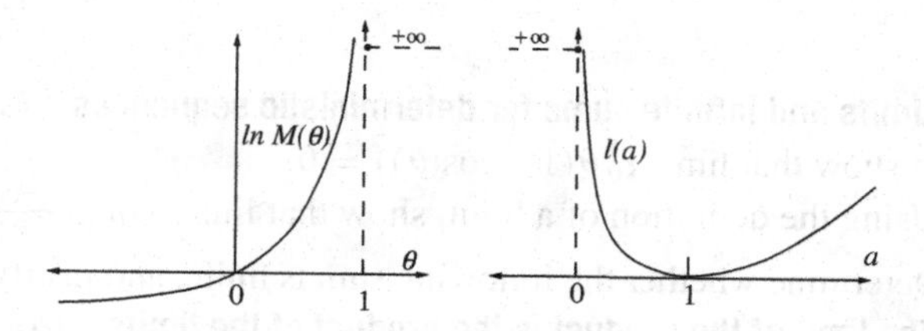

Figure 2.14 $\ln M(\theta)$ and $l(a)$ for an Exp(1) random variable.

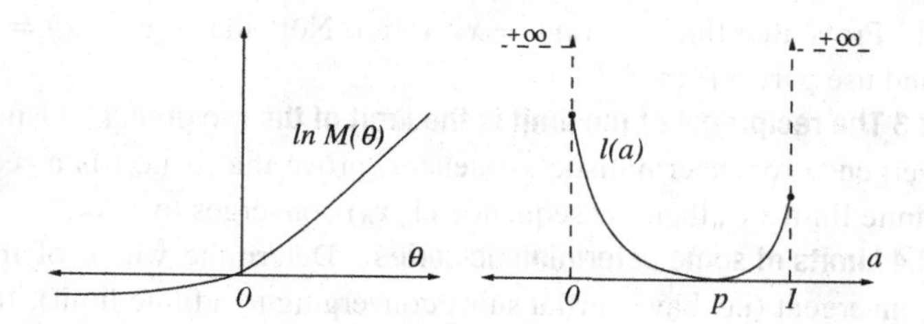

Figure 2.15 $\ln M(\theta)$ and $l(a)$ for a Bernoulli distribution.

Therefore, for any $a \in \mathbb{R}$,

$$\begin{aligned} l(a) &= \max_{\theta}\{a\theta - \ln M(\theta)\} \\ &= \max_{\theta<1}\{a\theta + \ln(1-\theta)\}. \end{aligned}$$

If $a \leq 0$ then $l(a) = +\infty$. On the other hand, if $a > 0$ then setting the derivative of $a\theta + \ln(1-\theta)$ to 0 yields the maximizing value $\theta = 1 - \frac{1}{a}$, and therefore

$$l(a) = \begin{cases} a - 1 - \ln(a) & a > 0 \\ +\infty & a \leq 0 \end{cases}.$$

The function l is shown in Figure 2.14.

Example 2.11 Let $X_1, X_2, \ldots$ be independent Bernoulli random variables with parameter p satisfying $0 < p < 1$. Thus S_n has the binomial distribution. Then $\ln M(\theta) = \ln(pe^\theta + (1-p))$, which has asymptotic slope 1 as $\theta \to +\infty$ and converges to a constant as $\theta \to -\infty$. Therefore, $l(a) = +\infty$ if $a > 1$ or if $a < 0$. For $0 \leq a \leq 1$, we find $a\theta - \ln M(\theta)$ is maximized by $\theta = \ln\left(\frac{a(1-p)}{p(1-a)}\right)$, leading to

$$l(a) = \begin{cases} a\ln\left(\frac{a}{p}\right) + (1-a)\ln\left(\frac{1-a}{1-p}\right) & 0 \leq a \leq 1 \\ +\infty & \text{else} \end{cases},$$

see Figure 2.15.

Problems

2.1 Limits and infinite sums for deterministic sequences (a) Using the definition of a limit, show that $\lim_{\theta\to 0}\theta(1+\cos(\theta))=0$.
(b) Using the definition of a limit, show that $\lim_{\theta\to 0,\theta>0}\frac{1+\cos(\theta)}{\theta}=+\infty$.
(c) Determine whether the following sum is finite, and justify your answer: $\sum_{n=1}^{\infty}\frac{1+\sqrt{n}}{1+n^2}$.
2.2 The limit of the product is the product of the limits Consider two (deterministic) sequences with finite limits: $\lim_{n\to\infty}x_n=x$ and $\lim_{n\to\infty}y_n=y$.
(a) Prove that the sequence (y_n) is bounded.
(b) Prove that $\lim_{n\to\infty}x_ny_n=xy$. (Hint: Note that $x_ny_n-xy=(x_n-x)y_n+x(y_n-y)$ and use part (a)).
2.3 The reciprocal of the limit is the limit of the reciprocal Using the definition of convergence for deterministic sequences, prove that if (x_n) is a sequence with a nonzero finite limit x_∞, then the sequence $(1/x_n)$ converges to $1/x_\infty$.
2.4 Limits of some deterministic series Determine which of the following series are convergent (i.e. have partial sums converging to a finite limit). Justify your answers.

$$\text{(a)}\ \sum_{n=0}^{\infty}\frac{3^n}{n!},\qquad \text{(b)}\ \sum_{n=1}^{\infty}\frac{(n+2)\ln n}{(n+5)^3},\qquad \text{(c)}\ \sum_{n=1}^{\infty}\frac{1}{(\ln(n+1))^5}.$$

2.5 On convergence of deterministic sequences and functions (a) Let $x_n=\frac{8n^2+n}{3n^2}$ for $n\geq 1$. Prove that $\lim_{n\to\infty}x_n=\frac{8}{3}$.
(b) Suppose f_n is a function on some set D for each $n\geq 1$, and suppose f is also a function on D. Then f_n is defined to converge to f *uniformly* if for any $\epsilon>0$, there exists an n_ϵ such that $|f_n(x)-f(x)|\leq\epsilon$ for all $x\in D$ whenever $n\geq n_\epsilon$. A key point is that n_ϵ does not depend on x. Show that the functions $f_n(x)=x^n$ on the semi-open interval $[0,1)$ do not converge uniformly to the zero function.
(c) The supremum of a function f on D, written $\sup_D f$, is the least upper bound of f. Equivalently, $\sup_D f$ satisfies $\sup_D f\geq f(x)$ for all $x\in D$, and given any $c<\sup_D f$, there is an $x\in D$ such that $f(x)>c$. Show that $|\sup_D f-\sup_D g|\leq\sup_D|f-g|$. Conclude that if f_n converges to f uniformly on D, then $\sup_D f_n$ converges to $\sup_D f$.
2.6 Convergence of alternating series Suppose $b_0\geq b_1\geq\ldots$ and that $b_k\to 0$ as $k\to\infty$. The purpose of this problem is to prove, using the Cauchy criteria, that the infinite sum $\sum_{k=0}^{\infty}(-1)^kb_k$ exists and is finite. By definition, the sum is equal to the limit of the partial sums $s_n=\sum_{k=0}^{n}(-1)^kb_k$ as $n\to\infty$, so it is to be proved that the sequence (s_n) has a finite limit. Please work to make your proof as simple and clean as possible.
(a) Show if $m\geq n$ then s_m is contained in the interval with endpoints s_n and s_{n+1}.
(b) Show that (s_n) is a Cauchy sequence. In particular, given $\epsilon>0$, specify how N_ϵ can be selected so that $|s_n-s_m|<\epsilon$ whenever $m\geq N_\epsilon$ and $n\geq N_\epsilon$.
2.7 On the Dirichlet criterion for convergence of a series Let (a_k) be a sequence with $a_k\geq 0$ for all $k\geq 0$ such that $\sum_{k=0}^{\infty}a_k$ is finite, and let L be a finite positive constant.
(a) Use the Cauchy criterion to show that if (d_k) is a sequence with $|d_k|\leq La_k$ for all k then the series $\sum_{k=0}^{\infty}d_k$ converges to a finite value.

Let $A_n = \sum_{k=n}^{\infty} a_k$. Then $a_k = A_k - A_{k+1}$ and the assumptions above about (a_k) are equivalent to the condition that (A_k) is a nonincreasing sequence converging to zero. Assume (B_k) is a sequence with $|B_k| \le L$ for all $k \ge 0$. Let $S_n = \sum_{k=0}^{n} A_k(B_k - B_{k-1})$, with the convention $B_{-1} = 0$.
(b) Prove the summation by parts formula: $S_n = \left(\sum_{k=0}^{n} a_k B_k\right) + A_{n+1}B_n$.
(c) Prove $\sum_{k=0}^{\infty} A_k(B_k - B_{k-1})$ converges to a finite limit.
(Note: If $B_n = 1$ for n even and $B_n = 0$ for n odd, the result of this problem provides an alternative proof of the result of the previous problem.)

2.8 Convergence of sequences of random variables Let Θ be uniformly distributed on the interval $[0, 2\pi]$. In which of the four senses (*a.s., m.s., p., d.*) do each of the following two sequences converge? Identify the limits, if they exist, and justify your answers.
(a) $(X_n : n \ge 1)$ defined by $X_n = \cos(n\Theta)$.
(b) $(Y_n : n \ge 1)$ defined by $Y_n = |1 - \frac{\Theta}{\pi}|^n$.

2.9 Convergence of a random sequence Suppose U_n for $n \ge 1$ are independent random variables, each uniformly distributed on the interval $[0, 1]$. Let $X_0 = 0$, and define X_n for $n \ge 1$ by the following recursion:

$$X_n = \max\left\{X_{n-1}, \frac{X_{n-1} + U_n}{2}\right\}.$$

(a) Does $\lim_{n\to\infty} X_n$ exist in the a.s. sense?
(b) Does $\lim_{n\to\infty} X_n$ exist in the m.s. sense?
(c) Identify the random variable Z such that $X_n \to Z$ in probability as $n \to \infty$. (Justify your answer.)

2.10 Convergence of random variables on (0,1] Let $\Omega = (0, 1]$, let $\mathcal{F}$ be the Borel σ algebra of subsets of $(0, 1]$, and let P be the probability measure on $\mathcal{F}$ such that $P([a, b]) = b - a$ for $0 < a \le b \le 1$. For the following sequences of random variables on $(\Omega, \mathcal{F}, P)$, determine in which of the four senses (a.s., p., m.s, d.), if any, each of the sequences converges. Justify your answers.
(a) $X_n(\omega) = n\omega - \lfloor n\omega \rfloor$, where $\lfloor x \rfloor$ is the largest integer less than or equal to x.
(b) $X_n(\omega) = n^2\omega$ if $0 < \omega < 1/n$, and $X_n(\omega) = 0$ otherwise.
(c) $X_n(\omega) = \frac{(-1)^n}{n\sqrt{\omega}}$.
(d) $X_n(\omega) = n\omega^n$.
(e) $X_n(\omega) = \omega \sin(2\pi n\omega)$. (Try at least for a heuristic justification.)

2.11 Convergence of some sequences of random variables Let V have the exponential distribution with parameter $\lambda = 3$. Determine in which of the four sense(s), a.s., m.s., p., or d., each of the following three sequences of random variables converges *to a finite limit random variable*.
(a) $X_n = \cos\left(\frac{V}{n}\right)$ for $n \ge 1$.
(b) $Y_n = \frac{V^n}{n}$ for $n \ge 1$.
(c) $Z_n = \left(1 + \frac{V}{n}\right)^n$ for $n \ge 1$.

2.12 A Gaussian sequence Suppose $W_1, W_2, \ldots$ are independent Gaussian random variables with mean zero and variance $\sigma^2 > 0$. Define the sequence $(X_n : n \ge 0)$ recursively by $X_0 = 0$ and $X_{k+1} = \frac{X_k + W_k}{2}$. Determine in which one(s) of the four senses, a.s., m.s., p., and d., the sequence (X_n) converges.

2.13 On the maximum of a random walk with negative drift Let $X_1, X_2, \ldots$ be independent, identically distributed random variables with mean $E[X_i] = -1$. Let $S_0 = 0$, and for $n \geq 1$, let $S_n = X_1 + \ldots + X_n$. Let $Z = \max\{S_n : n \geq 0\}$.
(a) Show that Z is well defined with probability one, and $P\{Z < +\infty\} = 1$.
(b) Does there exist a finite constant L, depending only on the above assumptions, such that $E[Z] \leq L$? Justify your answer. (Hint: $Z \geq \max\{S_0, S_1\} = \max\{0, X_1\}$.)

2.14 Convergence of a sequence of discrete random variables Let $X_n = X + (1/n)$ where $P\{X = i\} = 1/6$ for $i = 1, 2, 3, 4, 5$, or 6, and let F_n denote the distribution function of X_n.
(a) For what values of x does $F_n(x)$ converge to $F_X(x)$ as n tends to infinity?
(b) At what values of x is $F_X(x)$ continuous?
(c) Does the sequence (X_n) converge in distribution to X?

2.15 Convergence in distribution to a nonrandom limit Let $(X_n, n \geq 1)$ be a sequence of random variables and let X be a random variable such that $P\{X = c\} = 1$ for some constant c. Prove that if $\lim_{n\to\infty} X_n = X$ *d.*, then $\lim_{n\to\infty} X_n = X$ *p.* That is, prove that convergence in distribution to a constant implies convergence in probability to the same constant.

2.16 Convergence of a minimum Let $U_1, U_2, \ldots$ be a sequence of independent random variables, with each variable being uniformly distributed over the interval $[0, 1]$, and let $X_n = \min\{U_1, \ldots, U_n\}$ for $n \geq 1$.
(a) Determine in which of the senses (a.s., m.s., p., d.) the sequence (X_n) converges as $n \to \infty$, and identify the limit, if any. Justify your answers.
(b) Determine the value of the constant θ so that (Y_n) defined by $Y_n = n^\theta X_n$ converges in distribution as $n \to \infty$ to a nonzero limit, and identify the limit distribution.

2.17 Convergence of a product Let $U_1, U_2, \ldots$ be a sequence of independent random variables, with each variable being uniformly distributed over the interval $[0, 2]$, and let $X_n = U_1 U_2 \ldots U_n$ for $n \geq 1$.
(a) Determine in which of the senses (a.s., m.s., p., d.) the sequence (X_n) converges as $n \to \infty$, and identify the limit, if any. Justify your answers.
(b) Determine the value of the constant θ so that (Y_n) defined by $Y_n = n^\theta \ln(X_n)$ converges in distribution as $n \to \infty$ to a nonzero limit.

2.18 Limits of functions of random variables Let g and h be functions defined as follows:

$$g(x) = \begin{cases} -1 & \text{if } x \leq -1 \\ x & \text{if } -1 \leq x \leq 1 \\ 1 & \text{if } x \geq 1 \end{cases} \qquad h(x) = \begin{cases} -1 & \text{if } x \leq 0 \\ 1 & \text{if } x > 0 \end{cases}.$$

Thus, g represents a clipper and h represents a hard limiter. Suppose $(X_n : n \geq 0)$ is a sequence of random variables, and that X is also a random variable, all on the same underlying probability space. Give a yes or no answer to each of the four questions below. For each yes answer, identify the limit and give a justification. For each no answer, give a counterexample.

(a) If $\lim_{n\to\infty} X_n = X$ *a.s.*, then does $\lim_{n\to\infty} g(X_n)$ *a.s.* necessarily exist?
(b) If $\lim_{n\to\infty} X_n = X$ *m.s.*, then does $\lim_{n\to\infty} g(X_n)$ *m.s.* necessarily exist?
(c) If $\lim_{n\to\infty} X_n = X$ *a.s.*, then does $\lim_{n\to\infty} h(X_n)$ *a.s.* necessarily exist?
(d) If $\lim_{n\to\infty} X_n = X$ *m.s.*, then does $\lim_{n\to\infty} h(X_n)$ *m.s.* necessarily exist?

2.19 Sums of iid random variables, I A gambler repeatedly plays the following game. She bets one dollar and then there are three possible outcomes: she wins two dollars back with probability 0.4, she gets just the one dollar back with probability 0.1, and otherwise she gets nothing back. Roughly what is the probability that she is ahead after playing the game one hundred times?

2.20 Sums of iid random variables, II Let $X_1, X_2, \ldots$ be independent random variables with $P\{X_i = 1\} = P\{X_i = -1\} = 0.5$.
(a) Compute the characteristic function of the following: X_1, $S_n = X_1 + \ldots + X_n$, and $V_n = S_n/\sqrt{n}$.
(b) Find the pointwise limits of the characteristic functions of S_n and V_n as $n \to \infty$.
(c) In what sense(s), if any, do the sequences (S_n) and (V_n) converge?

2.21 Sums of iid random variables, III Fix $\lambda > 0$. For each integer $n > \lambda$, let $X_{1,n}, X_{2,n}, \ldots, X_{n,n}$ be independent random variables such that $P[X_{i,n} = 1] = \lambda/n$ and $P\{X_{i,n} = 0\} = 1 - (\lambda/n)$. Let $Y_n = X_{1,n} + X_{2,n} + \ldots + X_{n,n}$.
(a) Compute the characteristic function of Y_n for each n.
(b) Find the pointwise limit of the characteristic functions as $n \to \infty$. The limit is the characteristic function of what probability distribution?
(c) In what sense(s), if any, does the sequence (Y_n) converge?

2.22 Convergence and robustness of the sample median Suppose F is a CDF such that there is a unique value c^* such that $F(c^*) = 0.5$. Let $X_1, X_2, \ldots$ be independent random variables with CDF F. For $n \geq 1$, let Y_n denote the sample median of $X_1, \ldots, X_{2n+1}$. That is, for given $\omega \in \Omega$, if the numbers $X_1(\omega), \ldots, X_{2n+1}(\omega)$ are sorted in nondecreasing order, then $Y_n(\omega)$ is the $n+1th$ number.
(a) Show that Y_n converges almost surely (a.s.) as $n \to \infty$, and identify the limit. (It follows that Y_n also converges in the p. and d. senses.)
(b) Show that $P\{|Y_n| \geq c\} \leq 2^{2n+1} P\{|X_1| \geq c\}^{n+1}$ for all $c > 0$. This shows the tails of the distribution of Y_n are smaller than the tails of the distribution represented by F. (Hint: The union bound is sufficient. The event $\{|Y_n| \geq c\}$ is contained in the union of $\binom{2n+1}{n+1}$ overlapping events (what are they?), each having probability $P\{|X_1| \geq c\}^{n+1}$, and $\binom{2n+1}{n+1} \leq 2^{2n+1}$.)
(c) Show that if F is the CDF for the Cauchy distribution, with pdf $f(u) = \frac{1}{\pi(1+u^2)}$, then $E[|Y_1|]] < \infty$. So $E[Y_1]$ is well defined, and by symmetry, is equal to zero, even though $E[X_1]$ is not well defined. (Hint: Try finding a simple upper bound for $P\{|Y_n| \geq c\}$ and use the area rule for expectation: $E[|Y_1|] = \int_0^\infty P\{|Y_1| \geq c\}dc$.)

2.23 On the growth of the maximum of n independent exponentials Suppose that X_1, $X_2, \ldots$ are independent random variables, each with the exponential distribution with parameter $\lambda = 1$. For $n \geq 2$, let $Z_n = \frac{\max\{X_1,\ldots,X_n\}}{\ln(n)}$.
(a) Find a simple expression for the CDF of Z_n.
(b) Show that (Z_n) converges in distribution to a constant, and find the constant. (Note: It follows immediately that Z_n converges in p. to the same constant. It can also be shown that (Z_n) converges in the a.s. and m.s. senses to the same constant.)

2.24 Normal approximation for quantization error Suppose each of 100 real numbers is rounded to the nearest integer and then added. Assume the individual roundoff errors are independent and uniformly distributed over the interval $[-0.5, 0.5]$. Using the normal approximation suggested by the central limit theorem, find the approximate probability that the absolute value of the sum of the errors is greater than 5.

2.25 Limit behavior of a stochastic dynamical system Let $W_1, W_2, \ldots$ be a sequence of independent, $N(0, 0.5)$ random variables. Let $X_0 = 0$, and define $X_1, X_2, \ldots$ recursively by $X_{k+1} = X_k^2 + W_k$. Determine in which of the senses (a.s., m.s., p., d.) the sequence (X_n) converges as $n \to \infty$, and identify the limit, if any. Justify your answer.

2.26 Applications of Jensen's inequality Explain how each of the inequalities below follows from Jensen's inequality. Specifically, identify the convex function and random variable used.

(a) $E[\frac{1}{X}] \geq \frac{1}{E[X]}$, for a positive random variable X with finite mean.

(b) $E[X^4] \geq E[X^2]^2$, for a random variable X with finite second moment.

(c) $D(f|g) \geq 0$, where f and g are positive probability densities on a set A, and D is the divergence distance defined by $D(f|g) = \int_A f(x) \ln \frac{f(x)}{g(x)} dx$. (The base used in the logarithm is not relevant.)

2.27 Convergence analysis of successive averaging Let $U_1, U_2, \ldots$ be independent random variables, each uniformly distributed on the interval $[0,1]$. Let $X_0 = 0$ and $X_1 = 1$, and for $n \geq 1$ let $X_{n+1} = (1 - U_n)X_n + U_n X_{n-1}$. Note that given X_{n-1} and X_n, the variable X_{n+1} is uniformly distributed on the interval with endpoints X_{n-1} and X_n.

(a) Sketch a typical sample realization of the first few variables in the sequence.

(b) Find $E[X_n]$ for all n.

(c) Show that X_n converges in the a.s. sense as n goes to infinity. Explain your reasoning. (Hint: Let $D_n = |X_n - X_{n-1}|$. Then $D_{n+1} = U_n D_n$, and if $m > n$ then $|X_m - X_n| \leq D_n$.)

2.28 Understanding the Markov inequality Suppose X is a random variable with $E[X^4] = 30$.

(a) Derive an upper bound on $P\{|X| \geq 10\}$. Show your work.

(b) (Your bound in (a) must be the best possible in order to get both parts (a) and (b) correct.) Find a distribution for X such that the bound you found in part (a) holds with equality.

2.29 Mean square convergence of a random series The sum of infinitely many random variables, $X_1 + X_2 + \ldots$ is defined as the limit as n tends to infinity of the partial sums $X_1 + X_2 + \ldots + X_n$. The limit can be taken in the usual senses (in probability, in distribution, etc.). Suppose that the X_i are mutually independent with mean zero. Show that $X_1 + X_2 + \ldots$ exists in the mean square sense if and only if the sum of the variances, $\mathrm{Var}(X_1) + \mathrm{Var}(X_2) + \ldots$, is finite. (Hint: Apply the Cauchy criteria for mean square convergence.)

2.30 Portfolio allocation Suppose that you are given one unit of money (for example, a million dollars). Each day you bet a fraction α of it on a coin toss. If you win, you get double your money back, whereas if you lose, you get half of your money back. Let W_n denote the wealth you have accumulated (or have left) after n days. Identify in what sense(s) the limit $\lim_{n\to\infty} W_n$ exists, and when it does, identify the value of the limit:

(a) for $\alpha = 0$ (pure banking),
(b) for $\alpha = 1$ (pure betting),
(c) for general α.
(d) What value of α maximizes the expected wealth, $E[W_n]$? Would you recommend using that value of α?
(e) What value of α maximizes the long term growth rate of W_n (Hint: Consider $\ln(W_n)$ and apply the LLN.)

2.31 A large deviation Let $X_1, X_2, \ldots$ be independent, N(0,1) random variables. Find the constant b such that

$$P\{X_1^2 + X_2^2 + \ldots + X_n^2 \geq 2n\} = \exp(-n(b+\epsilon_n)),$$

where $\epsilon_n \to 0$ as $n \to \infty$. What is the numerical value of the approximation $\exp(-nb)$ if $n = 100$.

2.32 Some large deviations Let $U_1, U_2, \ldots$ be a sequence of independent random variables, each uniformly distributed on the interval [0, 1].
(a) For what values of $c \geq 0$ does there exist b with $b > 0$ (depending on c) so that $P\{U_1 + \ldots + U_n \geq cn\} \leq e^{-bn}$ for all $n \geq 1$?
(b) For what values of $c \geq 0$ does there exist b with $b > 0$ (depending on c) so that $P\{U_1 + \ldots + U_n \geq c(U_{n+1} + \ldots + U_{2n})\} \leq e^{-bn}$ for all $n \geq 1$?

2.33 Sums of independent Cauchy random variables Let $X_1, X_2, \ldots$ be independent, each with the standard Cauchy density function. The standard Cauchy density and its characteristic function are given by $f(x) = \frac{1}{\pi(1+x^2)}$ and $\Phi(u) = \exp(-|u|)$. Let $S_n = X_1 + X_2 + \cdots + X_n$.
(a) Find the characteristic function of $\frac{S_n}{n^\theta}$ for a constant θ.
(b) Does $\frac{S_n}{n}$ converge in distribution as $n \to \infty$? Justify your answer, and if the answer is yes, identify the limiting distribution.
(c) Does $\frac{S_n}{n^2}$ converge in distribution as $n \to \infty$? Justify your answer, and if the answer is yes, identify the limiting distribution.
(d) Does $\frac{S_n}{\sqrt{n}}$ converge in distribution as $n \to \infty$? Justify your answer, and if the answer is yes, identify the limiting distribution.

2.34 A rapprochement between the CLT and large deviations Let $X_1, X_2, \ldots$ be independent, identically distributed random variables with mean zero, variance σ^2, and probability density function f. Suppose the moment generating function $M(\theta)$ is finite for θ in an open interval I containing zero.
(a) Show that for $\theta \in I$, $(\ln M(\theta))''$ is the variance for the "tilted" density function f_θ defined by $f_\theta(x) = f(x)\exp(\theta x - \ln M(\theta))$. In particular, since $(\ln M(\theta))''$ is nonnegative, $\ln M$ is a convex function. (The interchange of expectation and differentiation with respect to θ can be justified for $\theta \in I$. You needn't give details.)
Let $b > 0$ and let $S_n = X_1 + \ldots + X_n$ for n any positive integer. By the central limit theorem, $P(S_n \geq b\sqrt{n}) \to Q(b/\sigma)$ as $n \to \infty$. An upper bound on the Q function is given by $Q(u) = \int_u^\infty \frac{1}{\sqrt{2\pi}} e^{-s^2/2} ds \leq \int_u^\infty \frac{s}{u\sqrt{2\pi}} e^{-s^2/2} ds = \frac{1}{u\sqrt{2\pi}} e^{-u^2/2}$. This bound is a good approximation if u is moderately large. Thus, $Q(b/\sigma) \approx \frac{\sigma}{b\sqrt{2\pi}} e^{-b^2/2\sigma^2}$ if b/σ is moderately large.

(b) The large deviations upper bound yields $P\{S_n \geq b\sqrt{n}\} \leq \exp(-n\ell(b/\sqrt{n}))$. Identify the limit of the large deviations upper bound as $n \to \infty$, and compare with the approximation given by the central limit theorem. (Hint: Approximate $\ln M$ near zero by its second order Taylor's approximation.)

2.35 Chernoff bound for Gaussian and Poisson random variables (a) Let X have the $N(\mu, \sigma^2)$ distribution. Find the optimized Chernoff bound on $P\{X \geq E[X] + c\}$ for $c \geq 0$.

(b) Let Y have the Poisson distribution with parameter λ. Find the optimized Chernoff bound on $P\{Y \geq E[Y] + c\}$ for $c \geq 0$.

(c) (The purpose of this problem is to highlight the similarity of the answers to parts (a) and (b).) Show that your answer to part (b) can be expressed as $P\{Y \geq E[Y] + c\} \leq \exp(-\frac{c^2}{2\lambda}\psi(\frac{c}{\lambda}))$ for $c \geq 0$, where $\psi(u) = 2g(1+u)/u^2$, with $g(s) = s(\ln s - 1) + 1$. (Note: Y has variance λ, so the essential difference between the normal and Poisson bounds is the ψ term. The function ψ is strictly positive and strictly decreasing on the interval $[-1, +\infty)$, with $\psi(-1) = 2$ and $\psi(0) = 1$. Also, $u\psi(u)$ is strictly increasing in u over the interval $[-1, +\infty)$.)

2.36 Large deviations of a mixed sum Let $X_1, X_2, \ldots$ have the $Exp(1)$ distribution, and $Y_1, Y_2, \ldots$ have the $Poi(1)$ distribution. Suppose all these random variables are mutually independent. Let $0 \leq f \leq 1$, and suppose $S_n = X_1 + \ldots + X_{nf} + Y_1 + \ldots + Y_{(1-f)n}$. Define $l(f, a) = \lim_{n\to\infty} \frac{1}{n} \ln P\{\frac{S_n}{n} \geq a\}$ for $a > 1$. Cramér's theorem can be extended to show that $l(f, a)$ can be computed by replacing the probability $P\{\frac{S_n}{n} \geq a\}$ by its optimized Chernoff bound. (For example, if $f = 1/2$, we simply view S_n as the sum of the $\frac{n}{2}$ iid random variables, $X_1 + Y_1, \ldots, X_{\frac{n}{2}} + Y_{\frac{n}{2}}$.) Compute $l(f, a)$ for $f \in \{0, \frac{1}{3}, \frac{2}{3}, 1\}$ and $a = 4$.

2.37 Large deviation exponent for a mixture distribution Problem 2.36 concerns an example such that $0 < f < 1$ and S_n is the sum of n independent random variables, such that a fraction f of the random variables have a CDF F_Y and a fraction $1 - f$ have a CDF F_Z. It is shown in the solutions that the large deviations exponent for $\frac{S_n}{n}$ is given by:

$$l(a) = \max_{\theta} \{\theta a - fM_Y(\theta) - (1-f)M_Z(\theta)\},$$

where $M_Y(\theta)$ and $M_Z(\theta)$ are the log moment generating functions for F_Y and F_Z respectively.

Consider the following variation. Let $X_1, X_2, \ldots, X_n$ be independent, and identically distributed, each with CDF given by $F_X(c) = fF_Y(c) + (1-f)F_Z(c)$. Equivalently, each X_i can be generated by flipping a biased coin with probability of heads equal to f, and generating X_i using CDF F_Y if heads shows and generating X_i with CDF F_Z if tails shows. Let $\widetilde{S}_n = X_1 + \ldots + X_n$, and let $\widetilde{l}$ denote the large deviations exponent for $\frac{\widetilde{S}_n}{n}$.

(a) Express the function $\widetilde{l}$ in terms of f, M_Y, and M_Z.

(b) Determine which is true and give a proof: $\widetilde{l}(a) \leq l(a)$ for all a, or $\widetilde{l}(a) \geq l(a)$ for all a. Can you also offer an intuitive explanation?

2.38 Bennett's inequality and Bernstein's inequality This problem illustrates that the proof of the Chernoff inequality is very easy to extend in many directions. Suppose it is known that $X_1, X_2, \ldots$ are independent with mean zero. Also, suppose that for some known positive constants L and d_i^2 for $i \geq 1$, $\mathrm{Var}(X_i) \leq d_i^2$ and $P\{|X_i| \leq L\} = 1$.

(a) Prove for $\theta > 0$ that $E[e^{\theta X_i}] \le \exp\left(\frac{d_i^2}{L^2}(e^{\theta L} - 1 - \theta L)\right)$. (Hint: Use the Taylor series expansion of e^u about $u = 0$, the fact $|X_i|^k \le |X_i|^2 L^{k-2}$ for $k \ge 2$, and the fact $1 + y \le e^y$ for all y.)
(b) For $\alpha > 0$, find θ that maximizes

$$\theta\alpha - \frac{\sum_{i=1}^n d_i^2}{L^2}(e^{\theta L} - 1 - \theta L).$$

(c) Prove Bennett's inequality: For $\alpha > 0$,

$$P\left\{\sum_{i=1}^n X_i \ge \alpha\right\} \le \exp\left(-\frac{\sum_{i=1}^n d_i^2}{L^2}\varphi\left(\frac{\alpha L}{\sum_i d_i^2}\right)\right),$$

where $\varphi(u) = (1 + u)\ln(1 + u) - u$.
(d) Show that $\varphi(u)/(u^2/2) \to 1$ as $u \to 0$ with $u \ge 0$. (Hint: Expand $\ln(1 + u)$ in a Taylor series about $u = 0$.)
(e) Using the fact $\varphi(u) \ge \frac{u^2}{2(1+\frac{u}{3})}$ for $u > 0$ (you needn't prove it), prove Bernstein's inequality:

$$P\left\{\sum_{i=1}^n X_i \ge \alpha\right\} \le \exp\left(-\frac{\frac{1}{2}\alpha^2}{\sum_{i=1}^n d_i^2 + \frac{\alpha L}{3}}\right).$$

2.39 Bernstein's inequality in various asymptotic regimes When the X_is are independent and identically distributed with variance σ^2 (and mean zero and there exists L so $P\{|X_1| \le L\} = 1$) Bernstein's inequality becomes $P\{S_n \ge \alpha\} \le \exp\left(-\frac{\frac{1}{2}\alpha^2}{n\sigma^2 + \frac{\alpha L}{3}}\right)$. See how the bound behaves for each of the following asymptotic regimes as $n \to \infty$:
(a) The values of σ^2 and L are fixed, and $\alpha = \theta\sqrt{n}$ for some fixed θ (i.e. the central limit theorem regime).
(b) The values of σ^2 and L are fixed, and $\alpha = cn$ for some fixed c (i.e. the large deviations regime).
(c) The values of L and α are fixed and $\sigma^2 = \frac{\gamma}{n}$ for some constant γ. (This regime is similar to the convergence of the binomial distribution with $p = \frac{\lambda}{n}$ to the Poisson distribution; the distribution of the Xs depends on n.)

2.40 The sum of products of a sequence of uniform random variables
Let $A_1, A_2, \ldots$ be a sequence of independent random variables, with $P(A_i = 1) = P(A_i = \frac{1}{2}) = \frac{1}{2}$ for all i. Let $B_k = A_1 \ldots A_k$.
(a) Does $\lim_{k\to\infty} B_k$ exist in the m.s. sense? Justify your answer.
(b) Does $\lim_{k\to\infty} B_k$ exist in the a.s. sense? Justify your answer.
(c) Let $S_n = B_1 + \ldots + B_n$. Show that $\lim_{m,n\to\infty} E[S_m S_n] = \frac{35}{3}$, which implies that $\lim_{n\to\infty} S_n$ exists in the m.s. sense.
(d) Find the mean and variance of the limit random variable.
(e) Does $\lim_{n\to\infty} S_n$ exist in the a.s. sense? Justify your answer.

2.41* Distance measures (metrics) for random variables For random variables X and Y, define

$$d_1(X,Y) = E[\mid X - Y \mid /(1+ \mid X - Y \mid)],$$
$$d_2(X,Y) = \min\{\epsilon \geq 0 : F_X(x+\epsilon) + \epsilon \geq F_Y(x) \text{ and } F_Y(x+\epsilon) + \epsilon \geq F_X(x) \text{ for all } x\},$$
$$d_3(X,Y) = (E[(X-Y)^2])^{1/2},$$

where in defining $d_3(X,Y)$ it is assumed that $E[X^2]$ and $E[Y^2]$ are finite.

(a) Show d_i is a metric for each i. Clearly $d_i(X,X) = 0$ and $d_i(X,Y) = d_i(Y,X)$. Verify in addition the triangle inequality. (The only other requirement of a metric is that $d_i(X,Y) = 0$ only if $X = Y$. For this to be true we must think of the metric as being defined on equivalence classes of random variables.)

(b) Let $X_1, X_2, \ldots$ be a sequence of random variables and let Y be a random variable. Show that X_n converges to Y:

(i) in probability if and only if $d_1(X,Y)$ converges to zero,
(ii) in distribution if and only if $d_2(X,Y)$ converges to zero,
(iii) in the mean square sense if and only if $d_3(X,Y)$ converges to zero (assume $E[Y^2] < \infty$).

(Hint for (i)): It helps to establish that

$$d_1(X,Y) - \epsilon/(1+\epsilon) \leq P\{\mid X - Y \mid \geq \epsilon\} \leq d_1(X,Y)(1+\epsilon)/\epsilon.$$

The "only if" part of (ii) is a little tricky. The metric d_2 is called the Levy metric.

2.42* Weak law of large numbers Let $X_1, X_2, \ldots$ be a sequence of random variables which are independent and identically distributed. Assume that $E[X_i]$ exists and is equal to zero for all i. If $\text{Var}(X_i)$ is finite, then Chebychev's inequality easily establishes that $(X_1 + \ldots + X_n)/n$ converges in probability to zero. Taking that result as a starting point, show that the convergence still holds even if $\text{Var}(X_i)$ is infinite. (Hint: Use "truncation" by defining $U_k = X_k I\{\mid X_k \mid \geq c\}$ and $V_k = X_k I\{\mid X_k \mid < c\}$ for some constant c. $E[\mid U_k \mid]$ and $E[V_k]$ don't depend on k and converge to zero as c tends to infinity. You might also find the previous problem helpful.)

2.43* Completing the proof of Cramér's theorem Prove Theorem 2.19 without the assumption that the random variables are bounded. To begin, select a large constant C and let $\widetilde{X}_i$ denote a random variable with the conditional distribution of X_i given that $|X_i| \leq C$. Let $\widetilde{S}_n = \widetilde{X}_1 + \ldots + \widetilde{X}_n$ and let $\widetilde{l}$ denote the large deviations exponent for $\widetilde{X}_i$. Then

$$P\left\{\frac{S_n}{n} \geq n\right\} \geq P\{|X_1| \leq C\}^n P\left\{\frac{\widetilde{S}_n}{n} \geq n\right\}.$$

One step is to show that $\widetilde{l}(a)$ converges to $l(a)$ as $C \to \infty$. It is equivalent to showing that if a pointwise monotonically increasing sequence of convex functions converges pointwise to a nonnegative convex function that is strictly positive outside some bounded set, then the minima of the convex functions converge to a nonnegative value.

3 Random vectors and minimum mean squared error estimation

Many of the concepts of random processes apply to the case in which there is only a finite number of observation times, leading to random vectors. In particular, we begin this chapter by seeing how the distribution of a random vector can be simplified by a linear change in coordinates – this same technique will be used in more general contexts in later chapters. This chapter also presents the geometric framework for estimation with the minimum mean squared error performance criterion, in which means and covariances of random variables come to the forefront. The framework includes the orthogonality principle that characterizes projections, and innovations sequences, in which new information is in a sense purified in order to simplify recursive estimation. We shall see that the multidimensional Gaussian distribution is particularly suitable for modeling systems with linear processing. An introduction to Kalman filtering, a flexible framework for estimation and tracking, is given, with an emphasis on the central role of the innovations sequence. The reader is encouraged to review the section on matrices in the Appendix before reading this chapter. □

3.1 Basic definitions and properties

A random vector X of dimension m has the form

$$X = \begin{pmatrix} X_1 \\ X_2 \\ \vdots \\ X_m \end{pmatrix},$$

where the X_is are random variables all on the same probability space. The expectation of X (also called the mean of X) is the vector $E[X]$ defined by

$$E[X] = \begin{pmatrix} E[X_1] \\ E[X_2] \\ \vdots \\ E[X_m]. \end{pmatrix}.$$

Suppose Y is another random vector on the same probability space as X, with dimension n. The cross correlation matrix of X and Y is the $m \times n$ matrix $E[XY^T]$, which has ijth entry $E[X_iY_j]$. The cross covariance matrix of X and Y, denoted by $\mathrm{Cov}(X, Y)$, is the matrix with ijth entry $\mathrm{Cov}(X_i, Y_j)$. Note that the correlation matrix is the matrix of correlations, and the covariance matrix is the matrix of covariances.

In the particular case that $n = m$ and $Y = X$, the cross correlation matrix of X with itself is simply called the correlation matrix of X, and is written as $E[XX^T]$, and it has ijth entry $E[X_iX_j]$. The cross covariance matrix of X with itself, $\mathrm{Cov}(X, X)$, has ijth entry $\mathrm{Cov}(X_i, X_j)$. This matrix is called the covariance matrix of X, and it is also denoted by $\mathrm{Cov}(X)$. So the notations $\mathrm{Cov}(X)$ and $\mathrm{Cov}(X, X)$ are interchangeable. While the notation $\mathrm{Cov}(X)$ is more concise, the notation $\mathrm{Cov}(X, X)$ is more indicative of the way the covariance matrix scales when X is multiplied by a constant.

Elementary properties of expectation, correlation, and covariance for vectors follow immediately from similar properties for ordinary scalar random variables. These properties include the following (here A and C are nonrandom matrices and b and d are nonrandom vectors):

1. $E[AX + b] = AE[X] + b$,
2. $\mathrm{Cov}(X, Y) = E[X(Y - E[Y])^T] = E[(X - E[X])Y^T] = E[XY^T] - (E[X])(E[Y])^T$,
3. $E[(AX)(CY)^T] = AE[XY^T]C^T$,
4. $\mathrm{Cov}(AX + b, CY + d) = A\mathrm{Cov}(X, Y)C^T$,
5. $\mathrm{Cov}(AX + b) = A\mathrm{Cov}(X)A^T$,
6. $\mathrm{Cov}(W + X, Y + Z) = \mathrm{Cov}(W, Y) + \mathrm{Cov}(W, Z) + \mathrm{Cov}(X, Y) + \mathrm{Cov}(X, Z)$.

In particular, the second property above shows the close connection between correlation matrices and covariance matrices. If the mean vector of either X or Y is zero, then the cross correlation and cross covariance matrices are equal.

Not every square matrix is a correlation matrix. For example, the diagonal elements must be nonnegative. Also, Schwarz's inequality (see Section 1.10) must be respected, so that $|\mathrm{Cov}(X_i, X_j)| \leq \sqrt{\mathrm{Cov}(X_i, X_i)\mathrm{Cov}(X_j, X_j)}$. Additional inequalities arise for consideration of three or more random variables at a time. Of course a square diagonal matrix is a correlation matrix if and only if its diagonal entries are nonnegative, because only vectors with independent entries need be considered. But if an $m \times m$ matrix is not diagonal, it is not a priori clear whether there are m random variables with all $m(m + 1)/2$ correlations matching the entries of the matrix. The following proposition neatly resolves these issues.

Proposition 3.1 *Correlation matrices and covariance matrices are symmetric positive semidefinite matrices. Conversely, if K is a symmetric positive semidefinite matrix, then K is the covariance matrix and correlation matrix for some mean zero random vector X.*

Proof If K is a correlation matrix, then $K = E[XX^T]$ for some random vector X. Given any vector α, $\alpha^T X$ is a scalar random variable, so

$$\alpha^T K\alpha = E[\alpha^T XX^T\alpha] = E[(\alpha^T X)(X^T\alpha)] = E[(\alpha^T X)^2] \geq 0.$$

Similarly, if $K = \mathrm{Cov}(X, X)$ then for any vector α,

$$\alpha^T K\alpha = \alpha^T \mathrm{Cov}(X, X)\alpha = \mathrm{Cov}(\alpha^T X, \alpha^T X) = \mathrm{Var}(\alpha^T X) \geq 0.$$

The first part of the proposition is proved.

For the converse part, suppose that K is an arbitrary symmetric positive semidefinite matrix. Let $\lambda_1, \ldots, \lambda_m$ and U be the corresponding set of eigenvalues and orthonormal matrix formed by the eigenvectors (see Section 11.7 in the Appendix). Let $Y_1, \ldots, Y_m$ be independent, mean 0 random variables with $\mathrm{Var}(Y_i) = \lambda_i$, and let Y be the random vector $Y = (Y_1, \ldots, Y_m)^T$. Then $\mathrm{Cov}(Y, Y) = \Lambda$, where Λ is the diagonal matrix with the λ_is on the diagonal. Let $X = UY$. Then $E[X] = 0$ and

$$\mathrm{Cov}(X, X) = \mathrm{Cov}(UY, UY) = U\Lambda U^T = K.$$

Therefore, K is both the covariance matrix and the correlation matrix of X. □

The *characteristic function* Φ_X of X is the function on $\mathbb{R}^m$ defined by

$$\Phi_X(u) = E[\exp(ju^T X)].$$

3.2 The orthogonality principle for minimum mean square error estimation

Let X be a random variable with some known distribution. Suppose X is not observed but that we wish to estimate X. If we use a constant b to estimate X, the estimation error will be $X - b$. The mean square error (MSE) is $E[(X - b)^2]$. Since $E[X - E[X]] = 0$ and $E[X] - b$ is constant,

$$\begin{aligned} E[(X-b)^2] &= E[((X - E[X]) + (E[X] - b))^2] \\ &= E[(X - E[X])^2 + 2(X - E[X])(E[X] - b) + (E[X] - b)^2] \\ &= \mathrm{Var}(X) + (E[X] - b)^2. \end{aligned}$$

From this expression it is easy to see that the mean square error is minimized with respect to b if and only if $b = E[X]$. The minimum possible value is $\mathrm{Var}(X)$.

Random variables X and Y are called orthogonal if $E[XY] = 0$. Orthogonality is denoted by "$X \perp Y$."

The essential fact $E[X - E[X]] = 0$ is equivalent to the following condition: $X - E[X]$ is orthogonal to constants: $(X - E[X]) \perp c$ for any constant c. Therefore, the choice of constant b yielding the minimum mean square error is the one that makes the error $X - b$ orthogonal to all constants. This result is generalized by the orthogonality principle, stated next.

Fix some probability space and let $L^2(\Omega, \mathcal{F}, P)$ be the set of all random variables on the probability space with finite second moments. Let X be a random variable in $L^2(\Omega, \mathcal{F}, P)$, and let $\mathcal{V}$ be a collection of random variables on the same probability space as X such that:

V.1 $\mathcal{V} \subset L^2(\Omega, \mathcal{F}, P)$,

V.2 $\mathcal{V}$ is a linear class: If $Z_1 \in \mathcal{V}$ and $Z_2 \in \mathcal{V}$ and a_1, a_2 are constants, then $a_1 Z_1 + a_2 Z_2 \in \mathcal{V}$,

V.3 $\mathcal{V}$ is closed in the mean square sense: If $Z_1, Z_2, \ldots$ is a sequence of elements of $\mathcal{V}$ and if $Z_n \to Z_\infty$ *m.s.* for some random variable Z_∞, then $Z_\infty \in \mathcal{V}$.

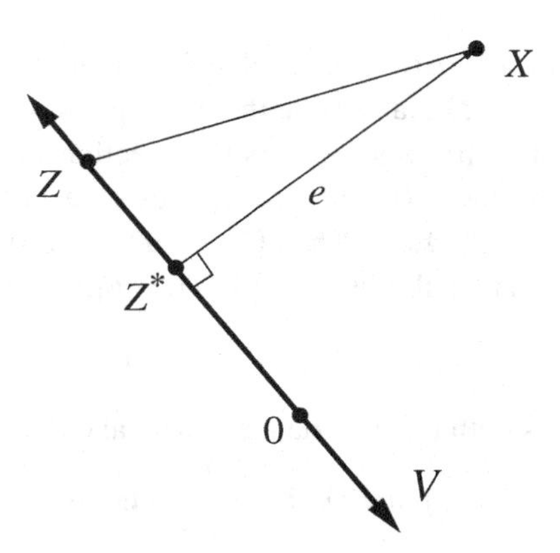

Figure 3.1 The orthogonality principle.

That is, $\mathcal{V}$ is a closed linear subspace of $L^2(\Omega, \mathcal{F}, P)$. The problem of interest is to find Z^* in $\mathcal{V}$ to minimize the mean square error, $E[(X-Z)^2]$, over all $Z \in \mathcal{V}$. That is, Z^* is the random variable in $\mathcal{V}$ that is closest to X in the minimum mean square error (MMSE) sense. We call it the *projection* of X onto $\mathcal{V}$ and denote it as $\Pi_{\mathcal{V}}(X)$.

Estimating a random variable by a constant corresponds to the case that $\mathcal{V}$ is the set of constant random variables: the projection of a random variable X onto the set of constant random variables is $E[X]$. The orthogonality principle stated next is illustrated in Figure 3.1.

Theorem 3.2 *(The orthogonality principle) Let $\mathcal{V}$ be a closed, linear subspace of $L^2(\Omega, \mathcal{F}, P)$, and let $X \in L^2(\Omega, \mathcal{F}, P)$, for some probability space $(\Omega, \mathcal{F}, P)$.*

(a) *(Existence and uniqueness) There exists a unique element Z^* (also denoted by $\Pi_{\mathcal{V}}(X)$) in $\mathcal{V}$ so that $E[(X - Z^*)^2] \leq E[(X - Z)^2]$ for all $Z \in \mathcal{V}$. (Here, we consider two elements Z and Z' of $\mathcal{V}$ to be the same if $P\{Z = Z'\} = 1$.)*

(b) *(Characterization) Let W be a random variable. Then $W = Z^*$ if and only if the following two conditions hold:*

(i) $W \in \mathcal{V}$,

(ii) $(X - W) \perp Z$ *for all Z in $\mathcal{V}$.*

(c) *(Error expression) The minimum mean square error (MMSE) is given by $E[(X-Z^*)^2] = E[X^2] - E[(Z^*)^2]$.*

Proof The proof of (a) is given in Problem 3.32. The technical condition V.3 on $\mathcal{V}$ is essential for the proof of existence. Here parts (b) and (c) are proved.

To establish the "if" half of (b), suppose W satisfies (i) and (ii) and let Z be an arbitrary element of $\mathcal{V}$. Then $W-Z \in \mathcal{V}$ because $\mathcal{V}$ is a linear class. Therefore, $(X-W) \perp (W-Z)$, which implies that

$$
\begin{aligned}
E[(X-Z)^2] &= E[(X-W+W-Z)^2] \\
&= E[(X-W)^2 + 2(X-W)(W-Z) + (W-Z)^2] \\
&= E[(X-W)^2] + E[(W-Z)^2].
\end{aligned}
$$

Thus $E[(X-W)^2] \leq E[(X-Z)^2]$. Since Z is an arbitrary element of $\mathcal{V}$, it follows that $W = Z^*$, and the "if" half of (b) is proved.

To establish the "only if" half of (b), note that $Z^* \in \mathcal{V}$ by the definition of Z^*. Let $Z \in \mathcal{V}$ and let $c \in \mathbb{R}$. Then $Z^* + cZ \in \mathcal{V}$, so that $E[(X-(Z^*+cZ))^2] \geq E[(X-Z^*)^2]$. But

$$
\begin{aligned}
E[(X-(Z^*+cZ))^2] &= E[(X-Z^*)-(cZ)^2] \\
&= E[(X-Z^*)^2] - 2cE[(X-Z^*)Z] + c^2E[Z^2],
\end{aligned}
$$

so that

$$-2cE[(X-Z^*)Z] + c^2E[Z^2] \geq 0. \tag{3.1}$$

As a function of c the left side of (3.1) is a parabola with value zero at $c = 0$. Hence its derivative with respect to c at 0 must be zero, which yields that $(X - Z^*) \perp Z$. The "only if" half of (b) is proved.

The expression of (c) is proved as follows. Since $X - Z^*$ is orthogonal to all elements of $\mathcal{V}$, including Z^* itself,

$$E[X^2] = E[((X-Z^*)+Z^*)^2] \;=\; E[(X-Z^*)^2] + E[(Z^*)^2].$$

This proves (c). □

The following propositions give some properties of the projection mapping $\Pi_{\mathcal{V}}$, with proofs based on the orthogonality principle.

Proposition 3.3 *(Linearity of projection) Suppose $\mathcal{V}$ is a closed linear subspace of $L^2(\Omega, \mathcal{F}, P)$, X_1 and X_2 are in $L^2(\Omega, \mathcal{F}, P)$, and a_1 and a_2 are constants. Then*

$$\Pi_{\mathcal{V}}(a_1X_1 + a_2X_2) = a_1\Pi_{\mathcal{V}}(X_1) + a_2\Pi_{\mathcal{V}}(X_2). \tag{3.2}$$

Proof By the characterization part of the orthogonality principle ((b) of Theorem 3.2), the projection $\Pi_{\mathcal{V}}(a_1X_1 + a_2X_2)$ is characterized by two properties. So, to prove (3.2), it suffices to show that $a_1\Pi_{\mathcal{V}_1}(X_1) + a_2\Pi_{\mathcal{V}_2}(X_2)$ satisfies these two properties. First, we must check that $a_1\Pi_{\mathcal{V}_1}(X_1) + a_2\Pi_{\mathcal{V}_2}(X_2) \in \mathcal{V}$. This follows immediately from the fact that $\Pi_{\mathcal{V}}(X_i) \in \mathcal{V}$, for $i = 1, 2$, and $\mathcal{V}$ is a linear subspace, so the first property is checked. Second, we must check that $e \perp Z$, where $e = a_1X_1 + a_2X_2 - (a_1\Pi_{\mathcal{V}}(X_1) + a_2\Pi_{\mathcal{V}}(X_2))$, and Z is an arbitrary element of $\mathcal{V}$. Now $e = a_1e_1 + a_2e_2$, where $e_i = X_i - \Pi_{\mathcal{V}}(X_i)$ for $i = 1, 2$, and $e_i \perp Z$ for $i = 1, 2$. So $E[eZ] = a_1E[e_1Z] + a_2E[e_2Z] = 0$, or equivalently, $e \perp Z$. Thus, the second property is also checked, and the proof is complete. □

Proposition 3.4 *(Projections onto nested subspaces) Suppose $\mathcal{V}_1$ and $\mathcal{V}_2$ are closed linear subspaces of $L^2(\Omega, \mathcal{F}, P)$ such that $\mathcal{V}_2 \subset \mathcal{V}_1$. Then for any $X \in L^2(\Omega, \mathcal{F}, P)$, $\Pi_{\mathcal{V}_2}(X) = \Pi_{\mathcal{V}_2}\Pi_{\mathcal{V}_1}(X)$. (In words, the projection of X onto $\mathcal{V}_2$ can be found by first projecting X onto $\mathcal{V}_1$, and then projecting the result onto $\mathcal{V}_2$.) Furthermore,*

$$E[(X - \Pi_{\mathcal{V}_2}(X))^2] = E[(X - \Pi_{\mathcal{V}_1}(X))^2] + E[(\Pi_{\mathcal{V}_1}(X) - \Pi_{\mathcal{V}_2}(X))^2]. \tag{3.3}$$

In particular, $E[(X - \Pi_{\mathcal{V}_2}(X))^2] \geq E[(X - \Pi_{\mathcal{V}_1}(X))^2]$.

Proof By the characterization part of the orthogonality principle (Theorem 3.2(b)), the projection $\Pi_{\mathcal{V}_2}(X)$ is characterized by two properties. So, to prove $\Pi_{\mathcal{V}_2}(X) = \Pi_{\mathcal{V}_2}\Pi_{\mathcal{V}_1}(X)$, it suffices to show that $\Pi_{\mathcal{V}_2}\Pi_{\mathcal{V}_1}(X)$ satisfies the two properties. First, we must check that $\Pi_{\mathcal{V}_2}\Pi_{\mathcal{V}_1}(X) \in \mathcal{V}_2$. This follows immediately from the fact that $\Pi_{\mathcal{V}_2}(X)$ maps into $\mathcal{V}_2$, so the first property is checked. Second, we must check that $e \perp Z$, where $e = X - \Pi_{\mathcal{V}_2}\Pi_{\mathcal{V}_1}(X)$, and Z is an arbitrary element of $\mathcal{V}_2$. Now $e = e_1 + e_2$, where $e_1 = X - \Pi_{\mathcal{V}_1}(X)$ and $e_2 = \Pi_{\mathcal{V}_1}(X) - \Pi_{\mathcal{V}_2}\Pi_{\mathcal{V}_1}(X)$. By the characterization of $\Pi_{\mathcal{V}_1}(X)$, e_1 is perpendicular to any random variable in $\mathcal{V}_1$. In particular, $e_1 \perp Z$, because $Z \in \mathcal{V}_2 \subset \mathcal{V}_1$. The characterization of the projection of $\Pi_{\mathcal{V}_1}(X)$ onto $\mathcal{V}_2$ implies that $e_2 \perp Z$. Since $e_i \perp Z$ for $i = 1, 2$, it follows that $e \perp Z$. Thus, the second property is also checked, so it is proved that $\Pi_{\mathcal{V}_2}(X) = \Pi_{\mathcal{V}_2}\Pi_{\mathcal{V}_1}(X)$.

As mentioned above, e_1 is perpendicular to any random variable in $\mathcal{V}_1$, which implies that $e_1 \perp e_2$. Thus, $E[e^2] = E[e_1^2] + E[e_2^2]$, which is equivalent to (3.3). Therefore, (3.3) is proved. The last inequality of the proposition follows, of course, from (3.3). The inequality is also equivalent to the inequality $\min_{W \in \mathcal{V}_2} E[(X - W)^2] \geq \min_{W \in \mathcal{V}_1} E[(X - W)^2]$, and this inequality is true because the minimum of a set of numbers cannot increase if more numbers are added to the set. □

The following proposition is closely related to the use of linear innovations sequences, discussed in Sections 3.5 and 3.6.

Proposition 3.5 *(Projection onto the span of orthogonal subspaces) Suppose $\mathcal{V}_1$ and $\mathcal{V}_2$ are closed linear subspaces of $L^2(\Omega, \mathcal{F}, P)$ such that $\mathcal{V}_1 \perp \mathcal{V}_2$, which means that $E[Z_1 Z_2] = 0$ for any $Z_1 \in \mathcal{V}_1$ and $Z_2 \in \mathcal{V}_2$. Let $\mathcal{V} = \mathcal{V}_1 \oplus \mathcal{V}_2 = \{Z_1 + Z_2 : Z_i \in \mathcal{V}_i\}$ denote the span of $\mathcal{V}_1$ and $\mathcal{V}_2$. Then for any $X \in L^2(\Omega, \mathcal{F}, P)$, $\Pi_{\mathcal{V}}(X) = \Pi_{\mathcal{V}_1}(X) + \Pi_{\mathcal{V}_2}(X)$. The minimum mean square error satisfies*

$$E[(X - \Pi_{\mathcal{V}}(X))^2] = E[X^2] - E[(\Pi_{\mathcal{V}_1}(X))^2] - E[(\Pi_{\mathcal{V}_2}(X))^2].$$

Proof The space $\mathcal{V}$ is also a closed linear subspace of $L^2(\Omega, \mathcal{F}, P)$ (see starred homework problem 3.33). By the characterization part of the orthogonality principle (Theorem 3.2(b)), the projection $\Pi_{\mathcal{V}}(X)$ is characterized by two properties. So to prove $\Pi_{\mathcal{V}}(X) = \Pi_{\mathcal{V}_1}(X) + \Pi_{\mathcal{V}_2}(X)$, it suffices to show that $\Pi_{\mathcal{V}_1}(X) + \Pi_{\mathcal{V}_2}(X)$ satisfies these two properties. First, we must check that $\Pi_{\mathcal{V}_1}(X) + \Pi_{\mathcal{V}_2}(X) \in \mathcal{V}$. This follows immediately from the fact that $\Pi_{\mathcal{V}_i}(X) \in \mathcal{V}_i$, for $i = 1, 2$, so the first property is checked. Second, we must check that $e \perp Z$, where $e = X - (\Pi_{\mathcal{V}_1}(X) + \Pi_{\mathcal{V}_2}(X))$, and Z is an arbitrary element of $\mathcal{V}$. Now any such Z can be written as $Z = Z_1 + Z_2$ where $Z_i \in \mathcal{V}_i$ for $i = 1, 2$. Observe that $\Pi_{\mathcal{V}_2}(X) \perp Z_1$ because $\Pi_{\mathcal{V}_2}(X) \in \mathcal{V}_2$ and $Z_1 \in \mathcal{V}_1$. Therefore,

$$\begin{aligned} E[eZ_1] &= E[(X - (\Pi_{\mathcal{V}_1}(X) + \Pi_{\mathcal{V}_2}(X))Z_1] \\ &= E[(X - \Pi_{\mathcal{V}_1}(X))Z_1] = 0, \end{aligned}$$

where the last equality follows from the characterization of $\Pi_{\mathcal{V}_1}(X)$. So, $e \perp Z_1$, and similarly $e \perp Z_2$, so $e \perp Z$. Thus, the second property is also checked, so $\Pi_{\mathcal{V}}(X) = \Pi_{\mathcal{V}_1}(X) + \Pi_{\mathcal{V}_2}(X)$ is proved.

Since $\Pi_{\mathcal{V}_i}(X) \in \mathcal{V}_i$ for $i = 1, 2$, $\Pi_{\mathcal{V}_1}(X) \perp \Pi_{\mathcal{V}_2}(X)$. Therefore, $E[(\Pi_{\mathcal{V}}(X))^2] = E[(\Pi_{\mathcal{V}_1}(X))^2] + E[(\Pi_{\mathcal{V}_2}(X))^2]$, and the expression for the MMSE in the proposition follows from the error expression in the orthogonality principle. □

3.3 Conditional expectation and linear estimators

In many applications, a random variable X is to be estimated based on observation of a random variable Y. Thus, an estimator is a function of Y. In applications, the two most frequently considered classes of functions of Y used in this context are essentially all functions, leading to the best unconstrained estimator, or all linear functions, leading to the best linear estimator. These two possibilities are discussed in this section.

3.3.1 Conditional expectation as a projection

Suppose a random variable X is to be estimated using an observed random vector Y of dimension m. Suppose $E[X^2] < +\infty$. Consider the most general class of estimators based on Y, by setting

$$\mathcal{V} = \{g(Y) : \ g : \mathbb{R}^m \to \mathbb{R}, \ E[g(Y)^2] < +\infty\}. \tag{3.4}$$

There is also the implicit condition that g is Borel measurable so that $g(Y)$ is a random variable. The projection of X onto this class $\mathcal{V}$ is the unconstrained minimum mean square error (MMSE) estimator of X given Y.

Let us first proceed to identify the optimal estimator by conditioning on the value of Y, thereby reducing this example to the estimation of a random variable by a constant, as discussed at the beginning of Section 3.2. For technical reasons we assume for now that X and Y have a joint pdf. Then, conditioning on Y,

$$E[(X - g(Y))^2] = \int_{\mathbb{R}^m} E[(X - g(Y))^2 | Y = y] f_Y(y) dy,$$

where

$$E[(X - g(Y))^2 | Y = y] = \int_{-\infty}^{\infty} (x - g(y))^2 f_{X|Y}(x|y) dx.$$

Since the mean is the MMSE estimator of a random variable among all constants, for each fixed y, the minimizing choice for $g(y)$ is

$$g^*(y) = E[X|Y = y] = \int_{-\infty}^{\infty} x f_{X|Y}(x|y) dx. \tag{3.5}$$

Therefore, the optimal estimator in $\mathcal{V}$ is $g^*(Y)$ which, by definition, is equal to the random variable $E[X|Y]$.

What does the orthogonality principle imply for this example? It implies that there exists an optimal estimator $g^*(Y)$ which is the unique element of $\mathcal{V}$ such that

$$(X - g^*(Y)) \perp g(Y)$$

for all $g(Y) \in \mathcal{V}$. If X, Y have a joint pdf then we can check that $E[X|Y]$ satisfies the required condition. Indeed,

$$\begin{aligned} E[(X - E[X|Y])g(Y)] &= \int\int (x - E[X|Y = y])g(y)f_{X|Y}(x|y)f_Y(y)dxdy \\ &= \int\left\{\int (x - E[X|Y = y])f_{X|Y}(x|y)dx\right\} g(y)f_Y(y)dy \\ &= 0, \end{aligned}$$

because the expression within the braces is zero.

In summary, if X and Y have a joint pdf (and similarly if they have a joint pmf) then the MMSE estimator of X given Y is $E[X|Y]$. Even if X and Y do not have a joint pdf or joint pmf, we *define* the *conditional expectation* $E[X|Y]$ to be the MMSE estimator of X given Y. By the orthogonality principle $E[X|Y]$ exists as long as $E[X^2] < \infty$, and it is the unique function of Y such that

$$E[(X - E[X|Y])g(Y)] = 0$$

for all $g(Y)$ in $\mathcal{V}$.

Estimation of a random variable has been discussed, but often we wish to estimate a random vector. A beauty of the MSE criterion is that it easily extends to estimation of random vectors, because the MSE for estimation of a random vector is the sum of the MSEs of the coordinates:

$$E[\| X - g(Y) \|^2] = \sum_{i=1}^{m} E[(X_i - g_i(Y))^2].$$

Therefore, for most sets of estimators $\mathcal{V}$ typically encountered, finding the MMSE estimator of a random vector X decomposes into finding the MMSE estimators of the coordinates of X separately.

Suppose a random vector X is to be estimated using estimators of the form g(Y), where here g maps $\mathbb{R}^n$ into $\mathbb{R}^m$. Assume $E[\|X\|^2] < +\infty$ and seek an estimator to minimize the MSE. As seen above, the MMSE estimator for each coordinate X_i is $E[X_i|Y]$, which is also the projection of X_i onto the set of unconstrained estimators based on Y, defined in (3.4). So the optimal estimator $g^*(Y)$ of the entire vector X is given by

$$g^*(Y) = E[X|Y] = \begin{pmatrix} E[X_1|Y] \\ E[X_2|Y] \\ \vdots \\ E[X_m|Y] \end{pmatrix}.$$

Let the estimation error be denoted by e, $e = X - E[X|Y]$. (Even though e is a random vector we use lower case for it for an obvious reason.)

The mean of the error is given by $Ee = 0$. As for the covariance of the error, note that $E[X_j|Y]$ is in $\mathcal{V}$ for each j, so $e_i \perp E[X_j|Y]$ for each i,j. Since $Ee_i = 0$, it follows that

$\text{Cov}(e_i, E[X_j|Y]) = 0$ for all i,j. Equivalently, $\text{Cov}(e, E[X|Y]) = 0$. Using this and the fact $X = E[X|Y] + e$ yields

$$\begin{aligned}\text{Cov}(X) &= \text{Cov}(E[X|Y] + e) \\ &= \text{Cov}(E[X|Y]) + \text{Cov}(e) + \text{Cov}(E[X|Y], e) + \text{Cov}(e, E[X|Y]) \\ &= \text{Cov}(E[X|Y]) + \text{Cov}(e).\end{aligned}$$

Thus, $\text{Cov}(e) = \text{Cov}(X) - \text{Cov}(E[X|Y])$.

In practice, computation of $E[X|Y]$ (for example, using (3.5) in case a joint pdf exists) may be too complex or may require more information about the joint distribution of X and Y than is available. For both of these reasons, it is worthwhile to consider classes of estimators that are constrained to smaller sets of functions of the observations. A widely used set is the set of all linear functions, leading to linear estimators, described next.

3.3.2 Linear estimators

Let X and Y be random vectors with $E[\|X\|^2] < +\infty$ and $E[\|Y\|^2] < +\infty$. Seek estimators of the form $AY + b$ to minimize the MSE. Such estimators are called linear estimators because each coordinate of $AY + b$ is a linear combination of $Y_1, Y_2, \ldots, Y_m$ and 1. Here "1" stands for the random variable that is always equal to 1.

To identify the optimal linear estimator we shall apply the orthogonality principle for each coordinate of X with

$$\mathcal{V} = \{c_0 + c_1 Y_1 + c_2 Y_2 + \ldots + c_n Y_n : c_0, c_1, \ldots, c_n \in \mathbb{R}\}.$$

Let e denote the estimation error $e = X-(AY+b)$. We must select A and b so that $e_i \perp Z$ for all $Z \in \mathcal{V}$. Equivalently, we must select A and b so that

$$\begin{aligned} e_i &\perp 1 \quad && \text{all } i \\ e_i &\perp Y_j \quad && \text{all } i,j. \end{aligned}$$

The condition $e_i \perp 1$, which means $Ee_i = 0$, implies that $E[e_i Y_j] = \text{Cov}(e_i, Y_j)$. Thus, the required orthogonality conditions on A and b become $Ee = 0$ and $\text{Cov}(e, Y) = 0$. The condition $Ee = 0$ requires that $b = E[X] - AE[Y]$, so we can restrict our attention to estimators of the form $E[X] + A(Y - E[Y])$, so that $e = X - E[X] - A(Y - E[Y])$. The condition $\text{Cov}(e, Y) = 0$ becomes $\text{Cov}(X, Y) - A\text{Cov}(Y, Y) = 0$. If $\text{Cov}(Y, Y)$ is not singular, then A must be given by $A = \text{Cov}(X, Y)\text{Cov}(Y, Y)^{-1}$. In this case the optimal linear estimator, denoted by $\widehat{E}[X|Y]$, is given by

$$\widehat{E}[X|Y] = E[X] + \text{Cov}(X, Y)\text{Cov}(Y, Y)^{-1}(Y - E[Y]). \tag{3.6}$$

Proceeding as in the case of unconstrained estimators of a random vector, we find that the covariance of the error vector satisfies

$$\text{Cov}(e) = \text{Cov}(X) - \text{Cov}(\widehat{E}[X|Y]),$$

which by (3.6) yields

$$\text{Cov}(e) = \text{Cov}(X) - \text{Cov}(X, Y)\text{Cov}(Y, Y)^{-1}\text{Cov}(Y, X). \tag{3.7}$$

3.3.3 Comparison of the estimators

As seen above, the expectation $E[X]$, the MMSE linear estimator $\widehat{E}[X|Y]$, and the conditional expectation $E[X|Y]$, are all instances of projection mappings $\Pi_{\mathcal{V}}$, for $\mathcal{V}$ consisting of constants, linear estimators based on Y, or unconstrained estimators based on Y, respectively. Hence, the orthogonality principle, and Propositions 3.3–3.5 all apply to these estimators.

Proposition 3.3 implies that these estimators are linear functions of X. In particular, $E[a_1X_1 + a_2X_2|Y] = a_1E[X_1|Y] + a_2E[X_2|Y]$, and the same is true with "E" replaced by "$\widehat{E}$."

Proposition 3.4, regarding projections onto nested subspaces, implies an ordering of the mean square errors:

$$E[(X - E[X|Y])^2] \leq E[(X - \widehat{E}[X|Y])^2] \leq \mathrm{Var}(X).$$

Furthermore, it implies that the best linear estimator of X based on Y is equal to the best linear estimator of the estimator $E[X|Y]$: that is, $\widehat{E}[X|Y] = \widehat{E}[E[X|Y]|Y]$. It follows, in particular, that $E[X|Y] = \widehat{E}[X|Y]$ if and only if $E[X|Y]$ has the linear form, $AX + b$. Similarly, $E[X]$, the best constant estimator of X, is also the best constant estimator of $\widehat{E}[X|Y]$ or of $E[X|Y]$. That is, $E[X] = E[\widehat{E}[X|Y]] = E[E[X|Y]]$. In fact, $E[X] = E[\widehat{E}[E[X|Y]|Y]]$.

Proposition 3.4 also implies relations among estimators based on different sets of observations. For example, suppose X is to be estimated and Y_1 and Y_2 are both possible observations. The space of unrestricted estimators based on Y_1 alone is a subspace of the space of unrestricted estimators based on both Y_1 and Y_2. Therefore, Proposition 3.4 implies that $E[E[X|Y_1, Y_2]|Y_1] = E[X|Y_1]$, a property that is sometimes called the *tower property* of conditional expectation. The same relation holds true for the same reason for the best linear estimators: $\widehat{E}[\widehat{E}[X|Y_1, Y_2]|Y_1] = \widehat{E}[X|Y_1]$.

Example 3.1 Let X, Y be jointly continuous random variables with the pdf

$$f_{XY}(x,y) = \begin{cases} x+y & 0 \leq x, y \leq 1 \\ 0 & \text{else} \end{cases}.$$

Let us find $E[X|Y]$ and $\widehat{E}[X|Y]$. To find $E[X|Y]$ we first identify $f_Y(y)$ and $f_{X|Y}(x|y)$:

$$f_Y(y) = \int_{-\infty}^{\infty} f_{XY}(x,y)dx = \begin{cases} \frac{1}{2} + y & 0 \leq y \leq 1 \\ 0 & \text{else} \end{cases}.$$

Therefore, $f_{X|Y}(x|y)$ is defined only for $0 \leq y \leq 1$, and for such y it is given by

$$f_{X|Y}(x|y) = \begin{cases} \frac{x+y}{\frac{1}{2}+y} & 0 \leq x \leq 1 \\ 0 & \text{else} \end{cases}.$$

So for $0 \leq y \leq 1$,

$$E[X|Y = y] = \int_0^1 x f_{X|Y}(x|y)dx = \frac{2+3y}{3+6y}.$$

Therefore, $E[X|Y] = \frac{2+3Y}{3+6Y}$. To find $\widehat{E}[X|Y]$ use $E[X] = E[Y] = \frac{7}{12}$, $\mathrm{Var}(Y) = \frac{11}{144}$ and $\mathrm{Cov}(X,Y) = -\frac{1}{144}$ so $\widehat{E}[X|Y] = \frac{7}{12} - \frac{1}{11}(Y - \frac{7}{12})$.

Example 3.2 Suppose that $Y = XU$, where X and U are independent random variables, X has the Rayleigh density

$$f_X(x) = \begin{cases} \frac{x}{\sigma^2} e^{-x^2/2\sigma^2} & x \geq 0 \\ 0 & \text{else} \end{cases}$$

and U is uniformly distributed on the interval $[0,1]$. We find $\widehat{E}[X|Y]$ and $E[X|Y]$. To compute $\widehat{E}[X|Y]$ we find

$$\begin{aligned} E[X] &= \int_0^\infty \frac{x^2}{\sigma^2} e^{-x^2/2\sigma^2} dx = \frac{1}{\sigma}\sqrt{\frac{\pi}{2}} \int_{-\infty}^\infty \frac{x^2}{\sqrt{2\pi\sigma^2}} e^{-x^2/2\sigma^2} dx = \sigma\sqrt{\frac{\pi}{2}}, \\ E[Y] &= E[X]E[U] = \frac{\sigma}{2}\sqrt{\frac{\pi}{2}}, \\ E[X^2] &= 2\sigma^2, \\ \mathrm{Var}(Y) &= E[Y^2] - E[Y]^2 = E[X^2]E[U^2] - E[X]^2E[U]^2 = \sigma^2\left(\frac{2}{3} - \frac{\pi}{8}\right), \\ \mathrm{Cov}(X,Y) &= E[U]E[X^2] - E[U]E[X]^2 = \frac{1}{2}\mathrm{Var}(X) = \sigma^2\left(1 - \frac{\pi}{4}\right). \end{aligned}$$

Thus

$$\widehat{E}[X|Y] = \sigma\sqrt{\frac{\pi}{2}} + \frac{(1-\frac{\pi}{4})}{(\frac{2}{3} - \frac{\pi}{8})}\left(Y - \frac{\sigma}{2}\sqrt{\frac{\pi}{2}}\right).$$

To find $E[X|Y]$ we first find the joint density and then the conditional density. Now

$$\begin{aligned} f_{XY}(x,y) &= f_X(x) f_{Y|X}(y|x) \\ &= \begin{cases} \frac{1}{\sigma^2} e^{-x^2/2\sigma^2} & 0 \leq y \leq x, \\ 0 & \text{else} \end{cases} \\ f_Y(y) &= \int_{-\infty}^\infty f_{XY}(x,y)dx = \begin{cases} \int_y^\infty \frac{1}{\sigma^2} e^{-x^2/2\sigma^2} dx = \frac{\sqrt{2\pi}}{\sigma} Q\left(\frac{y}{\sigma}\right) & y \geq 0 \\ 0 & y < 0 \end{cases}, \end{aligned}$$

where Q is the complementary CDF for the standard normal distribution. So for $y \geq 0$

$$\begin{aligned} E[X|Y=y] &= \int_{-\infty}^\infty x f_{XY}(x,y)dx / f_Y(y) \\ &= \frac{\int_y^\infty \frac{x}{\sigma^2} e^{-x^2/2\sigma^2} dx}{\frac{\sqrt{2\pi}}{\sigma} Q(\frac{y}{\sigma})} = \frac{\sigma \exp(-y^2/2\sigma^2)}{\sqrt{2\pi}Q(\frac{y}{\sigma})}. \end{aligned}$$

Thus,

$$E[X|Y] = \frac{\sigma \exp(-Y^2/2\sigma^2)}{\sqrt{2\pi}Q(\frac{Y}{\sigma})}.$$

Example 3.3 Suppose that Y is a random variable and f is a Borel measurable function such that $E[f(Y)^2] < \infty$. Let us show that $E[f(Y)|Y] = f(Y)$. By definition, $E[f(Y)|Y]$ is the random variable of the form $g(Y)$ which is closest to $f(Y)$ in the mean square sense. If we take $g(Y) = f(Y)$, then the mean square error is zero. No other estimator can have a smaller mean square error. Thus, $E[f(Y)|Y] = f(Y)$. Similarly, if Y is a random vector with $E[||Y||^2] < \infty$, and if A is a matrix and b a vector, then $\widehat{E}[AY + b|Y] = AY + b$.

3.4 Joint Gaussian distribution and Gaussian random vectors

Recall that a random variable X is Gaussian (or normal) with mean μ and variance $\sigma^2 > 0$ if X has pdf

$$f_X(x) = \frac{1}{\sqrt{2\pi\sigma^2}} e^{-\frac{(x-\mu)^2}{2\sigma^2}}.$$

As a degenerate case, we say X is Gaussian with mean μ and variance 0 if $P\{X = \mu\} = 1$. Equivalently, X is Gaussian with mean μ and variance $\sigma^2 \geq 0$ if its characteristic function is given by

$$\Phi_X(u) = \exp\left(-\frac{u^2\sigma^2}{2} + j\mu u\right).$$

Lemma 3.6 *Suppose $X_1, X_2, \ldots, X_n$ are independent Gaussian random variables. Then any linear combination $a_1X_1 + \ldots + a_nX_n$ is a Gaussian random variable.*

Proof By an induction argument on n, it is sufficient to prove the lemma for $n = 2$. Also, if X is a Gaussian random variable, then so is aX for any constant a, so we can assume without loss of generality that $a_1 = a_2 = 1$. It remains to prove that if X_1 and X_2 are independent Gaussian random variables, then the sum $X = X_1 + X_2$ is also a Gaussian random variable. Let $\mu_i = E[X_i]$ and $\sigma_i^2 = \mathrm{Var}(X_i)$. Then the characteristic function of X is given by

$$\begin{aligned}\Phi_X(u) &= E[e^{juX}] = E[e^{juX_1}e^{juX_2}] = E[e^{juX_1}]E[e^{juX_2}]\\ &= \exp\left(-\frac{u^2\sigma_1^2}{2} + j\mu_1 u\right)\exp\left(-\frac{u^2\sigma_2^2}{2} + j\mu_2 u\right) = \exp\left(-\frac{u^2\sigma^2}{2} + j\mu u\right),\end{aligned}$$

where $\mu = \mu_1 + \mu_2$ and $\sigma^2 = \sigma_1^2 + \sigma_2^2$. Thus, X is an $N(\mu, \sigma^2)$ random variable. □

Let $(X_i : i \in I)$ be a collection of random variables indexed by some set I, which possibly has infinite cardinality. A finite linear combination of $(X_i : i \in I)$ is a random variable of the form

$$a_1X_{i_1} + a_2X_{i_2} + \ \ldots \ + a_nX_{i_n},$$

where n is finite, $i_k \in I$ for each k, and $a_k \in \mathbb{R}$ for each k.

Definition 3.7 A collection $(X_i : i \in I)$ of random variables has a *joint Gaussian distribution* (and the random variables $X_i : i \in I$ themselves are said to be *jointly Gaussian*) if every finite linear combination of $(X_i : i \in I)$ is a Gaussian random variable. A random vector X is called a *Gaussian random vector* if its coordinate random variables are jointly Gaussian. A collection of random vectors is said to have a joint Gaussian distribution if all of the coordinate random variables of all of the vectors are jointly Gaussian.

We write that X is an $N(\mu, K)$ random vector if X is a Gaussian random vector with mean vector μ and covariance matrix K.

Proposition 3.8 (a) *If $(X_i : i \in I)$ has a joint Gaussian distribution, then each of the random variables itself is Gaussian.*

(b) *If the random variables $X_i : i \in I$ are each Gaussian and if they are independent, which means that $X_{i_1}, X_{i_2}, \ldots, X_{i_n}$ are independent for any finite number of indices $i_1, i_2, \ldots, i_n$, then $(X_i : i \in I)$ has a joint Gaussian distribution.*

(c) *(Preservation of joint Gaussian property under linear combinations and limits) Suppose $(X_i : i \in I)$ has a joint Gaussian distribution. Let $(Y_j : j \in J)$ denote a collection of random variables such that each Y_j is a finite linear combination of $(X_i : i \in I)$, and let $(Z_k : k \in K)$ denote a set of random variables such that each Z_k is a limit in probability (or in the m.s. or a.s. senses) of a sequence from $(Y_j : j \in J)$. Then $(Y_j : j \in J)$ and $(Z_k : k \in K)$ each have a joint Gaussian distribution.*

(c′) *(Alternative version of (c)) Suppose $(X_i : i \in I)$ has a joint Gaussian distribution. Let $\mathcal{Z}$ denote the smallest set of random variables that contains $(X_i : i \in I)$, that is a linear class, and is closed under taking limits in probability. Then $\mathcal{Z}$ has a joint Gaussian distribution.*

(d) *The characteristic function of an $N(\mu, K)$ random vector is given by $\Phi_X(u) = E[e^{ju^TX}] = e^{ju^T\mu - \frac{1}{2}u^TKu}$.*

(e) *If X is an $N(\mu, K)$ random vector and K a diagonal matrix (i.e. $Cov(X_i, X_j) = 0$ for $i \neq j$, or equivalently, the coordinates of X are uncorrelated) then the coordinates $X_1, \ldots, X_m$ are independent.*

(f) *An $N(\mu, K)$ random vector X such that K is nonsingular has a pdf given by*

$$f_X(x) = \frac{1}{(2\pi)^{\frac{m}{2}}|K|^{\frac{1}{2}}} \exp\left(-\frac{(x-\mu)^T K^{-1}(x-\mu)}{2}\right). \tag{3.8}$$

Any random vector X such that $Cov(X)$ is singular does not have a pdf.

(g) *If X and Y are jointly Gaussian vectors, then they are independent if and only if $Cov(X, Y) = 0$.*

Proof (a) Suppose $(X_i : i \in I)$ has a joint Gaussian distribution, so that all finite linear combinations of the X_is are Gaussian random variables. Each X_i for $i \in I$ is itself a finite linear combination of all the variables (with only one term). So each X_i is a Gaussian random variable.

(b) Suppose the variables $X_i : i \in I$ are mutually independent, and each is Gaussian. Then any finite linear combination of $(X_i : i \in I)$ is the sum of finitely many independent Gaussian random variables (by Lemma 3.6), and is hence also a Gaussian random variable. So $(X_i : i \in I)$ has a joint Gaussian distribution.

(c) Suppose the hypotheses of (c) are true. Let V be a finite linear combination of $(Y_j : j \in J) : V = b_1 Y_{j_1} + b_2 Y_{j_2} + \ldots + b_n Y_{j_n}$. Each Y_j is a finite linear combination of $(X_i : i \in I)$, so V can be written as a finite linear combination of $(X_i : i \in I)$:

$$V = b_1(a_{11}X_{i_{11}} + \ldots a_{1k_1}X_{i_{1k_1}}) + \ldots + b_n(a_{n1}X_{i_{n1}} + \ldots + a_{nk_n}X_{i_{nk_n}}).$$

Therefore V is thus a Gaussian random variable. Thus, any finite linear combination of $(Y_j : j \in J)$ is Gaussian, so that $(Y_j : j \in J)$ has a joint Gaussian distribution.

Let W be a finite linear combination of $(Z_k : k \in K)$: $W = a_1 Z_{k_1} + \ldots + a_m Z_{k_m}$. By assumption, for $1 \leq l \leq m$, there is a sequence $(j_{l,n} : n \geq 1)$ of indices from J such that $Y_{j_{l,n}} \stackrel{d.}{\to} Z_{k_l}$ as $n \to \infty$. Let $W_n = a_1 Y_{j_{1,n}} + \ldots + a_m Y_{j_{m,n}}$. Each W_n is a Gaussian random variable, because it is a finite linear combination of $(Y_j : j \in J)$. Also,

$$|W - W_n| \leq \sum_{l=1}^{m} a_l |Z_{k_l} - Y_{j_{l,n}}|. \tag{3.9}$$

Since each term on the righthand side of (3.9) converges to zero in probability, it follows that $W_n \stackrel{p.}{\to} W$ as $n \to \infty$. Since limits in probability of Gaussian random variables are also Gaussian random variables (Proposition 2.8), it follows that W is a Gaussian random variable. Thus, an arbitrary finite linear combination W of $(Z_k : k \in K)$ is Gaussian, so, by definition, $(Z_k : k \in K)$ has a joint Gaussian distribution.

(c$'$) Suppose $(X_i : i \in I)$ has a joint Gaussian distribution. Using the notation of (c), let $(Y_j : j \in J)$ denote the set of *all* finite linear combinations of $(X_i : i \in I)$ and let $(Z_k : k \in K)$ denote the set of *all* random variables that are limits in probability of random variables in $(Y_j : j \in I)$. We will show that $\mathcal{Z} = (Z_k : k \in K)$, which together with (c) already proved, will establish (c$'$). We begin by establishing that $(Z_k : k \in K)$ satisfies the three properties required of $\mathcal{Z}$:

(i) $(Z_k : k \in K)$ contains $(X_i : i \in I)$,
(ii) $(Z_k : k \in K)$ is a linear class,
(iii) $(Z_k : k \in K)$ is closed under taking limits in probability.

Property (i) follows from the fact that for any $i_o \in I$, the random variable X_{i_o} is trivially a finite linear combination of $(X_i : i \in I)$, and it is trivially the limit in probability of the sequence with all entries equal to itself. Property (ii) is true because a linear combination of the form $a_1 Z_{k_1} + a_2 Z_{k_2}$ is the limit in probability of a sequence of random variables of the form $a_1 Y_{j_{n,1}} + a_2 Y_{j_{n,2}}$, and, since $(Y_j : j \in J)$ is a linear class, $a_1 Y_{j_{n,1}} + a_2 Y_{j_{n_2}}$ is a random variable from $(Y_j : j \in J)$ for each n. To prove (iii), suppose $Z_{k_n} \stackrel{p.}{\to} Z_\infty$ as $n \to \infty$ for some sequence $k_1, k_2, \ldots$ from K. By passing to a subsequence if necessary, it can be assumed that $P\{|Z_\infty - Z_{k_n}| \geq 2^{-(n+1)}\} \leq 2^{-(n+1)}$ for all $n \geq 1$. Since each Z_{k_n} is the limit in probability of a sequence from $(Y_j : j \in J)$, for each n there is $j_n \in J$

so $P\{|Z_{k_n} - Y_{j_n}| \geq 2^{-(n+1)}\} \leq 2^{-(n+1)}$. Since $|Z_\infty - Y_{j_n}| \leq |Z_\infty - Z_{k_n}| + |Z_{k_n} - Y_{j_n}|$, it follows that $P\{|Z_\infty - Y_{j_n}| \geq 2^{-n}\} \leq 2^{-n}$. So $Y_{j_n} \xrightarrow{p} Z_\infty$. Therefore, Z_∞ is a random variable in $(Z_k : k \in K)$, so $(Z_k : k \in K)$ is closed under convergence in probability. In summary, $(Z_k : k \in K)$ has properties (i)–(iii). Any set of random variables with these three properties must contain $(Y_j : j \in J)$, and hence must contain $(Z_k : k \in K)$. So $(Z_k : k \in K)$ is indeed the smallest set of random variables with properties (i)–(iii). That is, $(Z_k : k \in K) = \mathcal{Z}$, as claimed.

(d) Let X be an $N(\mu, K)$ random vector. Then for any vector u with the same dimension as X, the random variable u^TX is Gaussian with mean $u^T\mu$ and variance given by

$$\mathrm{Var}(u^TX) = \mathrm{Cov}(u^TX, u^TX) = u^TKu.$$

Thus, we already know the characteristic function of u^TX. But the characteristic function of the vector X evaluated at u is the characteristic function of u^TX evaluated at 1:

$$\Phi_X(u) = E[e^{ju^TX}] = E[e^{j(u^TX)}] = \Phi_{u^TX}(1) = e^{ju^T\mu - \frac{1}{2}u^TKu},$$

which establishes (d) of the proposition.

(e) If X is an $N(\mu, K)$ random vector and K is a diagonal matrix, then

$$\Phi_X(u) = \prod_{i=1}^m \exp\left(ju_i\mu_i - \frac{k_{ii}u_i^2}{2}\right) = \prod_i \Phi_i(u_i),$$

where k_{ii} denotes the ith diagonal element of K, and Φ_i is the characteristic function of an $N(\mu_i, k_{ii})$ random variable. By uniqueness of distribution for a given joint characteristic function, it follows that $X_1, \ldots, X_m$ are independent random variables.

(f) Let X be an $N(\mu, K)$ random vector. Since K is positive semidefinite it can be written as $K = U\Lambda U^T$ where U is orthonormal (so $UU^T = U^TU = I$) and Λ is a diagonal matrix with the nonnegative eigenvalues $\lambda_1, \lambda_2, \ldots, \lambda_m$ of K along the diagonal (see Section 11.7 of the Appendix.) Let $Y = U^T(X - \mu)$. Then Y is a Gaussian vector with mean 0 and covariance matrix given by $\mathrm{Cov}(Y, Y) = \mathrm{Cov}(U^TX, U^TX) = U^TKU = \Lambda$. In summary, we have $X = UY + \mu$, and Y is a vector of independent Gaussian random variables, the ith one being $N(0, \lambda_i)$. Suppose further that K is nonsingular, meaning $\det(K) \neq 0$. Since $\det(K) = \lambda_1\lambda_2 \cdots \lambda_m$ this implies that $\lambda_i > 0$ for each i, so that Y has the joint pdf

$$f_Y(y) = \prod_{i=1}^m \frac{1}{\sqrt{2\pi\lambda_i}} \exp\left(-\frac{y_i^2}{2\lambda_i}\right) = \frac{1}{(2\pi)^{\frac{m}{2}}\sqrt{\det(K)}} \exp\left(-\frac{y^T\Lambda^{-1}y}{2}\right).$$

Since $|\det(U)| = 1$ and $U\Lambda^{-1}U^T = K^{-1}$, the joint pdf for the $N(\mu, K)$ random vector X is given by

$$f_X(x) = f_Y(U^T(x - \mu)) = \frac{1}{(2\pi)^{\frac{m}{2}}|K|^{\frac{1}{2}}} \exp\left(-\frac{(x-\mu)^TK^{-1}(x-\mu)}{2}\right).$$

Now suppose, instead, that X is any random vector with some mean μ and a singular covariance matrix K. That means that $\det K = 0$, or equivalently that $\lambda_i = 0$ for one of the eigenvalues of K, or equivalently, that there is a vector α such that $\alpha^T K\alpha = 0$ (such an α is an eigenvector of K for eigenvalue zero). But then $0 = \alpha^T K\alpha = \alpha^T \text{Cov}(X, X)\alpha = \text{Cov}(\alpha^T X, \alpha^T X) = \text{Var}(\alpha^T X)$. Therefore, $P\{\alpha^T X = \alpha^T \mu\} = 1$. That is, with probability one, X is in the subspace $\{x \in \mathbb{R}^m : \alpha^T(x - \mu) = 0\}$. Therefore, X does not have a pdf.

(g) Suppose X and Y are jointly Gaussian vectors and uncorrelated (i.e. $\text{Cov}(X, Y) = 0$). Let Z denote the dimension $m + n$ vector with coordinates $X_1, \ldots, X_m, Y_1, \ldots, Y_n$. Since $\text{Cov}(X, Y) = 0$, the covariance matrix of Z is block diagonal:

$$\text{Cov}(Z) = \begin{pmatrix} \text{Cov}(X) & 0 \\ 0 & \text{Cov}(Y) \end{pmatrix}.$$

Therefore, for $u \in \mathbb{R}^m$ and $v \in \mathbb{R}^n$,

$$\begin{aligned} \Phi_Z\left(\begin{pmatrix} u \\ v \end{pmatrix}\right) &= \exp\left(-\frac{1}{2}\begin{pmatrix} u \\ v \end{pmatrix}^T \text{Cov}(Z)\begin{pmatrix} u \\ v \end{pmatrix} + j\begin{pmatrix} u \\ v \end{pmatrix}^T EZ\right) \\ &= \Phi_X(u)\Phi_Y(v). \end{aligned}$$

Such factorization implies that X and Y are independent. The if part of (g) is proved. Conversely, if X and Y are jointly Gaussian and independent of each other, then the characteristic function of the joint density must factor, which implies that $\text{Cov}(Z)$ is block diagonal as above. That is, $\text{Cov}(X, Y) = 0$. □

Recall that in general, if X and Y are two random vectors on the same probability space, then the mean square error for the MMSE linear estimator $\widehat{E}[X|Y]$ is greater than or equal to the mean square error for the best unconstrained estimator, $E[X|Y]$. The tradeoff, however, is that $E[X|Y]$ can be much more difficult to compute than $\widehat{E}[X|Y]$, which is determined entirely by first and second moments. As shown in the next proposition, if X and Y are jointly Gaussian, the two estimators coincide. That is, the MMSE unconstrained estimator of Y is linear. We also know that $E[X|Y = y]$ is the mean of the conditional mean of X given $Y = y$. The proposition identifies not only the conditional mean, but the entire conditional distribution of X given $Y = y$, for the case X and Y are jointly Gaussian.

Proposition 3.9 *Let X and Y be jointly Gaussian vectors and $y \in \mathbb{R}$. The conditional distribution of X given $Y = y$ is $N(\widehat{E}[X|Y = y], Cov(e))$. In particular, the conditional mean $E[X|Y = y]$ is equal to $\widehat{E}[X|Y = y]$. That is, if X and Y are jointly Gaussian, $E[X|Y] = \widehat{E}[X|Y]$.*

If $Cov(Y)$ is nonsingular,

$$E[X|Y = y] = \widehat{E}[X|Y = y] - E[X] + \text{Cov}(X, Y)\text{Cov}(Y)^{-1}(y - E[Y]), \tag{3.10}$$

$$\text{Cov}(e) = \text{Cov}(X) - \text{Cov}(X, Y)\text{Cov}(Y)^{-1}\text{Cov}(Y, X), \tag{3.11}$$

and if Cov(e) is nonsingular,

$$f_{X|Y}(x|y) = \frac{1}{(2\pi)^{\frac{m}{2}}|\mathrm{Cov}(e)|^{\frac{1}{2}}} \times \exp\left(-\frac{1}{2}\left(x - \widehat{E}[X|Y=y]\right)^T \mathrm{Cov}(e)^{-1}\left(x - \widehat{E}[X|Y=y]\right)\right). \tag{3.12}$$

Proof Consider the MMSE linear estimator $\widehat{E}[X|Y]$ of X given Y, and let e denote the corresponding error vector: $e = X - \widehat{E}[X|Y]$. Recall that, by the orthogonality principle, $Ee = 0$ and $\mathrm{Cov}(e, Y) = 0$. Since Y and e are obtained from X and Y by linear transformations, they are jointly Gaussian. Since $\mathrm{Cov}(e, Y) = 0$, the random vectors e and Y are also independent. For the next part of the proof, the reader should keep in mind that if a is a deterministic vector of some dimension m, and Z is an $N(0, K)$ random vector, for a matrix K that is not a function of a, then $Z + a$ has the $N(a, K)$ distribution.

Focus on the following rearrangement of the definition of e:

$$X = e + \widehat{E}[X|Y]. \tag{3.13}$$

(Basically, the whole proof of the proposition hinges on (3.13).) Since $\widehat{E}[X|Y]$ is a function of Y and since e is independent of Y with distribution $N(0, \mathrm{Cov}(e))$, the following key observation can be made. Given $Y = y$, the conditional distribution of e is the $N(0, \mathrm{Cov}(e))$ distribution, which does not depend on y, while $\widehat{E}[X|Y = y]$ is completely determined by y. So, given $Y = y$, X can be viewed as the sum of the $N(0, \mathrm{Cov}(e))$ vector e and the determined vector $\widehat{E}[X|Y = y]$. So the conditional distribution of X given $Y = y$ is $N(\widehat{E}[X|Y = y], \mathrm{Cov}(e))$. In particular, $E[X|Y = y]$, which in general is the mean of the conditional distribution of X given $Y = y$, is therefore the mean of the $N(\widehat{E}[X|Y = y], \mathrm{Cov}(e))$ distribution. Hence $E[X|Y = y] = \widehat{E}[X|Y = y]$. Since this is true for all y, $E[X|Y] = \widehat{E}[X|Y]$.

Equations (3.10) and (3.11), respectively, are just the equations (3.6) and (3.7) derived for the MMSE linear estimator, $\widehat{E}[X|Y]$, and its associated covariance of error. Equation (3.12) is just the formula (3.8) for the pdf of an $N(\mu, K)$ vector, with $\mu = \widehat{E}[X|Y = y]$ and $K = \mathrm{Cov}(e)$. □

Example 3.4 Suppose X and Y are jointly Gaussian mean zero random variables such that the vector $\begin{pmatrix} X \\ Y \end{pmatrix}$ has covariance matrix $\begin{pmatrix} 4 & 3 \\ 3 & 9 \end{pmatrix}$. Let us find simple expressions for the two random variables $E[X^2|Y]$ and $P(X \geq c|Y)$. Note that if W is a random variable with the $N(\mu, \sigma^2)$ distribution, then $E[W^2] = \mu^2 + \sigma^2$ and $P\{W \geq c\} = Q(\frac{c-\mu}{\sigma})$, where Q is the standard Gaussian complementary CDF. The idea is to apply these facts to the conditional distribution of X given Y. Given $Y = y$, the conditional distribution of X is $N\left(\frac{\mathrm{Cov}(X,Y)}{\mathrm{Var}(Y)}y, \mathrm{Cov}(X) - \frac{\mathrm{Cov}(X,Y)^2}{\mathrm{Var}(Y)}\right)$, or $N(\frac{y}{3}, 3)$. Therefore, $E[X^2|Y = y] = (\frac{y}{3})^2 + 3$ and $P(X \geq c|Y = y) = Q\left(\frac{c-(y/3)}{\sqrt{3}}\right)$. Applying these two functions to the random variable Y yields $E[X^2|Y] = (\frac{Y}{3})^2 + 3$ and $P(X \geq c|Y) = Q\left(\frac{c-(Y/3)}{\sqrt{3}}\right)$.

3.5 Linear innovations sequences

Let $X, Y_1, \ldots, Y_n$ be random vectors with finite second moments, all on the same probability space. In general, computation of the joint projection $\widehat{E}[X|Y_1, \ldots, Y_n]$ is considerably more complicated than computation of the individual projections $\widehat{E}[X|Y_i]$, because it requires inversion of the covariance matrix of all the Ys. However, if $E[Y_i] = 0$ for all i and $E[Y_i Y_j^T] = 0$ for $i \neq j$ (i.e. all coordinates of Y_i are orthogonal to constants and to all coordinates of Y_j for $i \neq j$), then

$$\widehat{E}[X|Y_1, \ldots, Y_n] = \overline{X} + \sum_{i=1}^{n} \widehat{E}[X - \overline{X}|Y_i], \tag{3.14}$$

where we write $\overline{X}$ for $E[X]$. The orthogonality principle can be used to prove (3.14) as follows. It suffices to prove that the right side of (3.14) satisfies the two properties that together characterize the left side of (3.14). First, the right side is a linear combination of $1, Y_1, \ldots, Y_n$. Secondly, let e denote the error when the right side of (3.14) is used to estimate X:

$$e = X - \overline{X} - \sum_{i=1}^{n} \widehat{E}[X - \overline{X}|Y_i].$$

It must be shown that $E[e(Y_1^T c_1 + Y_2^T c_2 + \ldots + Y_n^T c_n + b)] = 0$ for any constant vectors $c_1, \ldots, c_n$ and constant b. It is enough to show that $E[e] = 0$ and $E[eY_j^T] = 0$ for all j. But $\widehat{E}[X - \overline{X}|Y_i]$ has the form $B_i Y_i$, because $X - \overline{X}$ and Y_i have mean zero. Thus, $E[e] = 0$. Furthermore,

$$E[eY_j^T] = E\left[\left(X - \widehat{E}[X|Y_j]\right) Y_j^T\right] - \sum_{i:i\neq j} E[B_i Y_i Y_j^T].$$

Each term on the right side of this equation is zero, so $E[eY_j^T] = 0$, and (3.14) is proved.

If $1, Y_1, Y_2, \ldots, Y_n$ have finite second moments but are not orthogonal, then (3.14) does not directly apply. However, by orthogonalizing this sequence we can obtain a sequence $1, \widetilde{Y}_1, \widetilde{Y}_2, \ldots, \widetilde{Y}_n$ that can be used instead. Let $\widetilde{Y}_1 = Y_1 - E[Y_1]$, and for $k \geq 2$ let

$$\widetilde{Y}_k = Y_k - \widehat{E}[Y_k|Y_1, \ldots, Y_{k-1}]. \tag{3.15}$$

Then $E[\widetilde{Y}_i] = 0$ for all i and $E[\widetilde{Y}_i \widetilde{Y}_j^T] = 0$ for $i \neq j$. In addition, by induction on k, we can prove that the set of all random variables obtained by linear transformation of $1, Y_1, \ldots, Y_k$ is equal to the set of all random variables obtained by linear transformation of $1, \widetilde{Y}_1, \ldots, \widetilde{Y}_k$.

Thus, for any random vector X with all components having finite second moments,

$$\widehat{E}[X|Y_1, \ldots, Y_n] = \widehat{E}[X|\widetilde{Y}_1, \ldots, \widetilde{Y}_n] = \overline{X} + \sum_{i=1}^{n} \widehat{E}[X - \overline{X}|\widetilde{Y}_i]$$

$$= \overline{X} + \sum_{i=1}^{n} \mathrm{Cov}(X, \widetilde{Y}_i)\mathrm{Cov}(\widetilde{Y}_i)^{-1}\widetilde{Y}_i.$$

(Since $E[\widetilde{Y}_i] = 0$ for $i \geq 1$, $\text{Cov}(X, \widetilde{Y}_i) = E[X\widetilde{Y}_i^T]$ and $\text{Cov}(\widetilde{Y}_i) = E[\widetilde{Y}_i\widetilde{Y}_i^T]$.) Moreover, this same result can be used to compute the innovations sequence recursively:

$$\widetilde{Y}_1 = Y_1 - E[Y_1], \text{ and}$$

$$\widetilde{Y}_k = Y_k - E[Y_k] - \sum_{i=1}^{k-1} \text{Cov}(X, \widetilde{Y}_i)\text{Cov}(\widetilde{Y}_i)^{-1}\widetilde{Y}_i \quad k \geq 2.$$

The sequence $\widetilde{Y}_1, \widetilde{Y}_2, \ldots, \widetilde{Y}_n$ is called the *linear innovations sequence* for $Y_1, Y_2, \ldots, Y_n$.

3.6 Discrete-time Kalman filtering

Kalman filtering is a state-space approach to the problem of estimating one random sequence from another. Recursive equations are found that are useful in many real-time applications. For notational convenience, because there are so many matrices in this section, lower case letters are used for random vectors. All the random variables involved are assumed to have finite second moments. The state sequence $x_0, x_1, \ldots$, is to be estimated from an observed sequence $y_0, y_1, \ldots$ These sequences of random vectors are assumed to satisfy the following state and observation equations:

$$\text{State:} x_{k+1} = F_k x_k + w_k \qquad k \geq 0$$

$$\text{Observation:} y_k = H_k^T x_k + v_k \qquad k \geq 0.$$

It is assumed that:

- $x_0, v_0, v_1, \ldots, w_0, w_1, \ldots$ are pairwise uncorrelated.
- $Ex_0 = \overline{x}_0$, $\text{Cov}(x_0) = P_0$, $Ew_k = 0$, $\text{Cov}(w_k) = Q_k$, $Ev_k = 0$, $\text{Cov}(v_k) = R_k$.
- F_k, H_k, Q_k, R_k for $k \geq 0$; P_0 are known matrices.
- $\overline{x}_0$ is a known vector.

See Figure 3.2 for a block diagram of the state and observation equations. The evolution of the state sequence $x_0, x_1, \ldots$ is driven by the random vectors $w_0, w_1, \ldots$, while the random vectors $v_0, v_1, \ldots$, represent observation noise.

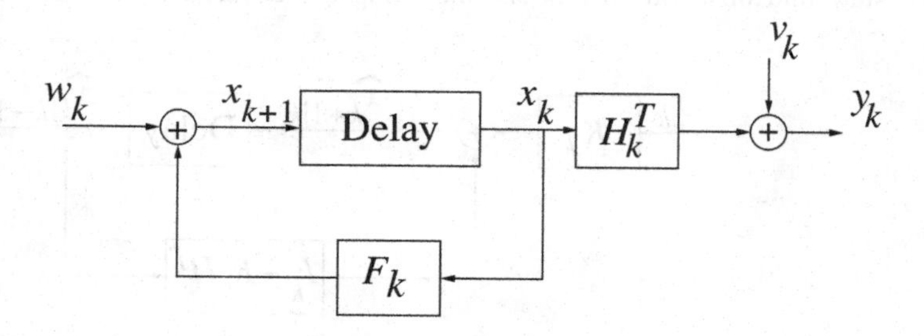

Figure 3.2 The state and observations equations.

Let $\bar{x}_k = E[x_k]$ and $P_k = \mathrm{Cov}(x_k)$. These quantities are recursively determined for $k \geq 1$ by

$$\bar{x}_{k+1} = F_k \bar{x}_k \text{ and } P_{k+1} = F_k P_k F_k^T + Q_k, \tag{3.16}$$

where the initial conditions $\bar{x}_0$ and P_0 are given as part of the state model. The idea of the Kalman filter equations is to recursively compute conditional expectations in a similar way.

Let $y^k = (y_0, y_1, \ldots, y_k)$ represent the observations up to time k. Define for nonnegative integers i, j

$$\widehat{x}_{i|j} = \widehat{E}[x_i | y^j]$$

and the associated covariance of error matrices

$$\Sigma_{i|j} = \mathrm{Cov}(x_i - \widehat{x}_{i|j}).$$

The goal is to compute $\widehat{x}_{k+1|k}$ for $k \geq 0$. The Kalman filter equations will first be stated, then briefly discussed, and then derived. The Kalman filter equations are given by

$$\begin{aligned} \widehat{x}_{k+1|k} &= \left[F_k - K_k H_k^T\right] \widehat{x}_{k|k-1} + K_k y_k \\ &= F_k \widehat{x}_{k|k-1} + K_k \left[y_k - H_k^T \widehat{x}_{k|k-1}\right], \end{aligned} \tag{3.17}$$

with the initial condition $\widehat{x}_{0|-1} = \bar{x}_0$, where the gain matrix K_k is given by

$$K_k = F_k \Sigma_{k|k-1} H_k \left[H_k^T \Sigma_{k|k-1} H_k + R_k\right]^{-1}, \tag{3.18}$$

and the covariance of error matrices are recursively computed by

$$\Sigma_{k+1|k} = F_k \left[\Sigma_{k|k-1} - \Sigma_{k|k-1} H_k \left(H_k^T \Sigma_{k|k-1} H_k + R_k\right)^{-1} H_k^T \Sigma_{k|k-1}\right] F_k^T + Q_k, \tag{3.19}$$

with the initial condition $\Sigma_{0|-1} = P_0$, see Figure 3.3 for the block diagram.

We comment briefly on the Kalman filter equations, before deriving them. First, observe what happens if H_k is the zero matrix, $H_k = 0$, for all k. Then the Kalman filter equations reduce to (3.16) with $\widehat{x}_{k|k-1} = \bar{x}_k$, $\Sigma_{k|k-1} = P_k$ and $K_k = 0$. Taking $H_k = 0$ for all k is equivalent to having no observations available.

In many applications, the sequence of gain matrices can be computed ahead of time according to (3.18) and (3.19). Then as the observations become available, the estimates can be computed using only (3.17). In some applications the matrices involved in the state and observation models, including the covariance matrices of the v_ks and w_ks,

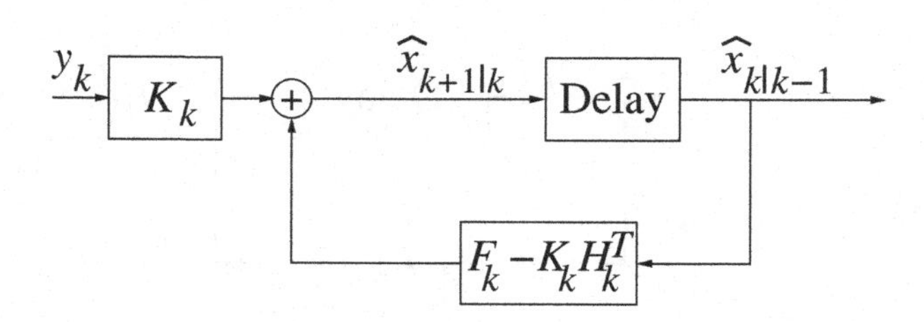

Figure 3.3 The Kalman filter.

do not depend on k. The gain matrices K_k could still depend on k due to the initial conditions, but if the model is stable in some sense, then the gains converge to a constant matrix K, so that in steady state the filter equation (3.17) becomes time invariant: $\widehat{x}_{k+1|k} = (F - KH^T)\widehat{x}_{k|k-1} + Ky_k$.

In other applications, particularly those involving feedback control, the matrices in the state and/or observation equations might not be known until just before they are needed.

The Kalman filter equations are now derived. Roughly speaking, there are two considerations for computing $\widehat{x}_{k+1|k}$ once $\widehat{x}_{k|k-1}$ is computed: (1) the *information update*, accounting for the availability of the new observation y_k, enabling the calculation of $\widehat{x}_{k|k}$; and (2) the *time update*, accounting for the change in state from x_k to x_{k+1}.

Information update: The observation y_k is not totally new because it can be predicted in part from the previous observations, or simply by its mean in the case $k = 0$. Specifically, we consider $\tilde{y}_k = y_k - \widehat{E}[y_k|y^{k-1}]$ to be the new part of the observation y_k. Here, $\tilde{y}_0, \tilde{y}_1, \ldots$ is the linear innovation sequence for the observation sequence $y_0, y_1, \ldots$, as defined in Section 3.5 (with the minor difference that here the vectors are indexed from time $k = 0$ on, rather than from time $k = 1$). Let $\tilde{y}^{k-1} = (\tilde{y}_0, \tilde{y}_1, \ldots, \tilde{y}_{k-1})$. Since the linear span of the random variables in $(1, y^{k-1}, y_k)$ is the same as the linear span of the random variables in $(1, y^{k-1}, \tilde{y}_k)$, for the purposes of incorporating the new observation we can pretend that $\tilde{y}_k$ is the new observation rather than y_k. From the observation equation, the fact $E[v_k] = 0$, and the fact w_k is orthogonal to all the random variables of y^{k-1}, it follows that

$$\begin{aligned}\widehat{E}[y_k|y^{k-1}] &= \widehat{E}\left[H_k^T x_k + w_k|y^{k-1}\right] \\ &= H_k^T \widehat{x}_{k|k-1},\end{aligned}$$

so $\tilde{y}_k = y_k - H_k^T \widehat{x}_{k|k-1}$. Since $(1, y^{k-1}, y_k)$ and $(1, \tilde{y}^{k-1}, \tilde{y}_k)$ have the same span and the random variables in $\tilde{y}^{k-1}$ are orthogonal to the random variables in $\tilde{y}_k$, and all these random variables have mean zero,

$$\begin{aligned}\widehat{x}_{k|k} &= \widehat{E}\left[x_k|\tilde{y}^{k-1}, \tilde{y}_k\right] \\ &= \widehat{E}\left[x_k|\tilde{y}^{k-1}\right] + \widehat{E}\left[x_k - \overline{x}_k|\tilde{y}^{k-1}, \tilde{y}_k\right] \\ &= \widehat{x}_{k|k-1} + \mathrm{Cov}(x_k, \tilde{y}^k)\mathrm{Cov}(\tilde{y}_k)^{-1}\tilde{y}_k. \end{aligned} \tag{3.20}$$

Furthermore, use of the new observation $\tilde{y}_k$ reduces the covariance of error for predicting x_k from $\Sigma_{k|k-1}$ by the covariance matrix of the innovative part of the estimator:

$$\Sigma_{k|k} = \Sigma_{k|k-1} - \mathrm{Cov}(x_k, \tilde{y}^k)\mathrm{Cov}(\tilde{y}_k)^{-1}\mathrm{Cov}(\tilde{y}^k, x_k). \tag{3.21}$$

Time update: In view of the state update equation and the fact that w_k is uncorrelated with the random variables of y^k and has mean zero,

$$\begin{aligned}\widehat{x}_{k+1|k} &= \widehat{E}[F_k x_k + w_k|y^k] \\ &= F_k\widehat{E}[x_k|y^k] + \widehat{E}[w_k|y^{k-1}] \\ &= F_k\widehat{x}_{k|k}.\end{aligned} \tag{3.22}$$

Thus, the time update consists of simply multiplying the estimate $\widehat{x}_{k|k}$ by F_k. Furthermore, the covariance of error matrix for predicting x_{k+1} by $\widehat{x}_{k+1|k}$, is given by

$$\begin{aligned}\Sigma_{k+1|k} &= \mathrm{Cov}(x_{k+1} - \widehat{x}_{k+1|k}) \\ &= \mathrm{Cov}(F_k(x_k - \widehat{x}_{k|k}) + w_k) \\ &= F_k \Sigma_{k|k} F_k^T + Q_k. \end{aligned} \tag{3.23}$$

Putting it all together: Combining (3.20) and (3.22) with the fact $\tilde{y}_k = y_k - H_k^T \widehat{x}_{k|k-1}$ yields the Kalman filter equation (3.17), if we set

$$K_k = F_k \mathrm{Cov}(x_k, \tilde{y}^k) \mathrm{Cov}(\tilde{y}_k)^{-1}. \tag{3.24}$$

Applying the facts:

$$\begin{aligned}\mathrm{Cov}(x_k, \tilde{y}_k) &= \mathrm{Cov}(x_k + w_k, H_k^T(x_k - \widehat{x}_{k|k-1}) + v_k) \\ &= \mathrm{Cov}(x_k, H_k^T(x_k - \widehat{x}_{k|k-1})) \\ &= \mathrm{Cov}(x_k - \widehat{x}_{k|k-1}, H_k^T(x_k - \widehat{x}_{k|k-1})) \quad (\text{since } \widehat{x}_{k|k-1} \perp x_k - \widehat{x}_{k|k-1}) \\ &= \Sigma_{k|k-1} H_k \end{aligned} \tag{3.25}$$

and

$$\begin{aligned}\mathrm{Cov}(\tilde{y}_k) &= \mathrm{Cov}(H_k^T(x_k - \widehat{x}_{k|k-1}) + v_k) \\ &= \mathrm{Cov}(H_k^T(x_k - \widehat{x}_{k|k-1})) + \mathrm{Cov}(v_k) \\ &= H_k^T \Sigma_{k|k-1} H_k + R_k \end{aligned} \tag{3.26}$$

to (3.24) yields (3.18) and to (3.21) yields

$$\Sigma_{k|k} = \Sigma_{k|k-1} - \Sigma_{k|k-1} H_k (H_k^T \Sigma_{k|k-1} H_k + R_k)^{-1} H_k^T \Sigma_{k|k-1}. \tag{3.27}$$

Finally, (3.23) and (3.27) yield (3.19). This completes the derivation of the Kalman filter equations.

Problems

3.1 Rotation of a joint normal distribution yielding independence Let X be a Gaussian vector with

$$E[X] = \begin{pmatrix} 10 \\ 5 \end{pmatrix} \quad \mathrm{Cov}(X) = \begin{pmatrix} 2 & 1 \\ 1 & 1 \end{pmatrix}.$$

(a) Write an expression for the pdf of X that does not use matrix notation.
(b) Find a vector b and orthonormal matrix U such that Y defined by $Y = U^T(X - b)$ is a mean zero Gaussian vector such that Y_1 and Y_2 are independent.

3.2 Linear approximation of the cosine function over an interval Let Θ be uniformly distributed on the interval $[0, \pi]$ (yes, $[0, \pi]$, not $[0, 2\pi]$). Suppose $Y = \cos(\Theta)$ is to be estimated by an estimator of the form $a + b\Theta$. What numerical values of a and b minimize the mean square error?

3.3 Calculation of some minimum mean square error estimators Let $Y = X+N$, where X has the exponential distribution with parameter λ, and N is Gaussian with mean 0 and variance σ^2. The variables X and N are independent, and the parameters λ and σ^2 are strictly positive. (Recall that $E[X] = \frac{1}{\lambda}$ and $\mathrm{Var}(X) = \frac{1}{\lambda^2}$.)
(a) Find $\widehat{E}[X|Y]$ and also find the mean square error for estimating X by $\widehat{E}[X|Y]$.
(b) Does $E[X|Y] = \widehat{E}[X|Y]$? Justify your answer. (Hint: Answer is yes if and only if there is no estimator for X of the form $g(Y)$ with a smaller MSE than $\widehat{E}[X|Y]$.)

3.4 Valid covariance matrix For what real values of a and b is the following matrix the covariance matrix of some real-valued random vector?

$$K = \begin{pmatrix} 2 & 1 & b \\ a & 1 & 0 \\ b & 0 & 1 \end{pmatrix}.$$

Hint: A symmetric $n \times n$ matrix is positive semidefinite if and only if the determinant of every matrix obtained by deleting a set of rows and the corresponding set of columns is nonnegative.

3.5 Conditional probabilities with joint Gaussians I Let $\begin{pmatrix} X \\ Y \end{pmatrix}$ be a mean zero Gaussian vector with correlation matrix $\begin{pmatrix} 1 & \rho \\ \rho & 1 \end{pmatrix}$, where $|\rho| < 1$.
(a) Express $P(X \le 1|Y)$ in terms of ρ, Y, and the standard normal CDF, Φ.
(b) Find $E[(X-Y)^2|Y=y]$ for real values of y.

3.6 Conditional probabilities with joint Gaussians II Let X, Y be jointly Gaussian random variables with mean zero and covariance matrix

$$\mathrm{Cov}\begin{pmatrix} X \\ Y \end{pmatrix} = \begin{pmatrix} 4 & 6 \\ 6 & 18 \end{pmatrix}.$$

Express your answers in terms of Φ defined by $\Phi(u) = \int_{-\infty}^{u} \frac{1}{\sqrt{2\pi}} e^{-s^2/2} ds$.
(a) Find $P\{|X-1| \ge 2\}$.
(b) What is the conditional density of X given that $Y = 3$? You can either write out the density in full, or describe it as a well known density with specified parameter values.
(c) Find $P\{|X - E[X|Y]| \ge 1\}$.

3.7 An estimation error bound Suppose the random vector $\binom{X}{Y}$ has mean vector $\binom{2}{-2}$ and covariance matrix $\begin{pmatrix} 8 & 3 \\ 3 & 2 \end{pmatrix}$. Let $e = X - E[X|Y]$.
(a) If possible, compute $E[e^2]$. If not, give an upper bound.
(b) For what joint distribution of X and Y (consistent with the given information) is $E[e^2]$ maximized? Is your answer unique?

3.8 An MMSE estimation problem (a) Let X and Y be jointly uniformly distributed over the triangular region in the $x-y$ plane with corners (0,0), (0,1), and (1,2). Find both the linear minimum mean square error (LMMSE) estimator of X given Y and the (possibly nonlinear) MMSE estimator of X given Y. Compute the mean square error for each estimator. What percentage reduction in MSE does the MMSE estimator provide over the LMMSE?

(b) Repeat (a) assuming Y is an $N(0,1)$ random variable and $X = |Y|$.

3.9 Comparison of MMSE estimators for an example Let $X = \frac{1}{1+U}$, where U is uniformly distributed over the interval $[0,1]$.

(a) Find $E[X|U]$ and calculate the MSE, $E[(X - E[X|U])^2]$.

(b) Find $\widehat{E}[X|U]$ and calculate the MSE, $E[(X - \widehat{E}[X|U])^2]$.

3.10 Conditional Gaussian comparison Suppose that X and Y are jointly Gaussian, mean zero, with $\mathrm{Var}(X) = \mathrm{Var}(Y) = 10$ and $\mathrm{Cov}(X,Y) = 8$. Express the following probabilities in terms of the Q function:

(a) $p_a \stackrel{\triangle}{=} P\{X \geq 2\}$.

(b) $p_b \stackrel{\triangle}{=} P(X \geq 2 | Y = 3)$.

(c) $p_c \stackrel{\triangle}{=} P(X \geq 2 | Y \geq 3)$. (Note: p_c can be expressed as an integral. You need not carry out the integration.)

(d) Indicate how p_a, p_b, and p_c are ordered, from smallest to largest.

3.11 Diagonalizing a two-dimensional Gaussian distribution Let $X = \binom{X_1}{X_2}$ be a mean zero Gaussian random vector with correlation matrix $\begin{pmatrix} 1 & \rho \\ \rho & 1 \end{pmatrix}$, where $|\rho| < 1$. Find an orthonormal 2 by 2 matrix U such that $X = UY$ for a Gaussian vector $Y = \binom{Y_1}{Y_2}$ such that Y_1 is independent of Y_2. Also, find the variances of Y_1 and Y_2.

Note: The following identity might be useful for some of the problems that follow. If A, B, C, and D are jointly Gaussian and mean zero, then $E[ABCD] = E[AB]E[CD] + E[AC]E[BD] + E[AD]E[BC]$. This implies that $E[A^4] = 3E[A^2]^2$, $\mathrm{Var}(A^2) = 2E[A^2]$, and $\mathrm{Cov}(A^2, B^2) = 2\mathrm{Cov}(A,B)^2$. Also, $E[A^2 B] = 0$.

3.12 An estimator of an estimator Let X and Y be square integrable random variables and let $Z = E[X|Y]$, so Z is the MMSE estimator of X given Y. Show that the LMMSE estimator of X given Y is also the LMMSE estimator of Z given Y. (Can you generalize this result?)

3.13 Projections onto nested linear subspaces (a) Use the orthogonality principle to prove the following statement: Suppose $\mathcal{V}_0$ and $\mathcal{V}_1$ are two closed linear spaces of second order random variables, such that $\mathcal{V}_0 \supset \mathcal{V}_1$, and suppose X is a random variable with finite second moment. Let Z_i^* be the random variable in $\mathcal{V}_i$ with the minimum mean square distance from X. Then Z_1^* is the variable in $\mathcal{V}_1$ with the minimum mean square distance from Z_0^*. (b) Suppose that X, Y_1, and Y_2 are random variables with finite second moments. For each of the following three statements, identify the choice of subspace $\mathcal{V}_0$ and $\mathcal{V}_1$ such that the statement follows from (a):

(i) $\widehat{E}[X|Y_1] = \widehat{E}[\ \widehat{E}[X|Y_1, Y_2]\ |Y_1]$.

(ii) $E[X|Y_1] = E[\ E[X|Y_1, Y_2]\ |Y_1]$ (sometimes called the "tower property").

(iii) $E[X] = E[\widehat{E}[X|Y_1]]$ (think of the expectation of a random variable as the constant closest to the random variable, in the m.s. sense).

3.14 Some identities for estimators Let X and Y be random variables with $E[X^2] < \infty$. For each of the following statements, determine if the statement is true. If yes, give a justification using the orthogonality principle. If no, give a counter example.

(a) $E[X\cos(Y)|Y] = E[X|Y]\cos(Y)$,
(b) $E[X|Y] = E[X|Y^3]$,
(c) $E[X^3|Y] = E[X|Y]^3$,
(d) $E[X|Y] = E[X|Y^2]$,
(e) $\widehat{E}[X|Y] = \widehat{E}[X|Y^3]$,
(f) If $E[(X - E[X|Y])^2] = \mathrm{Var}(X)$, then $E[X|Y] = \widehat{E}[X|Y]$.

3.15 Some identities for estimators, version 2 Let X, Y, and Z be random variables with finite second moments and suppose X is to be estimated. For each of the following, if true, give a brief explanation. If false, give a counter example.
(a) $E[(X - E[X|Y])^2] \leq E[(X - \widehat{E}[X|Y, Y^2])^2]$,
(b) $E[(X - E[X|Y])^2] = E[(X - \widehat{E}[X|Y, Y^2])^2]$ if X and Y are jointly Gaussian,
(c) $E[(X - E[E[X|Z]\,|Y])^2] \leq E[(X - E[X|Y])^2]$,
(d) If $E[(X - E[X|Y])^2] = \mathrm{Var}(X)$, then X and Y are independent.

3.16 Some simple examples Give an example of each of the following, and in each case, explain your reasoning.
(a) Two random variables X and Y such that $\widehat{E}[X|Y] = E[X|Y]$, and such that $E[X|Y]$ is not simply constant, and X and Y are not jointly Gaussian.
(b) A pair of random variables X and Y on some probability space such that X is Gaussian, Y is Gaussian, but X and Y are not jointly Gaussian.
(c) Three random variables X, Y, and Z, which are pairwise independent, but all three together are not independent.

3.17 The square root of a positive semidefinite matrix (a) True or false? If B is a matrix over the reals, then BB^T is positive semidefinite.
(b) True or false? If K is a symmetric positive semidefinite matrix over the reals, then there exists a symmetric positive semidefinite matrix S over the reals such that $K = S^2$. (Hint: What if K is also diagonal?)

3.18 Estimating a quadratic Let $\begin{pmatrix} X \\ Y \end{pmatrix}$ be a mean zero Gaussian vector with correlation matrix $\begin{pmatrix} 1 & \rho \\ \rho & 1 \end{pmatrix}$, where $|\rho| < 1$.
(a) Find $E[X^2|Y]$, the best estimator of X^2 given Y.
(b) Compute the mean square error for the estimator $E[X^2|Y]$.
(c) Find $\widehat{E}[X^2|Y]$, the best linear (actually, affine) estimator of X^2 given Y, and compute the mean square error.

3.19 A quadratic estimator Suppose Y has the $N(0, 1)$ distribution and that $X = |Y|$. Find the estimator for X of the form $\widehat{X} = a + bY + cY^2$ which minimizes the mean square error. (You can use the following numerical values: $E[|Y|] = 0.8$, $E[Y^4] = 3$, $E[|Y|Y^2] = 1.6$.)
(a) Use the orthogonality principle to derive equations for a, b, and c.
(b) Find the estimator $\widehat{X}$.
(c) Find the resulting minimum mean square error.

3.20 An innovations sequence and its application Let $\begin{pmatrix} Y_1 \\ Y_2 \\ Y_3 \\ X \end{pmatrix}$ be a mean zero random vector with correlation matrix $\begin{pmatrix} 1 & 0.5 & 0.5 & 0 \\ 0.5 & 1 & 0.5 & 0.25 \\ 0.5 & 0.5 & 1 & 0.25 \\ 0 & 0.25 & 0.25 & 1 \end{pmatrix}$.

(a) Let $\widetilde{Y}_1, \widetilde{Y}_2, \widetilde{Y}_3$ denote the innovations sequence. Find the matrix A so that $\begin{pmatrix} \widetilde{Y}_1 \\ \widetilde{Y}_2 \\ \widetilde{Y}_3 \end{pmatrix} = A \begin{pmatrix} Y_1 \\ Y_2 \\ Y_3 \end{pmatrix}$.

(b) Find the correlation matrix of $\begin{pmatrix} \widetilde{Y}_1 \\ \widetilde{Y}_2 \\ \widetilde{Y}_3 \end{pmatrix}$ and cross covariance matrix $\mathrm{Cov}\left(X, \begin{pmatrix} \widetilde{Y}_1 \\ \widetilde{Y}_2 \\ \widetilde{Y}_3 \end{pmatrix}\right)$.

(c) Find the constants a, b, and c to minimize $E[(X - a\widetilde{Y}_1 - b\widetilde{Y}_2 - c\widetilde{Y}_3)^2]$.

3.21 Estimation for an additive Gaussian noise model Assume x and n are independent Gaussian vectors with means $\bar{x}$, $\bar{n}$ and covariance matrices Σ_x and Σ_n. Let $y = x + n$. Then x and y are jointly Gaussian.

(a) Show that $E[x|y]$ is given by either $\bar{x} + \Sigma_x(\Sigma_x + \Sigma_n)^{-1}(y - (\bar{x} + \bar{n}))$ or $\Sigma_n(\Sigma_x + \Sigma_n)^{-1}\bar{x} + \Sigma_x(\Sigma_x + \Sigma_n)^{-1}(y - \bar{n})$.

(b) Show that the conditional covariance matrix of x given y is given by any of the three expressions: $\Sigma_x - \Sigma_x(\Sigma_x + \Sigma_n)^{-1}\Sigma_x = \Sigma_x(\Sigma_x + \Sigma_n)^{-1}\Sigma_n = (\Sigma_x^{-1} + \Sigma_n^{-1})^{-1}$. (Assume that the various inverses exist.)

3.22 A Kalman filtering example (a) Let $\sigma^2 > 0$, let f be a real constant, and let x_0 denote an $N(0, \sigma^2)$ random variable. Consider the state and observation sequences defined by:

$$\text{(state)} \quad x_{k+1} = fx_k + w_k,$$
$$\text{(observation)} \quad y_k = x_k + v_k,$$

where $w_1, w_2, \ldots; v_1, v_2, \ldots$ are mutually independent $N(0, 1)$ random variables. Write down the Kalman filter equations for recursively computing the estimates $\hat{x}_{k|k-1}$, the (scalar) gains K_k, and the sequence of the variances of the errors (for brevity write σ_k^2 for the covariance or error instead of $\Sigma_{k|k-1}$).

(b) For what values of f is the sequence of error variances bounded?

3.23 Steady state gains for one-dimensional Kalman filter This is a continuation of the previous problem.

(a) Show that $\lim_{k\to\infty} \sigma_k^2$ exists.

(b) Express the limit, σ_∞^2, in terms of f.
(c) Explain why $\sigma_\infty^2 = 1$ if $f = 0$.

3.24 A variation of Kalman filtering Let $\sigma^2 > 0$, let f be a real constant, and let x_0 denote an $N(0, \sigma^2)$ random variable. Consider the state and observation sequences defined by:

$$\text{(state)} \quad x_{k+1} = fx_k + w_k,$$
$$\text{(observation)} \quad y_k = x_k + w_k,$$

where $w_1, w_2, \ldots$ are mutually independent $N(0, 1)$ random variables. Note that the state and observation equations are driven by the same sequence, so that some of the Kalman filtering equations derived in the notes do not apply. Derive recursive equations needed to compute $\hat{x}_{k|k-1}$, including recursive equations for any needed gains or variances of error. (Hints: What modifications need to be made to the derivation for the standard model? Check that your answer is correct for $f = 1$.)

3.25 Estimation with jointly Gaussian random variables Suppose X and Y are jointly Gaussian random variables with $E[X] = 2$, $E[Y] = 4$, $\mathrm{Var}(X) = 9$, $\mathrm{Var}(Y) = 25$, and $\rho = 0.2$. (ρ is the correlation coefficient.) Let $W = X + 2Y + 3$.
(a) Find $E[W]$ and $\mathrm{Var}(W)$.
(b) Calculate the numerical value of $P\{W \geq 20\}$.
(c) Find the unconstrained estimator $g^*(W)$ of Y based on W with the minimum MSE, and find the resulting MSE.

3.26 An innovations problem Let $U_1, U_2, \ldots$ be a sequence of independent random variables, each uniformly distributed on the interval $[0, 1]$. Let $Y_0 = 1$, and $Y_n = U_1U_2 \ldots U_n$ for $n \geq 1$.
(a) Find the variance of Y_n for each $n \geq 1$.
(b) Find $E[Y_n|Y_0, \ldots, Y_{n-1}]$ for $n \geq 1$.
(c) Find $\widehat{E}[Y_n|Y_0, \ldots, Y_{n-1}]$ for $n \geq 1$.
(d) Find the linear innovations sequence $\widetilde{Y} = (\widetilde{Y}_0, \widetilde{Y}_1, \ldots)$.
(e) Fix a positive integer M and let $X_M = U_1 + \ldots + U_M$. Using the answer to (d), find $\widehat{E}[X_M|Y_0, \ldots, Y_M]$, the best linear estimator of X_M given $(Y_0, \ldots, Y_M)$.

3.27 Innovations and orthogonal polynomials for the normal distribution (a) Let X be an $N(0, 1)$ random variable. Show that for integers $n \geq 0$,

$$E[X^n] = \begin{cases} \frac{n!}{(n/2)!2^{n/2}} & n \text{ even} \\ 0 & n \text{ odd} \end{cases}.$$

Hint: One approach is to apply the power series expansion for e^x on each side of the identity $E[e^{uX}] = e^{u^2/2}$, and identify the coefficients of u^n.
(b) Let X be an $N(0, 1)$ random variable, and let $Y_n = X^n$ for integers $n \geq 0$. Note that $Y_0 \equiv 1$. Express the first five terms, $\widetilde{Y}_0$ through $\widetilde{Y}_4$, of the linear innovations sequence of Y in terms of U.

3.28 Linear innovations and orthogonal polynomials for the uniform distribution
(a) Let U be uniformly distributed on the interval $[-1, 1]$. Show that for $n \geq 0$,

$$E[U^n] = \begin{cases} \frac{1}{n+1} & n \text{ even} \\ 0 & n \text{ odd} \end{cases}.$$

(b) Let $Y_n = U^n$ for integers $n \geq 0$. Note that $Y_0 \equiv 1$. Express the first four terms, $\widetilde{Y}_1$ through $\widetilde{Y}_4$, of the linear innovations sequence of Y in terms of U.

3.29 Representation of three random variables with equal cross covariances Let K be a matrix of the form

$$K = \begin{pmatrix} 1 & a & a \\ a & 1 & a \\ a & a & 1 \end{pmatrix},$$

where $a \in \mathbb{R}$.
(a) For what values of a is K the covariance matrix of some random vector?
(b) Let a have one of the values found in (a). Fill in the missing entries of the matrix U,

$$U = \begin{pmatrix} * & * & \frac{1}{\sqrt{3}} \\ * & * & \frac{1}{\sqrt{3}} \\ * & * & \frac{1}{\sqrt{3}} \end{pmatrix},$$

to yield an orthonormal matrix, and find a diagonal matrix Λ with nonnegative entries, so that if Z is a three-dimensional random vector with $\mathrm{Cov}(Z) = I$, then $U\Lambda^{\frac{1}{2}}Z$ has covariance matrix K. (Hint: It happens that the matrix U can be selected independently of a. Also, $1 + 2a$ is an eigenvalue of K.)

3.30 Example of extended Kalman filter Often dynamical systems in engineering applications have nonlinearities in the state dynamics and/or observation model. If the nonlinearities are not too severe and if the rate of change of the state is not too large compared to the observation noise (so that tracking is accurate) then an effective extension of Kalman filtering is based on linearizing the nonlinearities about the current state estimate. For example, consider the following example

$$x_{k+1} = x_k + w_k, \qquad y_k = \sin(2\pi fk + x_k) + v_k,$$

where the w_ks are $N(0, q)$ random variables and the v_ks are $N(0, r)$ random variables with $q << 1$ and f is a constant frequency. Here the random process x can be viewed as the phase of a sinusoidal signal, and the goal of filtering is to track the phase. In communication systems such tracking is implemented using a phase lock loop, and in this instance we expect the extended Kalman filter to give similar equations. The equations for the extended Kalman filter are the same as for the ordinary Kalman filter with the variation that $\widetilde{y}_k = y_k - \sin(2\pi fk + \widehat{x}_{k|k-1})$ and, in the equations for the covariance of error and Kalman gains, $H_k = \left.\frac{d\sin(2\pi fk+x)}{dx}\right|_{x=\widehat{x}_{k|k-1}}$.
(a) Write down the equations for the update $\widehat{x}_{k|k-1} \rightarrow \widehat{x}_{k+1|k}$, including expressing the Kalman gain K_k in terms of $\Sigma_{k|k-1}$ and $\widehat{x}_{k|k-1}$. (You don't need to write out the equations for update of the covariance of error, which, intuitively, should be slowly varying in

steady state. Also, ignore the fact that the phase can only be tracked modulo 2π over the long run.)

(b) Verify/explain why, if the covariance of error is small, the extended Kalman filter adjusts the estimated phase in the right direction. That is, the change to $\widehat{x}$ in one step tends to have the opposite sign as the error $\widehat{x} - x$.

3.31 Kalman filter for a rotating state Consider the Kalman state and observation equations for the following matrices, where $\theta_o = 2\pi/10$ (the matrices don't depend on time, so the subscript k is omitted):

$$F = (0.99)\begin{pmatrix} \cos(\theta_o) & -\sin(\theta_o) \\ \sin(\theta_o) & \cos(\theta_o) \end{pmatrix} \quad H = \begin{pmatrix} 1 \\ 0 \end{pmatrix} \quad Q = \begin{pmatrix} 1 & 0 \\ 0 & 1 \end{pmatrix} \quad R = 1.$$

(a) Explain in words what successive iterates $F^n x_o$ are like, for a nonzero initial state x_o (this is the same as the state equation, but with the random term w_k left off).

(b) Write out the Kalman filter equations for this example, simplifying as much as possible (but no more than possible! The equations don't simplify all that much.)

3.32* Proof of the orthogonality principle Prove the seven statements lettered (a)–(g) in what follows.

Let X be a random variable and let $\mathcal{V}$ be a collection of random variables on the same probability space such that:

(i) $E[Z^2] < +\infty$ for each $Z \in \mathcal{V}$,

(ii) $\mathcal{V}$ is a linear class, i.e. if $Z, Z' \in \mathcal{V}$ then so is $aZ + bZ'$ for any real numbers a and b,

(iii) $\mathcal{V}$ is closed in the sense that if $Z_n \in \mathcal{V}$ for each n and Z_n converges to a random variable Z in the mean square sense, then $Z \in \mathcal{V}$.

The orthogonality principle is there is a unique $Z^* \in \mathcal{V}$ so $E[(X - Z^*)^2] \leq E[(X - Z)^2]$ for all $Z \in \mathcal{V}$. Furthermore, a random variable $W \in \mathcal{V}$ is equal to Z^* if and only if $(X - W) \perp Z$ for all $Z \in \mathcal{V}$. ($(X - W) \perp Z$ means $E[(X - W)Z] = 0$.)

The remainder of this problem is aimed at a proof. Let $d = \inf\{E[(X - Z)^2] : Z \in \mathcal{V}\}$. By definition of infimum there exists a sequence $Z_n \in \mathcal{V}$ so that $E[(X - Z_n)^2] \to d$ as $n \to +\infty$.

(a) The sequence Z_n is Cauchy in the mean square sense.
(Hint: Use the "parallelogram law": $E[(U - V)^2] + E[(U + V)^2] = 2(E[U^2] + E[V^2]$.)
Thus, by the Cauchy criteria, there is a random variable Z^* such that Z_n converges to Z^* in the mean square sense.

(b) Z^* satisfies the conditions advertised in the first sentence of the principle.

(c) The element Z^* satisfying the condition in the first sentence of the principle is unique. (Consider two random variables that are equal to each other with probability one to be the same.) This completes the proof of the first sentence.

(d) ("if" part of second sentence.) If $W \in \mathcal{V}$ and $(X - W) \perp Z$ for all $Z \in \mathcal{V}$, then $W = Z^*$.

(The "only if" part of the second sentence is divided into three parts:)

(e) $E[(X - Z^* - cZ)^2] \geq E[(X - Z^*)^2]$ for any real constant c.

(f) $-2cE[(X - Z^*)Z] + c^2 E[Z^2] \geq 0$ for any real constant c.

(g) $(X - Z^*) \perp Z$, and the principle is proved.

3.33* The span of two closed subspaces is closed Check that the span, $\mathcal{V}_1 \oplus \mathcal{V}_2$, of two closed orthogonal linear spaces (defined in Proposition 3.5) is also a closed linear space. A hint for showing that $\mathcal{V}$ is closed is to use the fact that if (Z_n) is a m.s. convergent sequence of random variables in $\mathcal{V}$, then each variable in the sequence can be represented as $Z_n = Z_{n,1} + Z_{n,2}$, where $Z_{n,i} \in \mathcal{V}_i$, and $E[(Z_n - Z_m)^2] = E[(Z_{n,1} - Z_{m,1})^2] + E[(Z_{n,2} - Z_{m,2})^2]$.

3.34* Von Neumann's alternating projections algorithm Let $\mathcal{V}_1$ and $\mathcal{V}_2$ be closed linear subspaces of $L^2(\Omega, \mathcal{F}, P)$, and let $X \in L^2(\Omega, \mathcal{F}, P)$. Define a sequence $(Z_n : n \geq 0)$ recursively, by alternating projections onto $\mathcal{V}_1$ and $\mathcal{V}_2$, as follows. Let $Z_0 = X$, and for $k \geq 0$, let $Z_{2k+1} = \Pi_{\mathcal{V}_1}(Z_{2k})$ and $Z_{2k+2} = \Pi_{\mathcal{V}_2}(Z_{2k+1})$. The goal of this problem is to show that $Z_n \stackrel{m.s.}{\to} \Pi_{\mathcal{V}_1 \cap \mathcal{V}_2}(X)$. The approach is to establish that (Z_n) converges in the m.s. sense, by verifying the Cauchy criteria, and using the orthogonality principle to identify the limit. Define $D(i,j) = E[(Z_i - Z_j)]^2$ for $i \geq 0$ and $j \geq 0$, and $\epsilon_i = D(i+1, i)$ for $i \geq 0$.

(a) Show that $\epsilon_i = E[(Z_i)^2] - E[(Z_{i+1})^2]$.

(b) Show that $\sum_{i=0}^{\infty} \epsilon_i \leq E[X^2] < \infty$.

(c) Use the orthogonality principle to show that for $n \geq 1$ and $k \geq 0$:

$$D(n, n+2k+1) = \epsilon_n + D(n+1, n+2k+1),$$
$$D(n, n+2k+2) = D(n, n+2k+1) - \epsilon_{n+2k+1}.$$

(d) Use the above equations to show that for $n \geq 1$ and $k \geq 0$,

$$D(n, n+2k+1) = \epsilon_n + \cdots + \epsilon_{n+k} - (\epsilon_{n+k+1} + \cdots + \epsilon_{n+2k}),$$
$$D(n, n+2k+2) = \epsilon_n + \cdots + \epsilon_{n+k} - (\epsilon_{n+k+1} + \cdots + \epsilon_{n+2k+1}).$$

Consequently, $D(n,m) \leq \sum_{i=n}^{m-1} \epsilon_i$ for $1 \leq n < m$, and therefore $(Z_n : n \geq 0)$ is a Cauchy sequence, so $Z_n \stackrel{m.s.}{\to} Z_\infty$ for some random variable Z_∞.

(e) Verify that $Z_\infty \in \mathcal{V}_1 \cap \mathcal{V}_2$.

(f) Verify that $(X - Z_\infty) \perp Z$ for any $Z \in \mathcal{V}_1 \cap \mathcal{V}_2$. (Hint: Explain why $(X - Z_n) \perp Z$ for all n, and let $n \to \infty$.)

By the orthogonality principle, (e) and (f) imply that $Z_\infty = \Pi_{\mathcal{V}_1 \cap \mathcal{V}_2}(X)$.

4 Random processes

After presenting the definition of a random process, this chapter discusses many of the most widely used examples and subclasses of random processes.

4.1 Definition of a random process

A random process X is an indexed collection $X = (X_t : t \in \mathbb{T})$ of random variables, all on the same probability space $(\Omega, \mathcal{F}, P)$. In many applications the index set $\mathbb{T}$ is a set of times. If $\mathbb{T} = \mathbb{Z}$, or more generally, if $\mathbb{T}$ is a set of consecutive integers, then X is called a discrete-time random process. If $\mathbb{T} = \mathbb{R}$ or if $\mathbb{T}$ is an interval of $\mathbb{R}$, then X is called a continuous-time random process. Three ways to view a random process $X = (X_t : t \in \mathbb{T})$ are as follows:

- For each t fixed, X_t is a function on Ω.
- X is a function on $\mathbb{T} \times \Omega$ with value $X_t(\omega)$ for given $t \in \mathbb{T}$ and $\omega \in \Omega$.
- For each ω fixed with $\omega \in \Omega$, $X_t(\omega)$ is a function of t, called the *sample path corresponding to ω.*

Example 4.1 Suppose $W_1, W_2, \ldots$ are independent random variables with $P\{W_k = 1\} = P\{W_k = -1\} = \frac{1}{2}$ for each k, $X_0 = 0$, and $X_n = W_1 + \ldots + W_n$ for positive integers n. Let $W = (W_k : k \geq 1)$ and $X = (X_n : n \geq 0)$. Then W and X are both discrete-time random processes. The index set $\mathbb{T}$ for X is $\mathbb{Z}_+$. A sample path of W and a corresponding sample path of X are shown in Figure 4.1.

The following notation is used:

$$\begin{aligned}
\mu_X(t) &= E[X_t], \\
R_X(s,t) &= E[X_s X_t], \\
C_X(s,t) &= \mathrm{Cov}(X_s, X_t), \\
F_{X,n}(x_1, t_1; \ldots; x_n, t_n) &= P\{X_{t_1} \leq x_1, \ldots, X_{t_n} \leq x_n\}.
\end{aligned}$$

μ_X is called the *mean function*, R_X is called the *correlation function*, C_X is called the *covariance function*, and $F_{X,n}$ is called the *nth order cumulative distribution function*

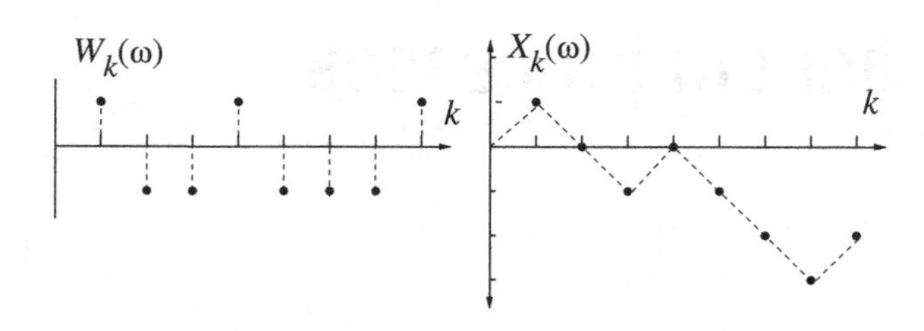

Figure 4.1 Typical sample paths.

(CDF). Sometimes the prefix "auto," meaning "self," is added to the words "correlation" and "covariance," to emphasize that only one random process is involved.

Definition 4.1 A *second order random process* is a random process $(X_t : t \in \mathbb{T})$ such that $E[X_t^2] < +\infty$ for all $t \in \mathbb{T}$.

The mean, correlation, and covariance functions of a second order random process are all well-defined and finite.

If X_t is a discrete random variable for each t, then the nth order pmf of X is defined by

$$p_{X,n}(x_1, t_1; \ldots; x_n, t_n) = P\{X_{t_1} = x_1, \ldots, X_{t_n} = x_n\}.$$

Similarly, if $X_{t_1}, \ldots, X_{t_n}$ are jointly continuous random variables for any distinct $t_1, \ldots, t_n$ in $\mathbb{T}$, then X has an nth order pdf $f_{X,n}$, such that for $t_1, \ldots, t_n$ fixed, $f_{X,n}(x_1, t_1; \ldots; x_n, t_n)$ is the joint pdf of $X_{t_1}, \ldots, X_{t_n}$.

Example 4.2 Let A and B be independent, $N(0, 1)$ random variables. Suppose $X_t = A+Bt+t^2$ for all $t \in \mathbb{R}$. Let us describe the sample functions, the mean, correlation, and covariance functions, and the first and second order pdfs of X.

Each sample function corresponds to some fixed ω in Ω. For ω fixed, $A(\omega)$ and $B(\omega)$ are numbers. The sample paths all have the same shape – they are parabolas with constant second derivative equal to 2. The sample path for ω fixed has $t = 0$ intercept $A(\omega)$, and minimum value $A(\omega) - \frac{B(\omega)^2}{4}$ achieved at $t = -\frac{B(w)}{2}$. Three typical sample paths are shown in Figure 4.2. The various moment functions are given by

$$\begin{aligned}
\mu_X(t) = E[A + Bt + t^2] &= t^2, \\
R_X(s, t) = E[(A + Bs + s^2)(A + Bt + t^2)] &= 1 + st + s^2t^2, \\
C_X(s, t) = R_X(s, t) - \mu_X(s)\mu_X(t) &= 1 + st.
\end{aligned}$$

As for the densities, for each t fixed, X_t is a linear combination of two independent Gaussian random variables, $\mu_X(t) = t^2$ and $\mathrm{Var}(X_t) = C_X(t, t) = 1 + t^2$. Thus, X_t is an

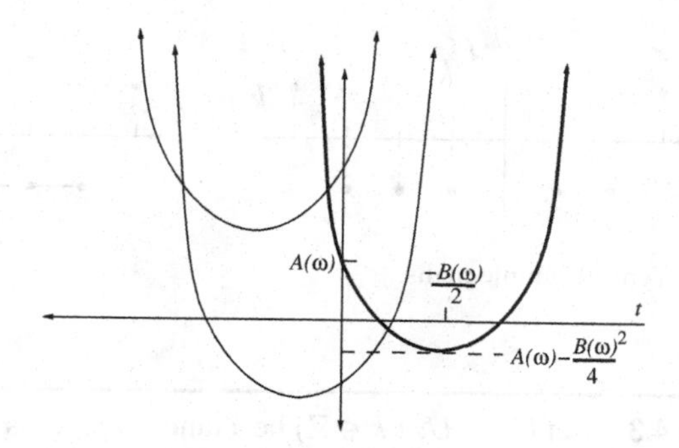

Figure 4.2 Typical sample paths.

$N(t^2, 1+t^2)$ random variable. That specifies the first order pdf $f_{X,1}$ well enough, but if one insists on writing it out in all detail it is given by

$$f_{X,1}(x,t) = \frac{1}{\sqrt{2\pi(1+t^2)}} \exp\left(-\frac{(x-t^2)^2}{2(1+t^2)}\right).$$

For distinct s and t, X_s and X_t are jointly Gaussian with

$$\mathrm{Cov}\begin{pmatrix} X_s \\ X_t \end{pmatrix} = \begin{pmatrix} 1+s^2 & 1+st \\ 1+st & 1+t^2 \end{pmatrix}.$$

The determinant of this matrix is $(s-t)^2$, which is nonzero. Thus X has a second order pdf $f_{X,2}$. For most purposes, we have already written enough about $f_{X,2}$ for this example, but in full detail $f_{X,2}(x,s;y,t)$ is given by

$$\frac{1}{2\pi|s-t|} \exp\left(-\frac{1}{2}\begin{pmatrix} x-s^2 \\ y-t^2 \end{pmatrix}^T \begin{pmatrix} 1+s^2 & 1+st \\ 1+st & 1+t^2 \end{pmatrix}^{-1} \begin{pmatrix} x-s^2 \\ y-t^2 \end{pmatrix}\right).$$

The nth order distributions of X are joint Gaussian distributions, but densities do not exist for $n \geq 3$ because the values of $\begin{pmatrix} X_{t_1} \\ X_{t_2} \\ X_{t_3} \end{pmatrix}$ for fixed t_1, t_2, t_3 are restricted to a plane embedded in $\mathbb{R}^3$.

A random process $(X_t : t \in \mathbb{T})$ is said to be *Gaussian* if the random variables $X_t : t \in \mathbb{T}$ comprising the process are jointly Gaussian. The process X in the above example is Gaussian. The finite order distributions of a Gaussian random process X are determined by the mean function μ_X and autocorrelation function R_X. Indeed, for any finite subset $\{t_1, t_2, \ldots, t_n\}$ of $\mathbb{T}$, $(X_{t_1}, \ldots, X_{t_n})^T$ is a Gaussian vector with mean $(\mu_X(t_1), \ldots, \mu_X(t_n))^T$ and covariance matrix with ijth element $C_X(t_i, t_j) = R_X(t_i, t_j) - \mu_X(t_i)\mu_X(t_j)$. Two or more random processes are said to be jointly Gaussian if all the random variables comprising the processes are jointly Gaussian.

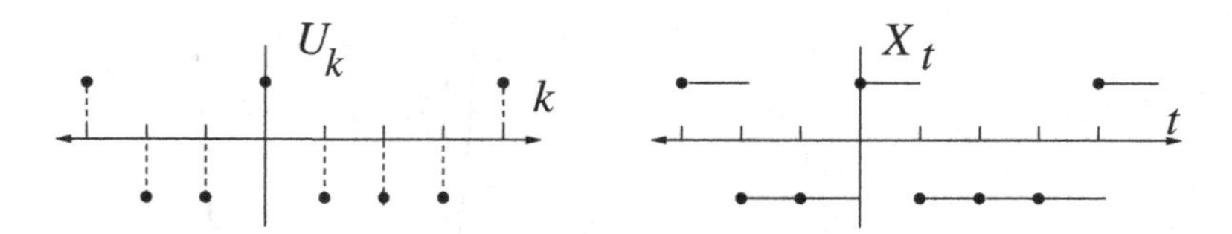

Figure 4.3 Typical sample paths.

Example 4.3 Let $U = (U_k : k \in \mathbb{Z})$ be a random process such that the random variables $U_k : k \in \mathbb{Z}$ are independent, and $P\{U_k = 1\} = P\{U_k = -1\} = \frac{1}{2}$ for all k. Let $X = (X_t : t \in \mathbb{R})$ be the random process obtained by letting $X_t = U_n$ for $n \le t < n+1$ for any n. Equivalently, $X_t = U_{\lfloor t \rfloor}$. A sample path of U and a corresponding sample path of X are shown in Figure 4.3. Both random processes have zero mean, so their covariance functions are equal to their correlation functions and are given by

$$R_U(k,l) = \begin{cases} 1 \text{ if } k = l \\ 0 \text{ else} \end{cases}, \qquad R_X(s,t) = \begin{cases} 1 \text{ if } \lfloor s \rfloor = \lfloor t \rfloor \\ 0 \text{ else} \end{cases}.$$

The random variables of U are discrete, so the nth order pmf of U exists for all n. It is given by

$$p_{U,n}(x_1, k_1; \ldots; x_n, k_n) = \begin{cases} 2^{-n} & \text{if } (x_1, \ldots, x_n) \in \{-1, 1\}^n \\ 0 & \text{else} \end{cases}$$

for distinct integers $k_1, \ldots, k_n$. The nth order pmf of X exists for the same reason, but it is a bit more difficult to write down. In particular, the joint pmf of X_s and X_t depends on whether $\lfloor s \rfloor = \lfloor t \rfloor$. If $\lfloor s \rfloor = \lfloor t \rfloor$ then $X_s = X_t$ and if $\lfloor s \rfloor \neq \lfloor t \rfloor$ then X_s and X_t are independent. Therefore, the second order pmf of X is given as follows:

$$p_{X,2}(x_1, t_1; x_2, t_2) = \begin{cases} \frac{1}{2} & \text{if } \lfloor t_1 \rfloor = \lfloor t_2 \rfloor \text{ and } x_1 = x_2 \in \{-1, 1\} \\ \frac{1}{4} & \text{if } \lfloor t_1 \rfloor \neq \lfloor t_2 \rfloor \text{ and } x_1, x_2 \in \{-1, 1\} \\ 0 & \text{else.} \end{cases}$$

4.2 Random walks and gambler's ruin

The topic of this section illustrates how interesting events concerning multiple random variables naturally arise in the study of random processes. Suppose p is given with $0 < p < 1$. Let $W_1, W_2, \ldots$ be independent random variables with $P\{W_i = 1\} = p$ and $P\{W_i = -1\} = 1 - p$ for $i \ge 1$. Suppose X_0 is an integer valued random variable independent of $(W_1, W_2, \ldots)$, and for $n \ge 1$, define X_n by $X_n = X_0 + W_1 + \ldots + W_n$. A sample path of $X = (X_n : n \ge 0)$ is shown in Figure 4.4. The random process X is called a random walk. Write P_k and E_k for conditional probabilities and conditional

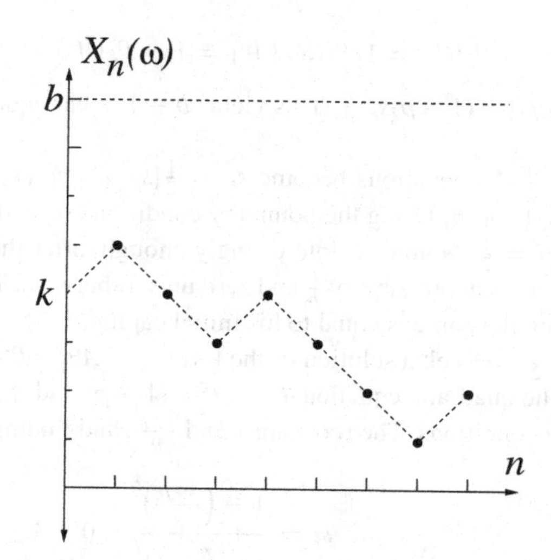

Figure 4.4 A typical sample path.

expectations given that $X_0 = k$. For example, $P_k(A) = P(A \mid X_0 = k)$ for any event A. Let us summarize some of the basic properties of X:

- $E_k[X_n] = k + n(2p - 1)$.
- $\mathrm{Var}_k(X_n) = \mathrm{Var}(k + W_1 + \cdots + W_n) = 4np(1 - p)$.
- $\lim_{n\to\infty} \frac{X_n}{n} = 2p - 1$ (a.s. and m.s. under P_k, k fixed).
- $\lim_{n\to\infty} P_k \left\{ \frac{X_n - n(2p-1)}{\sqrt{4np(1-p)}} \leq c \right\} = \Phi(c)$.
- $P_k\{X_n = k + j - (n - j)\} = \binom{n}{j} p^j (1-p)^{n-j}$ for $0 \leq j \leq n$.

Almost all the properties listed are properties of the one-dimensional distributions of X. In fact, only the strong law of large numbers, giving the a.s. convergence in the third property listed, depends on the joint distribution of the X_ns.

The so-called *gambler's ruin problem* is a nice example of the calculation of a probability involving the joint distributions of the random walk X. Interpret X_n as the number of units of money a gambler has at time n. Assume that the initial wealth k satisfies $k \geq 0$, and suppose the gambler has a goal of accumulating b units of money for some positive integer $b \geq k$. While the random walk $(X_n : n \geq 0)$ continues on forever, we are only interested in it until it hits either 0 (the gambler is ruined) or b (the gambler is successful). Let S_b denote the event that the gambler is successful, meaning the random walk reaches b or more without first reaching 0. The gambler's success probability is $P_k(S_b)$. A simple idea allows us to compute the success probability. The idea is to condition on the value of the first step W_1, and then to recognize that after the first step is taken, the conditional probability of success is the same as the unconditional probability of success for initial wealth $k + W_1$.

Let $s_k = P_k(S_b)$ for $0 \leq k \leq b$, so s_k is the success probability for the gambler with initial wealth k and target wealth b. Clearly $s_0 = 0$ and $s_b = 1$. For $1 \leq k \leq b - 1$, condition on W_1 to yield

$$s_k = P_k\{W_1 = 1\}P_k(S_b \mid W_1 = 1) + P_k\{W_1 = -1\}P_k(S_b \mid W_1 = -1)$$

or $s_k = ps_{k+1} + (1-p)s_{k-1}$. This yields $b-1$ linear equations for the $b-1$ unknowns $s_1, \ldots, s_{b-1}$.

If $p = \frac{1}{2}$ the equations become $s_k = \frac{1}{2}\{s_{k-1} + s_{k+1}\}$ so that $s_k = A + Bk$ for some constants A and B. Using the boundary conditions $s_0 = 0$ and $s_b = 1$, we find $s_k = \frac{k}{b}$ in the case $p = \frac{1}{2}$. Note that, interestingly enough, after the gambler stops playing, he'll have b units with probability $\frac{k}{b}$ and zero units otherwise. Thus, his expected wealth after completing the game is equal to his initial capital, k.

If $p \neq \frac{1}{2}$, we seek a solution of the form $s_k = A\theta_1^k + B\theta_2^k$, where θ_1 and θ_2 are the two roots of the quadratic equation $\theta = p\theta^2 + (1-p)$ and A, B are selected to meet the two boundary conditions. The roots are 1 and $\frac{1-p}{p}$, and finding A and B yields that, if $p \neq \frac{1}{2}$,

$$s_k = \frac{1 - \left(\frac{1-p}{p}\right)^k}{1 - \left(\frac{1-p}{p}\right)^b} \qquad 0 \leq k \leq b.$$

Now suppose $p > \frac{1}{2}$. By the law of large numbers, $\frac{X_n}{n} \to 2p-1$ a.s. as $n \to \infty$. This implies, in particular, that $X_n \to +\infty$ a.s. as $n \to \infty$. Thus, unless the gambler is ruined in finite time, his capital converges to infinity. Let S be the event that the gambler's wealth converges to infinity without ever reaching zero. The events S_b decrease with b because if b is larger the gambler has more possibilities to be ruined before accumulating b units of money: $S_1 \supset S_2 \supset \ldots$ and $S = \{X_n \to \infty\} \cap \left(\cap_{b=1}^{\infty} S_b\right)$. Therefore, by the fact $P\{X_n \to \infty\} = 1$ and the continuity of probability,

$$P_k(S) = P\left(\cap_{b=1}^{\infty} S_b\right) = \lim_{b\to\infty} P_k(S_b) \quad = \quad \lim_{b\to\infty} s_k \quad = \quad 1 - \left(\frac{1-p}{p}\right)^k.$$

Thus, the probability of eventual ruin decreases geometrically with the initial wealth k.

4.3 Processes with independent increments and martingales

The increment of a random process $X = (X_t : t \in \mathbb{T})$ over an interval $[a, b]$ is the random variable $X_b - X_a$. A random process has *independent increments* if for any positive integer n and any $t_0 < t_1 < \ldots < t_n$ in $\mathbb{T}$, the increments $X_{t_1} - X_{t_0}, \ldots, X_{t_n} - X_{t_{n-1}}$ are mutually independent.

A random process $(X_t : t \in \mathbb{T})$ is called a *martingale* if $E[X_t]$ is finite for all t and for any positive integer n and $t_1 < t_2 < \ldots < t_n < t_{n+1}$,

$$E[X_{t_{n+1}} \mid X_{t_1}, \ldots, X_{t_n}] = X_{t_n}$$

or, equivalently,

$$E[X_{t_{n+1}} - X_{t_n} \mid X_{t_1}, \ldots, X_{t_n}] = 0.$$

If t_n is interpreted as the present time, then t_{n+1} is a future time and the value of $(X_{t_1}, \ldots, X_{t_n})$ represents information about the past and present values of X. With this interpretation, the martingale property is that the future increments of X have conditional mean zero, given the past and present values of the process.

An example of a martingale is the following. Suppose a gambler has initial wealth X_0. Suppose the gambler makes bets with various odds, such that, as far as the past history of X can determine, the bets made are all for fair games in which the expected net gains are zero. Then if X_t denotes the wealth of the gambler at any time $t \geq 0$, then $(X_t : t \geq 0)$ is a martingale.

Suppose (X_t) is an independent increment process with index set $\mathbb{T} = \mathbb{R}_+$ or $\mathbb{T} = \mathbb{Z}_+$, with X_0 equal to a constant and with mean zero increments. Then X is a martingale, as we now show. Let $t_1 < \ldots < t_{n+1}$ be in $\mathbb{T}$. Then $(X_{t_1}, \ldots, X_{t_n})$ is a function of the increments $X_{t_1} - X_0, X_{t_2} - X_{t_1}, \ldots, X_{t_n} - X_{t_{n-1}}$, and hence it is independent of the increment $X_{t_{n+1}} - X_{t_n}$. Thus

$$E[X_{t_{n+1}} - X_{t_n} \mid X_{t_1}, \ldots, X_{t_n}] = E[X_{t_{n+1}} - X_{t_n}] \ = \ 0.$$

The random walk $(X_n : n \geq 0)$ arising in the gambler's ruin problem is an independent increment process, and if $p = \frac{1}{2}$ it is also a martingale.

The following proposition is stated, without proof, to give an indication of some of the useful deductions that follow from the martingale property.

Proposition 4.2 (a) *(Doob's maximal inequality) Let $X_0, X_1, X_2, \ldots$ be nonnegative random variables such that $E[X_{k+1} \mid X_0, \ldots, X_k] \leq X_k$ for $k \geq 0$ (such X is a nonnegative supermartingale). Then,*

$$P\left\{\left(\max_{0 \leq k \leq n} X_k\right) \geq \gamma\right\} \leq \frac{E[X_0]}{\gamma}.$$

(b) *(Doob's L^2 inequality) Let $X_0, X_1, \ldots$ be a martingale sequence with $E[X_n^2] < +\infty$ for some n. Then*

$$E\left[\left(\max_{0 \leq k \leq n} X_k\right)^2\right] \leq 4E[X_n^2].$$

Martingales can be used to derive concentration inequalities involving sums of dependent random variables, as shown next. A random sequence $X_1, X_2, \ldots$ is called a *martingale difference sequence* if the process of partial sums defined by $S_n = X_1 + \ldots + X_n$ (with $S_0 = 0$) is a martingale, or equivalently, if $E[X_n | X_1, \ldots, X_{n-1}] = 0$ for each $n \geq 1$. The following proposition shows that Bennett's inequality and Bernstein's inequality given in Problem 2.38 readily extend from the case of sums of independent random variables to sums of martingale difference random variables. A related analysis in Section 10.3 yields the Azuma–Hoeffding inequality.

Proposition 4.3 *(Bennett's and Bernstein's inequalities for martingale difference sequences) Suppose $X_1, X_2, \ldots$ is a martingale difference sequence such that for some constant L and constants d_i^2, $i \geq 1$: $P\{|X_i| \leq L\} = 1$ and $E[X_i^2 | X_1, \ldots, X_{i-1}] \leq d_i^2$ for $i \geq 1$. Then for $\alpha > 0$ and $n \geq 1$:*

$$P\left\{\sum_{i=1}^{n} X_i \geq \alpha\right\} \leq \exp\left(-\frac{\sum_{i=1}^{n} d_i^2}{L^2}\varphi\left(\frac{\alpha L}{\sum_i d_i^2}\right)\right) \quad \textit{(Bennett's inequality)}$$
$$\leq \exp\left(-\frac{\frac{1}{2}\alpha^2}{\sum_{i=1}^{n} d_i^2 + \frac{\alpha L}{3}}\right) \quad \textit{(Bernstein's inequality)},$$

where $\varphi(u) = (1+u)\ln(1+u) - u$.

Proof Problem 2.38(a) yields $E[e^{\theta X_i}|X_1, \ldots, X_{i-1}] \leq \exp\left(\frac{d_i^2(e^{\theta L}-1-\theta L)}{L^2}\right)$ for $\theta > 0$. Therefore,

$$\begin{aligned} E[e^{\theta S_n}] &= E[E[e^{\theta X_n} e^{\theta S_{n-1}}|X_1, \ldots, X_{n-1}]] \\ &= E[E[e^{\theta X_n}|X_1, \ldots, X_{n-1}]e^{\theta S_{n-1}}] \\ &\leq \exp\left(\frac{d_n^2(e^{\theta L} - 1 - \theta L)}{L^2}\right) E[e^{\theta S_{n-1}}], \end{aligned}$$

which by induction on n implies

$$E[e^{\theta S_n}] \leq \exp\left(\frac{\left(\sum_{i=1}^{n} d_i^2\right)(e^{\theta L} - 1 - \theta L)}{L^2}\right),$$

just as if the X_is were independent. The remainder of the proof is identical to the proof of the Chernoff bound. □

4.4 Brownian motion

A *Brownian motion*, also called a *Wiener process*, with parameter $\sigma^2 > 0$, is a random process $W = (W_t : t \geq 0)$ such that:

B.0 $P\{W_0 = 0\} = 1$,
B.1 W has independent increments,
B.2 $W_t - W_s$ has the $N(0, \sigma^2(t-s))$ distribution for $t \geq s$,
B.3 $P\{W_t$ is a continuous function of $t\} = 1$, or in other words, W is sample path continuous with probability one.

A typical sample path of a Brownian motion is shown in Figure 4.5. A Brownian motion, being a mean zero independent increment process with $P\{W_0 = 0\} = 1$, is a martingale.

The mean, correlation, and covariance functions of a Brownian motion W are given by

$$\mu_W(t) = E[W_t] = E[W_t - W_0] = 0$$

and, for $s \leq t$,

$$\begin{aligned} R_W(s,t) &= E[W_s W_t] \\ &= E[(W_s - W_0)(W_s - W_0 + W_t - W_s)] \\ &= E[(W_s - W_0)^2] = \sigma^2 s, \end{aligned}$$

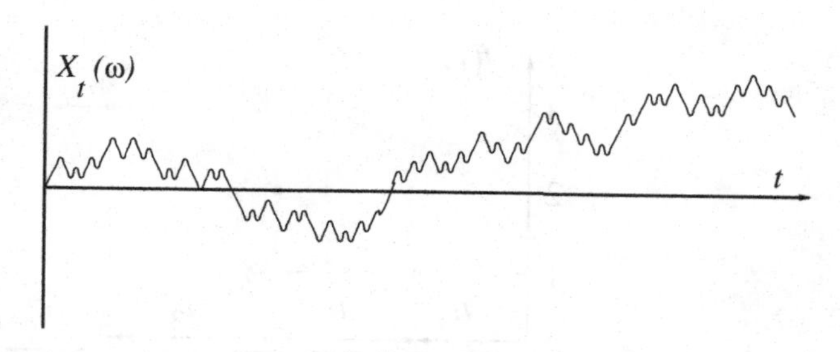

Figure 4.5 A typical sample path of Brownian motion.

so that, in general,

$$C_W(s,t) = R_W(s,t) \;\; = \;\; \sigma^2(s \wedge t).$$

A Brownian motion is Gaussian, because if $0 = t_0 \le t_1 \le \cdots \le t_n$, then each coordinate of the vector $(W_{t_1}, \ldots, W_{t_n})$ is a linear combination of the n independent Gaussian random variables $(W_{t_i} - W_{t_{i-1}} : 1 \le i \le n)$. Thus, properties B.0–B.2 imply that W is a Gaussian random process with $\mu_W = 0$ and $R_W(s,t) = \sigma^2(s \wedge t)$. In fact, the converse is also true. If $W = (W_t : t \ge 0)$ is a Gaussian random process with mean zero and $R_W(s,t) = \sigma^2(s \wedge t)$, then B.0–B.2 are true.

Property B.3 does not come automatically. For example, if W is a Brownian motion and if U is a Unif(0,1) distributed random variable independent of W, let $\widetilde{W}$ be defined by

$$\widetilde{W}_t = W_t + I_{\{U=t\}}.$$

Then $P\{\widetilde{W}_t = W_t\} = 1$ for each $t \ge 0$ and $\widetilde{W}$ also satisfies B.0–B.2, but $\widetilde{W}$ fails to satisfy B.3. Thus, $\widetilde{W}$ is not a Brownian motion. The difference between W and $\widetilde{W}$ is significant if events involving uncountably many values of t are investigated. For example,

$$P\{W_t \le 1 \text{ for } 0 \le t \le 1\} \ne P\{\widetilde{W}_t \le 1 \text{ for } 0 \le t \le 1\}.$$

4.5 Counting processes and the Poisson process

A function f on $\mathbb{R}_+$ is called a *counting function* if $f(0) = 0$, f is nondecreasing, f is right continuous, and f is integer valued. The interpretation is that $f(t)$ is the number of counts observed during the interval $(0,t]$. An increment $f(b) - f(a)$ is the number of counts in the interval $(a,b]$. If t_i denotes the time of the ith count for $i \ge 1$, then f can be described by the sequence (t_i). Or, if $u_1 = t_1$ and $u_i = t_i - t_{i-1}$ for $i \ge 2$, then f can be described by the sequence (u_i), see Figure 4.6. The numbers $t_1, t_2, \ldots$ are called the *count times* and

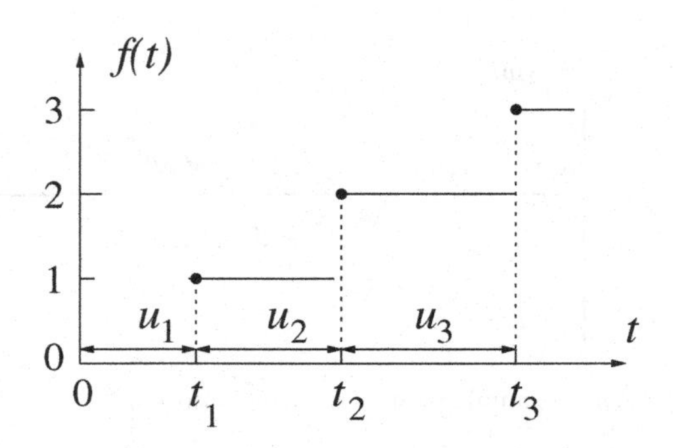

Figure 4.6 A counting function.

the numbers $u_1, u_2, \ldots$ are called the *intercount times*. The following equations clearly hold:

$$f(t) = \sum_{n=1}^{\infty} I_{\{t \geq t_n\}},$$
$$t_n = \min\{t : f(t) \geq n\},$$
$$t_n = u_1 + \ldots + u_n.$$

A random process is called a *counting process* if with probability one its sample path is a counting function. A counting process has two corresponding random sequences, the sequence of count times and the sequence of intercount times.

The most widely used example of a counting process is a Poisson process, defined next.

Definition 4.4 Let $\lambda \geq 0$. A *Poisson process* with rate λ is a random process $N = (N_t : t \geq 0)$ such that:

N.1 N is a counting process,
N.2 N has independent increments,
N.3 $N(t) - N(s)$ has the $Poi(\lambda(t-s))$ distribution for $t \geq s$.

Proposition 4.5 *Let N be a counting process and let $\lambda > 0$. The following are equivalent:*

(a) *N is a Poisson process with rate λ.*
(b) *The intercount times $U_1, U_2, \ldots$ are mutually independent, $Exp(\lambda)$ random variables.*
(c) *For each $\tau > 0$, N_τ is a Poisson random variable with parameter $\lambda\tau$, and given $\{N_\tau = n\}$, the times of the n counts during $[0, \tau]$ are the same as n independent, Unif $[0, \tau]$ random variables, reordered to be nondecreasing. That is, for any $n \geq 1$, the conditional density of the first n count times, $(T_1, \ldots, T_n)$, given the event $\{N_\tau = n\}$, is:*

$$f(t_1, \ldots, t_n | N_\tau = n) = \begin{cases} \frac{n!}{\tau^n} & 0 < t_1 < \cdots < t_n \leq \tau \\ 0 & else \end{cases}. \tag{4.1}$$

Proof It will be shown that (a) implies (b), (b) implies (c), and (c) implies (a).

(a) implies (b). Suppose N is a Poisson process. The joint pdf of the first n count times $T_1, \ldots, T_n$ can be found as follows. Let $0 < t_1 < t_2 < \ldots < t_n$. Select $\epsilon > 0$ so small that $(t_1 - \epsilon, t_1]$, $(t_2 - \epsilon, t_2]$, ..., $(t_n - \epsilon, t_n]$ are disjoint intervals of $\mathbb{R}_+$. Then the probability that $(T_1, \ldots, T_n)$ is in the n-dimensional cube with upper corner $t_1, \ldots, t_n$ and sides of length ϵ is given by

$$\begin{aligned} &P\{T_i \in (t_i - \epsilon, t_i] \text{ for } 1 \leq i \leq n\} \\ &= P\{N_{t_1 - \epsilon} = 0, N_{t_1} - N_{t_1 - \epsilon} = 1, N_{t_2 - \epsilon} - N_{t_1} = 0, \ldots, N_{t_n} - N_{t_n - \epsilon} = 1\} \\ &= (e^{-\lambda(t_1 - \epsilon)})(\lambda \epsilon e^{-\lambda \epsilon})(e^{-\lambda(t_2 - \epsilon - t_1)}) \cdots (\lambda \epsilon e^{-\lambda \epsilon}) \\ &= (\lambda \epsilon)^n e^{-\lambda t_n}. \end{aligned}$$

The volume of the cube is ϵ^n. Therefore $(T_1, \ldots, T_n)$ has the pdf

$$f_{T_1 \cdots T_n}(t_1, \ldots, t_n) = \begin{cases} \lambda^n e^{-\lambda t_n} & \text{if } 0 < t_1 < \cdots < t_n \\ 0 & \text{else} \end{cases}. \tag{4.2}$$

The vector $(U_1, \ldots, U_n)$ is the image of $(T_1, \ldots, T_n)$ under the mapping $(t_1, \ldots, t_n) \to (u_1, \ldots, u_n)$ defined by $u_1 = t_1$, $u_k = t_k - t_{k-1}$ for $k \geq 2$. The mapping is invertible, because $t_k = u_1 + \ldots + u_k$ for $1 \leq k \leq n$. The range of the mapping is $\mathbb{R}_+^n$, and the Jacobian, given by

$$\frac{\partial u}{\partial t} = \begin{pmatrix} 1 & & & & \\ -1 & 1 & & & \\ & -1 & 1 & & \\ & & \ddots & \ddots & \\ & & & -1 & 1 \end{pmatrix},$$

has unit determinant. Therefore, by the formula for the transformation of random vectors (see Section 1.11),

$$f_{U_1 \ldots U_n}(u_1, \ldots, u_n) = \begin{cases} \lambda^n e^{-\lambda(u_1 + \ldots + u_n)} & u \in \mathbb{R}_+^n \\ 0 & \text{else} \end{cases}. \tag{4.3}$$

The joint pdf in (4.3) factors into the product of n pdfs, with each pdf being for an *Exp*(λ) random variable. Thus the intercount times $U_1, U_2, \ldots$ are independent and each is exponentially distributed with parameter λ. So (a) implies (b).

(b) implies (c). Suppose that N is a counting process such that the intercount times $U_1, U_2, \ldots$ are independent, *Exp*(λ) random variables, for some $\lambda > 0$. Thus, for $n \geq 1$, the first n intercount times have the joint pdf given in (4.3). Equivalently, appealing to the transformation of random vectors in the reverse direction, the pdf of the first n count times, $(T_1, \ldots, T_n)$, is given by (4.2). Fix $\tau > 0$ and an integer $n \geq 1$. The event $\{N_\tau = n\}$ is equivalent to the event $(T_1, \ldots, T_{n+1}) \in A_{n,\tau}$, where

$$A_{n,\tau} = \{t \in \mathbb{R}_+^{n+1} : 0 < t_1 < \ldots < t_n \leq \tau < t_{n+1}\}.$$

The conditional pdf of $(T_1, \ldots, T_{n+1})$, given that $\{N_\tau = n\}$, is obtained by starting with the joint pdf of $(T_1, \ldots, T_{n+1})$, namely $\lambda^{n+1}e^{-\lambda(t_{n+1})}$ on the set $\{t \in \mathbb{R}^{n+1} : 0 < t_1 < \ldots < t_{n+1}\}$, setting it equal to zero on the complement of the set $A_{n,\tau}$, and scaling it up by the factor $1/P\{N_\tau = n\}$ on $A_{n,\tau}$:

$$f(t_1, \ldots, t_{n+1}|N_\tau = n) = \begin{cases} \frac{\lambda^{n+1}e^{-\lambda t_{n+1}}}{P\{N_\tau=n\}} & 0 < t_1 < \cdots < t_n \leq \tau < t_{n+1} \\ 0 & \text{else} \end{cases}. \tag{4.4}$$

The joint density of $(T_1, \ldots, T_n)$, given that $\{N_\tau = n\}$, is obtained for each $(t_1, \ldots, t_n)$ by integrating the density in (4.4) with respect to t_{n+1} over $\mathbb{R}$. If $0 < t_1 < \ldots < t_n \leq \tau$ does not hold, the density in (4.4) is zero for all values of t_{n+1}. If $0 < t_1 < \ldots < t_n \leq \tau$, then the density in (4.4) is nonzero for $t_{n+1} \in (\tau, \infty)$. Integrating (4.4) with respect to t_{n+1} over (τ, ∞) yields:

$$f(t_1, \ldots, t_n|N_\tau = n) = \begin{cases} \frac{\lambda^n e^{-\lambda\tau}}{P\{N_\tau=n\}} & 0 < t_1 < \cdots < t_n \leq \tau \\ 0 & \text{else} \end{cases}. \tag{4.5}$$

The conditional density in (4.5) is constant on $\{t \in \mathbb{R}_+^n : 0 < t_1 < \ldots < t_n \leq \tau\}$, and that constant must be the reciprocal of the n-dimensional volume of the set. The unit cube $[0, \tau]^n$ in $\mathbb{R}^n$ has volume τ^n. It can be partitioned into $n!$ equal volume subsets determined by the $n!$ possible orderings of the numbers $t_1, \ldots, t_n$. Therefore, the set $\{t \in \mathbb{R}_+^n : 0 \leq t_1 < \cdots < t_n \leq \tau\}$ has volume $\tau^n/n!$. Hence, (4.5) implies both that (4.1) holds and that $P\{N_\tau = n\} = \frac{(\lambda\tau)^n e^{-\lambda\tau}}{n!}$. These implications are for $n \geq 1$. Also, $P\{N_\tau = 0\} = P\{U_1 > \tau\} = e^{-\lambda\tau}$. Thus, N_τ is a Poi$(\lambda\tau)$ random variable.

(c) implies (a). Suppose $t_0 < t_1 < \ldots < t_k$ and let $n_1, \ldots, n_k$ be nonnegative integers. Set $n = n_1 + \ldots + n_k$ and $p_i = (t_i - t_{i-1})/t_k$ for $1 \leq i \leq k$. Suppose (c) is true. Given there are n counts in the interval $[0, \tau]$, by (c), the distribution of the numbers of counts in each subinterval is as if each of the n counts is thrown into a subinterval at random, falling into the ith subinterval with probability p_i. The probability that, for $1 \leq i \leq K$, n_i particular counts fall into the ith interval, is $p_1^{n_1} \cdots p_k^{n_k}$. The number of ways to assign n counts to the intervals such that there are n_i counts in the ith interval is $\binom{n}{n_1 \ldots n_k} = \frac{n!}{n_1! \ldots n_k!}$. This thus gives rise to what is known as a multinomial distribution for the numbers of counts per interval. We have

$$\begin{aligned}
&P\{N(t_i) - N(t_{i-1}) = n_i \text{ for } 1 \leq i \leq k\} \\
&= P\{N(t_k) = n\} P[N(t_i) - N(t_{i-1}) = n_i \text{ for } 1 \leq i \leq k \mid N(t_k) = n] \\
&= \frac{(\lambda t_k)^n e^{-\lambda t_k}}{n!} \binom{n}{n_1 \ \ldots \ n_k} p_1^{n_1} \ldots p_k^{n_k} \\
&= \prod_{i=1}^{k} \frac{(\lambda(t_i - t_{i-1}))^{n_i} e^{-\lambda(t_i - t_{i-1})}}{n_i!}.
\end{aligned}$$

Therefore the increments $N(t_i) - N(t_{i-1})$, $1 \leq i \leq k$, are independent, with $N(t_i) - N(t_{i-1})$ being a Poisson random variable with mean $\lambda(t_i - t_{i-1})$, for $1 \leq i \leq k$. So (a) is proved. □

A Poisson process is not a martingale. However, if $\widetilde{N}$ is defined by $\widetilde{N}_t = N_t - \lambda t$, then $\widetilde{N}$ is an independent increment process with mean 0 and $\widetilde{N}_0 = 0$. Thus, $\widetilde{N}$ is a martingale. Note that $\widetilde{N}$ has the same mean and covariance function as a Brownian motion with $\sigma^2 = \lambda$, which shows how little one really knows about a process from its mean function and correlation function alone.

4.6 Stationarity

Consider a random process $X = (X_t : t \in \mathbb{T})$ such that either $\mathbb{T} = \mathbb{Z}$ or $\mathbb{T} = \mathbb{R}$. Then X is said to be *stationary* if for any $t_1, \ldots, t_n$ and s in $\mathbb{T}$, the random vectors $(X_{t_1}, \ldots, X_{t_n})$ and $(X_{t_1+s}, \ldots, X_{t_n+s})$ have the same distribution. In other words, the joint statistics of X of all orders are unaffected by a shift in time. The condition of stationarity of X can also be expressed in terms of the CDFs of X: X is stationary if for any $n \geq 1, s, t_1, \ldots, t_n \in \mathbb{T}$, and $x_1, \ldots, x_n \in \mathbb{R}$,

$$F_{X,n}(x_1, t_1; \ldots; x_n, t_n) = F_{X,n}(x_1, t_1 + s; \ldots; x_n; t_n + s).$$

Suppose X is a stationary second order random process. (Recall that second order means that $E[X_t^2] < \infty$ for all t.) Then by the $n = 1$ part of the definition of stationarity, X_t has the same distribution for all t. In particular, $\mu_X(t)$ and $E[X_t^2]$ do not depend on t. Moreover, by the $n = 2$ part of the definition $E[X_{t_1}X_{t_2}] = E[X_{t_1+s}X_{t_2+s}]$ for any $s \in \mathbb{T}$. If $E[X_t^2] < +\infty$ for all t, then $E[X_{t+s}]$ and $R_X(t_1 + s, t_2 + s)$ are finite and both do not depend on s.

A second order random process $(X_t : t \in \mathbb{T})$ with $\mathbb{T} = \mathbb{Z}$ or $\mathbb{T} = \mathbb{R}$ is called *wide sense stationary* (WSS) if

$$\mu_X(t) = \mu_X(s + t) \text{ and } R_X(t_1, t_2) = R_X(t_1 + s, t_2 + s)$$

for all t, s, $t_1, t_2 \in \mathbb{T}$. As shown above, a stationary second order random process is WSS. Wide sense stationarity means that $\mu_X(t)$ is a finite number, not depending on t, and $R_X(t_1, t_2)$ depends on t_1, t_2 only through the difference $t_1 - t_2$. By a convenient and widely accepted abuse of notation, if X is WSS, we use μ_X to be the constant and R_X to be the function of one real variable such that

$$E[X_t] = \mu_X \qquad t \in \mathbb{T}$$
$$E[X_{t_1}X_{t_2}] = R_X(t_1 - t_2) \qquad t_1, t_2 \in \mathbb{T}.$$

The dual use of the notation R_X if X is WSS leads to the identity $R_X(t_1, t_2) = R_X(t_1 - t_2)$. As a practical matter, this means replacing a comma by a minus sign. Since one interpretation of R_X requires it to have two arguments, and the other interpretation requires only one argument, the interpretation is clear from the number of arguments. Some brave authors even skip mentioning that X is WSS when they write: "Suppose $(X_t : t \in \mathbb{R})$ has mean μ_X and correlation function $R_X(\tau)$," because it is implicit in this statement that X is WSS.

Since the covariance function C_X of a random process X satisfies

$$C_X(t_1, t_2) = R_X(t_1, t_2) - \mu_X(t_1)\mu_X(t_2),$$

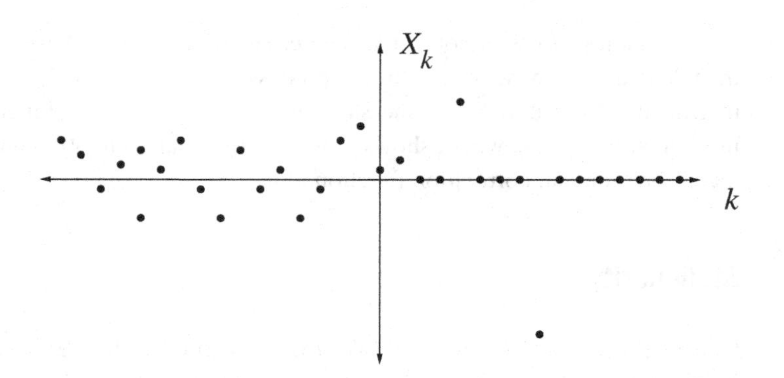

Figure 4.7 A typical sample path of an example of a WSS random process.

if X is WSS then $C_X(t_1, t_2)$ is a function of $t_1 - t_2$. The notation C_X is also used to denote the function of one variable such that $C_X(t_1 - t_2) = \text{Cov}(X_{t_1}, X_{t_2})$. Therefore, if X is WSS then $C_X(t_1 - t_2) = C_X(t_1, t_2)$. Also, $C_X(\tau) = R_X(\tau) - \mu_X^2$, where in this equation τ should be thought of as the difference of two times, $t_1 - t_2$.

In general, there is much more to know about a random vector or a random process than the first and second moments. Therefore, one can mathematically define WSS processes that are spectacularly different in appearance from any stationary random process. For example, any random process $(X_k : k \in \mathbb{Z})$ such that the X_k are independent with $E[X_k] = 0$ and $\text{Var}(X_k) = 1$ for all k is WSS. To be specific, we could take the X_k to be independent, with X_k being $N(0, 1)$ for $k \leq 0$ and with X_k having pmf

$$p_{X,1}(x, k) = P\{X_k = x\} \quad = \quad \begin{cases} \frac{1}{2k^2} & x \in \{k, -k\} \\ 1 - \frac{1}{k^2} & \text{if } x = 0 \\ 0 & \text{else} \end{cases}$$

for $k \geq 1$. A typical sample path is shown in Figure 4.7.

The situation is much different if X is a Gaussian process. Indeed, suppose X is Gaussian and WSS. Then for any $t_1, t_2, \ldots, t_n, s \in \mathbb{T}$, the vector $(X_{t_1+s}, X_{t_2+s}, \ldots, X_{t_n+s})^T$ is Gaussian with mean $(\mu, \mu, \ldots, \mu)^T$ and covariance matrix with ijth entry given by $C_X((t_i + s) - (t_j + s)) = C_X(t_i - t_j)$. This mean and covariance matrix do not depend on s. Thus, the distribution of the vector does not depend on s. Therefore, X is stationary.

In summary, if X is stationary then X is WSS, and if X is both Gaussian and WSS, then X is stationary.

Example 4.4 Let $X_t = A\cos(\omega_c t + \Theta)$, where ω_c is a nonzero constant, A and Θ are independent random variables with $P\{A > 0\} = 1$ and $E[A^2] < +\infty$. Each sample path of the random process $(X_t : t \in \mathbb{R})$ is a pure sinusoidal function at frequency ω_c radians per unit time, with amplitude A and phase Θ.

We address two questions. First, what additional assumptions, if any, are needed on the distributions of A and Θ to imply that X is WSS? Second, we consider two distributions for Θ which each make X WSS, and see if they make X stationary.

To address whether X is WSS, the mean and correlation functions can be computed. Since A and Θ are independent and $\cos(\omega_c t+\Theta) = \cos(\omega_c t)\cos(\Theta) - \sin(\omega_c t)\sin(\Theta)$,

$$\mu_X(t) = E[A]\left(E[\cos(\Theta)]\cos(\omega_c t) - E[\sin(\Theta)]\sin(\omega_c t)\right).$$

The function $\mu_X(t)$ is a linear combination of $\cos(\omega_c t)$ and $\sin(\omega_c t)$. The only way such a linear combination can be independent of t is if the coefficients of $\cos(\omega_c t)$ and $\sin(\omega_c t)$ are zero (in fact, it is enough to equate the values of $\mu_X(t)$ at $\omega_c t = 0, \frac{\pi}{2}$, and π). So, $\mu_X(t)$ does not depend on t if and only if $E[\cos(\Theta)] = E[\sin(\Theta)] = 0$.

Turning next to R_X, using the identity $\cos(a)\cos(b) = (\cos(a-b)+\cos(a+b))/2$ yields

$$\begin{aligned} R_X(s,t) &= E[A^2]E[\cos(\omega_c s+\Theta)\cos(\omega_c t+\Theta)] \\ &= \frac{E[A^2]}{2}\{\cos(\omega_c(s-t)) + E[\cos(\omega_c(s+t)+2\Theta)]\}. \end{aligned}$$

Since $s+t$ can be arbitrary for $s-t$ fixed, in order that $R_X(s,t)$ be a function of $s-t$ alone it is necessary that $E[\cos(\omega_c(s+t)+2\Theta)]$ be a constant, independent of the value of $s+t$. Arguing just as in the case of μ_X, with Θ replaced by 2Θ, yields that $R_X(s,t)$ is a function of $s-t$ if and only if $E[\cos(2\Theta)] = E[\sin(2\Theta)] = 0$.

Combining the findings for μ_X and R_X, yields that X is WSS, if and only if,

$$E[\cos(\Theta)] = E[\sin(\Theta)] = E[\cos(2\Theta)] = E[\sin(2\Theta)] = 0.$$

There are many distributions for Θ in $[0,2\pi]$ such that the four moments specified are zero. Two possibilities are: (a) Θ is uniformly distributed on the interval $[0,2\pi]$; or (b) Θ is a discrete random variable, taking the four values $0, \frac{\pi}{2}, \pi, \frac{3\pi}{2}$ with equal probability. Is X stationary for either possibility?

We shall show that X is stationary if Θ is uniformly distributed over $[0,2\pi]$. Stationarity means for any fixed constant s, the random processes $(X_t : t \in \mathbb{R})$ and $(X_{t+s} : t \in \mathbb{R})$ have the same finite order distributions. For this example,

$$X_{t+s} = A\cos(\omega_c(t+s)+\Theta) = A\cos(\omega_c t+\tilde{\Theta}),$$

where $\tilde{\Theta} = ((\omega_c s+\Theta) \bmod 2\pi)$. By Example 1.10, $\tilde{\Theta}$ is again uniformly distributed on the interval $[0,2\pi]$. Thus (A,Θ) and $(A,\tilde{\Theta})$ have the same joint distribution, so $A\cos(\omega_c t+\Theta)$ and $A\cos(\omega_c t+\tilde{\Theta})$ have the same finite order distributions. Hence, X is indeed stationary if Θ is uniformly distributed over $[0,2\pi]$.

Assume now that Θ takes on each of the values of $0, \frac{\pi}{2}, \pi$, and $\frac{3\pi}{2}$ with equal probability. Is X stationary? If X were stationary then X_t would have the same distribution for all t. On one hand, $P\{X_0 = 0\} = P\{\Theta = \frac{\pi}{2} \text{ or } \Theta = \frac{3\pi}{2}\} = \frac{1}{2}$. On the other hand, if $\omega_c t$ is not an integer multiple of $\frac{\pi}{2}$, then $\omega_c t+\Theta$ cannot be an integer multiple of $\frac{\pi}{2}$, so $P\{X_t = 0\} = 0$. Hence X is not stationary.

With more work it can be shown that X is stationary, if and only if, $(\Theta \bmod 2\pi)$ is uniformly distributed over the interval $[0,2\pi]$.

4.7 Joint properties of random processes

Two random processes X and Y are said to be jointly stationary if their parameter set $\mathbb{T}$ is either $\mathbb{Z}$ or $\mathbb{R}$, and if for any $t_1, \ldots, t_n, s \in \mathbb{T}$, the distribution of the random vector

$$(X_{t_1+s}, X_{t_2+s}, \ldots, X_{t_n+s}, Y_{t_1+s}, Y_{t_2+s}, \ldots, Y_{t_n+s})$$

does not depend on s.

The random processes X and Y are said to be jointly Gaussian if all the random variables comprising X and Y are jointly Gaussian.

If X and Y are second order random processes on the same probability space, the cross correlation function, R_{XY}, is defined by $R_{XY}(s,t) = E[X_s Y_t]$, and the cross covariance function, C_{XY}, is defined by $C_{XY}(s,t) = \mathrm{Cov}(X_s, Y_t)$.

The random processes X and Y are said to be jointly WSS, if X and Y are each WSS, and if $R_{XY}(s,t)$ is a function of $s - t$. If X and Y are jointly WSS, we use $R_{XY}(\tau)$ for $R_{XY}(s,t)$ where $\tau = s - t$, and similarly we use $C_{XY}(s-t) = C_{XY}(s,t)$. Note that $C_{XY}(s,t) = C_{YX}(t,s)$, so $C_{XY}(\tau) = C_{YX}(-\tau)$.

4.8 Conditional independence and Markov processes

Markov processes are naturally associated with the state space approach for modeling a system. The idea of a state space model for a given system is to define the state of the system at any given time t. The state of the system at time t should summarize everything about the system up to and including time t that is relevant to the future of the system. For example, the state of an aircraft at time t could consist of the position, velocity, and remaining fuel at time t. Think of t as the present time. The state at time t determines the possible future part of the aircraft trajectory. For example, it determines how much longer the aircraft can fly and where it could possibly land. The state at time t does not completely determine the entire past trajectory of the aircraft. Rather, the state summarizes enough about the system up to the present so that if the state is known, no more information about the past is relevant to the future possibilities. The concept of state is inherent in the Kalman filtering model discussed in Chapter 3. The notion of state is captured for random processes using the notions of conditional independence and the Markov property, which are discussed next.

Let X, Y, Z be random vectors. We shall define the condition that X and Z are conditionally independent given Y. Such a condition is denoted by $X - Y - Z$. If X, Y, Z are discrete, then $X - Y - Z$ is defined to hold if

$$P(X = i, Z = k \mid Y = j) = P(X = i \mid Y = j)P(Z = k \mid Y = j) \tag{4.6}$$

for all i, j, k with $P\{Y = j\} > 0$. Equivalently, $X - Y - Z$ if

$$P\{X = i, Y = j, Z = k\}P\{Y = j\} = P\{X = i, Y = j\}P\{Z = k, Y = j\} \tag{4.7}$$

for all i, j, k. Equivalently again, X–Y–Z if

$$P(Z = k \mid X = i, Y = j) = P(Z = k \mid Y = j) \tag{4.8}$$

for all i, j, k with $P\{X = i, Y = j\} > 0$. The forms (4.6) and (4.7) make it clear that the condition $X - Y - Z$ is symmetric in X and Z: thus $X - Y - Z$ is the same condition as $Z - Y - X$. The form (4.7) does not involve conditional probabilities, so no requirement about conditioning on events having positive probability is needed. The form (4.8) shows that $X - Y - Z$ means that knowing Y alone is as informative as knowing both X and Y, for the purpose of determining conditional probabilities of Z. Intuitively, the condition $X - Y - Z$ means that the random variable Y serves as a state.

If X, Y, and Z have a joint pdf, then the condition $X - Y - Z$ can be defined using the pdfs and conditional pdfs in a similar way. For example, the conditional independence condition $X - Y - Z$ holds by definition if

$$f_{XZ|Y}(x, z|y) = f_{X|Y}(x|y) f_{Z|Y}(z|y) \quad \text{whenever } f_Y(y) > 0.$$

An equivalent condition is

$$f_{Z|XY}(z|x, y) = f_{Z|Y}(z|y) \quad \text{whenever } f_{XY}(x, y) > 0. \tag{4.9}$$

Example 4.5 Suppose X, Y, Z are jointly Gaussian vectors. Let us see what the condition $X - Y - Z$ means in terms of the covariance matrices. Assume without loss of generality that the vectors have mean zero. Because X, Y, and Z are jointly Gaussian, the condition (4.9) is equivalent to the condition that $E[Z|X, Y] = E[Z|Y]$ (because given X, Y, or just given Y, the conditional distribution of Z is Gaussian, and in the two cases the mean and covariance of the conditional distribution of Z are the same). The idea of linear innovations applied to the length two sequence (Y, X) yields $E[Z|X, Y] = E[Z|Y] + E[Z|\tilde{X}]$ where $\tilde{X} = X - E[X|Y]$. Thus $X - Y - Z$ if and only if $E[Z|\tilde{X}] = 0$, or equivalently, if and only if $\text{Cov}(\tilde{X}, Z) = 0$. Since $\tilde{X} = X - \text{Cov}(X, Y)\text{Cov}(Y)^{-1}Y$, it follows that

$$\text{Cov}(\tilde{X}, Z) = \text{Cov}(X, Z) - \text{Cov}(X, Y)\text{Cov}(Y)^{-1}\text{Cov}(Y, Z).$$

Therefore, $X - Y - Z$ if and only if

$$\text{Cov}(X, Z) = \text{Cov}(X, Y)\text{Cov}(Y)^{-1}\text{Cov}(Y, Z). \tag{4.10}$$

In particular, if X, Y, and Z are jointly Gaussian random variables with nonzero variances, the condition $X - Y - Z$ holds if and only if the correlation coefficients satisfy $\rho_{XZ} = \rho_{XY}\rho_{YZ}$.

A general definition of conditional probabilities and conditional independence, based on the general definition of conditional expectation given in Chapter 3, is given next. Recall that $P(F) = E[I_F]$ for any event F, where I_F denotes the indicator function of F. If Y is a random vector, we define $P(F|Y)$ to equal $E[I_F|Y]$. This means that $P(F|Y)$ is the unique (in the sense that any two versions are equal with probability one) random variable such that:

(1) $P(F|Y)$ is a function of Y and it has finite second moments, and
(2) $E[g(Y)P(F|Y)] = E[g(Y)I_F]$ for any $g(Y)$ with finite second moment.

Given arbitrary random vectors, we define X and Z to be conditionally independent given Y, (written $X - Y - Z$) if for any Borel sets A and B,

$$P(\{X \in A\}\{Z \in B\}|Y) = P(X \in A|Y)P(Z \in B|Y).$$

Equivalently, $X - Y - Z$ if for any Borel set B, $P(Z \in B|X, Y) = P(Z \in B|Y)$.

Definition 4.6 A random process $X = (X_t : t \in \mathbb{T})$ is said to be a *Markov process* if for any $t_1, \ldots, t_{n+1}$ in $\mathbb{T}$ with $t_1 < \ldots < t_n$, the following conditional independence condition holds:

$$(X_{t_1}, \ldots, X_{t_n}) \; - \; X_{t_n} \; - \; X_{t_{n+1}}. \tag{4.11}$$

It turns out that the Markov property is equivalent to the following conditional independence property: For any $t_1, \ldots, t_{n+m}$ in $\mathbb{T}$ with $t_1 < \ldots < t_{n+m}$,

$$(X_{t_1}, \ldots, X_{t_n}) \; - \; X_{t_n} \; - \; (X_{t_n}, \ldots, X_{t_{n+m}}). \tag{4.12}$$

The definition (4.11) is easier to check than condition (4.12), but (4.12) is appealing because it is symmetric in time. In words, thinking of t_n as the present time, the Markov property means that the past and future of X are conditionally independent given the present state X_{t_n}.

Example 4.6 (*Markov property of independent increment processes*) Let $(X_t : t \geq 0)$ be an independent increment process such that X_0 is a constant. Then for any $t_1, \ldots, t_{n+1}$ with $0 \leq t_1 \leq \ldots \leq t_{n+1}$, the vector $(X_{t_1}, \ldots, X_{t_n})$ is a function of the n increments $X_{t_1} - X_0$, $X_{t_2} - X_{t_1}$, $X_{t_n} - X_{t_{n-1}}$, and is thus independent of the increment $V = X_{t_{n+1}} - X_{t_n}$. But $X_{t_{n+1}}$ is determined by V and X_{t_n}. Thus, X is a Markov process. In particular, random walks, Brownian motions, and Poisson processes are Markov processes.

Example 4.7 (*Gaussian Markov processes*) Suppose $X = (X_t : t \in \mathbb{T})$ is a Gaussian random process with $\mathrm{Var}(X_t) > 0$ for all t. By the characterization of conditional independence for jointly Gaussian vectors (4.10), the Markov property (4.11) is equivalent to

$$\mathrm{Cov}\left(\begin{pmatrix} X_{t_1} \\ X_{t_2} \\ \vdots \\ X_{t_n} \end{pmatrix}, X_{t_{n+1}}\right) = \mathrm{Cov}\left(\begin{pmatrix} X_{t_1} \\ X_{t_2} \\ \vdots \\ X_{t_n} \end{pmatrix}, X_{t_n}\right) \mathrm{Var}(X_{t_n})^{-1}\mathrm{Cov}(X_{t_n}, X_{t_{n+1}}),$$

which, letting $\rho(s,t)$ denote the correlation coefficient between X_s and X_t, is equivalent to the requirement

$$\begin{pmatrix} \rho(t_1,t_{n+1}) \\ \rho(t_2,t_{n+1}) \\ \vdots \\ \rho(t_n,t_{n+1}) \end{pmatrix} = \begin{pmatrix} \rho(t_1,t_n) \\ \rho(t_2,t_n) \\ \vdots \\ \rho(t_n,t_n) \end{pmatrix} \rho(t_n,t_{n+1}).$$

Therefore a Gaussian process X is Markovian if and only if

$$\rho(r,t) = \rho(r,s)\rho(s,t) \quad \text{whenever } r,s,t \in \mathbb{T} \text{ with } r < s < t. \tag{4.13}$$

If $X = (X_k : k \in Z)$ is a discrete-time stationary Gaussian process, then $\rho(s,t)$ may be written as $\rho(k)$, where $k = s - t$. Note that $\rho(k) = \rho(-k)$. Such a process is Markovian if and only if $\rho(k_1 + k_2) = \rho(k_1)\rho(k_2)$ for all positive integers k_1 and k_2. Therefore, X is Markovian if and only if $\rho(k) = b^{|k|}$ for all k, for some constant b with $|b| \leq 1$. Equivalently, a stationary Gaussian process $X = (X_k : k \in Z)$ with $\mathrm{Var}(X_k) > 0$ for all k is Markovian if and only if the covariance function has the form $C_X(k) = Ab^{|k|}$ for some constants A and b with $A > 0$ and $|b| \leq 1$.

Similarly, if $(X_t : t \in R)$ is a continuous-time stationary Gaussian process with $\mathrm{Var}(X_t) > 0$ for all t, X is Markovian if and only if $\rho(s+t) = \rho(s)\rho(t)$ for all $s,t \geq 0$. The only bounded real-valued functions satisfying such a multiplicative condition are exponential functions. Therefore, a stationary Gaussian process X with $\mathrm{Var}(X_t) > 0$ for all t is Markovian if and only if ρ has the form $\rho(\tau) = \exp(-\alpha|\tau|)$, for some constant $\alpha \geq 0$, or equivalently, if and only if C_X has the form $C_X(\tau) = A\exp(-\alpha|\tau|)$ for some constants $A > 0$ and $\alpha \geq 0$.

The following proposition should be intuitively clear, and it often applies in practice.

Proposition 4.7 *(Markov property of a sequence determined by a recursion driven by independent random variables) Suppose $X_0, U_1, U_2, \ldots$ are mutually independent random variables and suppose $(X_n : n \geq 1)$ is determined by a recursion of the form $X_{n+1} = h_{n+1}(X_n, U_{n+1})$ for $n \geq 0$. Then $(X_n : n \geq 0)$ is a Markov process.*

Proof The proposition will first be proved in the case when the random variables are all discrete type. Let $n \geq 1$, let $B \subset \mathbb{R}$, and let φ be the function defined by $\varphi(x_n) = P\{h_{n+1}(x_n, U_{n+1}) \in B\}$. The random vector $(X_0, \ldots, X_n)$ is determined by $(X_0, U_1, \ldots, U_n)$, and is therefore independent of U_{n+1}. Thus, for any possible value $(x_0, \ldots, x_n)$ of $(X_0, \ldots, X_n)$,

$$\begin{aligned} P(X_{n+1} &\in B | X_0 = x_o, \ldots, X_n = x_n) \\ &= P(h_{n+1}(x_n, U_{n+1}) \in B | X_0 = x_o, \ldots, X_n = x_n) \\ &= \varphi(x_n). \end{aligned}$$

So the conditional distribution of X_{n+1} given $(X_0, \ldots, X_n)$ depends only on X_n, establishing the Markov property.

For the general case use the general version of conditional probability. Let $n \geq 1$, let B be a Borel subset of $\mathbb{R}$, and let φ be defined as before. We will show that $P(X_{n+1} \in B|X_0, \ldots, X_n) = \varphi(X_n)$ by checking that $\varphi(X_n)$ has the two properties that characterize $P(X_{n+1} \in B|X_0, \ldots, X_n)$. First, $\varphi(X_n)$ is a function of $X_0, \ldots, X_n$ with finite second moments. Secondly, if g is an arbitrary Borel function such that $g(X_0, \ldots, X_n)$ has a finite second moment, then

$$
\begin{aligned}
&E\left[I_{\{X_{n+1} \in B\}} g(X_0, \ldots, X_n)\right] \\
&= \int_{\mathbb{R}^n} \int_{\{u: h_{n+1}(x_n, u) \in B\}} g(x_0, \ldots, x_n) dF_{U_{n+1}}(u) dF_{X_0, \ldots, X_n}(x_0, \ldots, x_n) \\
&= \int_{\mathbb{R}^n} \left(\int_{\{u: h_{n+1}(x_n, u) \in B\}} dF_{U_{n+1}}(u) \right) g(x_0, \ldots, x_n) dF_{X_0, \ldots, X_n}(x_0, \ldots, x_n) \\
&= \int_{\mathbb{R}^n} \varphi(x_n) g(x_0, \ldots, x_n) dF_{X_0, \ldots, X_n}(x_0, \ldots, x_n) \\
&= E\left[\varphi(X_n) g(X_0, \ldots, X_n)\right].
\end{aligned}
$$

Therefore, $P(X_{n+1} \in B|X_0, \ldots X_n) = \varphi(X_n)$. Hence, $P(X_{n+1} \in B|X_0, \ldots X_n)$ is a function of X_n so that $P(X_{n+1} \in B|X_0, \ldots X_n) = P(X_{n+1} \in B|X_n)$. Since B is arbitrary it implies $(X_0, \ldots, X_n) - X_n - X_{n+1}$, so $(X_n : n \geq 0)$ is a Markov process. □

For example, if the driving terms $w_k : k \geq 0$ used for discrete-time Kalman filtering are independent (rather than just being pairwise uncorrelated), then the state process of the Kalman filtering model has the Markov property.

4.9 Discrete-state Markov processes

This section delves further into the theory of Markov processes in the case of a discrete state space $\mathcal{S}$, assumed to be a finite or countably infinite set. Given a probability space $(\Omega, \mathcal{F}, P)$, an $\mathcal{S}$ valued random variable is defined to be a function Y mapping Ω to $\mathcal{S}$ such that $\{\omega : Y(\omega) = s\} \in \mathcal{F}$ for each $s \in \mathcal{S}$. Assume that the elements of $\mathcal{S}$ are ordered so that $\mathcal{S} = \{a_1, a_2, \ldots, a_n\}$ when $\mathcal{S}$ has finite cardinality, or $\mathcal{S} = \{a_1, a_2, a_3, \ldots\}$ when $\mathcal{S}$ has infinite cardinality. Given the ordering, an $\mathcal{S}$ valued random variable is equivalent to a positive integer valued random variable, so it is nothing exotic. Think of the probability distribution of an $\mathcal{S}$ valued random variable Y as a row vector called a probability vector: $p_Y = (P\{Y = a_1\}, P\{Y = a_2\}, \ldots)$. Similarly think of a deterministic function g on $\mathcal{S}$ as a column vector, $g = (g(a_1), g(a_2), \ldots)^T$. Since the elements of $\mathcal{S}$ may not even be numbers, it might not make sense to speak of the expected value of an $\mathcal{S}$ valued random variable. However, if g is a function mapping $\mathcal{S}$ to the reals, then $g(Y)$ is a real-valued random variable and its expectation is given by the inner product of the probability vector p_Y and the column vector g: $E[g(Y)] = \sum_{i \in \mathcal{S}} p_Y(i) g(i) = p_Y g$. A random process $X = (X_t : t \in \mathbb{T})$ is said to have state space $\mathcal{S}$ if X_t is an $\mathcal{S}$ valued random variable for each $t \in \mathbb{T}$, and the Markov property of such a random process is defined just as it is for a real-valued random process.

Let $(X_t : t \in \mathbb{T})$ be a Markov process with state space $\mathcal{S}$. For brevity denote the pmf of X at time t as $\pi(t) = (\pi_i(t) : i \in \mathcal{S})$. So $\pi_i(t) = p_X(i,t) = P\{X(t) = i\}$. The following notation is used to denote conditional probabilities:

$$P\left(X_{t_1} = j_1, \ldots, X_{t_n} = j_n \middle| X_{s_1} = i_1, \ldots, X_{s_m} = i_m\right) = p_X(j_1, t_1; \ldots; j_n, t_n | i_1, s_1; \ldots; i_m, s_m).$$

For brevity, conditional probabilities of the form $P(X_t = j|X_s = i)$ are written as $p_{ij}(s,t)$, and are called the *transition probabilities* of X.

The first order pmfs $\pi(t)$ and the transition probabilities $p_{ij}(s,t)$ determine all the finite order distributions of the Markov process as follows. Given

$$\begin{cases} t_1 < t_2 < \ldots < t_n \text{ in } \mathbb{T} \\ i_i, i_2, \ldots, i_n \in \mathcal{S} \end{cases}, \tag{4.14}$$

one writes

$$\begin{aligned} &p_X(i_1, t_1; \cdots; i_n, t_n) \\ &\quad = p_X(i_1, t_1; \cdots; i_{n-1}, t_{n-1}) p_X(i_n, t_n | i_1, t_1; \cdots; i_{n-1}, t_{n-1}) \\ &\quad = p_X(i_1, t_1; \cdots; i_{n-1}, t_{n-1}) p_{i_{n-1} i_n}(t_{n-1}, t_n). \end{aligned}$$

Application of this operation $n-2$ more times yields that

$$p_X(i_1, t_1; \cdots; i_n, t_n) = \pi_{i_1}(t_1) p_{i_1 i_2}(t_1, t_2) \cdots p_{i_{n-1} i_n}(t_{n-1}, t_n), \tag{4.15}$$

which shows that the finite order distributions of X are indeed determined by the first order pmfs and the transition probabilities. Equation (4.15) can be used to easily verify that the form (4.12) of the Markov property holds.

Given $s < t$, the collection $H(s,t)$ defined by $H(s,t) = (p_{ij}(s,t) : i,j \in \mathcal{S})$ should be thought of as a matrix, and it is called the transition probability matrix for the interval $[s,t]$. Let e denote the column vector with all ones, indexed by $\mathcal{S}$. Since $\pi(t)$ and the rows of $H(s,t)$ are probability vectors, it follows that $\pi(t)e = 1$ and $H(s,t)e = e$. Computing the distribution of X_t by summing over all possible values of X_s yields that $\pi_j(t) = \sum_i P(X_s = i, X_t = j) = \sum_i \pi_i(s) p_{ij}(s,t)$, which in matrix form yields that $\pi(t) = \pi(s)H(s,t)$ for $s,t \in \mathbb{T}, s \leq t$. Similarly, given $s < \tau < t$, computing the conditional distribution of X_t given X_s by summing over all possible values of X_τ yields

$$H(s,t) = H(s,\tau)H(\tau,t) \qquad s,\tau,t \in \mathbb{T},\ s < \tau < t. \tag{4.16}$$

The relations (4.16) are known as the Chapman–Kolmogorov equations.

A Markov process is *time-homogeneous* if the transition probabilities $p_{ij}(s,t)$ depend on s and t only through $t-s$. In that case we write $p_{ij}(t-s)$ instead of $p_{ij}(s,t)$, and $H_{ij}(t-s)$ instead of $H_{ij}(s,t)$. If the Markov process is time-homogeneous, then $\pi(s+\tau) = \pi(s)H(\tau)$ for $s, s+\tau \in \mathbb{T}$ and $\tau \geq 0$. A probability distribution π is called an *equilibrium* (or invariant) distribution if $\pi H(\tau) = \pi$ for all $\tau \geq 0$.

Recall that a random process is *stationary* if its finite order distributions are invariant with respect to translation in time. On one hand, referring to (4.15), we see that a time-homogeneous Markov process is stationary if and only if $\pi(t) = \pi$ for all t for some equilibrium distribution π. On the other hand, a Markov random process that is stationary is time-homogeneous.

Repeated application of the Chapman–Kolmogorov equations yields that $p_{ij}(s,t)$ can be expressed in terms of transition probabilities for s and t close together. For example, consider Markov processes with index set $\mathbb{Z}$. Then $H(n, k+1) = H(n,k)P(k)$ for $n \le k$, where $P(k) = H(k, k+1)$ is the one-step transition probability matrix. Fixing n and using forward recursion starting with $H(n,n) = I$, $H(n, n+1) = P(n)$, $H(n, n+2) = P(n)P(n+1)$, and so forth yields

$$H(n,l) = P(n)P(n+1)\cdots P(l-1).$$

In particular, if the chain is time-homogeneous then $H(k) = P^k$ for all k, where P is the time independent *one-step transition probability matrix,* and $\pi(l) = \pi(k)P^{l-k}$ for $l \ge k$. In this case a probability distribution π is an equilibrium distribution if and only if $\pi P = \pi$.

Example 4.8 Consider a two-stage pipeline through which packets flow, as pictured in Figure 4.8. Some assumptions about the pipeline will be made in order to model it as a simple discrete-time Markov process. Each stage has a single buffer. Normalize time so that in one unit of time a packet can make a single transition. Call the time interval between k and $k+1$ the kth "time slot," and assume that the pipeline evolves in the following way during a given slot.

If at the beginning of the slot, there are no packets in stage one, then a new packet arrives to stage one with probability a, independently of the past history of the pipeline and of the outcome at stage two.

If at the beginning of the slot, there is a packet in stage one and no packet in stage two, then the packet is transferred to stage two with probability d_1.

If at the beginning of the slot, there is a packet in stage two, then the packet departs from the stage and leaves the system with probability d_2, independently of the state or outcome of stage one.

These assumptions lead us to model the pipeline as a discrete-time Markov process with the state space $\mathcal{S} = \{00, 01, 10, 11\}$, *transition probability diagram* shown in Figure 4.9 (using the notation $\bar{x} = 1 - x$) and one-step transition probability matrix P given by

$$P = \begin{pmatrix} \bar{a} & 0 & a & 0 \\ \bar{a}d_2 & \bar{a}\bar{d}_2 & ad_2 & a\bar{d}_2 \\ 0 & d_1 & \bar{d}_1 & 0 \\ 0 & 0 & d_2 & \bar{d}_2 \end{pmatrix}.$$

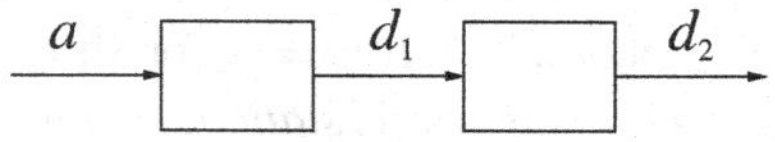

a d_1 d_2

Figure 4.8 A two-stage pipeline.

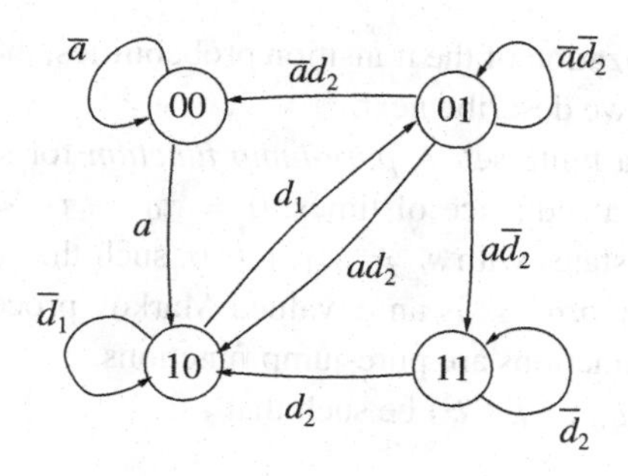

Figure 4.9 One-step transition probability diagram.

The rows of P are probability vectors. For example, the first row is the probability distribution of the state at the end of a slot, given that the state is 00 at the beginning of a slot. Now that the model is specified, let us determine the throughput rate of the pipeline.

The equilibrium probability distribution $\pi = (\pi_{00}, \pi_{01}, \pi_{10}, \pi_{11})$ is the probability vector satisfying the linear equation $\pi = \pi P$. Once π is found, the throughput rate η can be computed as follows. It is defined to be the rate (averaged over a long time) that packets transit the pipeline. Since at most two packets can be in the pipeline at a time, the following three quantities are all clearly the same, and can be taken to be the throughput rate:

The rate of arrivals to stage one;
The rate of departures from stage one (or rate of arrivals to stage two);
The rate of departures from stage two.

Focus on the first of these three quantities to obtain

$$\begin{aligned}\eta &= P\{\text{an arrival at stage 1}\}\\ &= P(\text{an arrival at stage 1}|\text{stage 1 empty at slot beginning})\\ &\quad \cdot P(\text{stage 1 empty at slot beginning})\\ &= a(\pi_{00} + \pi_{01}).\end{aligned}$$

Similarly, by focusing on departures from stage 1, obtain $\eta = d_1\pi_{10}$. Finally, by focusing on departures from stage 2, obtain $\eta = d_2(\pi_{01} + \pi_{11})$. These three expressions for η must agree.

Consider the numerical example $a = d_1 = d_2 = 0.5$. The equation $\pi = \pi P$ yields that π is proportional to the vector $(1, 2, 3, 1)$. Applying the fact that π is a probability distribution yields that $\pi = (1/7, 2/7, 3/7, 1/7)$. Therefore $\eta = 3/14 = 0.214\ldots$.

In the remainder of this section we assume that X is a continuous-time, finite-state Markov process. The transition probabilities for arbitrary time intervals can be described in terms of the transition probabilities over arbitrarily short time intervals. By saving only a linearization of the transition probabilities, the concept of generator matrix arises naturally, as we describe next.

Let $\mathcal{S}$ be a finite set. A *pure-jump function* for state space $\mathcal{S}$ is $x : \mathcal{R}_+ \to \mathcal{S}$ such that there is a sequence of times, $0 = \tau_0 < \tau_1 < \ldots$ with $\lim_{i\to\infty} \tau_i = \infty$, and a sequence of states with $s_i \neq s_{i+1}$, $i \geq 0$, such that $x(t) = s_i$ for $\tau_i \leq t < \tau_{i+1}$. A *pure-jump Markov process* is an $\mathcal{S}$ valued Markov process such that, with probability one, the sample functions are pure-jump functions.

Let $Q = (q_{ij} : i,j \in \mathcal{S})$ be such that

$$\begin{array}{ll} q_{ij} \geq 0 & i,j \in \mathcal{S},\ i \neq j \\ q_{ii} = -\sum_{j\in\mathcal{S},j\neq i} q_{ij} & i \in \mathcal{S}. \end{array} \quad (4.17)$$

An example for state space $\mathcal{S} = \{1, 2, 3\}$ is

$$Q = \begin{pmatrix} -1 & 0.5 & 0.5 \\ 1 & -2 & 1 \\ 0 & 1 & -1 \end{pmatrix},$$

and this matrix Q can be represented by the *transition rate diagram* shown in Figure 4.10. A pure-jump, time-homogeneous Markov process X has *generator matrix* Q if the transition probabilities $(p_{ij}(\tau))$ satisfy

$$\lim_{h\searrow 0} (p_{ij}(h) - I_{\{i=j\}})/h = q_{ij} \quad i,j \in \mathcal{S}, \quad (4.18)$$

or equivalently

$$p_{ij}(h) = I_{\{i=j\}} + hq_{ij} + o(h) \quad i,j \in \mathcal{S}, \quad (4.19)$$

where $o(h)$ represents a quantity such that $\lim_{h\to 0} o(h)/h = 0$. For the example this means that the transition probability matrix for a time interval of duration h is given by

$$\begin{pmatrix} 1-h & 0.5h & 0.5h \\ h & 1-2h & h \\ 0 & h & 1-h \end{pmatrix} + \begin{pmatrix} o(h) & o(h) & o(h) \\ o(h) & o(h) & o(h) \\ o(h) & o(h) & o(h) \end{pmatrix}.$$

For small enough h, the rows of the first matrix are probability distributions, owing to the assumptions on the generator matrix Q.

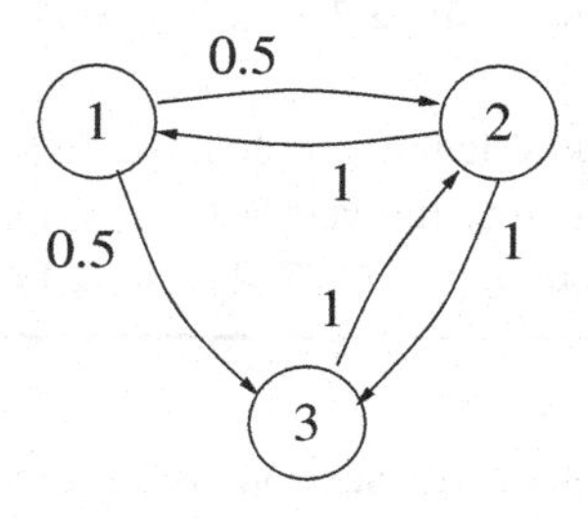

Figure 4.10 Transition rate diagram for a continuous-time Markov process.

Proposition 4.8 *Given a matrix Q satisfying (4.17), and a probability distribution $\pi(0) = (\pi_i(0) : i \in \mathcal{S})$, there is a pure-jump, time-homogeneous Markov process with generator matrix Q and initial distribution $\pi(0)$. The finite order distributions of the process are uniquely determined by $\pi(0)$ and Q.*

The first order distributions and the transition probabilities can be derived from Q and an initial distribution $\pi(0)$ by solving differential equations, derived as follows. Fix $t > 0$ and let h be a small positive number. The Chapman–Kolmogorov equations imply that

$$\frac{\pi_j(t+h) - \pi_j(t)}{h} = \sum_{i \in \mathcal{S}} \pi_i(t) \left(\frac{p_{ij}(h) - I_{\{i=j\}}}{h} \right). \tag{4.20}$$

Letting h converge to zero yields the differential equation:

$$\frac{\partial \pi_j(t)}{\partial t} = \sum_{i \in \mathcal{S}} \pi_i(t) q_{ij}, \tag{4.21}$$

or, in matrix notation, $\frac{\partial \pi(t)}{\partial t} = \pi(t)Q$. These equations, known as the Kolmogorov forward equations, can be rewritten as

$$\frac{\partial \pi_j(t)}{\partial t} = \sum_{i \in \mathcal{S}, i \neq j} \pi_i(t) q_{ij} - \sum_{i \in \mathcal{S}, i \neq j} \pi_j(t) q_{ji}, \tag{4.22}$$

which shows that the rate change of the probability of being at state j is the rate of *probability flow* into state j minus the rate of probability flow out of state j.

The Kolmogorov forward equations (4.21), or equivalently, (4.22), for $(\pi(t) : t \geq 0)$ take as input data the initial distribution $\pi(0)$ and the generator matrix Q. These equations include as special cases differential equations for the transition probability functions, $p_{i,j}(t)$. After all, for i_o fixed, $p_{i_o,j}(t) = P(X_t = j | X_0 = i_o) = \pi_j(t)$ if the initial distribution of $(\pi(t))$ is $\pi_i(0) = I_{\{i=i_o\}}$. Thus, (4.21) specializes to

$$\frac{\partial p_{i_o,j}(t)}{\partial t} = \sum_{i \in \mathcal{S}} p_{i_o,i}(t) q_{i,j} \quad p_{i_o,i}(0) = I_{\{i=i_o\}}. \tag{4.23}$$

Recall that $H(t)$ is the matrix with (i,j)th element equal to $p_{i,j}(t)$. Therefore, for any i_o fixed, the differential equation (4.23) determines the i_oth row of $(H(t); t \geq 0)$. The equations (4.23) for all choices of i_o can be written together in the following matrix form: $\frac{\partial H(t)}{\partial t} = H(t)Q$ with $H(0)$ equal to the identity matrix. An occasionally useful general expression for the solution is $H(t) = \exp(Qt) \triangleq \sum_{n=0}^{\infty} \frac{t^n Q^n}{n!}$.

Example 4.9 Consider the two-state, continuous-time Markov process with the transition rate diagram shown in Figure 4.11 for some positive constants α and β. The generator matrix is given by

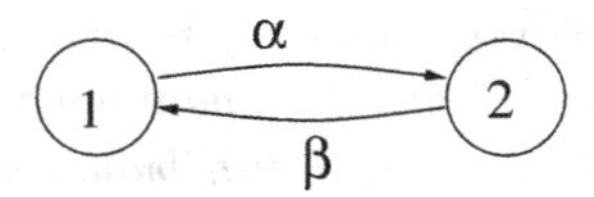

Figure 4.11 Transition rate diagram for a two-state continuous-time Markov process.

$$Q = \begin{bmatrix} -\alpha & \alpha \\ \beta & -\beta \end{bmatrix}.$$

Let us solve the forward Kolmogorov equation for a given initial distribution $\pi(0)$. The equation for $\pi_1(t)$ is

$$\frac{\partial \pi_1(t)}{\partial t} = -\alpha\pi_1(t) + \beta\pi_2(t); \quad \pi_1(0) \text{ given.}$$

But $\pi_1(t) = 1 - \pi_2(t)$, so

$$\frac{\partial \pi_1(t)}{\partial t} = -(\alpha+\beta)\pi_1(t) + \beta; \quad \pi_1(0) \text{ given.}$$

By differentiation we check that this equation has the solution

$$\begin{aligned} \pi_1(t) &= \pi_1(0)e^{-(\alpha+\beta)t} + \int_0^t e^{-(\alpha+\beta)(t-s)}\beta ds \\ &= \pi_1(0)e^{-(\alpha+\beta)t} + \frac{\beta}{\alpha+\beta}(1 - e^{-(\alpha+\beta)t}), \end{aligned}$$

so that

$$\pi(t) = \pi(0)e^{-(\alpha+\beta)t} + \left(\frac{\beta}{\alpha+\beta}, \frac{\alpha}{\alpha+\beta}\right)(1 - e^{-(\alpha+\beta)t}). \tag{4.24}$$

For any initial distribution $\pi(0)$,

$$\lim_{t\to\infty} \pi(t) = \left(\frac{\beta}{\alpha+\beta}, \frac{\alpha}{\alpha+\beta}\right).$$

The rate of convergence is exponential, with rate parameter $\alpha+\beta$, and the limiting distribution is the unique probability distribution satisfying $\pi Q = 0$.

By specializing (4.24) we determine $H(t)$. Specifically, $H(t)$ is a 2×2 matrix; its top row is $\pi(t)$ for the initial condition $\pi(0) = (1, 0)$; its bottom row is $\pi(t)$ for the initial condition $\pi(0) = (0, 1)$; the result is:

$$H(t) = \begin{pmatrix} \frac{\alpha e^{-(\alpha+\beta)t}+\beta}{\alpha+\beta} & \frac{\alpha(1-e^{-(\alpha+\beta)t})}{\alpha+\beta} \\ \frac{\beta(1-e^{-(\alpha+\beta)t})}{\alpha+\beta} & \frac{\alpha+\beta e^{-(\alpha+\beta)t}}{\alpha+\beta} \end{pmatrix}. \tag{4.25}$$

Note that $H(t)$ is a transition probability matrix for each $t \geq 0$, $H(0)$ is the 2×2 identity matrix; each row of $\lim_{t\to\infty} H(t)$ is equal to $\lim_{t\to\infty} \pi(t)$.

4.10 Space-time structure of discrete-state Markov processes

The previous section showed the distribution of a time-homogeneous, discrete-state Markov process can be specified by an initial probability distribution, and either a one-step transition probability matrix P (for discrete-time processes) or a generator matrix Q (for continuous-time processes). Another way to describe these processes is to specify the space-time structure, which is simply the sequences of states visited and how long each state is visited. The space-time structure is discussed first for discrete-time processes, and then for continuous-time processes. One benefit is to show how little difference there is between discrete-time and continuous-time processes.

Let $(X_k : k \in \mathbb{Z}_+)$ be a time-homogeneous Markov process with one-step transition probability matrix P. Let T_k denote the time that elapses between the kth and $k+1th$ jumps of X, and let $X^J(k)$ denote the state after k jumps, see Figure 4.12. More precisely, the *holding times* are defined by

$$T_0 = \min\{t \geq 0 : X(t) \neq X(0)\}, \tag{4.26}$$

$$T_k = \min\{t \geq 0 : X(T_0 + \ldots + T_{k-1} + t) \neq X(T_0 + \ldots + T_{k-1})\}, \tag{4.27}$$

and the *jump process* $X^J = (X^J(k) : k \geq 0)$ is defined by

$$X^J(0) = X(0) \text{ and } X^J(k) = X(T_0 + \ldots + T_{k-1}). \tag{4.28}$$

Clearly the holding times and jump process contain all the information needed to construct X, and vice versa. Thus, the following description of the joint distribution of the holding times and the jump process characterizes the distribution of X.

Proposition 4.9 *Let $X = (X(k) : k \in \mathbb{Z}_+)$ be a time-homogeneous Markov process with one-step transition probability matrix P.*

(a) The jump process X^J is itself a time-homogeneous Markov process, and its one-step transition probabilities are given by $p^J_{ij} = p_{ij}/(1-p_{ii})$ for $i \neq j$, and $p^J_{ii} = 0$, $i,j \in \mathcal{S}$.

(b) Given $X(0)$, $X^J(1)$ is conditionally independent of T_0.

(c) Given $(X^J(0), \ldots, X^J(n)) = (j_0, \ldots, j_n)$, the variables $T_0, \ldots, T_n$ are conditionally independent, and the conditional distribution of T_l is geometric with parameter $p_{j_l j_l}$:

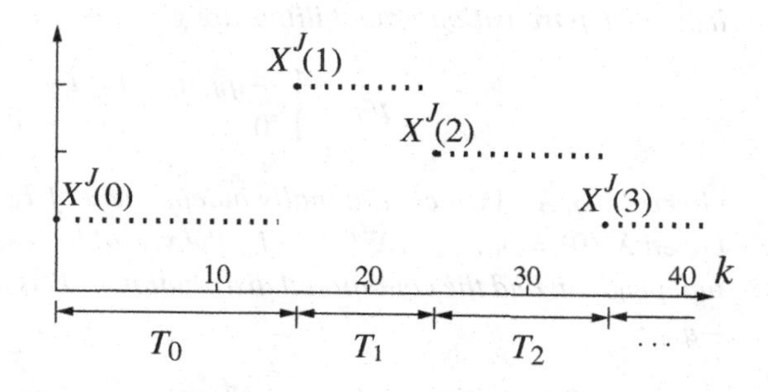

Figure 4.12 Illustration of jump process and holding times.

$$P(T_l = k | X^J(0) = j_0, \ldots, X^J(n) = j_n) = p_{j_l j_l}^{k-1}(1 - p_{j_l j_l}) \quad 0 \le l \le n,\ k \ge 1.$$

Proof Observe that if $X(0) = i$, then

$$\{T_0 = k, X^J(1) = j\} = \{X(1) = i, X(2) = i, \ldots, X(k-1) = i, X(k) = j\},$$

so

$$P(T_0 = k, X^J(1) = j | X(0) = i) = p_{ii}^{k-1} p_{ij} = \left[(1 - p_{ii}) p_{ii}^{k-1}\right] p_{ij}^J. \tag{4.29}$$

Because for i fixed the last expression in (4.29) displays the product of two probability distributions, conclude that *given* $X(0) = i$:

T_0 has distribution $((1 - p_{ii}) p_{ii}^{k-1} : k \ge 1)$, the geometric distribution of mean $1/(1 - p_{ii})$,
$X^J(1)$ has distribution $(p_{ij}^J : j \in \mathcal{S})$ (i fixed),
T_0 and $X^J(1)$ are independent.

More generally, check that

$$P\left(X^J(1) = j_1, \ldots, X^J(n) = j_n, T_o = k_0, \ldots, T_n = k_n \middle| X^J(0) = i\right)$$
$$= p_{ij_1}^J p_{j_1 j_2}^J \cdots p_{j_{n-1} j_n}^J \prod_{l=0}^{n} \left(p_{j_l j_l}^{k_l - 1}(1 - p_{j_l j_l})\right).$$

This establishes the proposition. □

Next we consider the space-time structure of time-homogeneous continuous-time pure-jump Markov processes. Essentially the only difference between the discrete- and continuous-time Markov processes is that the holding times for the continuous-time processes are exponentially distributed rather than geometrically distributed. Indeed, we define the holding times $T_k, k \ge 0$ and the jump process X^J using (4.26)–(4.28) as before.

Proposition 4.10 *Let $X = (X(t) : t \in \mathbb{R}_+)$ be a time-homogeneous, pure-jump Markov process with generator matrix Q. Then*

(a) *The jump process X^J is a discrete-time, time-homogeneous Markov process, and its one-step transition probabilities are given by*

$$p_{ij}^J = \begin{cases} -q_{ij}/q_{ii} & \text{for } i \ne j \\ 0 & \text{for } i = j \end{cases}. \tag{4.30}$$

(b) *Given $X(0)$, $X^J(1)$ is conditionally independent of T_0.*

(c) *Given $X^J(0) = j_0, \ldots, X^J(n) = j_n$, the variables $T_0, \ldots, T_n$ are conditionally independent, and the conditional distribution of T_l is exponential with parameter $-q_{j_l j_l}$:*

$$P(T_l \ge c | X^J(0) = j_0, \ldots, X^J(n) = j_n) = exp(c q_{j_l j_l}) \quad 0 \le l \le n.$$

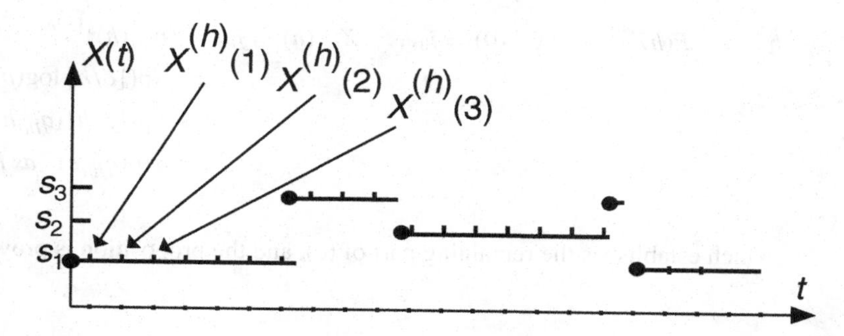

Figure 4.13 Sampling of a pure-jump function.

Proof Fix $h > 0$ and define the "sampled" process $X^{(h)}$ by $X^{(h)}(k) = X(hk)$ for $k \geq 0$, see Figure 4.13. Then $X^{(h)}$ is a discrete-time Markov process with one-step transition probabilities $p_{ij}(h)$ (the transition probabilities for the original process for an interval of length h). Let $(T_k^{(h)} : k \geq 0)$ denote the sequence of holding times and $(X^{J,h}(k) : k \geq 0)$ the jump process for the process $X^{(h)}$.

The assumption that with probability one the sample paths of X are pure-jump functions, implies that *with probability one:*

$$\lim_{h \to 0} (X^{J,h}(0), X^{J,h}(1), \ldots, X^{J,h}(n), hT_0^{(h)}, hT_1^{(h)}, \ldots, hT_n^{(h)})$$
$$= (X^J(0), X^J(1), \ldots, X^J(n), T_0, T_1, \ldots, T_n). \tag{4.31}$$

Since convergence with probability one implies convergence in distribution, the goal of identifying the distribution of the random vector on the righthand side of (4.31) can be accomplished by identifying the limit of the distribution of the vector on the left.

First, the limiting distribution of the process $X^{J,h}$ is identified. Since $X^{(h)}$ has one-step transition probabilities $p_{ij}(h)$, the formula for the jump process probabilities for discrete-time processes (see Proposition 4.9, part (a)) yields that the one-step transition probabilities $p_{ij}^{J,h}$ for $X^{(J,h)}$ are given by

$$\begin{aligned} p_{ij}^{J,h} &= \frac{p_{ij}(h)}{1 - p_{ii}(h)} \\ &= \frac{p_{ij}(h)/h}{(1 - p_{ii}(h))/h} \to \frac{q_{ij}}{-q_{ii}} \text{ as } h \to 0 \end{aligned} \tag{4.32}$$

for $i \neq j$, where the limit indicated in (4.32) follows from the definition (4.18) of the generator matrix Q. Thus, the limiting distribution of $X^{J,h}$ is that of a Markov process with one-step transition probabilities given by (4.30), establishing part (a) of the proposition. The conditional independence properties stated in (b) and (c) of the proposition follow in the limit from the corresponding properties for the jump process $X^{J,h}$ guaranteed by Proposition 4.9. Finally, since $\log(1 + \theta) = \theta + o(\theta)$ by Taylor's formula, we have for all $c \geq 0$ that

$$\begin{aligned}P(hT_l^{(h)} > c|X^{J,h}(0) = j_0, \ldots, X^{J,h}(n) = j_n) &= (p_{j_l j_l}(h))^{\lfloor c/h \rfloor} \\ &= \exp(\lfloor c/h \rfloor \log(p_{j_l j_l}(h))) \\ &= \exp(\lfloor c/h \rfloor (q_{j_l j_l} h + o(h))) \\ &\to \exp(q_{j_l j_l} c) \quad as\ h \to 0,\end{aligned}$$

which establishes the remaining part of (c), and the proposition is proved. □

Problems

4.1 Event probabilities for a simple random process
Define the random process X by $X_t = 2A + Bt$ where A and B are independent random variables with $P\{A = 1\} = P\{A = -1\} = P\{B = 1\} = P\{B = -1\} = 0.5$.
(a) Sketch the possible sample functions.
(b) Find $P\{X_t \geq 0\}$ for all t.
(c) Find $P\{X_t \geq 0 \text{ for all } t\}$.

4.2 Correlation function of a product Let Y and Z be independent random processes with $R_Y(s,t) = 2\exp(-|s-t|)\cos(2\pi f(s-t))$ and $R_Z(s,t) = 9 + \exp(-3|s-t|^4)$. Find the autocorrelation function $R_X(s,t)$ where $X_t = Y_t Z_t$.

4.3 A sinusoidal random process Let $X_t = A\cos(2\pi Vt + \Theta)$ where the amplitude A has mean 2 and variance 4, the frequency V in Hertz is uniform on $[0,5]$, and the phase Θ is uniform on $[0, 2\pi]$. Furthermore, suppose A, V and Θ are independent. Find the mean function $\mu_X(t)$ and autocorrelation function $R_X(s,t)$. Is X WSS?

4.4 Another sinusoidal random process Suppose that X_1 and X_2 are random variables such that $EX_1 = EX_2 = EX_1X_2 = 0$ and $\text{Var}(X_1) = \text{Var}(X_2) = \sigma^2$. Define $Y_t = X_1\cos(2\pi t) - X_2\sin(2\pi t)$.
(a) Is the random process Y necessarily wide-sense stationary?
(b) Give an example of random variables X_1 and X_2 satisfying the given conditions such that Y is stationary.
(c) Give an example of random variables X_1 and X_2 satisfying the given conditions such that Y is not (strict sense) stationary.

4.5 A random line Let $X = (X_t : t \in \mathbb{R})$ be a random process such that $X_t = R - St$ for all t, where R and S are independent random variables, having the Rayleigh distribution with positive parameters σ_R^2 and σ_S^2, respectively.
(a) Indicate three typical sample paths of X in a single sketch. Describe in words the set of possible sample paths of X.
(b) Is X a Markov process? Why or why not?
(c) Does X have independent increments? Why or why not?
(d) Let A denote the area of the triangle bounded by portions of the coordinate axes and the graph of X. Find $E[A]$. Simplify your answer as much as possible.

4.6 A random process corresponding to a random parabola Define a random process X by $X_t = A + Bt + t^2$, where A and B are independent, $N(0,1)$ random variables.
(a) Find $\hat{E}[X_5|X_1]$, the linear minimum mean square error (LMMSE) estimator of X_5 given X_1, and compute the mean square error.

(b) Find the MMSE (possibly nonlinear) estimator of X_5 given X_1, and compute the mean square error.

(c) Find $\hat{E}[X_5|X_0, X_1]$ and compute the mean square error. (Hint: Can do by inspection.)

4.7 Some probabilities for a Brownian motion Let $(W_t : t \geq 1)$ be a standard Brownian motion.

(a) Express $P\{W_3 \geq \frac{W_2+W_4}{2} + 1\}$ in terms of the Q function.

(b) Find the limit of the distribution of $\frac{W_t^2}{t}$ as $t \to \infty$.

4.8 Brownian motion: Ascension and smoothing Let W be a Brownian motion process and suppose $0 \leq r < s < t$.

(a) Find $P\{W_r \leq W_s \leq W_t\}$.

(b) Find $E[W_s|W_r, W_t]$. (This part is unrelated to part (a).)

4.9 Brownian bridge Let $W = (W_t : t \geq 0)$ be a standard Brownian motion (i.e. a Brownian motion with parameter $\sigma^2 = 1$). Let $B_t = W_t - tW_1$ for $0 \leq t \leq 1$. The process $B = (B_t : 0 \leq t \leq 1)$ is called a *Brownian bridge* process. Like W, B is a mean zero Gaussian random process.

(a) Sketch a typical sample path of W, and the corresponding sample path of B.

(b) Find the autocorrelation function of B.

(c) Is B a Markov process?

(d) Show that B is independent of the random variable W_1. (This means that for any finite collection, $t_1, \ldots, t_n \in [0, 1]$, the random vector $(B_{t_1}, \ldots, B_{t_n})^T$ is independent of W_1.)

(e) (Due to J.L. Doob.) Let $X_t = (1-t)W_{\frac{t}{1-t}}$, for $0 \leq t < 1$ and let $X_1 = 0$. Let X denote the random process $X = (X_t : 0 \leq t \leq 1)$. Like W, X is a mean zero, Gaussian random process. Find the autocorrelation function of X. Can you draw any conclusions?

4.10 Empirical distribution functions as random processes Let $X_1, X_2, \ldots$ be independent random variables, all with the same CDF F. For $n \geq 1$, the empirical CDF for n observations is defined by $\widehat{F}_n(t) = \frac{1}{n}\sum_{k=1}^{n} I_{\{X_k \leq t\}}$ for $t \in \mathbb{R}$.

(a) Find the mean function and autocovariance function of the random process $(\widehat{F}_n(t) : t \in \mathbb{R})$ for fixed n. (Hint: For computing the autocovariance, it may help to treat the cases $s \leq t$ and $s \geq t$ separately.)

(b) Explain why, for each $t \in \mathbb{R}$, $\lim_{n\to\infty} \widehat{F}_n(t) = F(t)$ almost surely.

(c) Let $D_n = \sup_{t\in\mathbb{R}} |\widehat{F}_n(t) - F(t)|$, so that D_n is a measure of distance between $\widehat{F}_n$ and F. Suppose the CDF F is continuous and strictly increasing. Show that the distribution of D_n is the same as it would be if the X_ns were all uniformly distributed on the interval $[0, 1]$. (Hint: Let $U_k = F(X_k)$. Show that the Us are uniformly distributed on the interval $[0, 1]$, let $\widehat{G}_n$ be the empirical CDF for the Us and let G be the CDF of the Us. Show that if $F(t) = v$, then $|\widehat{F}_n(t) - F(t)| = |\widehat{G}_n(v) - G(v)|$. Then complete the proof.)

(d) Let $X_n(t) = \sqrt{n}(\widehat{F}_n(t) - F(t))$ for $t \in \mathbb{R}$. Find the limit in distribution of $X_n(t)$ for t fixed as $n \to \infty$.

(e) (Note that $\sqrt{n}D_n = \sup_{t\in\mathbb{R}} |X_n(t)|$.) Show that in the case the Xs are uniformly distributed on the interval $[0, 1]$, the autocorrelation function of the process $(X_n(t) : 0 \leq t \leq 1)$ is the same as for a Brownian bridge (discussed in the previous problem). (Note: The distance D_n is known as the Kolmogorov–Smirnov statistic, and by pursuing the method of this problem further, the limiting distribution of $\sqrt{n}D_n$ can be found and it is equal to the distribution of the maximum magnitude of a Brownian bridge, a result due to J.L. Doob.)

4.11 Some Poisson process calculations Let $N = (N_t : t \geq 0)$ be a Poisson process with rate $\lambda > 0$.
(a) Give a simple expression for $P(N_1 \geq 1 | N_2 = 2)$ in terms of λ.
(b) Give a simple expression for $P(N_2 = 2 | N_1 \geq 1)$ in terms of λ.
(c) Let $X_t = N_t^2$. Is $X = (X_t : t \geq 0)$ a time-homogeneous Markov process? If so, give the transition probabilities $p_{ij}(\tau)$. If not, explain.

4.12 MMSE prediction for a Gaussian process based on two observations Let X be a mean zero stationary Gaussian process with $R_X(\tau) = 5\cos(\frac{\pi\tau}{2})3^{-|\tau|}$.
(a) Find the covariance matrix of $(X(2), X(3), X(4))^T$.
(b) Find $E[X(4)|X(2)]$.
(c) Find $E[X(4)|X(2), X(3)]$.

4.13 A simple discrete-time random process Let $U = (U_n : n \in \mathbb{Z})$ consist of independent random variables, each uniformly distributed on the interval $[0, 1]$. and let $X = (X_k : k \in \mathbb{Z}\}$ be defined by $X_k = \max\{U_{k-1}, U_k\}$.
(a) Sketch a typical sample path of the process X.
(b) Is X stationary?
(c) Is X Markov?
(d) Describe the first order distributions of X.
(e) Describe the second order distributions of X.

4.14 Poisson process probabilities Consider a Poisson process with rate $\lambda > 0$.
(a) Find the probability that there is (exactly) one count in each of the three intervals [0,1], [1,2], and [2,3].
(b) Find the probability that there are two counts in the interval $[0, 2]$ and two counts in the interval $[1, 3]$. (Note: your answer to part (b) should be larger than your answer to part (a)).
(c) Find the probability that there are two counts in the interval [1,2], given that there are two counts in the interval [0,2] and two counts in the interval [1,3].

4.15 Sliding function of an iid Poisson sequence Let $X = (X_k : k \in Z)$ be a random process such that the X_i are independent, Poisson random variables with mean λ, for some $\lambda > 0$. Let $Y = (Y_k : k \in Z)$ be the random process defined by $Y_k = X_k + X_{k+1}$.
(a) Show that Y_k is a Poisson random variable with parameter 2λ for each k.
(b) Show that X is a stationary random process.
(c) Is Y a stationary random process? Justify your answer.

4.16 Adding jointly stationary Gaussian processes Let X and Y be jointly stationary, jointly Gaussian random processes with mean zero, autocorrelation functions $R_X(t) = R_Y(t) = \exp(-|t|)$, and cross-correlation function $R_{XY}(t) = (0.5)\exp(-|t-3|)$.
(a) Let $Z(t) = (X(t) + Y(t))/2$ for all t. Find the autocorrelation function of Z.
(b) Is Z a stationary random process? Explain.
(c) Find $P\{X(1) \leq 5Y(2) + 1\}$. You may express your answer in terms of the standard normal cumulative distribution function Φ.

4.17 Invariance of properties under transformations Let $X = (X_n : n \in \mathbb{Z})$, $Y = (Y_n : n \in \mathbb{Z})$, and $Z = (Z_n : n \in \mathbb{Z})$ be random processes such that $Y_n = X_n^2$ for all n and $Z_n = X_n^3$ for all n. Determine whether each of the following statements is always true. If true, give a justification. If not, give a simple counter example.
(a) If X is Markov then Y is Markov.
(b) If X is Markov then Z is Markov.
(c) If Y is Markov then X is Markov.
(d) If X is stationary then Y is stationary.
(e) If Y is stationary then X is stationary.
(f) If X is wide sense stationary then Y is wide sense stationary.
(g) If X has independent increments then Y has independent increments.
(h) If X is a martingale then Z is a martingale.

4.18 A linear evolution equation with random coefficients Let the variables $A_k, B_k, k \geq 0$ be mutually independent with mean zero. Let A_k have variance σ_A^2 and let B_k have variance σ_B^2 for all k. Define a discrete-time random process Y by $Y = (Y_k : k \geq 0)$, such that $Y_0 = 0$ and $Y_{k+1} = A_k Y_k + B_k$ for $k \geq 0$.
(a) Find a recursive method for computing $P_k = E[(Y_k)^2]$ for $k \geq 0$.
(b) Is Y a Markov process? Explain.
(c) Does Y have independent increments? Explain.
(d) Find the autocorrelation function of Y. (You can use the second moments (P_k) in expressing your answer.)
(e) Find the corresponding linear innovations sequence $(\widetilde{Y}_k : k \geq 1)$.

4.19 On an $M/D/$infinity system Suppose customers enter a service system according to a Poisson point process on $\mathbb{R}$ of rate λ, meaning that the number of arrivals, $N(a, b]$, in an interval $(a, b]$, has the Poisson distribution with mean $\lambda(b-a)$, and the numbers of arrivals in disjoint intervals are independent. Suppose each customer stays in the system for one unit of time, independently of other customers. Because the arrival process is memoryless, because the service times are deterministic, and because the customers are served simultaneously, corresponding to infinitely many servers, this queueing system is called an $M/D/\infty$ queueing system. The number of customers in the system at time t is given by $X_t = N(t-1, t]$.
(a) Find the mean and autocovariance function of X.
(b) Is X stationary? Is X wide sense stationary?
(c) Is X a Markov process?
(d) Find a simple expression for $P\{X_t = 0 \text{ for } t \in [0, 1]\}$ in terms of λ.
(e) Find a simple expression for $P\{X_t > 0 \text{ for } t \in [0, 1]\}$ in terms of λ.

4.20 A Poisson spacing probability Let $N = (N_t : t \geq 0)$ be a Poisson process with some rate $\lambda > 0$. For $t \geq 0$, let A_t be the event that during the interval $[0, t]$ no two arrivals in the interval are closer than one unit of time apart. Let $x(t) = P(A_t)$.
(a) Find $x(t)$ for $0 \leq t \leq 1$.
(b) Derive a differential equation for $(x(t) : t \geq 1)$ which expresses $x'(t)$ as a function of $x(t)$ and $x(t-1)$. Begin by supposing $t \geq 1$ and h is a small positive constant, and writing an expression for $x(t+h)$ in terms of $x(t)$ and $x(t-1)$. (This is a linear differential

equation with a delay term. From the viewpoint of solving such differential equations, we view the initial condition of the equation as the waveform $(x(t) : 0 \leq t \leq 1)$. Since x is determined over $[0, 1]$ in part (a), the differential equation can then be used to solve, at least numerically, for x over the interval $[1, 2]$, then over the interval $[2, 3]$, and so on, to determine $x(t)$ for all $t \geq 0$. Moreover, this shows that the solution $(x(t) : t \geq 0)$ is an increasing function of its initial value, $(x(t) : 0 \leq t \leq 1)$. This monotonicity is different from monotonicity with respect to time.)
(c) Give equations that identify $\theta^* > 0$ and constants c_0 and c_1 so that $c_0 \leq x(t)e^{\theta^* t} \leq c_1$ for all $t \geq 0$. (Hint: Use the fact that there is a solution of the differential equation found in part (b), but not satisfying the initial condition over $[0, 1]$ found in part (a), of the form $y(t) = e^{-\theta^* t}$ for some $\theta^* > 0$, and use the monotonicity property identified in part (b).)
(d) The conditional probability of A_t, given there are exactly k arrivals during $[0, t]$, is $\left(\frac{t-k+1}{t}\right)^k$ for $0 \leq k \leq \lceil t \rceil$ (Why?). Use that fact to give a series representation for $(x(t) : t \geq 0)$.

4.21 Hitting the corners of a triangle Consider a discrete-time Markov process $(X_k : k \geq 0)$, with state space $\{1, 2, 3, 4, 5, 6\}$. Suppose the states are arranged in the triangle shown,

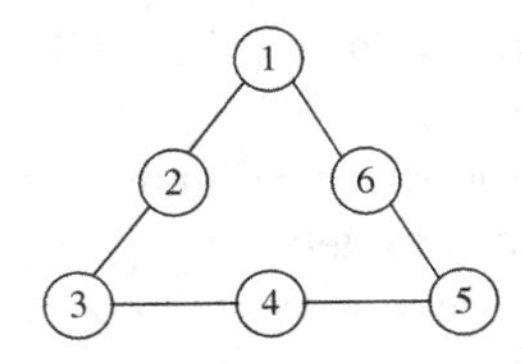

and given $X_k = i$, the next state X_{k+1} is one of the two neighbors of i, selected with probability 0.5 each. Suppose $P\{X_0 = 1\} = 1$.
(a) Let $\tau_B = \min\{k : X_k \in \{3, 4, 5\}\}$. So τ_B is the time the base of the triangle is first reached. Find $E[\tau_B]$.
(b) Let $\tau_3 = \min\{k : X_k = 3\}$. Find $E[\tau_3]$.
(c) Let τ_C be the first time $k \geq 1$ such that both states 3 and 5 have been visited by time k. Find $E[\tau_C]$. (Hint: Use results of (a) and (b) and symmetry.)
(d) Let τ_R denote the first time $k \geq \tau_C$ such that $X_k = 1$. That is, τ_R is the first time the process returns to vertex 1 of the triangle after reaching both of the other vertices. Find $E[\tau_R]$. (Hint: Use results of (c) and (b) and symmetry.)

4.22 A fly on a cube Consider a cube with vertices 000, 001, 010, 100, 110, 101, 011, 111. Suppose a fly walks along edges of the cube from vertex to vertex, and for any integer $t \geq 0$, let X_t denote which vertex the fly is at at time t. Assume $X = (X_t : t \geq 0)$ is a discrete-time Markov process, such that given X_t, the next state X_{t+1} is equally likely to be any one of the three vertices neighboring X_t.
(a) Sketch the one-step transition probability diagram for X.
(b) Let Y_t denote the distance of X_t, measured in number of hops, between vertex 000 and X_t. For example, if $X_t = 101$, then $Y_t = 2$. The process Y is a Markov process with states 0,1,2, and 3. Sketch the one-step transition probability diagram for Y.

(c) Suppose the fly begins at vertex 000 at time zero. Let τ be the first time that X returns to vertex 000 after time 0, or equivalently, the first time that Y returns to 0 after time 0. Find $E[\tau]$.

4.23 Time elapsed since Bernoulli renewals Let $U = (U_k : k \in \mathbb{Z})$ be such that for some $p \in (0, 1)$, the random variables U_k are independent, with each having the Bernoulli distribution with parameter p. Interpret $U_k = 1$ to mean that a renewal, or replacement, of some part takes place at time k. For $k \in \mathbb{Z}$, let $X_k = \min\{i \geq 1 : U_{k-i} = 1\}$. In words, X_k is the time elapsed since the last renewal strictly before time k.

(a) The process X is a time-homogeneous Markov process. Indicate a suitable state space, and describe the one-step transition probabilities.

(b) Find the distribution of X_k for k fixed.

(c) Is X a stationary random process? Explain.

(d) Find the k-step transition probabilities, $p_{i,j}(k) = P\{X_{n+k} = j | X_n = i\}$.

4.24 A random process created by interpolation Let $U = (U_k : k \in \mathbb{Z})$ such that the U_k are independent, and each is uniformly distributed on the interval $[0, 1]$. Let $X = (X_t : t \in \mathbb{R})$ denote the continuous time random process obtained by linearly interpolating between the Us. Specifically, $X_n = U_n$ for any $n \in \mathbb{Z}$, and X_t is affine on each interval of the form $[n, n + 1]$ for $n \in \mathbb{Z}$.

(a) Sketch a sample path of U and a corresponding sample path of X.

(b) Let $t \in \mathbb{R}$. Find and sketch the first order marginal density, $f_{X,1}(x, t)$. (Hint: Let $n = \lfloor t \rfloor$ and $a = t - n$, so that $t = n + a$. Then $X_t = (1 - a)U_n + aU_{n+1}$. It is helpful to consider the cases $0 \leq a \leq 0.5$ and $0.5 < a < 1$ separately. For brevity, you need only consider the case $0 \leq a \leq 0.5$.)

(c) Is the random process X WSS? Justify your answer.

(d) Find $P\{\max_{0 \leq t \leq 10} X_t \leq 0.5\}$.

4.25 Reinforcing samples (Due to G. Polya) Suppose at time $k = 2$, there is a bag with two balls in it, one orange and one blue. During each time step between k and $k+1$, one of the balls is selected from the bag at random, with all balls in the bag having equal probability. That ball, and a new ball of the same color, are both put into the bag. Thus, at time k there are k balls in the bag, for all $k \geq 2$. Let X_k denote the number of blue balls in the bag at time k.

(a) Is $X = (X_k : k \geq 2)$ a Markov process?

(b) Let $M_k = \frac{X_k}{k}$. Thus, M_k is the fraction of balls in the bag at time k that are blue. Determine whether $M = (M_k : k \geq 2)$ is a martingale.

(c) By the theory of martingales, since M is a bounded martingale, it converges a.s. to some random variable M_∞. Let $V_k = M_k(1 - M_k)$. Show that $E[V_{k+1}|V_k] = \frac{k(k+2)}{(k+1)^2} V_k$, and therefore that $E[V_k] = \frac{(k+1)}{6k}$. It follows that $\text{Var}(\lim_{k\to\infty} M_k) = \frac{1}{12}$.

(d) More concretely, find the distribution of M_k for each k, and then identify the distribution of the limit random variable, M_∞.

4.26 Restoring samples Suppose at time $k = 2$, there is a bag with two balls in it, one orange and one blue. During each time step between k and $k+1$, one of the balls is selected from the bag at random, with all balls in the bag having equal probability. That ball, and a new ball of the other color, are both put into the bag. Thus, at time k there are k balls in the bag, for all $k \geq 2$. Let X_k denote the number of blue balls in the bag at time k.

(a) Is $X = (X_k : k \geq 2)$ a Markov process? If so, describe the one-step transition probabilities.
(b) Compute $E[X_{k+1}|X_k]$ for $k \geq 2$.
(c) Let $M_k = \frac{X_k}{k}$. Thus, M_k is the fraction of balls in the bag at time k that are blue. Determine whether $M = (M_k : k \geq 2)$ is a martingale.
(d) Let $D_k = M_k - \frac{1}{2}$. Show that

$$E[D_{k+1}^2|X_k] = \frac{1}{(k+1)^2}\left\{k(k-2)D_k^2 + \frac{1}{4}\right\}.$$

(e) Let $v_k = E[D_k^2]$. Prove by induction on k that $v_k \leq \frac{1}{4k}$. What can you conclude about the limit of M_k as $k \to \infty$? (Be sure to specify what sense(s) of limit you mean.)

4.27 A space-time transformation of Brownian motion Suppose $(X_t : t \geq 0)$ is a real-valued, mean zero, independent increment process, and let $E[X_t^2] = \rho_t$ for $t \geq 0$. Assume $\rho_t < \infty$ for all t.
(a) Show that ρ must be nonnegative and nondecreasing over $[0, \infty)$.
(b) Express the autocorrelation function $R_X(s, t)$ in terms of the function ρ for all $s \geq 0$ and $t \geq 0$.
(c) Conversely, suppose a nonnegative, nondecreasing function ρ on $[0, \infty)$ is given. Let $Y_t = W(\rho_t)$ for $t \geq 0$, where W is a standard Brownian motion with $R_W(s, t) = \min\{s, t\}$. Explain why Y is an independent increment process with $E[Y_t^2] = \rho_t$ for all $t \geq 0$.
(d) Define a process Z in terms of a standard Brownian motion W by $Z_0 = 0$ and $Z_t = tW(\frac{1}{t})$ for $t > 0$. Does Z have independent increments? Justify your answer.

4.28 An M/M/1/B queueing system Suppose X is a continuous-time Markov process with the transition rate diagram shown, for a positive integer B and positive constant λ.

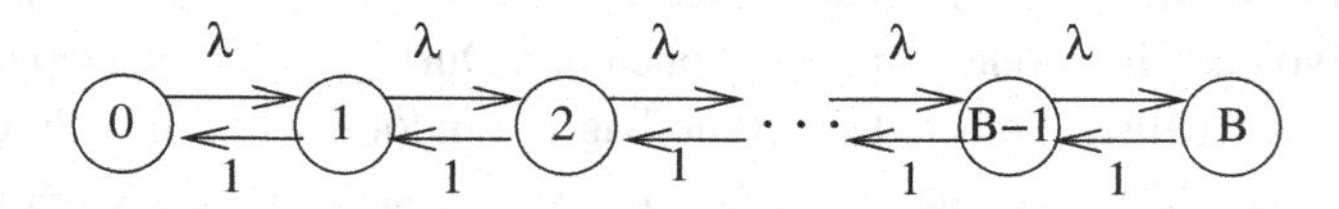

(a) Find the generator matrix, Q, of X for $B = 4$.
(b) Find the equilibrium probability distribution. (Note: The process X models the number of customers in a queueing system with a Poisson arrival process, exponential service times, one server, and a finite buffer.)

4.29 Identification of special properties of two discrete-time processes (I) Determine which of the properties:

(i) Markov property,
(ii) martingale property,
(iii) independent increment property,

are possessed by the following two random processes. Justify your answers.
(a) $X = (X_k : k \geq 0)$ defined recursively by $X_0 = 1$ and $X_{k+1} = (1 + X_k)U_k$ for $k \geq 0$, where $U_0, U_1, \ldots$ are independent random variables, each uniformly distributed on the interval $[0, 1]$.

(b) $Y = (Y_k : k \geq 0)$ defined by $Y_0 = V_0$, $Y_1 = V_0 + V_1$, and $Y_k = V_{k-2} + V_{k-1} + V_k$ for $k \geq 2$, where $V_k : k \in Z$ are independent Gaussian random variables with mean zero and variance one.

4.30 Identification of special properties of two discrete-time processes (II) Determine which of the properties:

(i) Markov property,
(ii) martingale property,
(iii) independent increment property,

are possessed by the following two random processes. Justify your answers.

(a) $(X_k : k \geq 0)$, where X_k is the number of cells alive at time k in a colony that evolves as follows. Initially, there is one cell, so $X_0 = 1$. During each discrete time step, each cell either dies or splits into two new cells, each possibility having probability one half. Suppose cells die or split independently. Let X_k denote the number of cells alive at time k.

(b) $(Y_k : k \geq 0)$, such that $Y_0 = 1$ and, for $k \geq 1$, $Y_k = U_1 U_2 \ldots U_k$, where $U_1, U_2, \ldots$ are independent random variables, each uniformly distributed over the interval $[0, 2]$.

4.31 Identification of special properties of two continuous-time processes (I) Answer as in the previous problem, for the following two random processes:

(a) $Z = (Z_t : t \geq 0)$, defined by $Z_t = \exp(W_t - \frac{\sigma^2 t}{2})$, where W is a Brownian motion with parameter σ^2. (Hint: Observe that $E[Z_t] = 1$ for all t.)

(b) $R = (R_t : t \geq 0)$ defined by $R_t = D_1 + D_2 + \cdots + D_{N_t}$, where N is a Poisson process with rate $\lambda > 0$ and $D_i : i \geq 1$ is an iid sequence of random variables, each having mean 0 and variance σ^2.

4.32 Identification of special properties of two continuous-time processes (II) Answer as in the previous problem, for the following two random processes:

(a) $Z = (Z_t : t \geq 0)$, defined by $Z_t = W_t^3$, where W is a Brownian motion with parameter σ^2.

(b) $R = (R_t : t \geq 0)$, defined by $R_t = \cos(2\pi t + \Theta)$, where Θ is uniformly distributed on the interval $[0, 2\pi]$.

4.33 A branching process Let $p = (p_i : i \geq 0)$ be a probability distribution on the nonnegative integers with mean m. Consider a population beginning with a single individual, comprising generation zero. The offspring of the initial individual form the first generation, and the offspring of the kth generation form the $k + 1$th generation. Suppose the number of offspring of any individual has the probability distribution p, independently of how many offspring other individuals have. Let $Y_0 = 1$, and for $k \geq 1$ let Y_k denote the number of individuals in the kth generation.

(a) Is $Y = (Y_k : k \geq 0)$ a Markov process? Briefly explain your answer.

(b) Find constants c_k so that $\frac{Y_k}{c_k}$ is a martingale.

(c) Let $a_m = P\{Y_m = 0\}$, the probability of extinction by the mth generation. Express a_{m+1} in terms of the distribution p and a_m. (Hint: condition on the value of Y_1, and note that the Y_1 subpopulations beginning with the Y_1 individuals in generation one are independent and statistically identical to the whole population.)

(d) Express the probability of eventual extinction, $a_\infty = \lim_{m\to\infty} a_m$, in terms of the distribution p. Under what condition is $a_\infty = 1$?

(e) Find a_∞ in terms of θ when $p_k = \theta^k(1-\theta)$ for $k \geq 0$ and $0 \leq \theta < 1$. (This distribution is similar to the geometric distribution, and it has mean $m = \frac{\theta}{1-\theta}$.)

4.34 Moving balls Consider the motion of three indistinguishable balls on a linear array of positions, indexed by the positive integers, such that one or more balls can occupy the same position. Suppose that at time $t = 0$ there is one ball at position one, one ball at position two, and one ball at position three. Given the positions of the balls at some integer time t, the positions at time $t+1$ are determined as follows. One of the balls in the left most occupied position is picked up, and one of the other two balls is selected at random (but not moved), with each choice having probability one half. The ball that was picked up is then placed one position to the right of the selected ball.
(a) Define a finite-state Markov process that tracks the relative positions of the balls. Try to use a small number of states. (Hint: Take the balls to be indistinguishable, and don't include the position numbers.) Describe the significance of each state, and give the one-step transition probability matrix for your process.
(b) Find the equilibrium distribution of your process.
(c) As time progresses, the balls all move to the right, and the average speed has a limiting value, with probability one. Find that limiting value. (You can use the fact that for a finite-state Markov process in which any state can eventually be reached from any other, the fraction of time the process is in a state i up to time t converges a.s. to the equilibrium probability for state i as $t \to \infty$.
(d) Consider the following continuous-time version of the problem. Given the current state at time t, a move as described above happens in the interval $[t, t+h]$ with probability $h + o(h)$. Give the generator matrix Q, find its equilibrium distribution, and identify the long term average speed of the balls.

4.35 Mean hitting time for a discrete-time Markov process
Let $(X_k : k \geq 0)$ be a time-homogeneous Markov process with the one-step transition probability diagram shown.

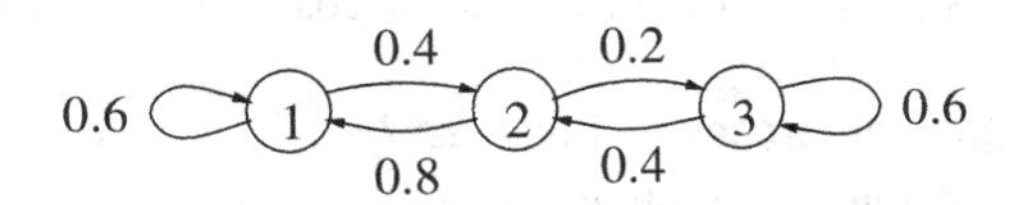

(a) Write down the one-step transition probability matrix P.
(b) Find the equilibrium probability distribution π.
(c) Let $\tau = \min\{k \geq 0 : X_k = 3\}$ and let $a_i = E[\tau | X_0 = i]$ for $1 \leq i \leq 3$. Clearly $a_3 = 0$. Derive equations for a_1 and a_2 by considering the possible values of X_1, in a way similar to the analysis of the gambler's ruin problem. Solve the equations to find a_1 and a_2.

4.36 Mean hitting time for a continuous-time Markov process Let $(X_t : t \geq 0)$ be a time-homogeneous Markov process with the transition rate diagram shown.

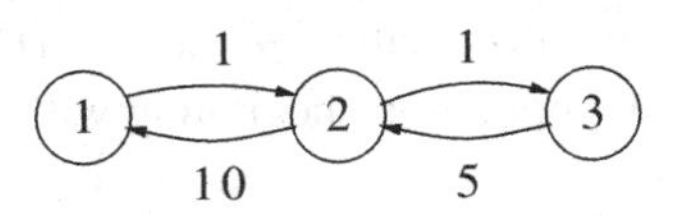

(a) Write down the rate matrix Q.
(b) Find the equilibrium probability distribution π.

(c) Let $\tau = \min\{t \geq 0 : X_t = 3\}$ and let $a_i = E[\tau|X_0 = i]$ for $1 \leq i \leq 3$. Clearly $a_3 = 0$. Derive equations for a_1 and a_2 by considering the possible values of $X_t(h)$ for small values of $h > 0$ and taking the limit as $h \to 0$. Solve the equations to find a_1 and a_2.

4.37 Poisson merger Summing counting processes corresponds to "merging" point processes. Show that the sum of K independent Poisson processes, having rates $\lambda_1, \ldots, \lambda_K$, respectively, is a Poisson process with rate $\lambda_1 + \ldots + \lambda_K$. (Hint: First formulate and prove a similar result for sums of random variables, and then think about what else is needed to get the result for Poisson processes. You can use the definition of a Poisson process or one of the equivalent descriptions given by Proposition 4.5 in the notes. Don't forget to check required independence properties.)

4.38 Poisson splitting Consider a stream of customers modeled by a Poisson process, and suppose each customer is one of K types. Let $(p_1, \ldots, p_K)$ be a probability vector, and suppose that for each k, the kth customer is type i with probability p_i. The types of the customers are mutually independent and also independent of the arrival times of the customers. Show that the stream of customers of a given type i is again a Poisson stream, and that its rate is λp_i. (Same hint as in the previous problem applies.) Show furthermore that the K substreams are mutually independent.

4.39 Poisson method for coupon collector's problem (a) Suppose a stream of coupons arrives according to a Poisson process $(A(t) : t \geq 0)$ with rate $\lambda = 1$, and suppose there are k types of coupons. (In network applications, the coupons could be pieces of a file to be distributed by some sort of gossip algorithm.) The type of each coupon in the stream is randomly drawn from the k types, each possibility having probability $\frac{1}{k}$, and the types of different coupons are mutually independent. Let $p(k, t)$ be the probability that at least one coupon of each type arrives by time t. (The letter "p" is used here because the number of coupons arriving by time t has the Poisson distribution.) Express $p(k, t)$ in terms of k and t.

(b) Find $\lim_{k\to\infty} p(k, k\ln k + kc)$ for an arbitrary constant c. That is, find the limit of the probability that the collection is complete at time $t = k\ln k + kc$. (Hint: If $a_k \to a$ as $k \to \infty$, then $(1 + \frac{a_k}{k})^k \to e^a$.)

(c) The rest of this problem shows that the limit found in part (b) also holds if the total number of coupons is deterministic, rather than Poisson distributed. One idea is that if t is large, then $A(t)$ is not too far from its mean with high probability. Show, specifically, that

$$\lim_{k\to\infty} P\{A(k\ln k + kc) \geq k\ln k + kc'\} = \begin{cases} 0 & \text{if } c < c' \\ 1 & \text{if } c > c' \end{cases}.$$

(d) Let $d(k, n)$ denote the probability that the collection is complete after n coupon arrivals. (The letter "d" is used here because the number of coupons, n, is deterministic.) Show that for any k, t, and n fixed, $d(k, n)P\{A(t) \geq n\} \leq p(k, t) \leq P\{A(t) \geq n\} + P\{A(t) \leq n\}d(k, n)$.

(e) Combine parts (c) and (d) to identify $\lim_{k\to\infty} d(k, k\ln k + kc)$.

4.40 Some orthogonal martingales based on Brownian motion Let $W = (W_t : t \geq 0)$ be a Brownian motion with $\sigma^2 = 1$ (called a standard Brownian motion), and let $M_t = \exp\left(\theta W_t - \frac{\theta^2 t}{2}\right)$ for an arbitrary constant θ.

(a) Show that $(M_t : t \geq 0)$ is a martingale. (Hint for parts (a) and (b): For notational brevity, let $\mathcal{W}_s$ represent $(W_u : 0 \leq u \leq s)$ for the purposes of conditioning. If Z_t

is a function of W_t for each t, then a sufficient condition for Z to be a martingale is that $E[Z_t|\mathcal{W}_s] = Z_s$ whenever $0 < s < t$, because then $E[Z_t|Z_u, 0 \le u \le s] = E[E[Z_t|\mathcal{W}_s]|Z_u, 0 \le u \le s] = E[Z_s|Z_u, 0 \le u \le s] = Z_s$.)
(b) By the power series expansion of the exponential function,

$$\exp\left(\theta W_t - \frac{\theta^2 t}{2}\right) = 1 + \theta W_t + \frac{\theta^2}{2}(W_t^2 - t) + \frac{\theta^3}{3!}(W_t^3 - 3tW_t) + \cdots$$
$$= \sum_{n=0}^{\infty} \frac{\theta^n}{n!} M_n(t),$$

where $M_n(t) = t^{n/2} H_n(\frac{W_t}{\sqrt{t}})$, and H_n is the nth Hermite polynomial. The fact that M is a martingale for any value of θ can be used to show that M_n is a martingale for each n (you don't need to supply details). Verify directly that $W_t^2 - t$ and $W_t^3 - 3tW_t$ are martingales.
(c) For fixed t, $(M_n(t) : n \ge 0)$ is a sequence of orthogonal random variables, because it is the linear innovations sequence for the variables $1, W_t, W_t^2, \ldots$ Use this fact and the martingale property of the M_n processes to show that if $n \ne m$ and $s, t \ge 0$, then $M_n(s) \perp M_m(t)$.

4.41 A state space reduction preserving the Markov property Consider a time-homogeneous, discrete-time Markov process $X = (X_k : k \ge 0)$ with state space $\mathcal{S} = \{1, 2, 3\}$, initial state $X_0 = 3$, and one-step transition probability matrix

$$P = \begin{pmatrix} 0.0 & 0.8 & 0.2 \\ 0.1 & 0.6 & 0.3 \\ 0.2 & 0.8 & 0.0 \end{pmatrix}.$$

(a) Sketch the transition probability diagram and find the equilibrium probability distribution $\pi = (\pi_1, \pi_2, \pi_3)$.
(b) Identify a function f on $\mathcal{S}$ so that $f(s) = a$ for two choices of s and $f(s) = b$ for the third choice of s, where $a \ne b$, such that the process $Y = (Y_k : k \ge 0)$ defined by $Y_k = f(X_k)$ is a Markov process with only two states, and give the one-step transition probability matrix of Y. Briefly explain your answer.

4.42* Autocorrelation function of a stationary Markov process Let $X = (X_k : k \in Z)$ be a Markov process such that the state space, $\{\rho_1, \rho_2, \ldots, \rho_n\}$, is a finite subset of the real numbers. Let $P = (p_{ij})$ denote the matrix of one-step transition probabilities. Let e be the column vector of all ones, and let $\pi(k)$ be the row vector $\pi(k) = (P\{X_k = \rho_1\}, \ldots, P\{X_k = \rho_n\})$.
(a) Show that $Pe = e$ and $\pi(k+1) = \pi(k)P$.
(b) Show that if the Markov chain X is a stationary random process then $\pi(k) = \pi$ for all k, where π is a vector such that $\pi = \pi P$.
(c) Prove the converse of part (b).
(d) Show that $P(X_{k+m} = \rho_j | X_k = \rho_i, X_{k-1} = s_1, \ldots, X_{k-m} = s_m) = p_{ij}^{(m)}$, where $p_{ij}^{(m)}$ is the i,jth element of the mth power of P, P^m, and $s_1, \ldots, s_m$ are arbitrary states.
(e) Assume that X is stationary. Express $R_X(k)$ in terms of P, (ρ_i), and the vector π of parts (b) and (c).

5 Inference for Markov models

This chapter gives a glimpse of the theory of iterative algorithms for graphical models, as well as an introduction to statistical estimation theory. It begins with a brief introduction to estimation theory: maximum likelihood and Bayes estimators are introduced, and an iterative algorithm, known as the expectation-maximization algorithm, for computation of maximum likelihood estimators in certain contexts, is described. This general background is then focused on three inference problems posed using Markov models.

5.1 A bit of estimation theory

The two most commonly used methods for producing estimates of unknown quantities are the maximum likelihood (ML) and Bayesian methods. These two methods are briefly described in this section, beginning with the ML method.

Suppose a parameter θ is to be estimated, based on observation of a random variable Y. An *estimator* of θ based on Y is a function $\widehat{\theta}$, which for each possible observed value y, gives the *estimate* $\widehat{\theta}(y)$. The ML method is based on the assumption that Y has a pmf $p_Y(y|\theta)$ (if Y is discrete type) or a pdf $f_Y(y|\theta)$ (if Y is continuous type), where θ is the unknown parameter to be estimated, and the family of functions $p_Y(y|\theta)$ or $f_Y(y|\theta)$, is known.

Definition 5.1 For a particular value y and parameter value θ, the *likelihood* of y for θ is $p_Y(y|\theta)$, if Y is discrete type, or $f_Y(y|\theta)$, if Y is continuous type. The *maximum likelihood estimate* of θ given $Y = y$ for a particular y is the value of θ that maximizes the likelihood of y. That is, the maximum likelihood estimator $\widehat{\theta}_{ML}$ is given by $\widehat{\theta}_{ML}(y) = \arg\max_\theta p_Y(y|\theta)$, or $\widehat{\theta}_{ML}(y) = \arg\max_\theta f_Y(y|\theta)$.

Note that the maximum likelihood estimator is not defined as one maximizing the likelihood of the parameter θ to be estimated. In fact, θ need not even be a random variable. Rather, the maximum likelihood estimator is defined by selecting the value of θ that maximizes the likelihood of the observation.

Example 5.1 Suppose Y is assumed to be an $N(\theta, \sigma^2)$ random variable, where σ^2 is known. Equivalently, we can write $Y = \theta + W$, where W is an $N(0, \sigma^2)$ random variable. Given that a value y is observed, the ML estimator is obtained by maximizing $f_Y(y|\theta) = \frac{1}{\sqrt{2\pi\sigma^2}} \exp\left(-\frac{(y-\theta)^2}{2\sigma^2}\right)$ with respect to θ. By inspection, $\widehat{\theta}_{ML}(y) = y$.

Example 5.2 Suppose Y is assumed to be a $Poi(\theta)$ random variable, for some $\theta > 0$. Given the observation $Y = k$ for some fixed $k \geq 0$, the ML estimator is obtained by maximizing $p_Y(k|\theta) = \frac{e^{-\theta}\theta^k}{k!}$ with respect to θ. Equivalently, dropping the constant $k!$ and taking the logarithm, θ is to be selected to maximize $-\theta + k \ln \theta$. The derivative is $-1 + k/\theta$, which is positive for $\theta < k$ and negative for $\theta > k$. Hence, $\widehat{\theta}_{ML}(k) = k$.

Note that in the ML method, the quantity to be estimated, θ, is not assumed to be random. This has the advantage that the modeler does not have to come up with a probability distribution for θ, and the modeler can still impose hard constraints on θ. But the ML method does not permit incorporation of soft probabilistic knowledge the modeler may have about θ before any observation is used.

The Bayesian method is based on estimating a random quantity. Thus, in the end, the variable to be estimated, say Z, and the observation, say Y, are jointly distributed random variables.

Definition 5.2 The Bayes estimator of Z given Y, for jointly distributed random variables Z and Y, and cost function $C(z, y)$, is the function $\widehat{Z} = g(Y)$ of Y which minimizes the average cost, $E[C(Z, \widehat{Z})]$.

The assumed distribution of Z is called the *prior* or *a priori* distribution, whereas the conditional distribution of Z given Y is called the *posterior* or *a posteriori* distribution. In particular, if Z is discrete, there is a prior pmf, p_Z, and a posterior pmf, $p_{Z|Y}$, or if Z and Y are jointly continuous, there is a prior pdf, f_Z, and a posterior pdf, $f_{Z|Y}$.

One of the most common choices of cost function is squared error, $C(z, \hat{z}) = (z - \hat{z})^2$, for which the Bayes estimators are the minimum mean squared error (MMSE) estimators, examined in Chapter 3. Recall that the MMSE estimators are given by the conditional expectation, $g(y) = E[Z|Y = y]$, which, given the observation $Y = y$, is the mean of the posterior distribution of Z given $Y = y$.

A commonly used choice of C when Z is a discrete random variable is $C(z, \hat{z}) = I_{\{z \neq \hat{z}\}}$. In this case, the Bayesian objective is to select $\hat{Z}$ to minimize $P\{Z \neq \hat{Z}\}$, or equivalently, to maximize $P\{Z = \hat{Z}\}$. For an estimator $\widehat{Z} = g(Y)$,

$$P\{Z = \hat{Z}\} = \sum_y P(Z = g(y)|Y = y)p_Y(y) = \sum_y p_{Z|Y}(g(y)|y)p_Y(y).$$

So a Bayes estimator for $C(z, \hat{z}) = I_{\{z \neq \hat{z}\}}$ maximizes $P(Z = g(y)|Y = y)$ for each y. That is, for each y, $g(y)$ is a maximizer of the posterior pmf of Z. The estimator, called the *maximum a posteriori probability* (MAP) estimator, can be written concisely as

$$\widehat{Z}_{\text{MAP}}(y) = \arg\max_z p_{Z|Y}(z|y).$$

Suppose there is a parameter θ to be estimated based on an observation Y, and suppose that the pmf of Y, $p_Y(y|\theta)$, is known for each θ. This is enough to determine the ML estimator, but determination of a Bayes estimator requires, in addition, a choice of cost function C and a prior probability distribution (i.e. a distribution for θ).

For example, if θ is a discrete variable, the Bayesian method would require that a prior pmf for θ be selected. In that case, we can view the parameter to be estimated as a random variable, which we might denote by the upper case symbol Θ, and the prior pmf could be denoted by $p_\Theta(\theta)$. Then, as required by the Bayesian method, the variable to be estimated, Θ, and the observation, Y, would be jointly distributed random variables. The joint pmf would be given by $p_{\Theta,Y}(\theta, Y) = p_\Theta(\theta)p_Y(y|\theta)$. The posterior probability distribution can be expressed as a conditional pmf, by Bayes' formula:

$$p_{\Theta|Y}(\theta|y) = \frac{p_\Theta(\theta)p_Y(y|\theta)}{p_Y(y)}, \tag{5.1}$$

where $p_Y(y) = \sum_{\theta'} p_{\Theta,Y}(\theta', y)$. Given y, the value of the MAP estimator is a value of θ that maximizes $p_{\Theta|Y}(\theta|y)$ with respect to θ. For that purpose, the denominator in the righthand side of (5.1) can be ignored, so that the MAP estimator is given by

$$\begin{aligned}\widehat{\Theta}_{\text{MAP}}(y) &= \arg\max_\theta p_{\Theta|Y}(\theta|y) \\ &= \arg\max_\theta p_\Theta(\theta)p_Y(y|\theta).\end{aligned} \tag{5.2}$$

The expression, (5.2), for $\widehat{\Theta}_{\text{MAP}}(y)$ is rather similar to the expression for the ML estimator, $\widehat{\theta}_{\text{ML}}(y) = \arg\max_\theta p_Y(y|\theta)$. In fact, the two estimators agree if the prior $p_\Theta(\theta)$ is *uniform*, meaning it is the same for all θ.

The MAP criterion for selecting estimators can be extended to the case that Y and θ are jointly continuous variables, leading to the following:

$$\begin{aligned}\widehat{\Theta}_{\text{MAP}}(y) &= \arg\max_\theta f_{\Theta|Y}(\theta|y) \\ &= \arg\max_\theta f_\Theta(\theta)f_Y(y|\theta).\end{aligned} \tag{5.3}$$

In this case, the probability that any estimator is exactly equal to θ is zero, but taking $\widehat{\Theta}_{\text{MAP}}(y)$ to maximize the posterior pdf maximizes the probability that the estimator is within ϵ of the true value of θ, in an asymptotic sense as $\epsilon \to 0$.

Example 5.3 Suppose Y is assumed to be a $N(\theta, \sigma^2)$ random variable, where the variance σ^2 is known and θ is to be estimated. Using the Bayesian method, suppose the prior density of θ is the $N(0, b^2)$ density for some known parameter b^2. Equivalently, we can write $Y = \Theta + W$, where Θ is an $N(0, b^2)$ random variable and W is an $N(0, \sigma^2)$ random variable, independent of Θ. By the properties of joint Gaussian densities given in Chapter 3, given $Y = y$, the posterior distribution (i.e. the conditional distribution of Θ given y) is the normal distribution with mean $\widehat{E}[\Theta|Y = y] = \frac{b^2 y}{b^2+\sigma^2}$ and variance $\frac{b^2\sigma^2}{b^2+\sigma^2}$. The mean and maximizing value of this conditional density are both equal to $\widehat{E}[\Theta|Y = y]$. Therefore, $\widehat{\Theta}_{\text{MMSE}}(y) = \widehat{\Theta}_{\text{MAP}}(y) = \widehat{E}[\Theta|Y = y]$. It is interesting to compare this example to Example 5.1. The Bayes estimators (MMSE and MAP) are both smaller in magnitude than $\widehat{\theta}_{\text{ML}}(y) = y$, by the factor $\frac{b^2}{b^2+\sigma^2}$. If b^2 is small compared to σ^2, the prior information indicates that $|\theta|$ is believed to be small, resulting in the Bayes estimators being smaller in magnitude than the ML estimator. As $b^2 \to \infty$, the prior distribution gets increasingly uniform, and the Bayes estimators converge to the ML estimator.

Example 5.4 Suppose Y is assumed to be a $Poi(\theta)$ random variable. Using the Bayesian method, suppose the prior distribution for θ is the uniform distribution over the interval $[0, \theta_{\max}]$, for some known value $\theta_{\max}$. Given the observation $Y = k$ for some fixed $k \geq 0$, the MAP estimator is obtained by maximizing

$$p_Y(k|\theta)f_\Theta(\theta) = \frac{e^{-\theta}\theta^k}{k!} \frac{I_{\{0 \leq \theta \leq \theta_{\max}\}}}{\theta_{\max}}$$

with respect to θ. As seen in Example 5.2, the term $\frac{e^{-\theta}\theta^k}{k!}$ is increasing in θ for $\theta < k$ and decreasing in θ for $\theta > k$. Therefore,

$$\widehat{\Theta}_{\mathrm{MAP}}(k) = \min\{k, \theta_{\max}\}.$$

It is interesting to compare this example to Example 5.2. Intuitively, the prior probability distribution indicates knowledge that $\theta \leq \theta_{\max}$, but no more than that, because the prior distribution restricted to $\theta \leq \theta_{\max}$ is uniform. If $\theta_{\max}$ is less than k, the MAP estimator is strictly smaller than $\widehat{\theta}_{\mathrm{ML}}(k) = k$. As $\theta_{\max} \to \infty$, the MAP estimator converges to the ML estimator. Actually, deterministic prior knowledge, such as $\theta \leq \theta_{\max}$, can also be incorporated into ML estimation as a hard constraint.

The next example makes use of the following lemma.

Lemma 5.3 *Suppose $c_i \geq 0$ for $1 \leq i \leq n$ and that $\bar{c} = \sum_{i=1}^n c_i > 0$. Then $\sum_{i=1}^n c_i \log p_i$ is maximized over all probability vectors $p = (p_1 \ldots, p_n)$ by $p_i = c_i/\bar{c}$.*

Proof If $c_j = 0$ for some j, then clearly $p_j = 0$ for the maximizing probability vector. By eliminating such terms from the sum, we can assume without loss of generality that $c_i > 0$ for all i. The function to be maximized is a strictly concave function of p over a region with linear constraints. The positivity constraints, namely $p_i \geq 0$, will be satisfied with strict inequality. The remaining constraint is $\sum_{i=1}^n p_i = 1$. We thus introduce a Lagrange multiplier λ for the equality constraint and seek the stationary point of the Lagrangian $L(p, \lambda) = \sum_{i=1}^n c_i \log p_i - \lambda((\sum_{i=1}^n p_i) - 1)$. By definition, the stationary point is the point at which the partial derivatives with respect to the variables p_i are all zero. Setting $\frac{\partial L}{\partial p_i} = \frac{c_i}{p_i} - \lambda = 0$ yields that $p_i = \frac{c_i}{\lambda}$ for all i. To satisfy the linear constraint, λ must equal $\bar{c}$. □

Example 5.5 Suppose $b = (b_1, b_2, \ldots, b_n)$ is a probability vector to be estimated by observing $Y = (Y_1, \ldots, Y_T)$. Assume $Y_1, \ldots, Y_T$ are independent, with each Y_t having probability distribution b: $P\{Y_t = i\} = b_i$ for $1 \leq t \leq T$ and $1 \leq i \leq n$. We shall determine the maximum likelihood estimate, $\widehat{b}_{\mathrm{ML}}(y)$, given a particular observation $y = (y_1, \ldots, y_T)$. The likelihood to be maximized with respect to b is $p(y|b) = b_{y_1} \cdots b_{y_T} =$

$\prod_{i=1}^n b_i^{k_i}$ where $k_i = |\{t : y_t = i\}|$. The log likelihood is $\ln p(y|b) = \sum_{i=1}^n k_i \ln(b_i)$. By Lemma 5.3, this is maximized by the empirical distribution of the observations, namely $b_i = \frac{k_i}{T}$ for $1 \le i \le n$. That is, $\widehat{b}_{\mathrm{ML}} = (\frac{k_1}{T}, \ldots, \frac{k_n}{T})$.

Example 5.6 This is a Bayesian version of the previous example. Suppose $b = (b_1, b_2, \ldots, b_n)$ is a probability vector to be estimated by observing $Y = (Y_1, \ldots, Y_T)$, and assume $Y_1, \ldots, Y_T$ are independent, with each Y_t having probability distribution b. For the Bayesian method, a distribution of the unknown distribution b must be assumed. That is right, a distribution of the distribution is needed. A convenient choice is the following. Suppose for some known numbers $d_i \ge 1$ that $(b_1, \ldots, b_{n-1})$ has the prior density:

$$f_B(b) = \begin{cases} \frac{\prod_{i=1}^n b_i^{d_i-1}}{Z(d)} & \text{if } b_i \ge 0 \text{ for } 1 \le i \le n-1, \text{ and } \sum_{i=1}^{n-1} b_i \le 1 \\ 0 & \text{else} \end{cases},$$

where $b_n = 1 - b_1 - \ldots - b_{n-1}$, and $Z(d)$ is a constant chosen so that f_B integrates to one. A larger value of d_i for a fixed i expresses an a priori guess that the corresponding value b_i may be larger. It can be shown, in particular, that if B has this prior distribution, then $E[B_i] = \frac{d_i}{d_1 + \cdots d_n}$. The MAP estimate, $\widehat{b}_{\mathrm{MAP}}(y)$, for a given observation vector y, is given by:

$$\begin{aligned} \widehat{b}_{\mathrm{MAP}}(y) &= \arg\max_b \ln\left(f_B(b) p(y|b)\right) \\ &= \arg\max_b \left\{ -\ln(Z(d)) + \sum_{i=1}^n (d_i - 1 + k_i) \ln(b_i) \right\}. \end{aligned}$$

By Lemma 5.3, $\widehat{b}_{\mathrm{MAP}}(y) = \left(\frac{d_1 - 1 + k_1}{\widetilde{T}}, \ldots, \frac{d_n - 1 + k_n}{\widetilde{T}}\right)$, where $\widetilde{T} = \sum_{i=1}^n (d_i - 1 + k_i) = T - n + \sum_{i=1}^n d_i$.

Comparison with Example 5.5 shows that the MAP estimate is the same as the ML estimate, except that $d_i - 1$ is added to k_i for each i. If the d_is are integers, the MAP estimate is the ML estimate with some prior observations mixed in, namely, $d_i - 1$ prior observations of outcome i for each i. A prior distribution such that the MAP estimate has the same algebraic form as the ML estimate is called a *conjugate prior*, and the specific density f_B for this example is called the *Dirichlet density* with parameter vector d.

Example 5.7 Suppose $Y = (Y_1, \ldots, Y_T)$ is observed, and it is assumed that the Y_i are independent, with the binomial distribution with parameters n and q. Suppose n is known, and q is an unknown parameter to be estimated from Y. Let us find the maximum likelihood estimate, $\widehat{q}_{\mathrm{ML}}(y)$, for a particular observation $y = (y_1, \ldots, y_T)$. The likelihood is

$$p(y|q) = \prod_{t=1}^T \left[\binom{n}{y_t} q^{y_t} (1-q)^{n-y_t} \right] = cq^s (1-q)^{nT-s},$$

where $s = y_1 + \cdots + y_T$, and c depends on y but not on q. The log likelihood is $\ln c + s\ln(q) + (nT - s)\ln(1-q)$. Maximizing over q yields $\hat{q}_{\mathrm{ML}} = \frac{s}{nT}$. An alternative way to think about this is to realize that each Y_t can be viewed as the sum of n independent Bernoulli(q) random variables, and s can be viewed as the observed sum of nT independent Bernoulli(q) random variables.

5.2 The expectation-maximization (EM) algorithm

The expectation-maximization algorithm is a computational method for computing maximum likelihood estimates in contexts where there are hidden random variables, in addition to observed data and unknown parameters. The following notation will be used:

θ, a parameter to be estimated,
X, the complete data,
$p_{cd}(x|\theta)$, the pmf of the complete data, which is a known function for each value of θ,
$Y = h(X)$, the observed random vector,
Z, the unobserved data. (This notation is used in the common case that X has the form $X = (Y, Z)$.)

We write $p(y|\theta)$ to denote the pmf of Y for a given value of θ. It can be expressed in terms of the pmf of the complete data by:

$$p(y|\theta) = \sum_{\{x:h(x)=y\}} p_{cd}(x|\theta). \tag{5.4}$$

In some applications, there can be a very large number of terms in the sum in (5.4), making it difficult to numerically maximize $p(y|\theta)$ with respect to θ (i.e. to compute $\widehat{\theta}_{\mathrm{ML}}(y)$).

Algorithm 5.4 *(Expectation-maximization (EM) algorithm) An observation y is given, along with an initial estimate $\theta^{(0)}$. The algorithm is iterative. Given $\theta^{(k)}$, the next value $\theta^{(k+1)}$ is computed in the following two steps:*

(Expectation step) Compute $Q(\theta|\theta^{(k)})$ for all θ, where

$$Q(\theta|\theta^{(k)}) = E[\ \log p_{cd}(X|\theta) \mid y, \theta^{(k)}]. \tag{5.5}$$

(Maximization step) Compute $\theta^{(k+1)} \in \arg\max_\theta Q(\theta|\theta^{(k)})$. In other words, find a value $\theta^{(k+1)}$ of θ that maximizes $Q(\theta|\theta^{(k)})$ with respect to θ.

Some intuition behind the algorithm is the following. If a vector of complete data x could be observed, it would be reasonable to estimate θ by maximizing the pmf of the complete data, $p_{cd}(x|\theta)$, with respect to θ. This plan is not feasible if the complete data are not observed. The idea is to estimate $\log p_{cd}(X|\theta)$ by its conditional expectation, $Q(\theta|\theta^{(k)})$, and then find θ to maximize this conditional expectation. The conditional expectation is well defined if some value of the parameter θ is fixed. For each iteration of the algorithm, the expectation step is completed using the latest value of θ, $\theta^{(k)}$, in computing the expectation of $\log p_{cd}(X|\theta)$.

In most applications there is some additional structure that helps in the computation of $Q(\theta|\theta^{(k)})$. This typically happens when p_{cd} factors into simple terms, such as in the case of hidden Markov models discussed in this chapter, or when p_{cd} has the form of the exponential of a low degree polynomial, such as the Gaussian or exponential distribution. In some cases there are closed form expressions for $Q(\theta|\theta^{(k)})$. In others, there may be an algorithm that generates samples of X with the desired pmf $p_{cd}(x|\theta^{(k)})$ using random number generators, and then $\log p_{cd}(X|\theta)$ is used as an approximation to $Q(\theta|\theta^{(k)})$.

Example 5.8 (Estimation of the variance of a signal) An observation Y is modeled as $Y = S+N$, where the signal S is assumed to be an $N(0,\theta)$ random variable, where θ is an unknown parameter, assumed to satisfy $\theta \geq 0$, and the noise N is an $N(0,\sigma^2)$ random variable where σ^2 is known and strictly positive. Suppose it is desired to estimate θ, the variance of the signal. Let y be a particular observed value of Y. We consider two approaches to finding $\widehat{\theta}_{\mathrm{ML}}$: a direct approach, and the EM algorithm.

For the direct approach, note that for θ fixed, Y is an $N(0,\theta+\sigma^2)$ random variable. Therefore, the pdf of Y evaluated at y, or likelihood of y, is given by

$$f(y|\theta) = \frac{\exp(-\frac{y^2}{2(\theta+\sigma^2)})}{\sqrt{2\pi(\theta+\sigma^2)}}.$$

The natural log of the likelihood is given by

$$\log f(y|\theta) = -\frac{\log(2\pi)}{2} - \frac{\log(\theta+\sigma^2)}{2} - \frac{y^2}{2(\theta+\sigma^2)}.$$

Maximizing over θ yields $\widehat{\theta}_{\mathrm{ML}} = (y^2-\sigma^2)_+$. While this one-dimensional case is fairly simple, the situation is different in higher dimensions, as explored in Problem 5.7. Thus, we examine use of the EM algorithm for this example.

To apply the EM algorithm for this example, take $X = (S,N)$ as the complete data. The observation is only the sum, $Y = S+N$, so the complete data are not observed. For given θ, S and N are independent, so the log of the joint pdf of the complete data is given as follows:

$$\log p_{cd}(s,n|\theta) = -\frac{\log(2\pi\theta)}{2} - \frac{s^2}{2\theta} - \frac{\log(2\pi\sigma^2)}{2} - \frac{n^2}{2\sigma^2}.$$

For the expectation step, we find

$$\begin{aligned} Q(\theta|\theta^{(k)}) &= E[\,\log p_{cd}(S,N|\theta)\,|y,\theta^{(k)}] \\ &= -\frac{\log(2\pi\theta)}{2} - \frac{E[S^2|y,\theta^{(k)}]}{2\theta} - \frac{\log(2\pi\sigma^2)}{2} - \frac{E[N^2|y,\theta^{(k)}]}{2\sigma^2}. \end{aligned}$$

For the maximization step, we find

$$\frac{\partial Q(\theta|\theta^{(k)})}{\partial\theta} = -\frac{1}{2\theta} + \frac{E[S^2|y,\theta^{(k)}]}{2\theta^2},$$

from which we see that $\theta^{(k+1)} = E[S^2|y,\theta^{(k)}]$. Computation of $E[S^2|y,\theta^{(k)}]$ is an exercise in conditional Gaussian distributions, similar to Example 3.4. The conditional second moment is the sum of the square of the conditional mean and the variance of the estimation error. Thus, the EM algorithm becomes the following recursion:

$$\theta^{(k+1)} = \left(\frac{\theta^{(k)}}{\theta^{(k)}+\sigma^2}\right)^2 y^2 + \frac{\theta^{(k)}\sigma^2}{\theta^{(k)}+\sigma^2}. \tag{5.6}$$

Problem 5.5 shows that if $\theta^{(0)} > 0$, then $\theta^{(k)} \to \widehat{\theta}_{\mathrm{ML}}$ as $k \to \infty$.

Proposition 5.7 below shows that the likelihood $p(y|\theta^{(k)})$ is nondecreasing in k. In the ideal case, the likelihood converges to the maximum possible value of the likelihood, and $\lim_{k\to\infty}\theta^{(k)} = \widehat{\theta}_{\mathrm{ML}}(y)$. However, the sequence could converge to a local, but not global, maximizer of the likelihood, or possibly even to an inflection point of the likelihood. This behavior is typical of gradient type nonlinear optimization algorithms, which the EM algorithm is similar to. Note that even if the parameter set is convex (as it is for the case of hidden Markov models), the corresponding sets of probability distributions on Y are not convex. It is the geometry of the set of probability distributions that really matters for the EM algorithm, rather than the geometry of the space of the parameters. Before the proposition is stated, the divergence between two probability vectors and some of its basic properties are discussed.

Definition 5.5 The *divergence* between probability vectors $p = (p_1,\ldots,p_n)$ and $q = (q_1,\ldots,q_n)$, denoted by $D(p||q)$, is defined by $D(p||q) = \sum_i p_i \log(p_i/q_i)$, with the understanding that $p_i\log(p_i/q_i) = 0$ if $p_i = 0$ and $p_i\log(p_i/q_i) = +\infty$ if $p_i > q_i = 0$.

Lemma 5.6 *(Basic properties of divergence)*

(i) $D(p||q) \geq 0$, *with equality if and only if* $p = q$;
(ii) D *is a convex function of the pair* (p,q).

Proof Property (i) follows from Lemma 5.3. Here is another proof. In proving (i), we can assume that $q_i > 0$ for all i. The function $\varphi(u) = \begin{cases} u\log u & u>0 \\ 0 & u=0 \end{cases}$ is convex. Thus, by Jensen's inequality,

$$D(p||q) = \sum_i \varphi\left(\frac{p_i}{q_i}\right) q_i \geq \varphi\left(\sum_i \frac{p_i}{q_i}\cdot q_i\right) = \varphi(1) = 0,$$

so (i) is proved.

The proof of (ii) is based on the *log-sum inequality*, which is the fact that for nonnegative numbers $a_1,\ldots,a_n,b_1,\ldots,b_n$:

$$\sum_i a_i \log\frac{a_i}{b_i} \geq \overline{a}\log\frac{\overline{a}}{\overline{b}}, \tag{5.7}$$

where $\overline{a} = \sum_i a_i$ and $\overline{b} = \sum_i b_i$. To verify (5.7), note that it is true if and only if it is true with each a_i replaced by ca_i, for any strictly positive constant c. So it can be assumed that $\overline{a} = 1$. Similarly, it can be assumed that $\overline{b} = 1$. For $\overline{a} = \overline{b} = 1$, (5.7) is equivalent to the fact $D(a||b) \geq 0$, already proved. So (5.7) is proved.

Let $0 < \alpha < 1$. Suppose $p^j = (p_1^j, \ldots, p_n^j)$ and $q^j = (q_1^j, \ldots, q_n^j)$ are probability distributions for $j = 1, 2$, and let $p_i = \alpha p_i^1 + (1-\alpha)p_i^2$ and $q_i = \alpha q_i^1 + (1-\alpha)q_i^2$, for $1 \leq i \leq n$. That is, (p^1, q^1) and (p^2, q^2) are two pairs of probability distributions, and $(p, q) = \alpha(p^1, q^1) + (1-\alpha)(p^2, q^2)$. For i fixed with $1 \leq i \leq n$, the log-sum inequality (5.7) with $(a_1, a_2, b_1, b_2) = (\alpha p_i^1, (1-\alpha)p_i^2, \alpha q_i^1, (1-\alpha)q_i^2)$ yields

$$\alpha p_i^1 \log \frac{p_i^1}{q_i^1} + (1-\alpha)p_i^2 \log \frac{p_i^2}{q_i^2} = \alpha p_i^1 \log \frac{\alpha p_i^1}{\alpha q_i^1} + (1-\alpha)p_i^2 \log \frac{(1-\alpha)p_i^2}{(1-\alpha)q_i^2}$$
$$\geq p_i \log \frac{p_i}{q_i}.$$

Summing each side of this inequality over i yields $\alpha D(p^1||q^1) + (1-\alpha)D(p^2||q^2) \geq D(p||q)$, so that $D(p||q)$ is a convex function of the pair (p, q). □

Proposition 5.7 *(Convergence of the EM algorithm) Suppose that the complete data pmf can be factored as $p_{cd}(x|\theta) = p(y|\theta)k(x|y, \theta)$ such that:*

(i) *$\log p(y|\theta)$ is differentiable in θ,*
(ii) *$E\left[\log k(X|y, \bar{\theta}) \mid y, \bar{\theta}\right]$ is finite for all $\bar{\theta}$,*
(iii) *$D(k(\cdot|y, \theta)||k(\cdot|y, \theta'))$ is differentiable with respect to θ' for fixed θ,*
(iv) *$D(k(\cdot|y, \theta)||k(\cdot|y, \theta'))$ is continuous in θ for fixed θ',*

and suppose that $p(y|\theta^{(0)}) > 0$. Then the likelihood $p(y|\theta^{(k)})$ is nondecreasing in k, and any limit point θ^* of the sequence $(\theta^{(k)})$ is a stationary point of the objective function $p(y|\theta)$, which by definition means

$$\frac{\partial p(y|\theta)}{\partial \theta}|_{\theta=\theta^*} = 0. \tag{5.8}$$

Proof Using the factorization $p_{cd}(x|\theta) = p(y|\theta)k(x|y, \theta)$,

$$\begin{aligned} Q(\theta|\theta^{(k)}) &= E[\log p_{cd}(X|\theta)|y, \theta^{(k)}] \\ &= \log p(y|\theta) + E[\log k(X|y, \theta) \; |y, \theta^{(k)}] \\ &= \log p(y|\theta) + E[\log \frac{k(X|y, \theta)}{k(X|y, \theta^{(k)})} \; |y, \theta^{(k)}] + R \\ &= \log p(y|\theta) - D(k(\cdot|y, \theta^{(k)})||k(\cdot|y, \theta)) + R, \end{aligned} \tag{5.9}$$

where

$$R = E[\log k(X|y, \theta^{(k)}) \; |y, \theta^{(k)}].$$

By assumption (ii), R is finite, and it depends on y and $\theta^{(k)}$, but not on θ. Therefore, the maximization step of the EM algorithm is equivalent to:

$$\theta^{(k+1)} = \arg\max_{\theta} \left[\log p(y|\theta) - D(k(\cdot|y, \theta^{(k)})||k(\cdot|y, \theta))\right]. \tag{5.10}$$

Thus, at each step, the EM algorithm attempts to maximize the log likelihood ratio $\log p(y|\theta)$ itself, minus a term which penalizes large differences between θ and $\theta^{(k)}$.

The definition of $\theta^{(k+1)}$ implies that $Q(\theta^{(k+1)}|\theta^{(k)}) \geq Q(\theta^{(k)}|\theta^{(k)})$. Therefore, using (5.9) and the fact $D(k(\cdot|y,\theta^{(k)})||k(\cdot|y,\theta^{(k)})) = 0$, yields

$$\log p(y|\theta^{(k+1)}) - D(k(\cdot|y,\theta^{(k)})||k(\cdot|y,\theta^{(k+1)})) \geq \log p(y|\theta^{(k)}). \tag{5.11}$$

In particular, since the divergence is nonnegative, $p(y|\theta^{(k)})$ is nondecreasing in k. Therefore, $\lim_{k\to\infty} \log p(y|\theta^{(k)})$ exists.

Suppose now that the sequence $(\theta^{(k)})$ has a limit point, θ^*. By continuity, implied by the differentiability assumption (i), $\lim_{k\to\infty} p(y|\theta^{(k)}) = p(y|\theta^*) < \infty$. For each k,

$$0 \leq \max_{\theta} \left[\log p(y|\theta) - D\left(k(\cdot|y,\theta^{(k)}) \mid\mid k(\cdot|y,\theta)\right)\right] - \log p(y|\theta^{(k)}) \tag{5.12}$$

$$\leq \log p(y|\theta^{(k+1)}) - \log p(y|\theta^{(k)}) \to 0 \quad \text{as } k \to \infty, \tag{5.13}$$

where (5.12) follows from the fact that $\theta^{(k)}$ is a possible value of θ in the maximization, and the inequality in (5.13) follows from (5.10) and the fact that the divergence is always nonnegative. Thus, the quantity on the righthand side of (5.12) converges to zero as $k \to \infty$. So by continuity, for any limit point θ^* of the sequence (θ_k),

$$\max_{\theta} \left[\log p(y|\theta) - D\left(k(\cdot|y,\theta^*) \mid\mid k(\cdot|y,\theta)\right)\right] - \log p(y|\theta^*) = 0$$

and therefore,

$$\theta^* \in \arg\max_{\theta} \left[\log p(y|\theta) - D\left(k(\cdot|y,\theta^*) \mid\mid k(\cdot|y,\theta)\right)\right].$$

So the derivative of $\log p(y|\theta) - D\left(k(\cdot|y,\theta^*) \mid\mid k(\cdot|y,\theta)\right)$ with respect to θ at $\theta = \theta^*$ is zero. The same is true of the term $D\left(k(\cdot|y,\theta^*) \mid\mid k(\cdot|y,\theta)\right)$ alone, because this term is nonnegative, it has value 0 at $\theta = \theta^*$, and it is assumed to be differentiable in θ. Therefore, the derivative of the first term, $\log p(y|\theta)$, must be zero at θ^*. □

Remark 5.1 In the above proposition and proof, we assume that θ^* is unconstrained. If there are inequality constraints on θ and if some of them are tight for θ^*, then we still find that if θ^* is a limit point of $\theta^{(k)}$, then it is a maximizer of $f(\theta) = \log p(y|\theta) - D\left(k(\cdot|y,\theta) \mid\mid k(\cdot|y,\theta^*)\right)$. Thus, under regularity conditions implying the existence of Lagrange multipliers, the Kuhn–Tucker optimality conditions are satisfied for the problem of maximizing $f(\theta)$. Since the derivatives of $D\left(k(\cdot|y,\theta) \mid\mid k(\cdot|y,\theta^*)\right)$ with respect to θ at $\theta = \theta^*$ are zero, and since the Kuhn–Tucker optimality conditions only involve the first derivatives of the objective function, those conditions for the problem of maximizing the true log likelihood function, $\log p(y|\theta)$, also hold at θ^*.

5.3 Hidden Markov models

A popular model of one-dimensional sequences with dependencies, explored especially in the context of speech processing, are the hidden Markov models. Suppose that:

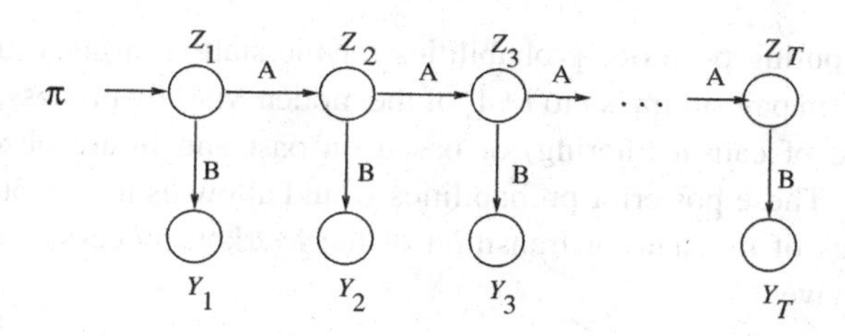

Figure 5.1 Structure of hidden Markov model.

$X = (Y, Z)$, where Z is unobserved data and Y is the observed data.

$Z = (Z_1, \ldots, Z_T)$ is a time-homogeneous Markov process, with one-step transition probability matrix $A = (a_{ij})$, and with Z_1 having the initial distribution π. Here, T, with $T \geq 1$, denotes the total number of observation times. The state-space of Z is denoted by $\mathcal{S}$, and the number of states of $\mathcal{S}$ is denoted by N_s.

$Y = (Y_1, \ldots, Y_T)$ are the observed data. They are such that given $Z = z$, for some $z = (z_1, \ldots, z_T)$, the variables $Y_1, \ldots, Y_T$ are conditionally independent with $P(Y_t = l|Z = z) = b_{z_t l}$, for a given observation generation matrix $B = (b_{il})$. The observations are assumed to take values in a set of size N_o, so that B is an $N_s \times N_o$ matrix and each row of B is a probability vector.

The parameter for this model is $\theta = (\pi, A, B)$. The model can be represented as a graph with nodes and directed edges, as illustrated in Figure 5.1. The pmf of the complete data, for a given choice of θ, is

$$p_{cd}(y, z|\theta) = \pi_{z_1} \prod_{t=1}^{T-1} a_{z_t z_{t+1}} \prod_{t=1}^{T} b_{z_t y_t}. \tag{5.14}$$

The correspondence between the pmf and the graph shown in Figure 5.1 is that each term on the righthand side of (5.14) corresponds to an edge in the graph.

In what follows we consider the following three estimation tasks associated with this model:

1. Given the observed data and θ, compute the conditional distribution of the state (solved by the forward-backward algorithm);
2. Given the observed data and θ, compute the most likely sequence for hidden states (solved by the Viterbi algorithm);
3. Given the observed data, compute the maximum likelihood (ML) estimate of θ (solved by the Baum–Welch/EM algorithm).

These problems are addressed in the next three subsections. As we will see, the first of these problems arises in solving the third problem. The second problem has some similarities to the first problem, but it can be addressed separately.

5.3.1 Posterior state probabilities and the forward-backward algorithm

In this subsection we assume that the parameter $\theta = (\pi, A, B)$ of the hidden Markov model is known and fixed. We shall describe computationally efficient methods for computing posterior probabilities for the state at a given time t, or for a transition at a given pair of times t to $t+1$, of the hidden Markov process, based on past observations (case of causal filtering) or based on past and future observations (case of smoothing). These posterior probabilities would allow us to compute, for example, MAP estimates of the state or transition of the Markov process at a given time. For example, we have:

$$\widehat{Z}_{t|t,\text{MAP}} = \arg\max_{i\in\mathcal{S}} P(Z_t = i | Y_1 = y_1, \ldots, Y_t = y_t, \theta), \tag{5.15}$$

$$\widehat{Z}_{t|T,\text{MAP}} = \arg\max_{i\in\mathcal{S}} P(Z_t = i | Y_1 = y_1, \ldots, Y_T = y_T, \theta), \tag{5.16}$$

$$\widehat{(Z_t, Z_{t+1})}_{|T,\text{MAP}} = \arg\max_{(i,j)\in\mathcal{S}\times\mathcal{S}} P(Z_t = i, Z_{t+1} = j | Y_1 = y_1, \ldots, Y_T = y_T, \theta), \tag{5.17}$$

where the convention for subscripts is similar to that used for Kalman filtering: "$t|T$" denotes that the state is to be estimated at time t based on the observations up to time T. The key to efficient computation is to recursively compute certain quantities through a recursion forward in time, and others through a recursion backward in time. We begin by deriving a forward recursion for the variables $\alpha_i(t)$ defined as follows:

$$\alpha_i(t) \stackrel{\triangle}{=} P(Y_1 = y_1, \cdots, Y_t = y_t, Z_t = i | \theta),$$

for $i \in \mathcal{S}$ and $1 \leq t \leq T$. The initial value is $\alpha_i(1) = \pi_i b_{iy_1}$. By the law of total probability, the update rule is:

$$\begin{aligned}
\alpha_j(t+1) &= \sum_{i\in\mathcal{S}} P(Y_1 = y_1, \cdots, Y_{t+1} = y_{t+1}, Z_t = i, Z_{t+1} = j | \theta) \\
&= \sum_{i\in\mathcal{S}} P(Y_1 = y_1, \cdots, Y_t = y_t, Z_t = i | \theta) \\
&\quad \times P(Z_{t+1} = j, Y_{t+1} = y_{t+1} | Y_1 = y_1, \cdots, Y_t = y_t, Z_t = i, \theta) \\
&= \sum_{i\in\mathcal{S}} \alpha_i(t) a_{ij} b_{jy_{t+1}}.
\end{aligned}$$

The righthand side of (5.15) can be expressed in terms of the αs as follows:

$$\begin{aligned}
P(Z_t = i | Y_1 = y_1, \ldots, Y_t = y_t, \theta) &= \frac{P(Z_t = i, Y_1 = y_1, \ldots, Y_t = y_t | \theta)}{P(Y_1 = y_1, \ldots, Y_t = y_t | \theta)} \\
&= \frac{\alpha_i(t)}{\sum_{j\in\mathcal{S}} \alpha_j(t)}.
\end{aligned} \tag{5.18}$$

The computation of the αs and the use of (5.18) is an alternative to, and very similar to, the Kalman filtering equations. The difference is that for Kalman filtering equations, the distributions involved are all Gaussian, so it suffices to compute means and variances, and also the normalization in (5.18), which is done once after the αs are computed, is more or less done at each step in the Kalman filtering equations.

To express the posterior probabilities involving both past and future observations used in (5.16), the following β variables are introduced:

$$\beta_i(t) \stackrel{\triangle}{=} P(Y_{t+1} = y_{t+1}, \cdots, Y_T = y_T | Z_t = i, \theta),$$

for $i \in \mathcal{S}$ and $1 \leq t \leq T$. The definition is not quite the time reversal of the definition of the αs, because the event $Z_t = i$ is being conditioned upon in the definition of $\beta_i(t)$. This asymmetry is introduced because the presentation of the model itself is not symmetric in time. The backward equation for the βs is as follows. The initial condition for the backward equations is $\beta_i(T) = 1$ for all i. By the law of total probability, the update rule is

$$\begin{aligned}
\beta_i(t-1) &= \sum_{j \in \mathcal{S}} P(Y_t = y_t, \cdots, Y_T = y_T, Z_t = j | Z_{t-1} = i, \theta) \\
&= \sum_{j \in \mathcal{S}} P(Y_t = y_t, Z_t = j | Z_{t-1} = i, \theta) \\
&\quad \cdot P(Y_{t+1} = y_{t+1}, \cdots, Y_T = y_T | Z_t = j, Y_t = y_t, Z_{t-1} = i, \theta) \\
&= \sum_{j \in \mathcal{S}} a_{ij} b_{jy_t} \beta_j(t).
\end{aligned}$$

Note that

$$\begin{aligned}
&P(Z_t = i, Y_1 = y_1, \ldots, Y_T = y_T | \theta) \\
&= P(Z_t = i, Y_1 = y_1, \ldots, Y_t = y_t | \theta) \\
&\quad \times P(Y_{t+1} = y_{t+1}, \ldots, Y_T = y_T | \theta, Z_t = i, Y_1 = y_1, \ldots, Y_t = y_t) \\
&= P(Z_t = i, Y_1 = y_1, \ldots, Y_t = y_t | \theta) \\
&\quad \times P(Y_{t+1} = y_{t+1}, \ldots, Y_T = y_T | \theta, Z_t = i) \\
&= \alpha_i(t) \beta_i(t),
\end{aligned}$$

from which we derive the smoothing equation for the conditional distribution of the state at a time t, given all the observations:

$$\begin{aligned}
\gamma_i(t) &\stackrel{\triangle}{=} P(Z_t = i | Y_1 = y_1, \ldots, Y_T = y_T, \theta) \\
&= \frac{P(Z_t = i, Y_1 = y_1, \ldots, Y_T = y_T | \theta)}{P(Y_1 = y_1, \ldots, Y_T = y_T | \theta)} \\
&= \frac{\alpha_i(t) \beta_i(t)}{\sum_{j \in \mathcal{S}} \alpha_j(t) \beta_j(t)}.
\end{aligned}$$

The variable $\gamma_i(t)$ defined here is the same as the probability in the righthand side of (5.16), so that we have an efficient way to find the MAP smoothing estimator defined in (5.16). For later use, we note from the above that for any i such that $\gamma_i(t) > 0$,

$$P(Y_1 = y_1, \ldots, Y_T = y_T|\theta) = \frac{\alpha_i(t)\beta_i(t)}{\gamma_i(t)}. \tag{5.19}$$

Similarly,

$$\begin{aligned}
&P(Z_t = i, Z_{t+1} = j, Y_1 = y_1, \ldots, Y_T = y_T|\theta) \\
&= P(Z_t = i, Y_1 = y_1, \ldots, Y_t = y_t|\theta) \\
&\quad \times P(Z_{t+1} = j, Y_{t+1} = y_{t+1}|\theta, Z_t = i, Y_1 = y_1, \ldots, Y_t = y_t) \\
&\quad \times P(Y_{t+2} = y_{t+2}, \ldots, Y_T = y_T|\theta, Z_t = i, Z_{t+1} = j, Y_1 = y_1, \ldots, Y_{t+1} = y_{t+1}) \\
&= \alpha_i(t) a_{ij} b_{jy_{t+1}} \beta_j(t+1),
\end{aligned}$$

from which we derive the smoothing equation for the conditional distribution of a state-transition for some pair of consecutive times t and $t+1$, given all the observations:

$$\begin{aligned}
\xi_{ij}(t) &\stackrel{\triangle}{=} P(Z_t = i, Z_{t+1} = j|Y_1 = y_1, \ldots, Y_T = y_T, \theta) \\
&= \frac{P(Z_t = i, Z_{t+1} = j, Y_1 = y_1, \ldots, Y_T = y_T|\theta)}{P(Y_1 = y_1, \ldots, Y_T = y_T|\theta)} \\
&= \frac{\alpha_i(t) a_{ij} b_{jy_{t+1}} \beta_j(t+1)}{\sum_{i',j'} \alpha_{i'}(t) a_{i'j'} b_{j'y_{t+1}} \beta_{j'}(t+1)} \\
&= \frac{\gamma_i(t) a_{ij} b_{jy_{t+1}} \beta_j(t+1)}{\beta_i(t)},
\end{aligned}$$

where the final expression is derived using (5.19). The variable $\xi_{ij}(t)$ defined here is the same as the probability in the righthand side of (5.17), so that we have an efficient way to find the MAP smoothing estimator of a state transition, defined in (5.17).

Summarizing, the forward-backward or α–β algorithm for computing the posterior distribution of the state or a transition is given by:

Algorithm 5.8 *(The forward-backward algorithm) The αs can be recursively computed forward in time, and the βs recursively computed backward in time, using:*

$$\begin{aligned}
\alpha_j(t+1) &= \sum_{i \in \mathcal{S}} \alpha_i(t) a_{ij} b_{jy_{t+1}}, \quad \textit{with initial condition } \alpha_i(1) = \pi_i b_{iy_1}, \\
\beta_i(t-1) &= \sum_{j \in \mathcal{S}} a_{ij} b_{jy_t} \beta_j(t), \quad \textit{with initial condition } \beta_i(T) = 1.
\end{aligned}$$

Then the posterior probabilities can be found:

$$P(Z_t = i|Y_1 = y_1, \ldots, Y_t = y_t, \theta) = \frac{\alpha_i(t)}{\sum_{j \in \mathcal{S}} \alpha_j(t)}, \tag{5.20}$$

$$\gamma_i(t) \stackrel{\triangle}{=} P(Z_t = i|Y_1 = y_1, \ldots, Y_T = y_T, \theta) = \frac{\alpha_i(t)\beta_i(t)}{\sum_{j \in \mathcal{S}} \alpha_j(t)\beta_j(t)} \tag{5.21}$$

$$
\begin{aligned}
\xi_{ij}(t) &\stackrel{\triangle}{=} P(Z_t = i, Z_{t+1} = j | Y_1 = y_1, \ldots, Y_T = y_T, \theta) \\
&= \frac{\alpha_i(t) a_{ij} b_{jy_{t+1}} \beta_j(t+1)}{\sum_{i',j'} \alpha_{i'}(t) a_{i'j'} b_{j'y_{t+1}} \beta_{j'}(t+1)} \qquad (5.22) \\
&= \frac{\gamma_i(t) a_{ij} b_{jy_{t+1}} \beta_j(t+1)}{\beta_i(t)}. \qquad (5.23)
\end{aligned}
$$

Remark 5.2 If the number of observations runs into the hundreds or thousands, the αs and βs can become so small that underflow problems can be encountered in numerical computation. However, the formulas (5.20), (5.21), and (5.22) for the posterior probabilities in the forward-backward algorithm are still valid if the αs and βs are multiplied by time dependent (but state independent) constants (for this purpose, (5.22) is more convenient than (5.23), because (5.23) involves βs at two different times). Then, the αs and βs can be renormalized after each time step of computation to have a sum equal to one. Moreover, the sum of the logarithms of the normalization factors for the αs can be stored in order to recover the log of the likelihood, $\log p(y|\theta) = \log \sum_{i=0}^{N_s-1} \alpha_i(T)$.

5.3.2 Most likely state sequence – Viterbi algorithm

Suppose the parameter $\theta = (\pi, A, B)$ is known, and that $Y = (Y_1, \ldots, Y_T)$ is observed. In some applications one wishes to have an estimate of the entire sequence Z. Since θ is known, Y and Z can be viewed as random vectors with a known joint pmf, namely $p_{cd}(y, z|\theta)$. For the remainder of this section, let y denote a fixed observed sequence, $y = (y_1, \ldots, y_T)$. We will seek the MAP estimate, $\widehat{Z}_{MAP}(y, \theta)$, of the entire state sequence $Z = (Z_1, \ldots, Z_T)$, given $Y = y$. By definition, it is the z that maximizes the posterior pmf $p(z|y, \theta)$, and as shown in Section 5.1, it is also equal to the maximizer of the joint pmf of Y and Z:

$$
\widehat{Z}_{MAP}(y, \theta) = \arg\max_z p_{cd}(y, z|\theta).
$$

The Viterbi algorithm (a special case of dynamic programming), described next, is a computationally efficient algorithm for simultaneously finding the maximizing sequence $z^* \in \mathcal{S}^T$ and computing $p_{cd}(y, z^*|\theta)$. It uses the variables:

$$
\delta_i(t) \stackrel{\triangle}{=} \max_{(z_1,\ldots,z_{t-1})} P(Z_1 = z_1, \ldots, Z_{t-1} = z_{t-1}, Z_t = i, Y_1 = y_1, \cdots, Y_t = y_t|\theta).
$$

These variables have a simple graphical representation. Note that, by (5.14), the complete data probability $p(y, z|\theta)$ is the product of terms encountered along the path determined by z through a trellis based on the Markov structure, as illustrated in Figure 5.2. Then $\delta_i(t)$ is the maximum, over all partial paths $(z_1, \ldots, z_t)$ going from stage 1 to stage t, of the product of terms encountered along the partial path.

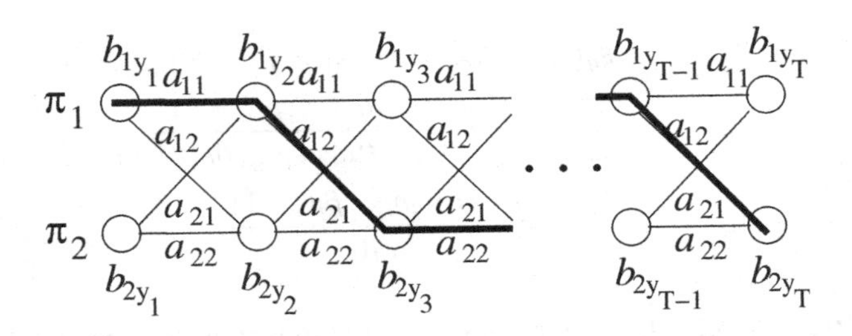

Figure 5.2 A sample path, $z = (1, 1, 2, 2, \ldots, 1, 1)$, of the hidden Markov process.

The δs can be computed by a recursion forward in time, using the initial values $\delta_i(1) = \pi(i)b_{iy_1}$ and the recursion derived as follows:

$$\begin{aligned}
\delta_j(t) &= \max_i \max_{\{z_1,\ldots,z_{t-2}\}} P(Z_1 = z_1, \ldots, Z_{t-1} = i, Z_t = j, Y_1 = y_1, \cdots, Y_t = y_t|\theta) \\
&= \max_i \max_{\{z_1,\ldots,z_{t-2}\}} P(Z_1 = z_1, \ldots, Z_{t-1} = i, Y_1 = y_1, \ldots, Y_{t-1} = y_{t-1}|\theta)a_{ij}b_{jy_t} \\
&= \max_i \left\{\delta_i(t-1)a_{ij}b_{jy_t}\right\}.
\end{aligned}$$

Note that $\delta_i(T) = \max_{z:z_T=i} p_{cd}(y, z|\theta)$. Thus, the following algorithm correctly finds $\widehat{Z}_{MAP}(y, \theta)$.

Algorithm 5.9 *(Viterbi algorithm) Compute the δs and associated back pointers by a recursion forward in time:*

$$\begin{aligned}
&\textit{(initial condition)} \quad \delta_i(1) = \pi(i)b_{iy_1}, \\
&\textit{(recursive step)} \quad \delta_j(t) = \max_i\{\delta_i(t-1)a_{ij}b_{jy_t}\}, \qquad (5.24) \\
&\textit{(storage of back pointers)} \quad \varphi_j(t) \stackrel{\triangle}{=} \arg\max_i\{\delta_i(t-1)a_{ij}b_{jy_t}\}.
\end{aligned}$$

Then $z^* = \widehat{Z}_{\mathrm{MAP}}(y, \theta)$ satisfies $p_{cd}(y, z^*|\theta) = \max_i \delta_i(T)$, and z^* is given by tracing backward in time:

$$z_T^* = \arg\max_i \delta_i(T) \quad \textit{and} \quad z_{t-1}^* = \varphi_{z_t^*}(t) \textit{ for } 2 \leq t \leq T. \qquad (5.25)$$

5.3.3 The Baum–Welch algorithm, or EM algorithm for HMM

The EM algorithm, introduced in Section 5.2, can be usefully applied to many parameter estimation problems with hidden data. This section shows how to apply it to the problem of estimating the parameter of a hidden Markov model from an observed output sequence. This results in the Baum–Welch algorithm, which was developed earlier than the EM algorithm, in the particular context of HMMs.

The parameter to be estimated is $\theta = (\pi, A, B)$. The complete data consist of (Y, Z) whereas the observed, incomplete data consist of Y alone. The initial parameter $\theta^{(0)} = (\pi^{(0)}, A^{(0)}, B^{(0)})$ should have all entries strictly positive, because any entry that is zero will remain zero at the end of an iteration. Suppose $\theta^{(k)}$ is given. The first half of an iteration of the EM algorithm is to compute, or determine in closed form, $Q(\theta|\theta^{(k)})$. Taking logarithms in the expression (5.14) for the pmf of the complete data yields

$$\log p_{cd}(y, z|\theta) = \log \pi_{z_1} + \sum_{t=1}^{T-1} \log a_{z_t z_{t+1}} + \sum_{t=1}^{T} \log b_{z_t y_t}.$$

Taking the expectation yields

$$\begin{aligned}Q(\theta|\theta^{(k)}) &= E[\log p_{cd}(y, Z|\theta)|y, \theta^{(k)}] \\ &= \sum_{i \in \mathcal{S}} \gamma_i(1) \log \pi_i + \sum_{t=1}^{T-1} \sum_{i,j} \xi_{ij}(t) \log a_{ij} + \sum_{t=1}^{T} \sum_{i \in \mathcal{S}} \gamma_i(t) \log b_{iy_t},\end{aligned}$$

where the variables $\gamma_i(t)$ and $\xi_{ij}(t)$ are defined using the model with parameter $\theta^{(k)}$. In view of this closed form expression for $Q(\theta|\theta^{(k)})$, the expectation step of the EM algorithm essentially comes down to computing the γs and the ξs. This computation can be done using the forward-backward algorithm, Algorithm 5.8, with $\theta = \theta^{(k)}$.

The second half of an iteration of the EM algorithm is to find the value of θ that maximizes $Q(\theta|\theta^{(k)})$, and set $\theta^{(k+1)}$ equal to that value. The parameter $\theta = (\pi, A, B)$ for this problem can be viewed as a set of probability vectors. Namely, π is a probability vector, and, for each i fixed, a_{ij} as j varies, and b_{il} as l varies, are probability vectors. Therefore, Example 5.5 and Lemma 5.3 will be of use. Motivated by these, we rewrite the expression found for $Q(\theta|\theta^{(k)})$ to get

$$\begin{aligned}Q(\theta|\theta^{(k)}) &= \sum_{i \in \mathcal{S}} \gamma_i(1) \log \pi_i + \sum_{i,j} \sum_{t=1}^{T-1} \xi_{ij}(t) \log a_{ij} + \sum_{i \in \mathcal{S}} \sum_{t=1}^{T} \gamma_i(t) \log b_{iy_t} \\ &= \sum_{i \in \mathcal{S}} \gamma_i(1) \log \pi_i + \sum_{i,j} \left(\sum_{t=1}^{T-1} \xi_{ij}(t) \right) \log a_{ij} \\ &\quad + \sum_{i \in \mathcal{S}} \sum_{l} \left(\sum_{t=1}^{T} \gamma_i(t) I_{\{y_t = l\}} \right) \log b_{il}. \end{aligned} \tag{5.26}$$

The first summation in (5.26) has the same form as the sum in Lemma 5.3. Similarly, for each i fixed, the sum over j involving a_{ij}, and the sum over l involving b_{il}, also have the same form as the sum in Lemma 5.3. Therefore, the maximization step of the EM algorithm can be written in the following form:

$$\pi_i^{(k+1)} = \gamma_i(1), \tag{5.27}$$

$$a_{ij}^{(k+1)} = \frac{\sum_{t=1}^{T-1} \xi_{ij}(t)}{\sum_{t=1}^{T-1} \gamma_i(t)}, \tag{5.28}$$

$$b_{il}^{(k+1)} = \frac{\sum_{t=1}^{T} \gamma_i(t) I_{\{y_t = l\}}}{\sum_{t=1}^{T} \gamma_i(t)}. \tag{5.29}$$

The update equations (5.27)–(5.29) have a natural interpretation. Equation (5.27) means that the new value of the distribution of the initial state, $\pi^{(k+1)}$, is simply the posterior distribution of the initial state, computed assuming $\theta^{(k)}$ is the true parameter value. The other two update equations are similar, but are more complicated because the transition matrix A and observation generation matrix B do not change with time. The denominator of (5.28) is the posterior expected number of times the state is equal to i up to time $T-1$, and the numerator is the posterior expected number of times two consecutive states are i,j. Thus, if we think of the time of a jump as being random, the righthand side of (5.28) is the time-averaged posterior conditional probability that, given the state at the beginning of a transition is i at a typical time, the next state will be j. Similarly, the right-hand side of (5.29) is the time-averaged posterior conditional probability that, given the state is i at a typical time, the observation will be l.

Algorithm 5.10 *(Baum–Welch algorithm, or EM algorithm for HMM) Select the state space $\mathcal{S}$, and in particular, the cardinality, N_s, of the state space, and let $\theta^{(0)}$ denote a given initial choice of parameter. Given $\theta^{(k)}$, compute $\theta^{(k+1)}$ by using the forward-backward algorithm (Algorithm 5.8) with $\theta = \theta^{(k)}$ to compute the γs and ξs. Then use (5.27)–(5.29) to compute $\theta^{(k+1)} = (\pi^{(k+1)}, A^{(k+1)}, B^{(k+1)})$.*

5.4 Notes

The EM algorithm is due to Dempster, Laird, and Rubin (1977). The paper includes examples and a proof that the likelihood is increased with each iteration of the algorithm. An article on the convergence of the EM algorithm is given in Wu (1983). Earlier related work includes that of Baum *et al.* (1970), giving the Baum–Welch algorithm. A tutorial on inference for HMMs and applications to speech recognition is given in Rabiner (1989).

Problems

5.1 Estimation of a Poisson parameter Suppose Y is assumed to be a $Poi(\theta)$ random variable. Using the Bayesian method, suppose the prior distribution of θ is the exponential distribution with some known parameter $\lambda > 0$.
(a) Find $\widehat{\Theta}_{\text{MAP}}(k)$, the MAP estimate of θ given that $Y = k$ is observed, for some $k \geq 0$.
(b) For what values of λ is $\widehat{\Theta}_{\text{MAP}}(k) \approx \widehat{\theta}_{\text{ML}}(k)$? (The ML estimator was found in Example 5.2.) Why should that be expected?

5.2 A variance estimation problem with Poisson observation The input voltage to an optical device is X and the number of photons observed at a detector is N. Suppose X is a Gaussian random variable with mean zero and variance σ^2, and that given X, the random variable N has the Poisson distribution with mean X^2. (Recall that the Poisson distribution with mean λ has probability mass function $\lambda^n e^{-\lambda}/n!$ for $n \geq 0$.)
(a) Express $P\{N = n\}$ in terms of σ^2. You can express this as an integral, which you do not have to evaluate.
(b) Find the maximum likelihood estimator of σ^2 given N. (Caution: Estimate σ^2, not X. Be as explicit as possible – the final answer has a simple form. Hint: You can

first simplify your answer to part (a) by using the fact that if X is an $N(0,\widetilde{\sigma}^2)$ random variable, then $E[X^{2n}] = \frac{\widetilde{\sigma}^{2n}(2n)!}{n!2^n}$.)

5.3 ML estimation of covariance matrix Suppose n independently generated p-dimensional random vectors $X_1, \ldots, X_n$, are observed, each assumed to have the $N(0,K)$ distribution for some unknown positive semidefinite matrix K. Let S denote the sample covariance function, defined by $S = \frac{1}{n}\sum_{i=1}^n X_i X_i^T$. The goal of this problem is to prove that S is the ML estimator of K. Let the observations be fixed for the remainder of this problem, and for simplicity, assume S has full rank. Therefore S is symmetric and positive definite.

(a) First, show that $\ln f(X_1, \ldots, X_n|K) = -\frac{n}{2}(p\ln(2\pi) + \ln\det(K) + \mathrm{Tr}(SK^{-1}))$, where Tr denotes the trace function.

(b) Then, using the diagonalization of S, explain why there is a symmetric positive definite matrix $S^{\frac{1}{2}}$ so that $S = S^{\frac{1}{2}}S^{\frac{1}{2}}$.

(c) Complete the proof by using the change of variables $\widetilde{K} = S^{-\frac{1}{2}}KS^{-\frac{1}{2}}$ and finding the value of $\widetilde{K}$ that maximizes the likelihood. Since the transformation from K to $\widetilde{K}$ is invertible, applying the inverse mapping to the maximizing value of $\widetilde{K}$ yields the ML estimator for K. (At some point you may need to use the fact that for matrices A and B such that AB is a square matrix, $\mathrm{Tr}(AB) = \mathrm{Tr}(BA)$.)

5.4 Estimation of Bernoulli parameter in Gaussian noise by EM algorithm Suppose $Y = (Y_1, \ldots, Y_T)$, $W = (W_1, \ldots, W_T)$, and $Z = (Z_1, \ldots, Z_T)$. Let $\theta \in [0,1]$ be a parameter to be estimated. Suppose $W_1, \ldots, W_T$ are independent, $N(0,1)$ random variables, and $Z_1, \ldots Z_T$ are independent random variables with $P\{Z_t = 1\} = \theta$ and $P\{Z_t = -1\} = 1-\theta$ for $1 \le t \le T$. Suppose $Y_t = Z_t + W_t$.

(a) Find a simple formula for $\varphi(t,\theta)$ defined by $\varphi(u,\theta) = E[Z_1|Y_1 = u, \theta]$.

(b) Using the function φ found in part (a) in your answer, derive the EM algorithm for calculation of $\widehat{\theta}_{ML}(y)$.

5.5 Convergence of the EM algorithm for an example The purpose of this exercise is to verify for Example 5.8 that if $\theta^{(0)} > 0$, then $\theta^{(k)} \to \widehat{\theta}_{\mathrm{ML}}$ as $k \to \infty$. As shown in the example, $\widehat{\theta}_{\mathrm{ML}} = (y^2 - \sigma^2)_+$. Let $F(\theta) = \left(\frac{\theta}{\theta+\sigma^2}\right)^2 y^2 + \frac{\theta\sigma^2}{\theta+\sigma^2}$ so that the recursion (5.6) has the form $\theta^{(k+1)} = F(\theta^{(k)})$. Clearly, over $\mathbb{R}_+$, F is increasing and bounded.

(a) Show that 0 is the only nonnegative solution of $F(\theta) = \theta$ if $y \le \sigma^2$ and that 0 and $y - \sigma^2$ are the only nonnegative solutions of $F(\theta) = \theta$ if $y > \sigma^2$.

(b) Show that for small $\theta > 0$, $F(\theta) = \theta + \frac{\theta^2(y^2-\sigma^2)}{\sigma^4} + o(\theta^3)$. (Hint: For $0 < \theta < \sigma^2$, $\frac{\theta}{\theta+\sigma^2} = \frac{\theta}{\sigma^2}\frac{1}{1+\theta/\sigma^2} = \frac{\theta}{\sigma^2}(1 - \frac{\theta}{\sigma^2} + (\frac{\theta}{\sigma^2})^2 - \ldots)$

(c) Sketch F and argue, using the above properties of F, that if $\theta^{(0)} > 0$, then $\theta^{(k)} \to \widehat{\theta}_{\mathrm{ML}}$.

5.6 Transformation of estimators and estimators of transformations Consider estimating a parameter $\theta \in [0,1]$ from an observation Y. A prior density of θ is available for the Bayes estimators, MAP and MMSE, and the conditional density of Y given θ is known. Answer the following questions and briefly explain your answers:

(a) Does $3 + 5\widehat{\theta}_{ML} = \widehat{(3+5\theta)}_{ML}$?

(b) Does $(\widehat{\theta}_{ML})^3 = \widehat{(\theta^3)}_{ML}$?

(c) Does $3 + 5\widehat{\theta}_{MAP} = \widehat{(3+5\theta)}_{MAP}$?

(d) Does $(\widehat{\theta}_{MAP})^3 = \widehat{(\theta^3)}_{MAP}$?
(e) Does $3 + 5\widehat{\theta}_{MMSE} = \widehat{(3 + 5\theta)}_{MMSE}$?
(f) Does $(\widehat{\theta}_{MMSE})^3 = \widehat{(\theta^3)}_{MMSE}$?

5.7 Using the EM algorithm for estimation of a signal variance This problem generalizes Example 5.8 to vector observations. Suppose the observation is $Y = S + N$, such that the signal S and noise N are independent random vectors in $\mathbb{R}^d$. Assume that S is $N(0, \theta I)$, and N is $N(0, \Sigma_N)$, where θ, with $\theta > 0$, is the parameter to be estimated, I is the identity matrix, and Σ_N is known.
(a) Suppose θ is known. Find the MMSE estimate of S, $\widehat{S}_{\text{MMSE}}$, and find an expression for the covariance matrix of the error vector, $S - \widehat{S}_{\text{MMSE}}$.
(b) Suppose now that θ is unknown. Describe a direct approach to computing $\widehat{\theta}_{\text{ML}}(Y)$.
(c) Describe how $\widehat{\theta}_{\text{ML}}(Y)$ can be computed using the EM algorithm.
(d) Consider how your answers to parts (b) and (c) simplify when $d = 2$ and the covariance matrix of the noise, Σ_N, is the identity matrix.

5.8 Finding a most likely path Consider an HMM with state space $\mathcal{S} = \{0, 1\}$, observation space $\{0, 1, 2\}$, and parameter $\theta = (\pi, A, B)$ given by:

$$\pi = (a, a^3), \quad A = \begin{pmatrix} a & a^3 \\ a^3 & a \end{pmatrix}, \quad B = \begin{pmatrix} ca & ca^2 & ca^3 \\ ca^2 & ca^3 & ca \end{pmatrix}.$$

Here a and c are positive constants. Their actual numerical values are not important, other than the fact that $a < 1$. Find the MAP state sequence for the observation sequence 021201, using the Viterbi algorithm. Show your work.

5.9 State estimation for an HMM with conditionally Gaussian observations Consider a discrete-time Markov process $Z = (Z_1, Z_2, Z_3, Z_4)$ with state-space $\{0, 1, 2\}$, initial distribution (i.e. distribution of Z_1) $\pi = (c2^{-3}, c, c2^{-5})$ (where $c > 0$ and its numerical value is not relevant), and transition probability diagram shown.

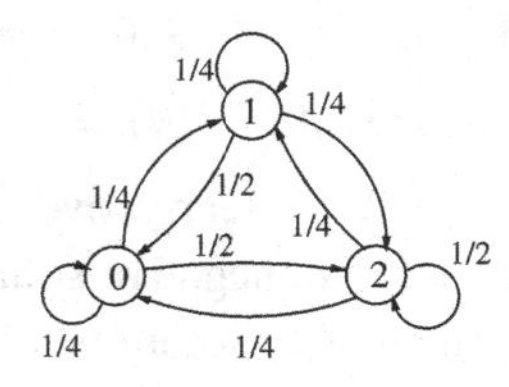

(a) Place weights on the edges of the trellis below so that the *minimum* sum of weights along a path in the trellis corresponds to the most likely state sequence of length four. That is, you are to use the Viterbi algorithm approach to find $z^* = (z_1^*, z_2^*, z_3^*, z_4^*)$ that maximizes $P\{(Z_1, Z_2, Z_3, Z_4) = (z_1, z_2, z_3, z_4)\}$ over all choices of (z_1, z_2, z_3, z_4). Also, find $\mathbf{z}^*$. (A weight i can represent a probability 2^{-i}, for example.)
(b) Using the same statistical model for the process Z as in part (a), suppose there is an observation sequence $(Y_t : 1 \le t \le 4)$ with $Y_t = Z_t + W_t$, where W_1, W_2, W_3, W_4 are $N(0, \sigma^2)$ random variables with $\frac{1}{2\sigma^2} = \ln 2$. (This choice of σ^2 simplifies the problem.) Suppose Z, W_1, W_2, W_3, W_4 are mutually independent. Find the MAP estimate $\widehat{Z}_{\text{MAP}}(y)$ of (Z_1, Z_2, Z_3, Z_4) for the observation sequence $y = (2, 0, 1, -2)$. Use an approach

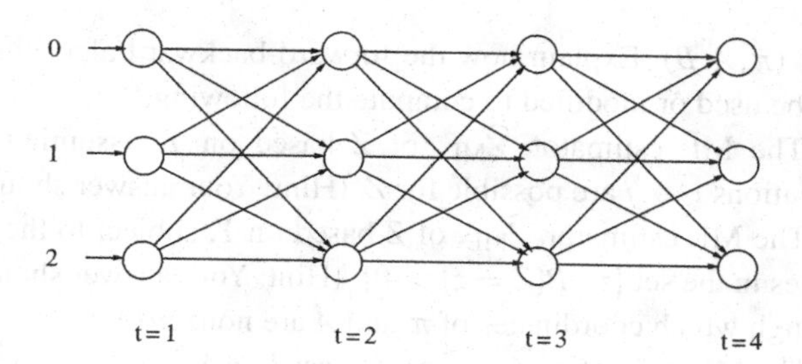

similar to part (a), by placing weights on the nodes and edges of the same trellis so that the MAP estimate is the minimum weight path in the trellis.

5.10 Estimation of the parameter of an exponential in exponential noise
Suppose an observation Y has the form $Y = Z + N$, where Z and N are independent, Z has the exponential distribution with parameter θ, N has the exponential distribution with parameter one, and $\theta > 0$ is an unknown parameter. We consider two approaches to finding $\widehat{\theta}_{\mathrm{ML}}(y)$.

(a) Show that $f_{cd}(y, z|\theta) = \begin{cases} \theta e^{-y+(1-\theta)z} & 0 \le z \le y \\ 0 & \text{else} \end{cases}$.

(b) Find $f(y|\theta)$. The direct approach to finding $\widehat{\theta}_{\mathrm{ML}}(y)$ is to maximize $f(y|\theta)$ (or its log) with respect to θ. You need not attempt the maximization.

(c) Derive the EM algorithm for finding $\widehat{\theta}_{\mathrm{ML}}(y)$. You may express your answer in terms of the function φ defined by:

$$\varphi(y, \theta) = E[Z|y, \theta] = \begin{cases} \frac{1}{\theta-1} - \frac{y}{\exp((\theta-1)y)-1} & \theta \neq 1 \\ \frac{y}{2} & \theta = 1 \end{cases}$$

You need not implement the algorithm.

(d) Suppose an observation $Y = (Y_1, \ldots, Y_T)$ has the form $Y = Z + N$, where $Z = (Z_1, \ldots, Z_T)$ and $N = (N_1, \ldots, N_T)$, such that $N_1, \ldots, N_T, Z_1, \ldots, Z_T$ are mutually independent, and for each t, Z_t has the exponential distribution with parameter θ, and N_t has the exponential distribution with parameter one, and $\theta > 0$ is an unknown parameter. Note that θ does not depend on t. Derive the EM algorithm for finding $\widehat{\theta}_{\mathrm{ML}}(y)$.

5.11 Estimation of a critical transition time of hidden state in HMM
Consider an HMM with unobserved data $Z = (Z_1, \ldots, Z_T)$, observed data $Y = (Y_1, \ldots, Y_T)$, and parameter vector $\theta = (\pi, A, B)$. Let $F \subset \mathcal{S}$, where $\mathcal{S}$ is the statespace of the hidden Markov process Z, and let τ_F be the first time t such that $Z_t \in F$ with the convention that $\tau_F = T + 1$ if ($Z_t \notin F$ for $1 \le t \le T$).

(a) Describe how to find the conditional distribution of τ_F given Y, under the added assumption that ($a_{ij} = 0$ for all (i,j) such that $i \in F$ and $j \notin F$), i.e. under the assumption that F is an absorbing set for Z.

(b) Describe how to find the conditional distribution of τ_F given Y, without the added assumption made in part (a).

5.12 Maximum likelihood estimation for HMMs Consider an HMM with unobserved data $Z = (Z_1, \ldots, Z_T)$, observed data $Y = (Y_1, \ldots, Y_T)$, and parameter vector

$\theta = (\pi, A, B)$. Explain how the forward-backward algorithm or the Viterbi algorithm can be used or modified to compute the following:
(a) The ML estimator, $\widehat{Z}_{\mathrm{ML}}$, of Z based on Y, assuming any initial state and any transitions $i \to j$ are possible for Z. (Hint: Your answer should not depend on π or A.)
(b) The ML estimator, $\widehat{Z}_{\mathrm{ML}}$, of Z based on Y, subject to the constraints that $\widehat{Z}_{\mathrm{ML}}$ takes values in the set $\{z : P\{Z = z\} > 0\}$. (Hint: Your answer should depend on π and A only through which coordinates of π and A are nonzero.)
(c) The ML estimator, $\widehat{Z}_{1,\mathrm{ML}}$, of Z_1 based on Y.
(d) The ML estimator, $\widehat{Z}_{t_o,\,\mathrm{ML}}$, of Z_{t_o} based on Y, for some fixed t_o with $1 \leq t_o \leq T$.

5.13 An underconstrained estimation problem Suppose the parameter $\theta = (\pi, A, B)$ for an HMM is unknown, but that it is assumed that the number of states N_s in the statespace $\mathcal{S}$ for (Z_t) is equal to the number of observations, T. Describe a trivial choice of the ML estimator $\widehat{\theta}_{\mathrm{ML}}(y)$ for a given observation sequence $y = (y_1, \ldots, y_T)$. What is the likelihood of y for this choice of θ?

5.14 Specialization of Baum–Welch algorithm for no hidden data (a) Determine how the Baum–Welch algorithm simplifies in the special case that B is the identity matrix, so that $X_t = Y_t$ for all t.
(b) Still assuming that B is the identity matrix, suppose that $\mathcal{S} = \{0, 1\}$ and the observation sequence is 0001110001110001110001. Find the ML estimator for π and A.

5.15 Bayes estimation for a simple product form distribution Let A be the three by three matrix with entries $a_{ij} = \begin{cases} 2 & i = j \\ 1 & i \neq j \end{cases}$. Suppose X, Y_1, Y_2, Y_3 have the joint pmf $P\{X = i, Y_1 = j, Y_2 = k, Y_3 = l\} = \frac{a_{ij}a_{ik}a_{il}}{Z}$, where Z is a normalizing constant so that the sum of $P\{X = i, Y_1 = j, Y_2 = k, Y_3 = l\}$ over all $i, j, k, l \in \{1, 2, 3\}$ is equal to one.
(a) Find the maximum a-posteriori (MAP) estimate of X given $(Y_1, Y_2, Y_3) = 122$.
(b) Find the conditional probability distribution of X given $(Y_1, Y_2, Y_3) = 122$.

5.16 Extending the forward-backward algorithm The forward-backward algorithm is a form of belief propagation (or message passing) algorithm for the special case of a graph structure that is a one-dimensional chain. It is easy to generalize the algorithm when the graph structure is a tree. For even more general graphs, with cycles, it is often useful to ignore the cycles and continue to use the same local computations, resulting in general belief propagation algorithms. To help explain how belief propagation equations can be derived for general graphs without a given linear ordering of nodes, this problem focuses on a symmetric version of the forward-backward algorithm. If the initial distribution π is uniform, then the complete probability distribution function can be written as

$$p_{cd}(y, z|\theta) = \frac{\prod_{t=1}^{T-1} a_{z_t z_{t+1}} \prod_{t=1}^{T} b_{iy_t}}{G}, \tag{5.30}$$

where G is the number of states in $\mathcal{S}$. Taking $\theta = (A, B)$, and dropping the requirement that the row sums of A and B be normalized to one, (5.30) still defines a valid joint distribution for Y and Z, with the understanding that the constant G is selected to make the sum over all pairs (y, z) sum to one. Note that G depends on θ. This representation of joint probability distributions for (Y, Z) is symmetric forward and backward in time.

(a) Assuming the distribution in (5.30), derive a symmetric variation of the forward-backward algorithm for computation of $\gamma_i(t) = P(Z_t = i|y, \theta)$. Instead of αs and βs, use variables of the form $\mu_i(t, t+1)$ to replace the αs; these are messages passed to the right, and variables of the form $\mu_i(t+1, t)$ to replace the βs; these are messages passed to the left. Here the notation $u(s, t)$ for two adjacent times s and t is for a message to be passed from node s to node t. A better notation might be $u(s \to t)$. The message $u(s, t)$ is a vector $u(s, t) = (u_i(s, t) : i \in S)$ of likelihoods, about the distribution of Z_t that has been collected from the direction s is from t. Give equations for calculating the μs and an equation to calculate the γs from the μs. (Hint: The backward variable $\mu(t+1, t)$ can be taken to be essentially identical to $\beta(t)$ for all t, whereas the forward variable $\mu(t, t+1)$ will be somewhat different from $\alpha(t)$ for all t. Note that $\alpha(t)$ depends on y_t but $\beta(t)$ does not. This asymmetry is used when $\alpha(t)$ and $\beta(t)$ are combined to give $\gamma(t)$.)
(b) Give expressions for $\mu_i(t, t+1)$ and $\mu(t+1, t)$ for $1 \le t \le T$ that involve multiple summations but no recursion. (These expressions can be verified by induction.)
(c) Explain using your answer to part (b) the correctness of your algorithm in part (a).

5.17 Free energy and the Boltzmann distribution Let S denote a finite set of possible states of a physical system, and suppose the (internal) energy of any state $s \in S$ is given by $V(s)$ for some function V on S. Let $T > 0$. The Helmholtz free energy of a probability distribution Q on S is defined to be the average (internal) energy minus the temperature times entropy: $F(Q) = \sum_i Q(i)V(i) + T\sum_i Q(i) \log Q(i)$. Note that F is a convex function of Q. (We're assuming Boltzmann's constant is normalized to one, so that T should actually be in units of energy, but by abuse of notation we will call T the temperature.)
(a) Use the method of Lagrange multipliers to show that the Boltzmann distribution defined by $B_T(i) = \frac{1}{Z(T)} \exp(-V(i)/T)$ minimizes $F(Q)$. Here $Z(T)$ is the normalizing constant required to make B_T a probability distribution.
(b) Describe the limit of the Boltzmann distribution as $T \to \infty$.
(c) Describe the limit of the Boltzmann distribution as $T \to 0$. If it is possible to simulate a random variable with the Boltzmann distribution, does this suggest an application?
(d) Show that $F(Q) = TD(Q||B_T)$ + (term not depending on Q). Therefore, given an energy function V on S and temperature $T > 0$, minimizing free energy over Q in some set is equivalent to minimizing the divergence $D(Q||B_T)$ over Q in the same set.

5.18 Baum–Welch saddlepoint Suppose that the Baum–Welch algorithm is run on a given data set with initial parameter $\theta^{(0)} = (\pi^{(0)}, A^{(0)}, B^{(0)})$ such that $\pi^{(0)} = \pi^{(0)}A^{(0)}$ (i.e. the initial distribution of the state is an equilibrium distribution of the state) and every row of $B^{(0)}$ is identical. Explain what happens, assuming that an ideal computer with infinite precision arithmetic is used.

5.19 Inference for a mixture model (a) An observed random vector Y is distributed as a mixture of Gaussian distributions in d dimensions. The parameter of the mixture distribution is $\theta = (\theta_1, \ldots, \theta_J)$, where θ_j is a d-dimensional vector for $1 \le j \le J$. Specifically, to generate Y a random variable Z, called the class label for the observation, is generated. The variable Z is uniformly distributed on $\{1, \ldots, J\}$, and the conditional distribution of Y given (θ, Z) is Gaussian with mean vector θ_Z and covariance the $d \times d$ identity matrix. The class label Z is not observed. Assuming that θ is known, find the

posterior pmf $p(z|y,\theta)$. Give a geometrical interpretation of the MAP estimate $\widehat{Z}$ for a given observation $Y = y$.
(b) Suppose now that the parameter θ is random with the uniform prior distribution over a very large region and suppose that given θ, n random variables are each generated as in part (a), independently, to produce $(Z^{(1)}, Y^{(1)}, Z^{(2)}, Y^{(2)}, \ldots, Z^{(n)}, Y^{(n)})$. Give an explicit expression for the joint distribution $P(\theta, z^{(1)}, y^{(1)}, z^{(2)}, y^{(2)}, \ldots, z^{(n)}, y^{(n)})$.
(c) The *iterative conditional modes* (ICM) algorithm for this example corresponds to taking turns maximizing $P(\widehat{\theta}, \widehat{z}^{(1)}, y^{(1)}, \widehat{z}^{(2)}, y^{(2)}, \ldots, \widehat{z}^{(n)}, y^{(n)})$ with respect to $\widehat{\theta}$ for $\widehat{z}$ fixed and with respect to $\widehat{z}$ for $\widehat{\theta}$ fixed. Give a simple geometric description of how the algorithm works and suggest a method to initialize the algorithm (there is no unique answer for the latter).
(d) Derive the EM algorithm for this example, in an attempt to compute the maximum likelihood estimate of θ given $y^{(1)}, y^{(2)}, \ldots, y^{(n)}$.

5.20 Constraining the Baum–Welch algorithm The Baum–Welch algorithm as presented placed no prior assumptions on the parameters π, A, B, other than the number of states N_s in the state space of (Z_t). Suppose matrices $\overline{A}$ and $\overline{B}$ are given with the same dimensions as the matrices A and B to be estimated, with all elements of $\overline{A}$ and $\overline{B}$ having values 0 and 1. Suppose that A and B are constrained to satisfy $A \le \overline{A}$ and $B \le \overline{B}$, in the element-by-element ordering (for example, $a_{ij} \le \overline{a}_{ij}$ for all i,j.) Explain how the Baum–Welch algorithm can be adapted to this situation.

5.21 MAP estimation of parameters of a Markov process Let Z be a Markov process with state space $\mathcal{S} = \{0, 1\}$, initial time $t = 1$, initial distribution π, and one-step transition probability matrix A.
(a) Suppose it is known that $A = \begin{pmatrix} 2/3 & 1/3 \\ 1/3 & 2/3 \end{pmatrix}$ and it is observed that $(Z(1), Z(4)) = (0, 1)$. Find the MAP estimate of $Z(2)$.
(b) Suppose instead $\theta = (\pi, A)$ and θ is unknown, and three independent observations of $(Z(1), Z(2), Z(3), Z(4))$ are generated using θ. Assuming the observations are 0001, 1011, 1110, find $\widehat{\theta}_{ML}$.

5.22* Implementation of algorithms Write a computer program:
(a) To simulate a HMM on a computer for a specified value of the parameter $\theta = (\pi, A, B)$;
(b) To run the forward-backward algorithm and compute the αs, βs, γs, and ξs;
(c) To run the Baum–Welch algorithm. Experiment a bit and describe your results. For example, if T observations are generated, and then if the Baum–Welch algorithm is used to estimate the parameter, how large does T need to be to insure that the estimates of θ are pretty accurate.

6 Dynamics of countable-state Markov models

Markov processes are useful for modeling a variety of dynamical systems. Often questions involving the long-time behavior of such systems are of interest, such as whether the process has a limiting distribution, or whether time averages constructed using the process are asymptotically the same as statistical averages.

6.1 Examples with finite state space

Recall that a probability distribution π on $\mathcal{S}$ is an *equilibrium probability distribution* for a time-homogeneous Markov process X if $\pi = \pi H(t)$ for all t. In the discrete-time case, this condition reduces to $\pi = \pi P$. We shall see in this section that under certain natural conditions, the existence of an equilibrium probability distribution is related to whether the distribution of $X(t)$ converges as $t \to \infty$. Existence of an equilibrium distribution is also connected to the mean time needed for X to return to its starting state. To motivate the conditions that will be imposed, we begin by considering four examples of finite state processes. Then the relevant definitions are given for finite or countably infinite state space, and propositions regarding convergence are presented.

Example 6.1 Consider the discrete-time Markov process with the one-step probability diagram shown in Figure 6.1. Note that the process can't escape from the set of states $\mathcal{S}_1 = \{a, b, c, d, e\}$, so that if the initial state $X(0)$ is in $\mathcal{S}_1$ with probability one, then the limiting distribution is supported by $\mathcal{S}_1$. Similarly if the initial state $X(0)$ is in $\mathcal{S}_2 = \{f, g, h\}$ with probability one, then the limiting distribution is supported by $\mathcal{S}_2$. Thus, the limiting distribution is not unique for this process. The natural way to deal with this problem is to decompose the original problem into two problems. That is, consider a Markov process on $\mathcal{S}_1$, and then consider a Markov process on $\mathcal{S}_2$.

Does the distribution of $X(0)$ necessarily converge if $X(0) \in \mathcal{S}_1$ with probability one? The answer is no. For example, note that if $X(0) = a$, then $X(k) \in \{a, c, e\}$ for all even values of k, whereas $X(k) \in \{b, d\}$ for all odd values of k. That is, $\pi_a(k) + \pi_c(k) + \pi_e(k)$ is one if k is even and is zero if k is odd. Therefore, if $\pi_a(0) = 1$, then $\pi(k)$ does not converge as $k \to \infty$.

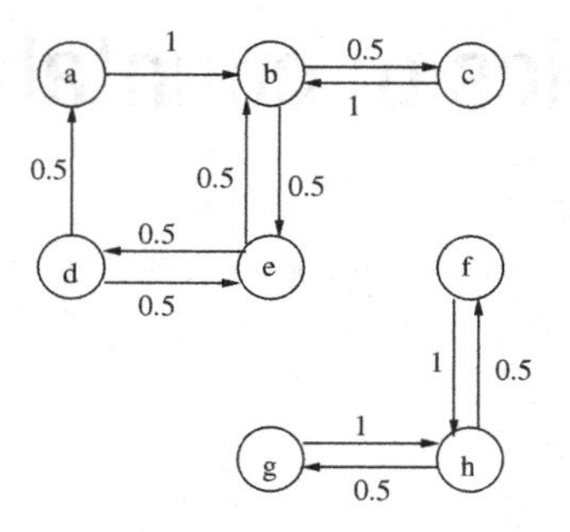

Figure 6.1 A one-step transition probability diagram with eight states.

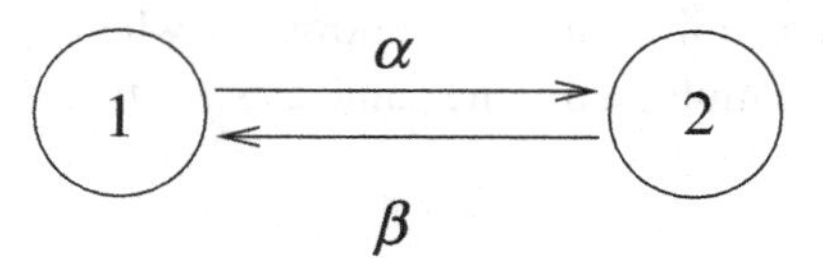

Figure 6.2 A transition rate diagram with two states.

Figure 6.3 A transition rate diagram with four states.

Basically speaking, the Markov process of Example 6.1 fails to have a unique limiting distribution independent of the initial state for two reasons: (i) the process is not irreducible; and (ii) the process is not aperiodic.

Example 6.2 Consider the two-state, continuous-time Markov process with the transition rate diagram shown in Figure 6.2 for some positive constants α and β. This was already considered in Example 4.9, where we found that for any initial distribution $\pi(0)$,

$$\lim_{t\to\infty} \pi(t) = \lim_{t\to\infty} \pi(0)H(t) = \left(\frac{\beta}{\alpha+\beta}, \frac{\alpha}{\alpha+\beta}\right).$$

The rate of convergence is exponential, with rate parameter $\alpha+\beta$, which happens to be the nonzero eigenvalue of Q. Note that the limiting distribution is the unique probability distribution satisfying $\pi Q = 0$. The periodicity problem of Example 6.1 does not arise for continuous-time processes.

Example 6.3 Consider the continuous-time Markov process with the transition rate diagram in Figure 6.3. The Q matrix is the block-diagonal matrix given by

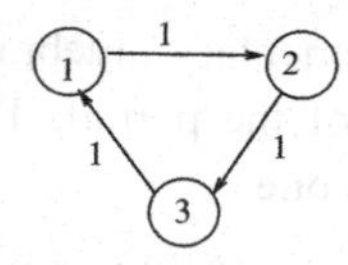

Figure 6.4 A one-step transition probability diagram with three states.

$$Q = \begin{bmatrix} -\alpha & \alpha & 0 & 0 \\ \beta & -\beta & 0 & 0 \\ 0 & 0 & -\alpha & \alpha \\ 0 & 0 & \beta & -\beta \end{bmatrix}.$$

This process is not irreducible, but rather the transition rate diagram can be decomposed into two parts, each equivalent to the diagram for Example 6.2. The equilibrium probability distributions are the probability distributions having the form $\pi = \left(\lambda\frac{\beta}{\alpha+\beta}, \lambda\frac{\alpha}{\alpha+\beta}, (1-\lambda)\frac{\beta}{\alpha+\beta}, (1-\lambda)\frac{\alpha}{\alpha+\beta}\right)$, where λ is the probability placed on the subset $\{1,2\}$.

Example 6.4 Consider the discrete-time Markov process with the transition probability diagram in Figure 6.4. The one-step transition probability matrix P is given by

$$P = \begin{bmatrix} 0 & 1 & 0 \\ 0 & 0 & 1 \\ 1 & 0 & 0 \end{bmatrix}.$$

Solving the equation $\pi = \pi P$ we find there is a unique equilibrium probability vector, namely $\pi = (\frac{1}{3}, \frac{1}{3}, \frac{1}{3})$. On the other hand, if $\pi(0) = (1,0,0)$, then

$$\pi(k) = \pi(0)P^k = \begin{cases} (1,0,0) & \text{if } k \equiv 0 \; mod \; 3 \\ (0,1,0) & \text{if } k \equiv 1 \; mod \; 3 \\ (0,0,1) & \text{if } k \equiv 2 \; mod \; 3 \end{cases}.$$

Therefore, $\pi(k)$ does not converge as $k \to \infty$.

6.2 Classification and convergence of discrete-time Markov processes

The following definition applies for either discrete time or continuous time.

Definition 6.1 Let X be a time-homogeneous Markov process on the countable state space $\mathcal{S}$. The process is said to be *irreducible* if for all $i, j \in \mathcal{S}$, there exists $s > 0$ so that $p_{ij}(s) > 0$.

The next definition is relevant only for discrete-time processes.

Definition 6.2 The *period* of a state i is GCD$\{k \geq 0 : p_{ii}(k) > 0\}$, where GCD stands for greatest common divisor. The set $\{k \geq 0 : p_{ii}(k) > 0\}$ is closed under addition, which by a result in elementary algebra[1] implies that the set contains all sufficiently large integer multiples of the period. The Markov process is called *aperiodic* if the period of all the states is one.

Proposition 6.3 *If X is irreducible, all states have the same period.*

Proof Let i and j be two states. By irreducibility, there are integers k_1 and k_2 so that $p_{ij}(k_1) > 0$ and $p_{ji}(k_2) > 0$. For any integer n, $p_{ii}(n + k_1 + k_2) \geq p_{ij}(k_1)p_{jj}(n)p_{ji}(k_2)$, so $\{k \geq 0 : p_{ii}(k) > 0\}$ contains $\{k \geq 0 : p_{jj}(k) > 0\}$ translated up by $k_1 + k_2$. Thus the period of i is less than or equal to the period of j. Since i and j were arbitrary states, the proposition follows. □

For a fixed state i, define $\tau_i = \min\{k \geq 1 : X(k) = i\}$, where we adopt the convention that the minimum of an empty set is $+\infty$. Let $M_i = E[\tau_i|X(0) = i]$. If $P(\tau_i < +\infty|X(0) = i) < 1$, state i is called *transient* (and by convention, $M_i = +\infty$). Otherwise $P(\tau_i < +\infty|X(0) = i) = 1$, and i is said to be *positive recurrent* if $M_i < +\infty$ and to be *null recurrent* if $M_i = +\infty$.

Proposition 6.4 *Suppose X is irreducible and aperiodic.*

(a) *All states are transient, or all are positive recurrent, or all are null recurrent.*
(b) *For any initial distribution $\pi(0)$, $\lim_{t\to\infty} \pi_i(t) = 1/M_i$, with the understanding that the limit is zero if $M_i = +\infty$.*
(c) *An equilibrium probability distribution π exists if and only if all states are positive recurrent.*
(d) *If it exists, the equilibrium probability distribution π is given by $\pi_i = 1/M_i$. (In particular, if it exists, the equilibrium probability distribution is unique.)*

Proof (a) Suppose state i is recurrent. Given $X(0) = i$, after leaving i the process returns to state i at time τ_i. The process during the time interval $\{0, \ldots, \tau_i\}$ is the first excursion of X from state 0. From time τ_i onward, the process behaves just as it did initially. Thus there is a second excursion from i, third excursion from i, and so on. Let T_k for $k \geq 1$ denote the length of the kth excursion. Then the T_ks are independent, and each has the same distribution as $T_1 = \tau_i$. Let j be another state and let ϵ denote the probability that X visits state j during one excursion from i. Since X is irreducible, $\epsilon > 0$. The excursions are independent, so state j is visited during the kth excursion with probability ϵ, independently of whether j was visited in earlier excursions. Thus, the number of excursions needed until state j is reached has the geometric distribution with parameter ϵ, which has mean $1/\epsilon$. In particular, state j is eventually visited with probability one. After j is visited the process eventually returns to state i, and then within an average of $1/\epsilon$ additional excursions, it will return to state j again. Thus, state j is also recurrent. Hence, if one state is recurrent, all states are recurrent.

The same argument shows that if i is positive recurrent, then j is positive recurrent. Given $X(0) = i$, the mean time needed for the process to visit j and then return to i is M_i/ϵ, since on average $1/\epsilon$ excursions of mean length M_i are needed. Thus, the

[1] Such as the Euclidean algorithm, Chinese remainder theorem, or Bezout theorem.

mean time to hit j starting from i, and the mean time to hit i starting from j, are both finite. Thus, j is positive recurrent. Hence, if one state is positive recurrent, all states are positive recurrent.

(b) Part (b) of the proposition follows by an application of the renewal theorem, which can be found in Asmussen (2003).

(c) Suppose all states are positive recurrent. By the law of large numbers, for any state j, the long run fraction of time the process is in state j is $1/M_j$ with probability one. Similarly, for any states i and j, the long run fraction of time the process is in state j is γ_{ij}/M_i, where γ_{ij} is the mean number of visits to j in an excursion from i. Therefore $1/M_j = \gamma_{ij}/M_i$. This implies that $\sum_i 1/M_i = 1$. That is, π defined by $\pi_i = 1/M_i$ is a probability distribution. The convergence for each i separately given in part (b), together with the fact that π is a probability distribution, imply that $\sum_i |\pi_i(t) - \pi_i| \to 0$. Thus, taking s to infinity in the equation $\pi(s)H(t) = \pi(s+t)$ yields $\pi H(t) = \pi$, so that π is an equilibrium probability distribution.

Conversely, if there is an equilibrium probability distribution π, consider running the process with initial state π. Then $\pi(t) = \pi$ for all t. So by part (b), for any state i, $\pi_i = 1/M_i$. Taking a state i such that $\pi_i > 0$, it follows that $M_i < \infty$. So state i is positive recurrent. By part (a), all states are positive recurrent.

(d) Part (d) was proved in the course of proving part (c). □

We conclude this section by describing a technique to establish a rate of convergence to the equilibrium distribution for finite-state Markov processes. Define $\delta(P)$ for a one-step transition probability matrix P by

$$\delta(P) = \min_{i,k} \sum_j p_{ij} \wedge p_{kj},$$

where $a \wedge b = \min\{a, b\}$. The number $\delta(P)$ is known as Dobrushin's coefficient of ergodicity. Since $a + b - 2(a \wedge b) = |a - b|$ for $a, b \geq 0$, we also have

$$1 - 2\delta(P) = \min_{i,k} \sum_j |p_{ij} - p_{kj}|.$$

Let $\|\mu\|_1$ for a vector μ denote the L_1 norm: $\|\mu\|_1 = \sum_i |\mu_i|$.

Proposition 6.5 *For any probability vectors π and σ, $\|\pi P - \sigma P\|_1 \leq (1 - \delta(P))\|\pi - \sigma\|_1$. Furthermore, if $\delta(P) > 0$ there is a unique equilibrium distribution π^∞, and for any other probability distribution π on $\mathcal{S}$, $\|\pi P^l - \pi^\infty\|_1 \leq 2(1 - \delta(P))^l$.*

Proof Let $\tilde{\pi}_i = \pi_i - \pi_i \wedge \sigma_i$ and $\tilde{\sigma}_i = \sigma_i - \pi_i \wedge \sigma_i$. Note that if $\pi_i \geq \sigma_i$ then $\tilde{\pi}_i = \pi_i - \sigma_i$ and $\tilde{\sigma}_i = 0$, and if $\pi_i \leq \sigma_i$ then $\tilde{\sigma}_i = \sigma_i - \pi_i$ and $\tilde{\pi}_i = 0$. Also, $\|\tilde{\pi}\|_1$ and $\|\tilde{\sigma}\|_1$ are both equal to $1 - \sum_i \pi_i \wedge \sigma_i$. Therefore, $\|\pi - \sigma\|_1 = \|\tilde{\pi} - \tilde{\sigma}\|_1 = 2\|\tilde{\pi}\|_1 = 2\|\tilde{\sigma}\|_1$. Furthermore,

$$\begin{aligned}\|\pi P - \sigma P\|_1 &= \|\tilde{\pi} P - \tilde{\sigma} P\|_1 \\ &= \sum_j \| \sum_i \tilde{\pi}_i P_{ij} - \sum_k \tilde{\sigma}_k P_{kj} \|_1 \\ &= (1/\|\tilde{\pi}\|_1) \sum_j \left| \sum_{i,k} \tilde{\pi}_i \tilde{\sigma}_k (P_{ij} - P_{kj}) \right| \\ &\le (1/\|\tilde{\pi}\|_1) \sum_{i,k} \tilde{\pi}_i \tilde{\sigma}_k \sum_j |P_{ij} - P_{kj}| \\ &\le \|\tilde{\pi}\|_1 (2 - 2\delta(P)) = \|\pi - \sigma\|_1 (1 - \delta(P)),\end{aligned}$$

which proves the first part of the proposition. Iterating the inequality just proved yields that

$$\|\pi P^l - \sigma P^l\|_1 \le (1 - \delta(P))^l \|\pi - \sigma\|_1 \le 2(1 - \delta(P))^l. \tag{6.1}$$

This inequality for $\sigma = \pi P^n$ yields that $\|\pi P^l - \pi P^{l+n}\|_1 \le 2(1 - \delta(P))^l$. Thus the sequence πP^l is a Cauchy sequence and has a limit π^∞, and $\pi^\infty P = \pi^\infty$. Finally, taking σ in (6.1) equal to π^∞ yields the last part of the proposition. □

Proposition 6.5 typically does not yield the exact asymptotic rate $\|\pi^l - \pi^\infty\|_1$ tends to zero. The asymptotic behavior can be investigated by computing $(I - zP)^{-1}$, and matching powers of z in the identity $(I - zP)^{-1} = \sum_{n=0}^\infty z^n P^n$.

6.3 Classification and convergence of continuous-time Markov processes

Chapter 4 discusses Markov processes in continuous time with a finite number of states. Here we extend the coverage of continuous-time Markov processes to include countably infinitely many states. For example, the state of a simple queue could be the number of customers in the queue, and if there is no upper bound on the number of customers that can be waiting in the queue, the state space is $\mathbb{Z}_+$. One possible complication, that rarely arises in practice, is that a continuous-time process can make infinitely many jumps in a finite amount of time.

Let $\mathcal{S}$ be a finite or countably infinite set with $\triangle \notin \mathcal{S}$. A *pure-jump function* is a function $x : \mathbb{R}_+ \to \mathcal{S} \cup \{\triangle\}$ such that for some sequence of times, $0 = \tau_0 < \tau_1 < \ldots$, and sequence of states, $s_0, s_1, \ldots$ with $s_i \in \mathcal{S}$, and $s_i \neq s_{i+1}$, $i \ge 0$, it holds that

$$x(t) = \begin{cases} s_i & \text{if } \tau_i \le t < \tau_{i+1} \quad i \ge 0 \\ \triangle & \text{if } t \ge \tau^* \end{cases}, \tag{6.2}$$

where $\tau^* = \lim_{i\to\infty} \tau_i$. If τ^* is finite it is said to be the *explosion time* of the function x, and if $\tau^* = +\infty$ the function is said to be nonexplosive. An example with $\mathcal{S} = \{1, 2, \ldots\}$, $s_i = i + 1$ for all i, and τ^* finite, is pictured in Figure 6.5.

Definition 6.6 A *pure-jump Markov process* $(X_t : t \ge 0)$ is a Markov process such that, with probability one, its sample paths are pure-jump functions. Such a process is said to be *nonexplosive* if its sample paths are nonexplosive, with probability one.

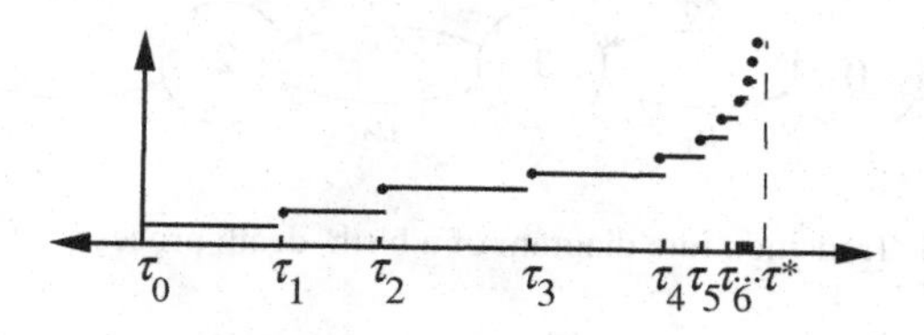

Figure 6.5 A pure-jump function with an explosion time.

Generator matrices are defined for countable-state Markov processes just as they are for finite-state Markov processes. A pure-jump, time-homogeneous Markov process X has *generator matrix* $Q = (q_{ij} : i,j \in \mathcal{S})$ if

$$\lim_{h \searrow 0} (p_{ij}(h) - I_{\{i=j\}})/h = q_{ij} \qquad i,j \in \mathcal{S} \tag{6.3}$$

or equivalently

$$p_{ij}(h) = I_{\{i=j\}} + hq_{ij} + o(h) \qquad i,j \in \mathcal{S}, \tag{6.4}$$

where $o(h)$ represents a quantity such that $\lim_{h\to 0} o(h)/h = 0$.

The space–time properties for continuous-time Markov processes with a countably infinite number of states are the same as for a finite number of states. There is a discrete-time jump process, and the holding times, given the jump process, are exponentially distributed. Also, the following holds.

Proposition 6.7 *Given a matrix $Q = (q_{ij} : i,j \in \mathcal{S})$ satisfying $q_{ij} \geq 0$ for distinct states i and j, and $q_{ii} = -\sum_{j\in\mathcal{S}, j\neq i} q_{ij}$ for each state i, and a probability distribution $\pi(0) = (\pi_i(0) : i \in \mathcal{S})$, there is a pure-jump, time-homogeneous Markov process with generator matrix Q and initial distribution $\pi(0)$. The finite-dimensional distributions of the process are uniquely determined by $\pi(0)$ and Q. The Chapman–Kolmogorov equations, $H(s,t) = H(s,\tau)H(\tau,t)$, and the Kolmogorov forward equations, $\frac{\partial \pi_j(t)}{\partial t} = \sum_{i\in\mathcal{S}} \pi_i(t) q_{ij}$, hold.*

Example 6.5 (Birth–death processes) A useful class of countable-state Markov processes is the set of birth–death processes. A (continuous-time) birth–death process with parameters $(\lambda_0, \lambda_2, \ldots)$ and $(\mu_1, \mu_2, \ldots)$ (also set $\lambda_{-1} = \mu_0 = 0$) is a pure-jump Markov process with state space $\mathcal{S} = \mathbb{Z}_+$ and generator matrix Q defined by $q_{kk+1} = \lambda_k$, $q_{kk} = -(\mu_k + \lambda_k)$, and $q_{kk-1} = \mu_k$ for $k \geq 0$, and $q_{ij} = 0$ if $|i-j| \geq 2$. The transition rate diagram is shown in Figure 6.6. The space-time structure, as defined in Section 4.10, of such a process is as follows. Given the process is in state k at time t, the next state visited is $k+1$ with probability $\lambda_k/(\lambda_k + \mu_k)$ and $k-1$ with probability $\mu_k/(\lambda_k + \mu_k)$. The holding time of state k is exponential with parameter $\lambda_k + \mu_k$. The

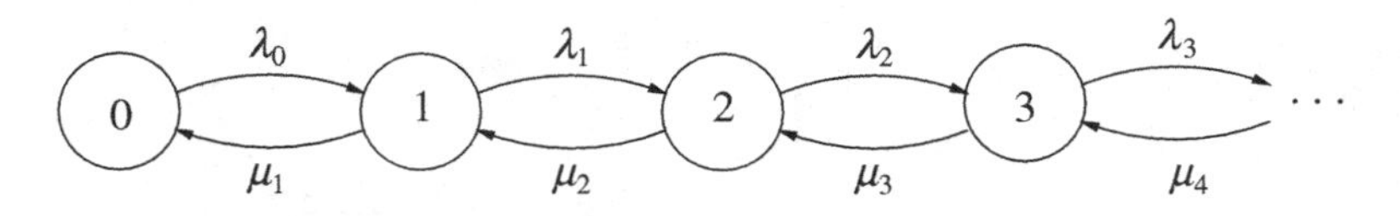

Figure 6.6 Transition rate diagram of a birth–death process.

Kolmogorov forward equations for birth–death processes are

$$\frac{\partial \pi_k(t)}{\partial t} = \lambda_{k-1}\pi_{k-1}(t) - (\lambda_k + \mu_k)\pi_k(t) + \mu_{k+1}\pi_{k+1}(t). \tag{6.5}$$

Example 6.6 (Description of a Poisson process as a Markov process) Let $\lambda > 0$ and consider a birth–death process N with $\lambda_k = \lambda$ and $\mu_k = 0$ for all k, with initial state zero. The space–time structure of this Markov process is rather simple. Each transition is an upward jump of size one, so the jump process is deterministic: $N^J(k) = k$ for all k. Ordinarily, the holding times are only conditionally independent given the jump process, but since the jump process is deterministic, the holding times are independent. Also, since $q_{k,k} = -\lambda$ for all k, each holding time is exponentially distributed with parameter λ. Therefore, N satisfies condition (b) of Proposition 4.5, so that N is a Poisson process with rate λ.

Define $\tau_i^o = \min\{t > 0 : X(t) \neq i\}$ and $\tau_i = \min\{t > \tau_i^o : X(t) = i\}$, for $i \in \mathcal{S}$. If $X(0) = i$, τ_i is the first time the process returns to state i, with the exception that $\tau_i = +\infty$ if the process never returns to state i. The following definitions are the same as when X is a discrete-time process. Let $M_i = E[\tau_i | X(0) = i]$. If $P\{\tau_i < +\infty\} < 1$, state i is called *transient*. Otherwise $P\{\tau_i < +\infty\} = 1$, and i is said to be *positive recurrent* if $M_i < +\infty$ and to be *null recurrent* if $M_i = +\infty$. The following propositions are analogous to those for discrete-time Markov processes. Proofs can be found in Asmussen (2003) and Norris (1997).

Proposition 6.8 *Suppose X is irreducible.*

(a) *All states are transient, or all are positive recurrent, or all are null recurrent.*
(b) *For any initial distribution* $\pi(0)$, $\lim_{t\to+\infty} \pi_i(t) = 1/(-q_{ii}M_i)$, *with the understanding that the limit is zero if* $M_i = +\infty$.

Proposition 6.9 *Suppose X is irreducible and nonexplosive.*

(a) *A probability distribution* π *is an equilibrium distribution if and only if* $\pi Q = 0$.
(b) *An equilibrium probability distribution exists if and only if all states are positive recurrent.*
(c) *If all states are positive recurrent, the equilibrium probability distribution is given by* $\pi_i = 1/(-q_{ii}M_i)$. *(In particular, if it exists, the equilibrium probability distribution is unique.)*

The assumption that X be nonexplosive is needed for Proposition 6.9(a) (per Problem 6.14), but the following proposition shows that the Markov processes encountered in most applications are nonexplosive.

Proposition 6.10 *Suppose X is irreducible. Fix a state i_o and for $k \geq 1$ let S_k denote the set of states reachable from i_o in k jumps. Suppose for each $k \geq 1$ there is a constant γ_k so that the jump intensities on S_k are bounded by γ_k, that is, suppose $-q_{ii} \leq \gamma_k$ for $i \in S_k$. If $\sum_{k=1}^{\infty} \frac{1}{\gamma_k} = +\infty$, the process X is nonexplosive.*

6.4 Classification of birth–death processes

The classification of birth–death processes, introduced in Example 6.5, is relatively simple. To avoid trivialities, consider a birth–death process such that the birth rates, $(\lambda_i : i \geq 0)$ and death rates $(\mu_i : i \geq 1)$ are all strictly positive. Then the process is irreducible.

First, investigate whether the process is nonexplosive, because this is a necessary condition for both recurrence and positive recurrence. This is usually a simple matter, because if the rates are bounded or grow at most linearly, the process is nonexplosive by Proposition 6.10. In some cases, even if Proposition 6.10 does not apply, it can be shown by some other means that the process is nonexplosive. For example, a test is given below for the process to be recurrent, and if it is recurrent, it is not explosive.

Next, investigate whether X is positive recurrent. Suppose we already know that the process is nonexplosive. Then the process is positive recurrent if and only if $\pi Q = 0$ for some probability distribution π, and if it is positive recurrent, π is the equilibrium distribution. Now $\pi Q = 0$ if and only if flow balance holds for any state k:

$$(\lambda_k + \mu_k)\pi_k = \lambda_{k-1}\pi_{k-1} + \mu_{k+1}\pi_{k+1}. \tag{6.6}$$

Equivalently, flow balance must hold for all sets of the form $\{0, \ldots, n-1\}$ (just sum each side of (6.6) over $k \in \{1, \ldots, n-1\}$). Therefore, $\pi Q = 0$ if and only if $\pi_{n-1}\lambda_{n-1} = \pi_n\mu_n$ for $n \geq 1$, which holds if and only if there is a probability distribution π with $\pi_n = \pi_0\lambda_0 \ldots \lambda_{n-1}/(\mu_1 \ldots \mu_n)$ for $n \geq 1$. Thus, a probability distribution π with $\pi Q = 0$ exists if and only if $S_1 < +\infty$, where

$$S_1 = \sum_{i=0}^{\infty} \frac{\lambda_0 \ldots \lambda_{i-1}}{\mu_1 \ldots \mu_i}, \tag{6.7}$$

with the understanding that the $i = 0$ term in the sum defining S_1 is one. Thus, under the assumption that X is nonexplosive, X is positive recurrent if and only if $S_1 < \infty$, and if X is positive recurrent, the equilibrium distribution is given by $\pi_n = (\lambda_0 \ldots \lambda_{n-1})/(S_1\mu_1 \ldots \mu_n)$.

Finally, investigate whether X is recurrent. This step is not necessary if we already know that X is positive recurrent, because a positive recurrent process is recurrent. The following test for recurrence is valid whether or not X is explosive. Since all states have the same classification, the process is recurrent if and only if state 0 is recurrent.

Thus, the process is recurrent if the probability the process never hits 0, for initial state 1, is zero. We shall first find the probability of never hitting state zero for a modified process, which stops upon reaching a large state n, and then let $n \to \infty$ to find the probability the original process never hits state 0. Let b_{in} denote the probability, for initial state i, the process does not reach zero before reaching n. Set the boundary conditions, $b_{0n} = 0$ and $b_{nn} = 1$. Fix i with $1 \le i \le n-1$, and derive an expression for b_{in} by first conditioning on the state reached by the first jump of the process, starting from state i. By the space–time structure, the probability the first jump is up is $\lambda_i/(\lambda_i + \mu_i)$ and the probability the first jump is down is $\mu_i/(\lambda_i + \mu_i)$. Thus,

$$b_{in} = \frac{\lambda_i}{\lambda_i + \mu_i} b_{i+1,n} + \frac{\mu_i}{\lambda_i + \mu_i} b_{i-1,n}$$

which can be rewritten as $\mu_i(b_{in} - b_{i-1,n}) = \lambda_i(b_{i+1,n} - b_{i,n})$. In particular, $b_{2n} - b_{1n} = b_{1n}\mu_1/\lambda_1$ and $b_{3n} - b_{2n} = b_{1n}\mu_1\mu_2/(\lambda_1\lambda_2)$, and so on, which upon summing yields the expression

$$b_{kn} = b_{1n} \sum_{i=0}^{k-1} \frac{\mu_1\mu_2 \dots \mu_i}{\lambda_1\lambda_2 \dots \lambda_i},$$

with the convention that the $i = 0$ term in the sum is one. Finally, the condition $b_{nn} = 1$ yields the solution

$$b_{1n} = \frac{1}{\sum_{i=0}^{n-1} \frac{\mu_1\mu_2 \dots \mu_i}{\lambda_1\lambda_2 \dots \lambda_i}}. \tag{6.8}$$

Note that b_{1n} is the probability, for initial state 1, of the event B_n that state n is reached without an earlier visit to state 0. Since $B_{n+1} \subset B_n$ for all $n \ge 1$,

$$P(\cap_{n \ge 1} B_n | X(0) = 1) = \lim_{n \to \infty} b_{1n} = 1/S_2, \tag{6.9}$$

where

$$S_2 = \sum_{i=0}^{\infty} \frac{\mu_1\mu_2 \dots \mu_i}{\lambda_1\lambda_2 \dots \lambda_i},$$

with the understanding that the $i = 0$ term in the sum defining S_2 is one. Due to the definition of pure jump processes used, whenever X visits a state in $\mathcal{S}$ the number of jumps up until that time is finite. Thus, on the event $\cap_{n \ge 1} B_n$, state zero is never reached. Conversely, if state zero is never reached, either the process remains bounded (which has probability zero) or $\cap_{n \ge 1} B_n$ is true. Thus, $P(\text{zero is never reached}|X(0) = 1) = 1/S_2$. Consequently, X is recurrent if and only if $S_2 = \infty$.

In summary, the following proposition is proved.

Proposition 6.11 *Suppose X is a continuous-time birth–death process with strictly positive birth rates and death rates. If X is nonexplosive (for example, if the rates are bounded or grow at most linearly with n, or if $S_2 = \infty$) then X is positive recurrent if and only if $S_1 < +\infty$. If X is positive recurrent the equilibrium probability distribution is given by $\pi_n = (\lambda_0 \dots \lambda_{n-1})/(S_1\mu_1 \dots \mu_n)$. The process X is recurrent if and only if $S_2 = \infty$.*

Discrete-time birth–death processes have a similar characterization. They are discrete-time, time-homogeneous Markov processes with state space equal to the set of nonnegative integers. Let nonnegative birth probabilities $(\lambda_k : k \geq 0)$ and death probabilities $(\mu_k : k \geq 1)$ satisfy $\lambda_0 \leq 1$, and $\lambda_k + \mu_k \leq 1$ for $k \geq 1$. The one-step transition probability matrix $P = (p_{ij} : i,j \geq 0)$ is given by

$$p_{ij} = \begin{cases} \lambda_i & \text{if } j = i+1 \\ \mu_i & \text{if } j = i-1 \\ 1 - \lambda_i - \mu_i & \text{if } j = i \geq 1 \\ 1 - \lambda_0 & \text{if } j = i = 0 \\ 0 & \text{else.} \end{cases} \tag{6.10}$$

Implicit in the specification of P is that births and deaths cannot happen simultaneously. If the birth and death probabilities are strictly positive, Proposition 6.11 holds as before, with the exception that the discrete-time process cannot be explosive.[2]

6.5 Time averages vs. statistical averages

Let X be a positive recurrent, irreducible, time-homogeneous Markov process with equilibrium probability distribution π. To be definite, suppose X is a continuous-time process, with pure-jump sample paths and generator matrix Q. The results of this section apply with minor modifications to the discrete-time setting as well. Above it is noted that $\lim_{t\to\infty} \pi_i(t) = \pi_i = 1/(-q_{ii}M_i)$, where M_i is the mean "cycle time" of state i. A related consideration is convergence of the *empirical* distribution of the Markov process, where the empirical distribution is the distribution observed over a (usually large) time interval.

For a fixed state i, the fraction of time the process spends in state i during $[0, t]$ is

$$\frac{1}{t}\int_0^t I_{\{X(s)=i\}}ds.$$

Let T_0 denote the time that the process is first in state i, and let T_k for $k \geq 1$ denote the time that the process jumps to state i for the kth time after T_0. The cycle times $T_{k+1}-T_k$, $k \geq 0$ are independent and identically distributed, with mean M_i. Therefore, by the law of large numbers, with probability one,

$$\lim_{k\to\infty} T_k/(kM_i) = \lim_{k\to\infty} \frac{1}{kM_i}\sum_{l=0}^{k-1}(T_{l+1} - T_l) = 1.$$

Furthermore, during the kth cycle interval $[T_k, T_{k+1})$, the amount of time spent by the process in state i is exponentially distributed with mean $-1/q_{ii}$, and the times spent in the state during disjoint cycles are independent. Thus, with probability one,

[2] If in addition $\lambda_i + \mu_i = 1$ for all i, the discrete-time process has period 2.

$$\lim_{k\to\infty}\frac{1}{kM_i}\int_0^{T_k} I_{\{X(s)=i\}}ds = \lim_{k\to\infty}\frac{1}{kM_i}\sum_{l=0}^{k-1}\int_{T_l}^{T_{l+1}} I_{\{X(s)=i\}}ds$$
$$= \frac{1}{M_i}E\left[\int_{T_0}^{T_1} I_{\{X(s)=i\}}ds\right]$$
$$= 1/(-q_{ii}M_i).$$

Combining these two observations yields that

$$\lim_{t\to\infty}\frac{1}{t}\int_0^t I_{\{X(s)=i\}}ds = 1/(-q_{ii}M_i) = \pi_i \tag{6.11}$$

with probability one. In short, the limit (6.11) is expected, because the process spends on average $-1/q_{ii}$ time units in state i per cycle from state i, and the cycle rate is $1/M_i$. Of course, since state i is arbitrary, if j is any other state,

$$\lim_{t\to\infty}\frac{1}{t}\int_0^t I_{\{X(s)=j\}}ds = 1/(-q_{jj}M_j) = \pi_j. \tag{6.12}$$

By considering how the time in state j is distributed among the cycles from state i, it follows that the mean time spent in state j per cycle from state i is $M_i\pi_j$.

So for any nonnegative function φ on $\mathcal{S}$,

$$\lim_{t\to\infty}\frac{1}{t}\int_0^t \varphi(X(s))ds = \lim_{k\to\infty}\frac{1}{kM_i}\int_0^{T_k}\varphi(X(s))ds$$
$$= \frac{1}{M_i}E\left[\int_{T_0}^{T_1}\varphi(X(s))ds\right]$$
$$= \frac{1}{M_i}E\left[\sum_{j\in\mathcal{S}}\varphi(j)\int_{T_0}^{T_1} I_{\{X(s)=j\}}ds\right]$$
$$= \frac{1}{M_i}\sum_{j\in\mathcal{S}}\varphi(j)E\left[\int_{T_0}^{T_1} I_{\{X(s)=j\}}\right]ds$$
$$= \sum_{j\in\mathcal{S}}\varphi(j)\pi_j. \tag{6.13}$$

Finally, for an arbitrary function φ on $\mathcal{S}$, (6.13) holds for both φ_+ and φ_-. So if either $\sum_{j\in\mathcal{S}}\varphi_+(j)\pi_j < \infty$ or $\sum_{j\in\mathcal{S}}\varphi_-(j)\pi_j < \infty$, then (6.13) holds for φ itself.

6.6 Queueing systems, M/M/1 queue and Little's law

Some basic terminology of queueing theory will now be explained. A simple type of queueing system is pictured in Figure 6.7. Notice that the *system* comprises a *queue* and a *server*. Ordinarily, whenever the system is not empty there is a customer in the server, and any other customers in the system are waiting in the queue. When the service of a customer is complete it departs from the server and then another customer from the queue, if any, immediately enters the server. The choice of which customer to be

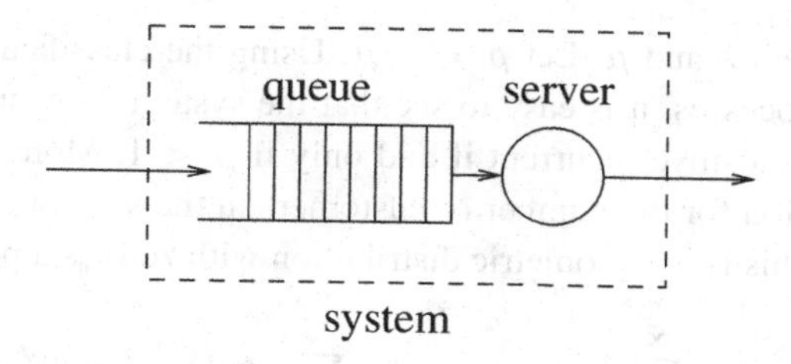

Figure 6.7 A single server queueing system.

served next depends on the *service discipline*. Common service disciplines are first-come first-served (FCFS) in which customers are served in the order of their arrival, or last-come first-served (LCFS) in which the customer that arrived most recently is served next. Some of the more complicated service disciplines involve priority classes, or the notion of "processor sharing" in which all customers present in the system receive equal attention from the server.

Often models of queueing systems involve a stochastic description. For example, given positive parameters λ and μ, we may declare that the arrival process is a Poisson process with rate λ, and that the service times of the customers are independent and exponentially distributed with parameter μ. Many queueing systems are given labels of the form A/B/s, where "A" is chosen to denote the type of arrival process, "B" is used to denote the type of departure process, and s is the number of servers in the system. In particular, the system just described is called an M/M/1 queueing system, so-named because the arrival process is memoryless (i.e. a Poisson arrival process), the service times are memoryless (i.e. are exponentially distributed), and there is a single server. Other labels for queueing systems have a fourth descriptor and thus have the form A/B/s/b, where b denotes the maximum number of customers that can be in the system. Thus, an M/M/1 system is also an M/M/1/∞ system, because there is no finite bound on the number of customers in the system.

A second way to specify an M/M/1 queueing system with parameters λ and μ is to let $A(t)$ and $D(t)$ be independent Poisson processes with rates λ and μ respectively. Process A marks customer arrival times and process D marks *potential* customer departure times. The number of customers in the system, starting from some initial value $N(0)$, evolves as follows. Each time there is a jump of A, a customer arrives to the system. Each time there is a jump of D, there is a potential departure, meaning that if there is a customer in the server at the time of the jump then the customer departs. If a potential departure occurs when the system is empty then the potential departure has no effect on the system. The number of customers in the system N can thus be expressed as

$$N(t) = N(0) + A(t) + \int_0^t I_{\{N(s-)\geq 1\}} dD(s).$$

It is easy to verify that the resulting process N is Markov, which leads to a third specification of an M/M/1 queueing system.

A third way to specify an M/M/1 queuing system is that the number of customers in the system $N(t)$ is a birth–death process with $\lambda_k = \lambda$ and $\mu_k = \mu$ for all k, for some parameters λ and μ. Let $\rho = \lambda/\mu$. Using the classification criteria derived for birth–death processes, it is easy to see that the system is recurrent if and only if $\rho \leq 1$, and that it is positive recurrent if and only if $\rho < 1$. Moreover, if $\rho < 1$ the equilibrium distribution for the number of customers in the system is given by $\pi_k = (1-\rho)\rho^k$ for $k \geq 0$. This is the geometric distribution with zero as a possible value, and with mean

$$\overline{N} = \sum_{k=0}^{\infty} k\pi_k = (1-\rho)\rho \sum_{k=1}^{\infty} \rho^{k-1}k = (1-\rho)\rho(\frac{1}{1-\rho})' = \frac{\rho}{1-\rho}.$$

The probability the server is busy, which is also the mean number of customers in the server, is $1 - \pi_0 = \rho$. The mean number of customers in the queue is thus given by $\rho/(1-\rho) - \rho = \rho^2/(1-\rho)$. This third specification is the most commonly used way to define an M/M/1 queueing process.

Since the M/M/1 process $N(t)$ is positive recurrent, the Markov ergodic convergence theorem implies that the statistical averages just computed, such as $\overline{N}$, are also equal to the limit of the time-averaged number of customers in the system as the averaging interval tends to infinity.

An important performance measure for a queueing system is the mean time spent in the system or the mean time spent in the queue. Little's law, described next, is a quite general and useful relationship that aids in computing mean transit time.

Little's law can be applied in a great variety of circumstances involving flow through a system with delay. In the context of queueing systems we speak of a flow of customers, but the same principle applies to a flow of water through a pipe. Little's law is that $\lambda T = \overline{N}$ where λ is the mean flow rate, T is the mean delay in the system, and $\overline{N}$ is the mean content of the system. For example, if water flows through a pipe with volume one cubic meter at the rate of two cubic meters per minute, the mean time (averaged over all drops of water) that water spends in the pipe is $T = \overline{N}/\lambda = 1/2$ minute. This is clear if water flows through the pipe without mixing, because the transit time of each drop of water is 1/2 minute. However, mixing within the pipe does not affect the average transit time.

Little's law is actually a set of results, each with somewhat different mathematical assumptions. The following version is quite general. Figure 6.8 pictures the cumulative number of arrivals ($\alpha(t)$) and the cumulative number of departures ($\delta(t)$) versus time,

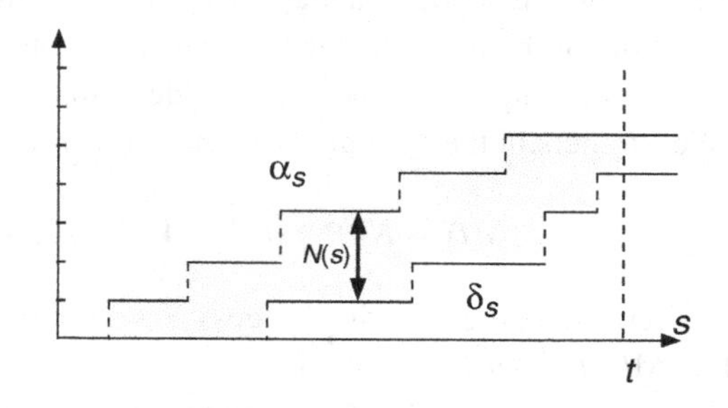

Figure 6.8 Cumulative arrival and departure processes and their difference.

for a queueing system assumed to be initially empty. Note that the number of customers in the system at any time s is given by the difference $N(s) = \alpha(s) - \delta(s)$, which is the vertical distance between the arrival and departure graphs in the figure. On the other hand, assuming that customers are served in first-come first-served order, the horizontal distance between the graphs gives the times in system for the customers. Given a (usually large) $t > 0$, let γ_t denote the area of the region between the two graphs over the interval $[0, t]$. This is the shaded region indicated in the figure. It is natural to define the time-averaged values of arrival rate and system content as

$$\overline{\lambda}_t = \alpha(t)/t \quad \text{and} \quad \overline{N}_t = \frac{1}{t}\int_0^t N(s)ds = \gamma_t/t.$$

Finally, the average, over the $\alpha(t)$ customers that arrive during the interval $[0, t]$, of the time spent in the system up to time t, is given by

$$\overline{T}_t = \gamma_t/\alpha(t).$$

Once these definitions are accepted, we have the following obvious proposition.

Proposition 6.12 *(Little's law, expressed using averages over time) For any $t > 0$,*

$$\overline{N}_t = \overline{\lambda}_t \overline{T}_t. \tag{6.14}$$

Furthermore, if any two of the three variables in (6.14) converge to a positive finite limit as $t \to \infty$, then so does the third variable, and the limits satisfy $\overline{N}_\infty = \overline{\lambda}_\infty \overline{T}_\infty$.

For example, the number of customers in an M/M/1 queue is a positive recurrent Markov process so that

$$\lim_{t\to\infty} \overline{N}_t = \overline{N} = \rho/(1-\rho),$$

where calculation of the statistical mean $\overline{N}$ was previously discussed. Also, by the law of large numbers applied to interarrival times, we have that the Poisson arrival process for an M/M/1 queue satisfies $\lim_{t\to\infty} \overline{\lambda}_t = \lambda$ with probability one. Thus, with probability one,

$$\lim_{t\to\infty} \overline{T}_t = \overline{N}/\lambda = \frac{1}{\mu - \lambda}.$$

In this sense, the average waiting time in an M/M/1 system is $1/(\mu - \lambda)$. The average time in service is $1/\mu$ (this follows from the third description of an M/M/1 queue, or also from Little's law applied to the server alone) so that the average waiting time in queue is given by $W = 1/(\mu - \lambda) - 1/\mu = \rho/(\mu - \lambda)$. This final result also follows from Little's law applied to the queue alone.

6.7 Mean arrival rate, distributions seen by arrivals, and PASTA

The mean arrival rate for the M/M/1 system is λ, the parameter of the Poisson arrival process. However, for some queueing systems the arrival rate depends on the number of customers in the system. In such cases the mean arrival rate is still typically meaningful, and it can be used in Little's law.

Suppose the number of customers in a queuing system is modeled by a birth–death process with arrival rates (λ_k) and departure rates (μ_k). Suppose in addition that the process is positive recurrent. Intuitively, the process spends a fraction of time π_k in state k and while in state k the arrival rate is λ_k. Therefore, the average arrival rate is

$$\overline{\lambda} = \sum_{k=0}^{\infty} \pi_k \lambda_k.$$

Similarly the average departure rate is

$$\overline{\mu} = \sum_{k=1}^{\infty} \pi_k \mu_k$$

and of course $\overline{\lambda} = \overline{\mu}$ because both are equal to the throughput of the system.

Often the distribution of a system at particular system-related sampling times is more important than the distribution in equilibrium. For example, the distribution seen by arriving customers may be the most relevant distribution, as far as the customers are concerned. If the arrival rate depends on the number of customers in the system then the distribution seen by arrivals need not be the same as the equilibrium distribution. Intuitively, $\pi_k \lambda_k$ is the long-term frequency of arrivals which occur when there are k customers in the system, so that the fraction of customers that see k customers in the system upon arrival is given by

$$r_k = \frac{\pi_k \lambda_k}{\overline{\lambda}}.$$

The following is an example of a system with variable arrival rate.

Example 6.7 (Single-server, discouraged arrivals) Suppose $\lambda_k = \alpha/(k+1)$ and $\mu_k = \mu$ for all k, where μ and α are positive constants. Then

$$S_2 = \sum_{k=0}^{\infty} \frac{(k+1)!\mu^k}{\alpha^k} = \infty \text{ and } S_1 = \sum_{k=0}^{\infty} \frac{\alpha^k}{k!\mu^k} = \exp\left(\frac{\alpha}{\mu}\right) < \infty$$

so that the number of customers in the system is a positive recurrent Markov process, with no additional restrictions on α and μ. Moreover, the equilibrium probability distribution is given by $\pi_k = (\alpha/\mu)^k \exp(-\alpha/\mu)/k!$, which is the Poisson distribution with mean $\overline{N} = \alpha/\mu$. The mean arrival rate is

$$\begin{aligned}\overline{\lambda} &= \sum_{k=0}^{\infty} \frac{\pi_k \alpha}{k+1} = \mu \exp(-\alpha/\mu) \sum_{k=0}^{\infty} \frac{(\alpha/\mu)^{k+1}}{(k+1)!} \\ &= \mu \exp(-\alpha/\mu)(\exp(\alpha/\mu) - 1) = \mu(1 - \exp(-\alpha/\mu)).\end{aligned}$$

This expression derived for $\overline{\lambda}$ is clearly equal to $\overline{\mu}$, because the departure rate is μ with probability $1 - \pi_0$ and zero otherwise. The distribution of the number of customers in the system seen by arrivals, (r_k) is given by

$$r_k = \frac{\pi_k \alpha}{(k+1)\overline{\lambda}} = \frac{(\alpha/\mu)^{k+1}\exp(-\alpha/\mu)}{(k+1)!(1-\exp(-\alpha/\mu))} \quad \text{for } k \geq 0,$$

which in words can be described as the result of removing the probability mass at zero in the Poisson distribution, shifting the distribution down by one, and then renormalizing. The mean number of customers in the queue seen by a typical arrival is therefore $(\alpha/\mu - 1)/(1 - \exp(-\alpha/\mu))$. This mean is somewhat less than $\overline{N}$ because, roughly speaking, the customer arrival rate is higher when the system is more lightly loaded.

The equivalence of time averages and statistical averages for computing the mean arrival rate and the distribution seen by arrivals can be shown by application of ergodic properties of the processes involved. The associated formal approach is described next, in slightly more generality. Let X denote an irreducible, positive-recurrent pure-jump Markov process. If the process makes a jump from state i to state j at time t, say that a transition of type (i,j) occurs. The sequence of transitions of X forms a new Markov process, Y. The process Y is a discrete-time Markov process with state space $\{(i,j) \in \mathcal{S} \times \mathcal{S} : q_{ij} > 0\}$, and it can be described in terms of the jump process for X, by $Y(k) = (X^J(k-1), X^J(k))$ for $k \geq 0$. (Let $X^J(-1)$ be defined arbitrarily.)

The one-step transition probability matrix of the jump process X^J is given by $\pi^J_{ij} = q_{ij}/(-q_{ii})$, and X^J is recurrent because X is recurrent. Its equilibrium distribution π^J (if it exists) is proportional to $-\pi_i q_{ii}$ (see Problem 6.3), and X^J is *positive* recurrent if and only if this distribution can be normalized to make a probability distribution, i.e. if and only if $R = -\sum_i \pi_i q_{ii} < \infty$. Assume for simplicity that X^J is positive recurrent. Then $\pi^J_i = -\pi_i q_{ii}/R$ is the equilibrium probability distribution of X^J. Furthermore, Y is positive recurrent and its equilibrium distribution is given by

$$\begin{aligned}\pi^Y_{ij} &= \pi^J_i p^J_{ij} \\ &= \frac{-\pi_i q_{ii}}{R}\frac{q_{ij}}{-q_{ii}} \\ &= \frac{\pi_i q_{ij}}{R}.\end{aligned}$$

Since limiting time averages equal statistical averages for Y,

$$\lim_{n\to\infty} (\text{number of first } n \text{ transitions of } X \text{ that are type } (i,j))/n = \pi_i q_{ij}/R$$

with probability one. Therefore, if $A \subset \mathcal{S} \times \mathcal{S}$, and if $(i,j) \in A$,

$$\lim_{n\to\infty} \frac{\text{number of first } n \text{ transitions of X that are type } (i,j)}{\text{number of first } n \text{ transitions of X with type in } A} = \frac{\pi_i q_{ij}}{\sum_{(i',j')\in A} \pi_{i'} q_{i'j'}}.$$

To apply this setup to the special case of a queueing system in which the number of customers in the system is a Markov birth–death process, let the set A be the set of transitions of the form $(i, i+1)$. Then deduce that the fraction of the first n arrivals that see i customers in the system upon arrival converges to $\pi_i \lambda_i / \sum_j \pi_j \lambda_j$ with probability one.

Note that if $\lambda_i = \lambda$ for all i, then $\overline{\lambda} = \lambda$ and $\pi = r$. The condition $\lambda_i = \lambda$ also implies that the arrival process is Poisson. This situation is called "Poisson Arrivals See Time Averages" (PASTA).

6.8 More examples of queueing systems modeled as Markov birth–death processes

For each of the four examples of this section it is assumed that new customers are offered to the system according to a Poisson process with rate λ, so that the PASTA property holds. Also, when there are k customers in the system then the service rate is μ_k for some given numbers μ_k. The number of customers in the system is a Markov birth–death process with $\lambda_k = \lambda$ for all k. Since the number of transitions of the process up to any given time t is at most twice the number of customers that arrived by time t, the Markov process is not explosive. Therefore the process is positive recurrent if and only if S_1 is finite, where

$$S_1 = \sum_{k=0}^{\infty} \frac{\lambda^k}{\mu_1 \mu_2 \ldots \mu_k}.$$

Special cases of this example are presented in the next four examples.

Example 6.8 (M/M/m systems) An M/M/m queueing system consists of a single queue and m servers. The arrival process is Poisson with some rate λ and the customer service times are independent and exponentially distributed with mean μ for some $\mu > 0$. The total number of customers in the system is a birth–death process with $\mu_k = \mu \min(k, m)$. Let $\rho = \lambda/(m\mu)$. Since $\mu_k = m\mu$ for all k large enough it is easy to check that the process is positive recurrent if and only if $\rho < 1$. Assume now that $\rho < 1$. Then the equilibrium distribution is given by

$$\pi_k = \frac{(\lambda/\mu)^k}{S_1 k!} \quad \text{for } 0 \leq k \leq m,$$

$$\pi_{m+j} = \pi_m \rho^j \quad \text{for } j \geq 1,$$

where S_1 makes the probabilities sum to one (use $1 + \rho + \rho^2 \ldots = 1/(1-\rho)$):

$$S_1 = \left(\sum_{k=0}^{m-1} \frac{(\lambda/\mu)^k}{k!} \right) + \frac{(\lambda/\mu)^m}{m!(1-\rho)}.$$

An arriving customer must join the queue (rather than go directly to a server) if and only if the system has m or more customers in it. By the PASTA property, this is the same as the equilibrium probability of having m or more customers in the system:

$$P_Q = \sum_{j=0}^{\infty} \pi_{m+j} = \pi_m/(1-\rho).$$

This formula is called the *Erlang C* formula for probability of queueing.

Example 6.9 (M/M/*m*/*m* systems) An M/M/*m*/*m* queueing system consists of m servers. The arrival process is Poisson with some rate λ and the customer service times are independent and exponentially distributed with mean μ for some $\mu > 0$. Since there is no queue, if a customer arrives when there are already m customers in the system, the arrival is blocked and cleared from the system. The total number of customers in the system is a birth–death process, but with the state space reduced to $\{0, 1, \ldots, m\}$, and with $\mu_k = k\mu$ for $1 \le k \le m$. The unique equilibrium distribution is given by

$$\pi_k = \frac{(\lambda/\mu)^k}{S_1 k!} \quad \text{for } 0 \le k \le m,$$

where S_1 is chosen to make the probabilities sum to one.

An arriving customer is blocked and cleared from the system if and only if the system already has m customers in it. By the PASTA property, this is the same as the equilibrium probability of having m customers in the system:

$$P_B = \pi_m = \frac{\frac{(\lambda/\mu)^m}{m!}}{\sum_{j=0}^{m} \frac{(\lambda/\mu)^j}{j!}}.$$

This formula is called the *Erlang B* formula for probability of blocking.

Example 6.10 (A system with a discouraged server) The number of customers in this system is a birth–death process with constant birth rate λ and death rates $\mu_k = 1/k$. It is easy to check that all states are transient for any positive value of λ (to verify this it suffices to check that $S_2 < \infty$). It is not difficult to show that $N(t)$ converges to $+\infty$ with probability one as $t \to \infty$.

Example 6.11 (A barely stable system) The number of customers in this system is a birth–death process with constant birth rate λ and death rates $\mu_k = \frac{\lambda(1+k^2)}{1+(k-1)^2}$ for all $k \ge 1$. Since the departure rates are barely larger than the arrival rates, this system is near the borderline between recurrence and transience. However,

$$S_1 = \sum_{k=0}^{\infty} \frac{1}{1+k^2} < \infty,$$

so $N(t)$ is positive recurrent with equilibrium distribution $\pi_k = 1/(S_1(1+k^2))$. The mean number of customers in the system is

$$\overline{N} = \sum_{k=0}^{\infty} \frac{k}{S_1(1+k^2)} = \infty.$$

By Little's law the mean time customers spend in the system is also infinite. It is debatable whether this system should be thought of as "stable" even though all states are positive recurrent and all waiting times are finite with probability one.

6.9 Foster–Lyapunov stability criterion and moment bounds

Communication network models can become quite complex, especially when dynamic scheduling, congestion, and physical layer effects such as fading wireless channel models are included. It is thus useful to have methods to give approximations or bounds on key performance parameters. The criteria for stability and related moment bounds discussed in this chapter are useful for providing such bounds.

Aleksandr Mikhailovich Lyapunov (1857–1918) contributed significantly to the theory of stability of dynamical systems. Although a dynamical system may evolve on a complicated, multiple dimensional state space, a recurring theme of dynamical systems theory is that stability questions can often be settled by studying the potential of a system for some nonnegative potential function V. Potential functions used for stability analysis are widely called Lyapunov functions. Similar stability conditions have been developed by many authors for stochastic systems. Below we present the well known criteria due to Foster (1953) for recurrence and positive recurrence. In addition we present associated bounds on the moments, which are expectations of some functions on the state space, computed with respect to the equilibrium probability distribution.[3]

Subsection 6.9.1 discusses the discrete-time tools, and presents examples involving load balancing routeing, and input queued crossbar switches. Subsection 6.9.2 presents the continuous-time tools, and an example.

6.9.1 Stability criteria for discrete-time processes

Consider an irreducible discrete-time Markov process X on a countable state space $\mathcal{S}$, with one-step transition probability matrix P. If f is a function on $\mathcal{S}$, Pf represents the function obtained by multiplication of the vector f by the matrix P: $Pf(i) = \sum_{j\in\mathcal{S}} p_{ij}f(j)$. If f is nonnegative, Pf is well defined, with the understanding that $Pf(i) = +\infty$ is possible for some, or all, values of i. An important property of Pf is that $Pf(i) = E[f(X(t+1))|X(t) = i]$. Let V be a nonnegative function on $\mathcal{S}$, to serve as the Lyapunov function. The *drift vector* of $V(X(t))$ is defined by $d(i) = E[V(X(t+1))|X(t) = i] - V(i)$. That is, $d = PV - V$. Note that $d(i)$ is always well-defined, if the value $+\infty$ is permitted. The drift vector is also given by

$$d(i) = \sum_{j:j\neq i} p_{ij}(V(j) - V(i)). \tag{6.15}$$

Proposition 6.13 *(Foster–Lyapunov stability criterion) Suppose $V : \mathcal{S} \to \mathbb{R}_+$ and C is a finite subset of $\mathcal{S}$.*

[3] A version of these moment bounds was given by Tweedie (1983), and a version of the moment bound method was used by Kingman (1962) in a queueing context. As noted in Meyn & Tweedie (1993), the moment bound method is closely related to Dynkin's formula. The works (Tassiulas & Ephremides 1992, Tassiulas & Ephremides 1993, Kumar & Meyn 1995, Tassiulas 1997), and many others, have demonstrated the wide applicability of the stability methods in various queueing network contexts, using quadratic Lyapunov functions.

(a) If $\{i : V(i) \leq K\}$ is finite for all K, and if $PV - V \leq 0$ on $\mathcal{S} - C$, then X is recurrent.
(b) If $\epsilon > 0$ and b is a constant such that $PV - V \leq -\epsilon + bI_C$, then X is positive recurrent.

Proposition 6.14 *(Moment bound) Suppose V, f, and g are nonnegative functions on $\mathcal{S}$ and suppose*

$$PV(i) - V(i) \leq -f(i) + g(i) \quad \text{for all } i \in \mathcal{S}. \tag{6.16}$$

In addition, suppose X is positive recurrent, so that the means $\bar{f} = \pi f$ and $\bar{g} = \pi g$ are well-defined. Then $\bar{f} \leq \bar{g}$. (In particular, if g is bounded, then $\bar{g}$ is finite, and therefore $\bar{f}$ is finite.)

Corollary 6.15 *(Combined Foster–Lyapunov stability criterion and moment bound) Suppose V, f, and g are nonnegative functions on $\mathcal{S}$ such that*

$$PV(i) - V(i) \leq -f(i) + g(i) \quad \text{for all } i \in \mathcal{S}. \tag{6.17}$$

In addition, suppose for some $\epsilon > 0$ that C defined by $C = \{i : f(i) < g(i) + \epsilon\}$ is finite. Then X is positive recurrent and $\bar{f} \leq \bar{g}$. (In particular, if g is bounded, then $\bar{g}$ is finite, and therefore $\bar{f}$ is finite.)

Proof Let $b = \max\{g(i) + \epsilon - f(i) : i \in C\}$. Then V, C, b, and ϵ satisfy the hypotheses of Proposition 6.13(b), so that X is positive recurrent. Therefore the hypotheses of Proposition 6.14 are satisfied, so that $\bar{f} \leq \bar{g}$. □

The assumptions in Propositions 6.13 and 6.14 and Corollary 6.15 do not imply that $\bar{V}$ is finite. Even so, since V is nonnegative, for a given initial state $X(0)$, the long term average drift of $V(X(t))$ is nonnegative. This gives an intuitive reason why the mean downward part of the drift, $\bar{f}$, must be less than or equal to the mean upward part of the drift, $\bar{g}$.

Example 6.12 (Probabilistic routeing to two queues) Consider the routeing scenario with two queues, queue 1 and queue 2, fed by a single stream of packets, as pictured in Figure 6.9. Here, $0 \leq a, u, d_1, d_2 \leq 1$, and $\bar{u} = 1 - u$. The state space for the process is $\mathcal{S} = \mathbb{Z}_+^2$, where the state $x = (x_1, x_2)$ denotes x_1 packets in queue 1 and x_2 packets in queue 2. In each time slot, a new arrival is generated with probability a, and then is routed to queue 1 with probability u and to queue 2 with probability $\bar{u}$. Then

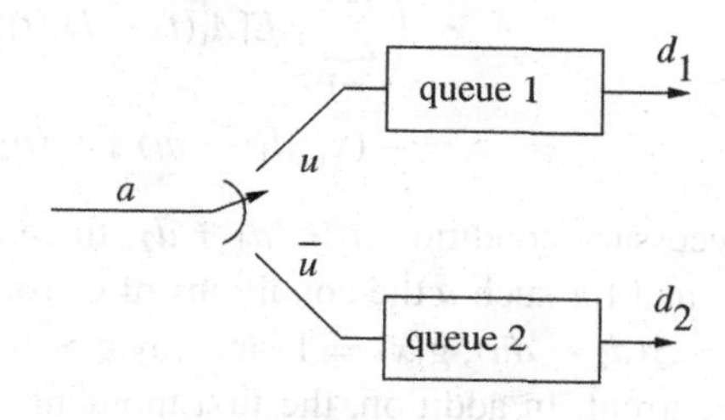

Figure 6.9 Two queues fed by a single arrival stream.

each queue i, if not empty, has a departure with probability d_i. Note that we allow a packet to arrive and depart in the same slot. Thus, if $X_i(t)$ is the number of packets in queue i at the beginning of slot t, then the system dynamics can be described as follows:

$$X_i(t+1) = X_i(t) + A_i(t) - D_i(t) + L_i(t) \quad \text{for } i \in \{0, 1\}, \tag{6.18}$$

where:

- $A(t) = (A_1(t), A_2(t))$ is equal to $(1, 0)$ with probability au, $(0, 1)$ with probability $a\bar{u}$, and $A(t) = (0, 0)$ otherwise.
- $D_i(t) : t \geq 0$, are *Bernoulli*(d_i) random variables, for $i \in \{0, 1\}$.
- All the $A(t)$s, $D_1(t)$s, and $D_2(t)$s are mutually independent.
- $L_i(t) = (-(X_i(t) + A_i(t) - D_i(t)))_+$ (see explanation next).

If $X_i(t) + A_i(t) = 0$, there can be no actual departure from queue i. However, we still allow $D_i(t)$ to equal one. To keep the queue length process from going negative, we add the random variable $L_i(t)$ in (6.18). Thus, $D_i(t)$ is the *potential* number of departures from queue i during the slot, and $D_i(t) - L_i(t)$ is the actual number of departures. This completes the specification of the one-step transition probabilities of the Markov process.

A necessary condition for positive recurrence is, for any routeing policy, $a < d_1 + d_2$, because the total arrival rate must be less than the total departure rate. We seek to show that this necessary condition is also sufficient, under the random routeing policy.

Let us calculate the drift of $V(X(t))$ for the choice $V(x) = (x_1^2 + x_2^2)/2$. Note that $(X_i(t+1))^2 = (X_i(t) + A_i(t) - D_i(t) + L_i(t))^2 \leq (X_i(t) + A_i(t) - D_i(t))^2$, because addition of the variable $L_i(t)$ can only push $X_i(t) + A_i(t) - D_i(t)$ closer to zero. Thus,

$$\begin{aligned} PV(x) - V(x) &= E[V(X(t+1))|X(t) = x] - V(x) \\ &\leq \frac{1}{2}\sum_{i=1}^{2} E[(x_i + A_i(t) - D_i(t))^2 - x_i^2|X(t) = x] \\ &= \sum_{i=1}^{2} x_i E[A_i(t) - D_i(t)|X(t) = x] \\ &\quad + \frac{1}{2} E[(A_i(t) - D_i(t))^2|X(t) = x] \qquad (6.19) \\ &\leq \left(\sum_{i=1}^{2} x_i E[A_i(t) - D_i(t)|X(t) = x] \right) + 1 \\ &= -(x_1(d_1 - au) + x_2(d_2 - a\bar{u})) + 1. \qquad (6.20) \end{aligned}$$

Under the necessary condition $a < d_1 + d_2$, there are choices of u so that $au < d_1$ and $a\bar{u} < d_2$, and for such u the conditions of Corollary 6.15 are satisfied, with $f(x) = x_1(d_1 - au) + x_2(d_2 - a\bar{u})$, $g(x) = 1$, and any $\epsilon > 0$, implying that the Markov process is positive recurrent. In addition, the first moments under the equilibrium distribution satisfy:

$$(d_1 - au)\overline{X}_1 + (d_2 - a\overline{u})\overline{X}_2 \leq 1. \tag{6.21}$$

In order to deduce an upper bound on $\overline{X}_1 + \overline{X}_2$, we select u^* to maximize the minimum of the two coefficients in (6.21). Intuitively, this entails selecting u to minimize the absolute value of the difference between the two coefficients. We find:

$$\begin{aligned}\epsilon &= \max_{0 \leq u \leq 1} \min\{d_1 - au, d_2 - a\overline{u}\} \\ &= \min\{d_1, d_2, \frac{d_1 + d_2 - a}{2}\}\end{aligned}$$

and the corresponding value u^* of u is given by

$$u^* = \begin{cases} 0 & \text{if } d_1 - d_2 < -a \\ \frac{1}{2} + \frac{d_1 - d_2}{2a} & \text{if } |d_1 - d_2| \leq a \\ 1 & \text{if } d_1 - d_2 > a \end{cases}.$$

For the system with $u = u^*$, (6.21) yields

$$\overline{X}_1 + \overline{X}_2 \leq \frac{1}{\epsilon}. \tag{6.22}$$

We note that, in fact,

$$\overline{X}_1 + \overline{X}_2 \leq \frac{2}{d_1 + d_2 - a}. \tag{6.23}$$

If $|d_1 - d_2| \leq a$ then the bounds (6.22) and (6.23) coincide, and otherwise, the bound (6.23) is strictly tighter. If $d_1 - d_2 < -a$ then $u^* = 0$, so that $\overline{X}_1 = 0$, and (6.21) becomes $(d_2 - a)\overline{X}_2 \leq 1$, which implies (6.23). Similarly, if $d_1 - d_2 > a$, then $u^* = 1$, so that $\overline{X}_2 = 0$, and (6.21) becomes $(d_1 - a)\overline{X}_1 \leq 1$, which implies (6.23). Thus, (6.23) is proved.

Example 6.13 (Route-to-shorter policy) Consider a variation of the previous example such that when a packet arrives, it is routed to the shorter queue. To be definite, in case of a tie, the packet is routed to queue 1. Then the evolution equation (6.18) still holds, but with the description of the arrival variables changed to the following:

- Given $X(t) = (x_1, x_2)$, $A(t) = (I_{\{x_1 \leq x_2\}}, I_{\{x_1 > x_2\}})$ with probability a, and $A(t) = (0, 0)$ otherwise.

Let P^{RS} denote the one-step transition probability matrix when the route-to-shorter policy is used. Proceeding as in (6.19) yields:

$$\begin{aligned}P^{RS}V(x) - V(x) &\leq \sum_{i=1}^{2} x_i E[A_i(t) - D_i(t)|X(t) = x] + 1 \\ &= a\left(x_1 I_{\{x_1 \leq x_2\}} + x_2 I_{\{x_1 > x_2\}}\right) - d_1 x_1 - d_2 x_2 + 1.\end{aligned}$$

Note that $x_1 I_{\{x_1 \leq x_2\}} + x_2 I_{\{x_1 > x_2\}} \leq u x_1 + \overline{u} x_2$ for $u \in [0, 1]$, with equality if $u = I_{\{x_1 \leq x_2\}}$. Therefore, the drift bound for V under the route-to-shorter policy is less than or equal to the drift bound (6.20), for V for any choice of probabilistic splitting. In fact, route-to-shorter routeing can be viewed as a controlled version of the independent splitting model, for which the control policy is selected to minimize the bound on the drift of V in each state. It follows that the route-to-shorter process is positive recurrent as long as $a < d_1 + d_2$, and (6.21) holds for any value of u such that $au < d_1$ and $a\overline{u} \leq d_2$. In particular, (6.22) holds for the route-to-shorter process.

We remark that the stronger bound (6.23) is not always true for the route-to-shorter policy. The problem is that even if $d_1 - d_2 < -a$, the route-to-shorter policy can still route to queue 1, and so $\overline{X}_1 \neq 0$. In fact, if a and d_2 are fixed with $0 < a < d_2 < 1$, then $\overline{X}_1 \to \infty$ as $d_1 \to 0$ for the route-to-shorter policy. Intuitively, that is because occasionally there will be a large number of customers in the system due to statistical fluctuations, and then there will be many customers in queue 1. But if $d_2 << 1$, those customers will remain in queue 2 for a very long time.

Example 6.14 (An input queued switch with probabilistic switching)[4] Consider a packet switch with N inputs and N outputs, as pictured in Figure 6.10. Suppose there are N^2 queues – N at each input – with queue i, j containing packets that arrived at input i and are destined for output j, for $i,j \in E$, where $E = \{1, \ldots, N\}$. Suppose the packets are all the same length, and adopt a discrete-time model, so that during one time slot, a transfer of packets can occur, such that at most one packet can be transferred from each input, and at most one packet can be transferred to each output. A permutation σ of E has the form $\sigma = (\sigma_1, \ldots, \sigma_N)$, where $\sigma_1, \ldots, \sigma_N$ are distinct elements of E. Let Π denote the set of all $N!$ such permutations. Given $\sigma \in \Pi$, let $R(\sigma)$ be the $N \times N$ *switching matrix* defined by $R_{ij} = I_{\{\sigma_i = j\}}$. Thus, $R_{ij}(\sigma) = 1$ means that under permutation σ, input i is connected to output j, or, equivalently, a packet in queue i,j is to depart, if there is any such packet. A state x of the system has the form $x = (x_{ij} : i,j \in E)$, where x_{ij} denotes the number of packets in queue i,j.

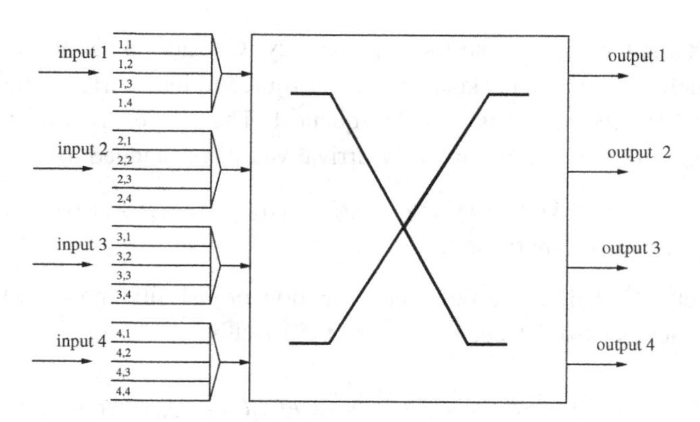

Figure 6.10 A 4 × 4 input queued switch.

[4] Tassiulas (1997) originally developed the results of Examples 6.14 and 6.15, in the context of wireless networks. The paper (McKeown, Mekkittikul, Anantharam & Walrand 1999) presents similar results in the context of a packet switch.

The evolution of the system over a time slot $[t, t+1)$ is described as follows:

$$X_{ij}(t+1) = X_{ij}(t) + A_{ij}(t) - R_{ij}(\sigma(t)) + L_{ij}(t),$$

where:

- $A_{ij}(t)$ is the number of packets arriving at input i, destined for output j, in the slot. Assume that the variables $(A_{ij}(t) : i,j \in E, t \geq 0)$ are mutually independent, and for each i,j, the random variables $(A_{ij}(t) : t \geq 0)$ are independent, identically distributed, with mean λ_{ij} and $E[A_{ij}^2] \leq K_{ij}$, for some constants λ_{ij} and K_{ij}. Let $\Lambda = (\lambda_{ij} : i,j \in E)$.
- $\sigma(t)$ is the switch state used during the slot.
- $L_{ij} = -(X_{ij}(t) + A_{ij}(t) - R_{ij}(\sigma(t)))_+$, which takes value one if there was an unused potential departure at queue ij during the slot, and is zero otherwise.

The number of packets at input i at the beginning of the slot is given by the row sum $\sum_{j\in E} X_{ij}(t)$, its mean is given by the row sum $\sum_{j\in E} \lambda_{ij}$, and at most one packet at input i can be served in a time slot. Similarly, the set of packets waiting for output j, called the *virtual queue* for output j, has size given by the column sum $\sum_{i\in E} X_{ij}(t)$. The mean number of arrivals to the virtual queue for output j is $\sum_{i\in E} \lambda_{ij}(t)$, and at most one packet in the virtual queue can be served in a time slot. These considerations lead us to impose the following restrictions on Λ:

$$\sum_{j\in E} \lambda_{ij} < 1 \text{ for all } i \quad \text{and} \quad \sum_{i\in E} \lambda_{ij} < 1 \text{ for all } j. \tag{6.24}$$

Except for trivial cases involving deterministic arrival sequences, the conditions (6.24) are necessary for stable operation, for any choice of the switch schedule $(\sigma(t) : t \geq 0)$.

Let us first explore random, independent, and identically distributed (iid) switching. That is, given a probability distribution u on Π, let $(\sigma(t) : t \geq 0)$ be independent with common probability distribution u. Once the distributions of the A_{ij}s and u are fixed, we have a discrete-time Markov process model. Given Λ satisfying (6.24), we wish to determine a choice of u so that the process with iid switch selection is positive recurrent.

Some standard background from switching theory is given in this paragraph. A *line sum* of a matrix M is either a row sum, $\sum_j M_{ij}$, or a column sum, $\sum_i M_{ij}$. A square matrix M is called *doubly stochastic* if it has nonnegative entries and if all of its line sums are one. Birkhoff's theorem, celebrated in the theory of switching, states that any doubly stochastic matrix M is a convex combination of switching matrices. That is, such an M can be represented as $M = \sum_{\sigma\in\Pi} R(\sigma)u(\sigma)$, where $u = (u(\sigma) : \sigma \in \Pi)$ is a probability distribution on Π. If $\widetilde{M}$ is a nonnegative matrix with all line sums less than or equal to one, then if some of the entries of $\widetilde{M}$ are increased appropriately, a doubly stochastic matrix can be obtained. That is, there exists a doubly stochastic matrix M so that $\widetilde{M}_{ij} \leq M_{ij}$ for all i,j. Applying Birkhoff's theorem to M yields that there is a probability distribution u so that $\widetilde{M}_{ij} \leq \sum_{\sigma\in\Pi} R(\sigma)u(\sigma)$ for all i,j.

Suppose Λ satisfies the necessary conditions (6.24). That is, suppose that all the line sums of Λ are less than one. Then with ϵ defined by

$$\epsilon = \frac{1 - (\text{maximum line sum of } \Lambda)}{N},$$

each line sum of $(\lambda_{ij} + \epsilon : i,j \in E)$ is less than or equal to one. Thus, by the observation at the end of the previous paragraph, there is a probability distribution u^* on Π so that $\lambda_{ij} + \epsilon \leq \mu_{ij}(u^*)$, where

$$\mu_{ij}(u) = \sum_{\sigma \in \Pi} R_{ij}(\sigma) u(\sigma).$$

We consider the system using probability distribution u^* for the switch states. That is, let $(\sigma(t) : t \geq 0)$ be independent, each with distribution u^*. Then for each ij, the random variables $R_{ij}(\sigma(t))$ are independent, Bernoulli$(\mu_{ij}(u^*))$ random variables.

Consider the quadratic Lyapunov function V given by $V(x) = \frac{1}{2}\sum_{i,j} x_{ij}^2$. As in (6.19),

$$PV(x) - V(x) \leq \sum_{i,j} x_{ij} E[A_{ij}(t) - R_{ij}(\sigma(t)) | X_{ij}(t) = x]$$
$$+ \frac{1}{2}\sum_{i,j} E[(A_{ij}(t) - R_{ij}(\sigma(t)))^2 | X(t) = x].$$

Now

$$E[A_{ij}(t) - R_{ij}(\sigma(t)) | X_{ij}(t) = x] = E[A_{ij}(t) - R_{ij}(\sigma(t))] = \lambda_{ij} - \mu_{ij}(u^*) \leq -\epsilon$$

and

$$\frac{1}{2}\sum_{i,j} E[(A_{ij}(t) - R_{ij}(\sigma(t)))^2 | X(t) = x] \leq \frac{1}{2}\sum_{i,j} E[(A_{ij}(t))^2 + (R_{ij}(\sigma(t)))^2] \leq K$$

where $K = \frac{1}{2}(N + \sum_{i,j} K_{ij})$. Thus,

$$PV(x) - V(x) \leq -\epsilon \left(\sum_{ij} x_{ij} \right) + K. \tag{6.25}$$

Therefore, by Corollary 6.15, the process is positive recurrent, and

$$\sum_{ij} \overline{X}_{ij} \leq \frac{K}{\epsilon}. \tag{6.26}$$

That is, the necessary condition (6.24) is also sufficient for positive recurrence and finite mean queue length in equilibrium, under iid random switching, for an appropriate probability distribution u^* on the set of permutations.

Example 6.15 (An input queued switch with maximum weight switching) The random switching policy used in Example 6.14 depends on the arrival rate matrix Λ, which may be unknown a priori. Also, the policy allocates potential departures to a given queue ij, whether or not the queue is empty, even if other queues could be served instead.

This suggests using a dynamic switching policy, such as the *maximum weight* switching policy, defined by $\sigma(t) = \sigma^{MW}(X(t))$, where for a state x,

$$\sigma^{MW}(x) = \arg\max_{\sigma\in\Pi} \sum_{ij} x_{ij}R_{ij}(\sigma). \tag{6.27}$$

The use of "arg max" here means that $\sigma^{MW}(x)$ is selected to be a value of σ that maximizes the sum on the righthand side of (6.27), which is the weight of permutation σ with edge weights x_{ij}. In order to obtain a particular Markov model, we assume that the set of permutations Π is numbered from 1 to $N!$ in some fashion, and in case there is a tie between two or more permutations for having the maximum weight, the lowest numbered permutation is used. Let P^{MW} denote the one-step transition probability matrix when the route-to-shorter policy is used.

Letting V and K be as in Example 6.14, we find under the maximum weight policy,

$$P^{MW}V(x) - V(x) \le \sum_{ij} x_{ij}(\lambda_{ij} - R_{ij}(\sigma^{MW}(x))) + K.$$

The maximum of a function is greater than or equal to the average of the function, so that for any probability distribution u on Π

$$\begin{aligned}\sum_{ij} x_{ij}R_{ij}(\sigma^{MW}(t)) &\ge \sum_{\sigma} u(\sigma)\sum_{ij} x_{ij}R_{ij}(\sigma) \qquad (6.28)\\ &= \sum_{ij} x_{ij}\mu_{ij}(u),\end{aligned}$$

with equality in (6.28) if and only if u is concentrated on the set of maximum weight permutations. In particular, the choice $u = u^*$ shows that

$$\sum_{ij} x_{ij}R_{ij}(\sigma^{MW}(t)) \ge \sum_{ij} x_{ij}\mu_{ij}(u*) \ge \sum_{ij} x_{ij}(\lambda_{ij} + \epsilon).$$

Therefore, if P is replaced by P^{MW}, (6.25) still holds. Therefore, by Corollary 6.15, the process is positive recurrent, and the same moment bound, (6.26), holds, as for the randomized switching strategy of Example 6.14. On one hand, implementing the maximum weight algorithm does not require knowledge of the arrival rates, but on the other hand, it requires that queue length information be shared, and that a maximization problem be solved for each time slot. Much recent work has gone towards reduced-complexity dynamic switching algorithms.

6.9.2 Stability criteria for continuous-time processes

Here is a continuous-time version of the Foster–Lyapunov stability criteria and the moment bounds. Suppose X is a time-homogeneous, irreducible, continuous-time Markov process with generator matrix Q. The drift vector of $V(X(t))$ is the vector QV. This definition is motivated by the fact that the mean drift of X for an interval of duration h is given by

$$
\begin{aligned}
d_h(i) &= \frac{E[V(X(t+h))|X(t)=i] - V(i)}{h} \\
&= \sum_{j\in\mathcal{S}} \left(\frac{p_{ij}(h) - \delta_{ij}}{h}\right) V(j) \\
&= \sum_{j\in\mathcal{S}} \left(q_{ij} + \frac{o(h)}{h}\right) V(j),
\end{aligned} \tag{6.29}
$$

so that if the limit as $h \to 0$ can be taken inside the summation in (6.29), then $d_h(i) \to QV(i)$ as $h \to 0$. The following useful expression for QV follows from the fact that the row sums of Q are zero:

$$
QV(i) = \sum_{j:j\neq i} q_{ij}(V(j) - V(i)). \tag{6.30}
$$

Formula (6.30) is quite similar to the formula (6.15) for the drift vector for a discrete-time process. The proof of the following proposition can be found in Hajek (2006).

Proposition 6.16 *(Foster–Lyapunov stability criterion – continuous time) Suppose $V : \mathcal{S} \to \mathbb{R}_+$ and C is a finite subset of $\mathcal{S}$.*
(a) If $QV \le 0$ on $\mathcal{S}$–C, and $\{i : V(i) \le K\}$ is finite for all K then X is recurrent.
(b) Suppose for some $b > 0$ and $\epsilon > 0$ that

$$
QV(i) \le -\epsilon + bI_C(i) \qquad \text{for all } i \in \mathcal{S}. \tag{6.31}
$$

Suppose further that $\{i : V(i) \le K\}$ is finite for all K, or that X is nonexplosive. Then X is positive recurrent.

Example 6.16 Suppose X has state space $\mathcal{S} = \mathbb{Z}_+$, with $q_{i0} = \mu$ for all $i \ge 1$, $q_{ii+1} = \lambda_i$ for all $i \ge 0$, and all other off-diagonal entries of the rate matrix Q equal to zero, where $\mu > 0$ and $\lambda_i > 0$ such that $\sum_{i\ge 0} \frac{1}{\lambda_i} < +\infty$. Let $C = \{0\}$, $V(0) = 0$, and $V(i) = 1$ for $i \ge 0$. Then $QV = -\mu + (\lambda_0 + \mu)I_C$, so that (6.31) is satisfied with $\epsilon = \mu$ and $b = \lambda_0 + \mu$. However, X is not positive recurrent. In fact, X is explosive. To see this, note that $p^J_{ii+1} = \frac{\lambda_i}{\mu+\lambda_i} \ge \exp(-\frac{\mu}{\lambda_i})$. Let δ be the probability that, starting from state 0, the jump process does not return to zero. Then $\delta = \prod_{i=0}^{\infty} p^J_{ii+1} \ge \exp(-\mu \sum_{i=0}^{\infty} \frac{1}{\lambda_i}) > 0$. Thus, X^J is transient. After the last visit to state zero, all the jumps of X^J are up one. The corresponding mean holding times of X are $\frac{1}{\lambda_i+\mu}$ which have a finite sum, so that the process X is explosive. This example illustrates the need for the assumption just after (6.31) in Proposition 6.16.

As for the case of discrete time, the drift conditions imply moment bounds.

Proposition 6.17 *(Moment bound–continuous time) Suppose $V, f,$ and g are nonnegative functions on $\mathcal{S}$, and suppose $QV(i) \le -f(i) + g(i)$ for all $i \in \mathcal{S}$. In addition, suppose X is positive recurrent, so that the means $\bar{f} = \pi f$ and $\bar{g} = \pi g$ are well-defined. Then $\bar{f} \le \bar{g}$.*

Corollary 6.18 *(Combined Foster–Lyapunov stability criterion and moment bound–continuous time) Suppose V, f, and g are nonnegative functions on $\mathcal{S}$ such that $QV(i) \leq -f(i)+g(i)$ for all $i \in \mathcal{S}$, and, for some $\epsilon > 0$, the set C defined by $C = \{i : f(i) < g(i)+\epsilon\}$ is finite. Suppose also that $\{i : V(i) \leq K\}$ is finite for all K. Then X is positive recurrent and $\bar{f} \leq \bar{g}$.*

Example 6.17 (Random server allocation with two servers) Consider the system shown in Figure 6.11. Suppose that each queue i is fed by a Poisson arrival process with rate λ_i, and suppose there are two potential departure processes, D_1 and D_2, which are Poisson processes with rates m_1 and m_2, respectively. The five Poisson processes are assumed to be independent. No matter how the potential departures are allocated to the permitted queues, the following conditions are necessary for stability:

$$\lambda_1 < m_1, \qquad \lambda_3 < m_2, \qquad \text{and } \lambda_1 + \lambda_2 + \lambda_3 < m_1 + m_2. \tag{6.32}$$

That is because server 1 is the only one that can serve queue 1, server 2 is the only one that can serve queue 3, and the sum of the potential service rates must exceed the sum of the potential arrival rates for stability. A vector $x = (x_1, x_2, x_2) \in \mathbb{Z}_+^3$ corresponds to x_i packets in queue i for each i. Consider random selection, so that when D_i has a jump, the queue served is chosen at random, with probabilities determined by $u = (u_1, u_2)$. As indicated in Figure 6.11, a potential service by server 1 is given to queue 1 with probability u_1, and to queue 2 with probability $\overline{u}_1$. Similarly, a potential service by server 2 is given to queue 2 with probability u_2, and to queue 3 with probability $\overline{u}_2$. The rates of potential service at the three stations are given by

$$\begin{aligned}
\mu_1(u) &= u_1 m_1 \\
\mu_2(u) &= \overline{u}_1 m_1 + u_2 m_2 \\
\mu_3(u) &= \overline{u}_2 m_2.
\end{aligned}$$

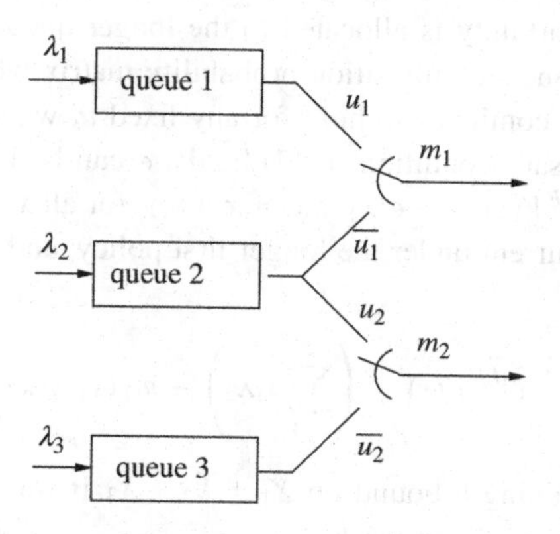

Figure 6.11 A system of three queues with two servers.

Let $V(x) = \frac{1}{2}(x_1^2 + x_2^2 + x_3^2)$. Using (6.30), we find that the drift vector QV is given by

$$QV(x) = \frac{1}{2}\left(\sum_{i=1}^{3}((x_i+1)^2 - x_i^2)\lambda_i\right) + \frac{1}{2}\left(\sum_{i=1}^{3}((x_i-1)_+^2 - x_i^2)\mu_i(u)\right).$$

Now $(x_i - 1)_+^2 \le (x_i - 1)^2$, so that

$$QV(x) \le \left(\sum_{i=1}^{3} x_i(\lambda_i - \mu_i(u))\right) + \frac{\gamma}{2}, \tag{6.33}$$

where γ is the total rate of events, given by $\gamma = \lambda_1 + \lambda_2 + \lambda_3 + \mu_1(u) + \mu_2(u) + \mu_3(u)$, or equivalently, $\gamma = \lambda_1 + \lambda_2 + \lambda_3 + m_1 + m_2$. Suppose that the necessary condition (6.32) holds. Then there exists some $\epsilon > 0$ and choice of u so that

$$\lambda_i + \epsilon \le \mu_i(u) \quad \text{for} \quad 1 \le i \le 3$$

and the largest such choice of ϵ is $\epsilon = \min\{m_1 - \lambda_1, m_2 - \lambda_3, \frac{m_1+m_2-\lambda_1-\lambda_2-\lambda_3}{3}\}$ (showing this is Problem 6.25). So $QV(x) \le -\epsilon(x_1 + x_2 + x_3) + \gamma$ for all x, so Corollary 6.18 implies that X is positive recurrent and $\overline{X}_1 + \overline{X}_2 + \overline{X}_3 \le \frac{\gamma}{2\epsilon}$.

Example 6.18 (Longer first server allocation with two servers) This is a continuation of Example 6.17, concerned with the system shown in Figure 6.11. Examine the righthand side of (6.33). Rather than taking a fixed value of u, suppose that the choice of u could be specified as a function of the state x. The maximum of a function is greater than or equal to the average of the function, so that for any probability distribution u,

$$\begin{aligned}\sum_{i=1}^{3} x_i\mu_i(u) &\le \max_{u'} \sum_i x_i\mu_i(u') \\ &= \max_{u'} m_1(x_1u_1' + x_2\overline{u}_1') + m_2(x_2u_2' + x_3\overline{u}_2') \\ &= m_1(x_1 \vee x_2) + m_2(x_2 \vee x_3),\end{aligned} \tag{6.34}$$

with equality in (6.34) for a given state x if and only if a longer first policy is used: each service opportunity is allocated to the longer queue connected to the server. Let Q^{LF} denote the one-step transition probability matrix when the longest first policy is used. Then (6.33) continues to hold for any fixed u, when Q is replaced by Q^{LF}. Therefore if the necessary condition (6.32) holds, ϵ can be taken as in Example 6.17, and then we have $Q^{LF}V(x) \le -\epsilon(x_1 + x_2 + x_3) + \gamma$ for all x. So Corollary 6.18 implies that X is positive recurrent under the longer first policy, and $\overline{X}_1 + \overline{X}_2 + \overline{X}_3 \le \frac{\gamma}{2\epsilon}$. (Note: We see that

$$Q^{LF}V(x) \le \left(\sum_{i=1}^{3} x_i\lambda_i\right) - m_1(x_1 \vee x_2) - m_2(x_2 \vee x_3) + \frac{\gamma}{2},$$

but for obtaining a bound on $\overline{X}_1 + \overline{X}_2 + \overline{X}_3$ it was simpler to compare to the case of random service allocation.)

Problems

6.1 Mean hitting time for a simple Markov process Let $(X(n) : n \geq 0)$ denote a discrete-time, time-homogeneous Markov chain with state space $\{0, 1, 2, 3\}$ and one-step transition probability matrix

$$P = \begin{pmatrix} 0 & 1 & 0 & 0 \\ 1-a & 0 & a & 0 \\ 0 & 0.5 & 0 & 0.5 \\ 0 & 0 & 1 & 0 \end{pmatrix}$$

for some constant a with $0 \leq a \leq 1$.
(a) Sketch the transition probability diagram for X and give the equilibrium probability vector. If the equilibrium vector is not unique, describe all the equilibrium probability vectors.
(b) Compute $E[\min\{n \geq 1 : X(n) = 3\}|X(0) = 0]$.

6.2 A two station pipeline in continuous time This is a continuous-time version of Example 4.8. Consider a pipeline consisting of two single-buffer stages in series. Model the system as a continuous-time Markov process. Suppose new packets are offered to the first stage according to a rate λ Poisson process. A new packet is accepted at stage one if the buffer in stage one is empty at the time of arrival. Otherwise the new packet is lost. If at a fixed time t there is a packet in stage one and no packet in stage two, then the packet is transferred during $[t, t+h)$ to stage two with probability $h\mu_1 + o(h)$. Similarly, if at time t the second stage has a packet, then the packet leaves the system during $[t, t+h)$ with probability $h\mu_2 + o(h)$, independently of the state of stage one. Finally, the probability of two or more arrival, transfer, or departure events during $[t, t+h)$ is $o(h)$.
(a) What is an appropriate state-space for this model?
(b) Sketch a transition rate diagram.
(c) Write down the Q matrix.
(d) Derive the throughput, assuming that $\lambda = \mu_1 = \mu_2 = 1$.
(e) Still assuming $\lambda = \mu_1 = \mu_2 = 1$, suppose the system starts with one packet in each stage. What is the expected time until both buffers are empty?

6.3 Equilibrium distribution of the jump chain Suppose that π is the equilibrium distribution for a time-homogeneous Markov process with transition rate matrix Q. Suppose that $B^{-1} = \sum_i -q_{ii}\pi_i$, where the sum is over all i in the state space, is finite. Show that the equilibrium distribution for the jump chain $(X^J(k) : k \geq 0)$ (defined in Section 4.10) is given by $\pi_i^J = -Bq_{ii}\pi_i$. (So π and π^J are identical if and only if q_{ii} is the same for all i.)

6.4 A simple Poisson process calculation Let $(N(t) : t \geq 0)$ be a Poisson random process with rate $\lambda > 0$. Compute $P(N(s) = i|N(t) = k)$ where $0 < s < t$ and i and k are nonnegative integers. (Caution: note order of s and t carefully.)

6.5 A simple question of periods Consider a discrete-time Markov process with the nonzero one-step transition probabilities indicated by the following graph:
(a) What is the period of state 4?
(b) What is the period of state 6?

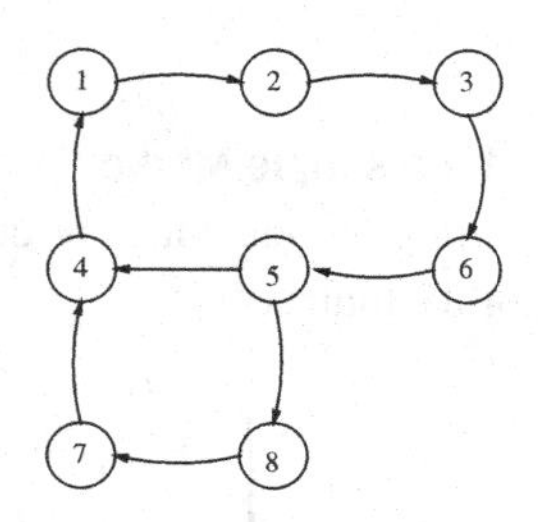

6.6 On distributions of three discrete-time Markov processes For each of the Markov processes with indicated one-step transition probability diagrams, determine the set of equilibrium distributions and whether $\lim_{t\to\infty} \pi_n(t)$ exists for all choices of the initial distribution, $\pi(0)$, and all states n.

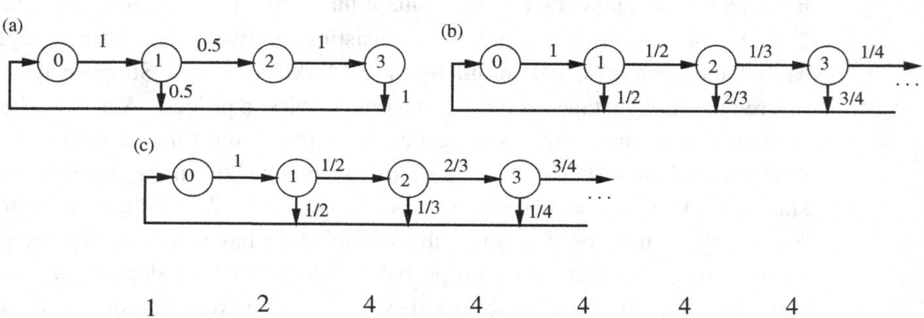

6.7 A simple birth–death Markov process Consider a continuous-time Markov process with the transition rate diagram shown.

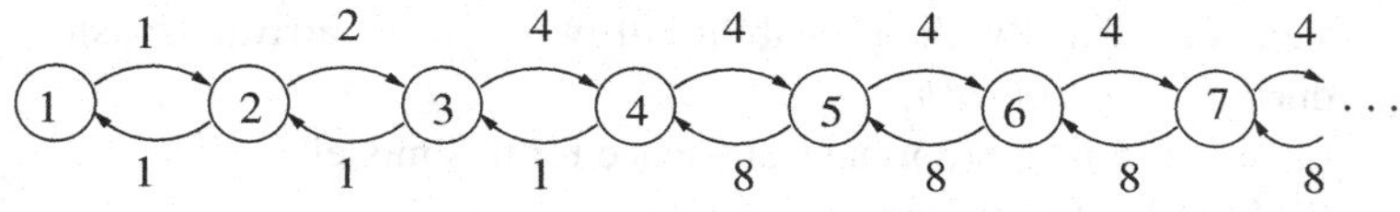

(a) What is the generator matrix Q?

(b) What is the equilibrium distribution?

(c) What is the mean time to reach state 1 starting in state 2?

6.8 A Markov process on a ring Consider a continuous-time Markov process with the transition rate diagram shown, where a, b, and c are strictly positive constants.

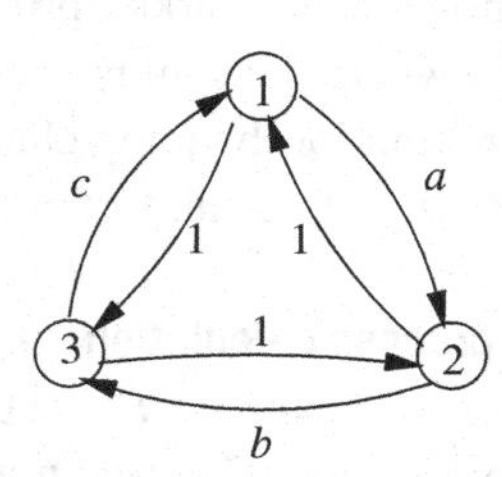

(a) Write down the Q matrix and verify that the equilibrium probability distribution π is proportional to $(1 + c + cb, 1 + a + ac, 1 + b + ba)$.

(b) Depending on the values of a, b and c, the process may tend to cycle clockwise, cycle counter clockwise, or tend to be cycle neutral. For example, it is cycle neutral if

$a = b = c = 1$. Let θ denote the long term rate of cycles per second in the clockwise direction per unit time. (A negative value indicates a long term rate of rotation in the counter clockwise direction.) For example, if $a = b = c$ then $\theta = (a-1)/3$. Give a simple expression for θ in terms of π, a, b, and c.
(c) Express θ in terms of a, b,, and c. What condition on a, b, and c is equivalent to the mean net cycle rate being zero?

6.9 Generating a random spanning tree Let $G = (V, E)$ be an undirected, connected graph with n vertices and m edges (so $|V| = n$ and $|E| = m$). Suppose that $m \geq n$, so the graph has at least one cycle. A *spanning tree* of G is a subset T of E with cardinality $n-1$ and no cycles. Let $\mathcal{S}$ denote the set of all spanning trees of G. We shall consider a Markov process with state space $\mathcal{S}$; the one-step transition probabilities are described as follows. Given a state T, an edge e is selected at random from among the $m-n+1$ edges in $E-T$, with all such edges having equal probability. The set $T \cup \{e\}$ then has a single cycle. One of the edges in the cycle (possibly edge e) is selected at random, with all edges in the cycle having equal probability of being selected, and is removed from $T \cup \{e\}$ to produce the next state, T'.
(a) Is the Markov process irreducible (for any choice of G satisfying the conditions given)? Justify your answer.
(b) Is the Markov process aperiodic (for any graph G satisfying the conditions given)?
(c) Show that the one-step transition probability matrix $P = (p_{T,T'} : T, T' \in \mathcal{S})$ is symmetric.
(d) Show that the equilibrium distribution assigns equal probability to all states in $\mathcal{S}$. Hence, a method for generating an approximately uniformly distributed spanning tree is to run the Markov process a long time and occasionally sample it.

6.10 A mean hitting time problem Let $(X(t) : t \geq 0)$ be a time-homogeneous, pure-jump Markov process with state space $\{0, 1, 2\}$ and Q matrix

$$Q = \begin{pmatrix} -4 & 2 & 2 \\ 1 & -2 & 1 \\ 2 & 0 & -2 \end{pmatrix}.$$

(a) Write down the state transition diagram and compute the equilibrium distribution.
(b) Compute $a_i = E[\min\{t \geq 0 : X(t) = 1\}|X(0) = i]$ for $i = 0, 1, 2$. If possible, use an approach that can be applied to larger state spaces.
(c) Derive a variation of the Kolmogorov forward differential equations for the quantities: $\alpha_i(t) = P(X(s) \neq 2 \text{ for } 0 \leq s \leq t \text{ and } X(t) = i|X(0) = 0)$ for $0 \leq i \leq 2$. (You need not solve the equations.)
(d) The forward Kolmogorov equations describe the evolution of the state probability distribution going forward in time, given an initial distribution. In other problems, a boundary condition is given at a final time, and a differential equation working backwards in time is called for (called Kolmogorov backward equations). Derive a backward differential equation for: $\beta_j(t) = P(X(s) \neq 2 \text{ for } t \leq s \leq t_f|X(t) = j)$, for $0 \leq j \leq 2$ and $t \leq t_f$ for some fixed time t_f. (Hint: Express $\beta_i(t-h)$ in terms of the $\beta_j(t)$s for $t \leq t_f$, and let $h \to 0$. You need not solve the equations.)

6.11 A birth–death process with periodic rates Consider a single server queueing system in which the number in the system is modeled as a continuous time birth–death process with the transition rate diagram shown, where $\lambda_a, \lambda_b, \mu_a$, and μ_b are strictly positive constants.

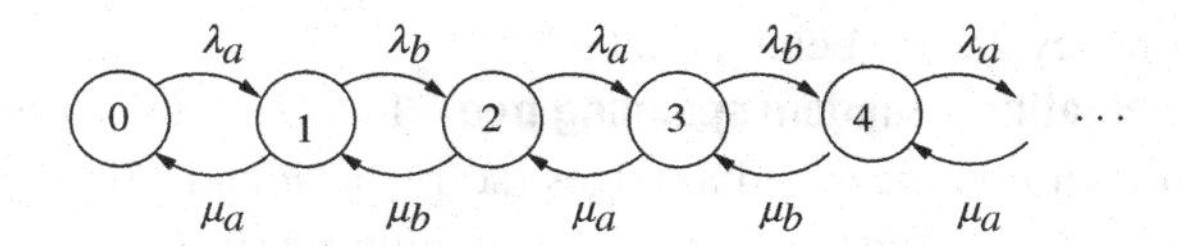

(a) Under what additional assumptions on these four parameters is the process positive recurrent?
(b) Assuming the system is positive recurrent, under what conditions on $\lambda_a, \lambda_b, \mu_a$, and μ_b is it true that the distribution of the number in the system at the time of a typical arrival is the same as the equilibrium distribution of the number in the system?

6.12 Markov model for a link with resets Suppose that a regulated communication link resets at a sequence of times forming a Poisson process with rate μ, as shown in the diagram. Packets are offered to the link according to a Poisson process with rate λ. Suppose the link shuts down after three packets pass in the absence of resets. Once the link is shut down, additional offered packets are dropped, until the link is reset again, at which time the process begins anew.

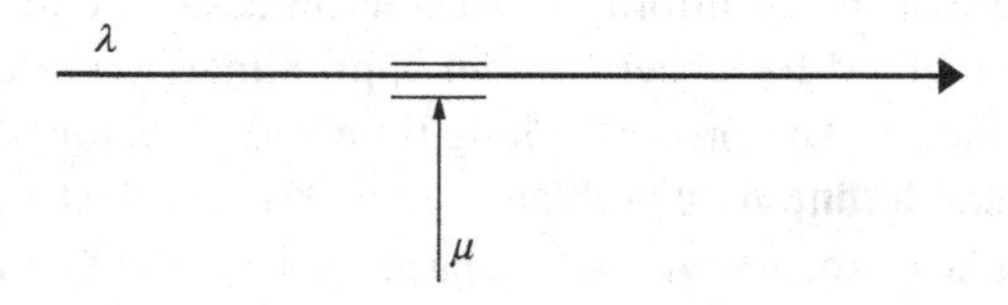

(a) Sketch a transition rate diagram for a finite state Markov process describing the system state.
(b) Express the dropping probability (same as the long term fraction of packets dropped) in terms of λ and μ.

6.13 An unusual birth–death process Consider the birth–death process X with arrival rates $\lambda_k = (p/(1-p))^k/a_k$ and death rates $\mu_k = (p/(1-p))^{k-1}/a_k$, where $.5 < p < 1$, and $a = (a_0, a_1, \ldots)$ is a probability distribution on the nonnegative integers with $a_k > 0$ for all k.
(a) Classify the states for the process X as transient, null recurrent or positive recurrent.
(b) Check that $aQ = 0$. Is a an equilibrium distribution for X? Explain.
(c) Find the one-step transition probabilities for the jump-chain, X^J.
(d) Classify the states for the process X^J as transient, null recurrent or positive recurrent.

6.14 A queue with decreasing service rate Consider a queueing system in which the arrival process is a Poisson process with rate λ. Suppose the instantaneous completion rate is μ when there are K or fewer customers in the system, and $\mu/2$ when there are $K+1$ or more customers in the system. The number in the system is modeled as a birth–death Markov process.

(a) Sketch the transition rate diagram.
(b) Under what condition on λ and μ are all states positive recurrent? Under this condition, give the equilibrium distribution.
(c) Suppose that $\lambda = (2/3)\mu$. Describe in words the typical behavior of the system, given that it is initially empty.

6.15 Limit of a discrete-time queueing system Model a queue by a discrete-time Markov chain by recording the queue state after intervals of q seconds each. Assume the queue evolves during one of the atomic intervals as follows: There is an arrival during the interval with probability αq, and no arrival otherwise. If there is a customer in the queue at the beginning of the interval then a single departure will occur during the interval with probability βq. Otherwise no departure occurs. Suppose that it is impossible to have an arrival and a departure in a single atomic interval.
(a) Find $a_k =$ P(an interarrival time is kq) and $b_k =$ P(a service time is kq).
(b) Find the equilibrium distribution, $p = (p_k : k \geq 0)$, of the number of customers in the system at the end of an atomic interval. What happens as $q \to 0$?

6.16 An M/M/1 queue with impatient customers Consider an M/M/1 queue with parameters λ and μ with the following modification. Each customer in the queue will defect (i.e. depart without service) with probability $\alpha h + o(h)$ in an interval of length h, independently of the other customers in the queue. Once a customer makes it to the server it no longer has a chance to defect and simply waits until its service is completed and then departs from the system. Let $N(t)$ denote the number of customers in the system (queue plus server) at time t.
(a) Give the transition rate diagram and generator matrix Q for the Markov chain $N = (N(t) : t \geq 0)$.
(b) Under what conditions are all states positive recurrent? Under this condition, find the equilibrium distribution for N. (You need not explicitly sum the series.)
(c) Suppose that $\alpha = \mu$. Find an explicit expression for p_D, the probability that a typical arriving customer defects instead of being served. Does your answer make sense as λ/μ converges to zero or to infinity?

6.17 Statistical multiplexing Consider the following scenario regarding a one-way link in a store-and-forward packet communication network. Suppose that the link supports eight connections, each generating traffic at 5 kilobits per second (kbps). The data for each connection are assumed to be in packets exponentially distributed in length with mean packet size 1 kilobit. The packet lengths are assumed mutually independent and the packets for each stream arrive according to a Poisson process. Packets are queued at the beginning of the link if necessary, and queue space is unlimited. Compute the mean delay (queueing plus transmission time – neglect propagation delay) for each of the following three scenarios. Compare your answers.
(a) (Full multiplexing) The link transmit speed is 50 kbps.
(b) The link is replaced by two 25 kbps links, and each of the two links carries four sessions. (Of course the delay would be larger if the sessions were not evenly divided.)
(c) (Multiplexing over two links) The link is replaced by two 25 kbps links. Each packet is transmitted on one link or the other, and neither link is idle whenever a packet from any session is waiting.

6.18 A queue with blocking (M/M/1/5 system) Consider an M/M/1 queue with service rate μ, arrival rate λ, and the modification that at any time, at most five customers can be in the system (including the one in service, if any). If a customer arrives and the system is full (i.e. already has five customers in it) then the customer is dropped, and is said to be blocked. Let $N(t)$ denote the number of customers in the system at time t. Then $(N(t) : t \geq 0)$ is a Markov chain.
(a) Indicate the transition rate diagram of the chain and find the equilibrium probability distribution.
(b) What is the probability, p_B, that a typical customer is blocked?
(c) What is the mean waiting time in queue, W, of a typical customer that is not blocked?
(d) Give a simple method to numerically calculate, or give a simple expression for, the mean length of a busy period of the system. (A busy period begins with the arrival of a customer to an empty system and ends when the system is again empty.)
6.19 Three queues and an autonomously traveling server Consider three stations that are served by a single rotating server, as pictured. Customers arrive to station i according

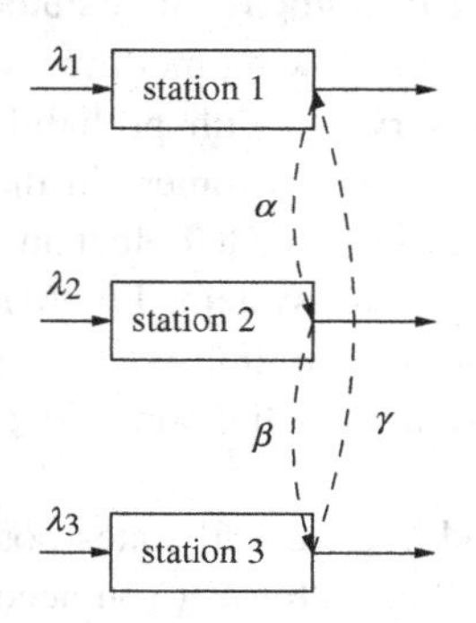

to a Poisson process of rate λ_i for $1 \leq i \leq 3$, and the total service requirement of each customer is exponentially distributed, with mean one. The rotation of the server is modeled by a three state Markov process with the transition rates α, β, and γ as indicated by the dashed lines. When at a station, the server works at unit rate, or is idle if the station is empty. If the service to a customer is interrupted because the server moves to the next station, the service is resumed when the server returns.
(a) Under what condition is the system stable? Briefly justify your answer.
(b) Identify a method for computing the mean customer waiting time at station one.
6.20 On two distributions seen by customers Consider a queueing system in which the number in the system only changes in steps of plus one or minus one. Let $D(k, t)$ denote the number of customers that depart in the interval [0,t] that leaves behind exactly k customers, and let R(k,t) denote the number of customers that arrive in the interval [0,t] to find exactly k customers already in the system.
(a) Show that $|D(k, t) - R(k, t)| \leq 1$ for all k and t.
(b) Let α_t (respectively δ_t) denote the number of arrivals (departures) up to time t. Suppose that $\alpha_t \rightarrow \infty$ and $\alpha_t/\delta_t \rightarrow 1$ as $t \rightarrow \infty$. Show that if the following two limits exist for a given value k, then they are equal: $r_k = \lim_{t\rightarrow\infty} R(k, t)/\alpha_t$ and $d_k = \lim_{t\rightarrow\infty} D(k, t)/\delta_t$.

6.21 Recurrence of mean zero random walks Suppose $B_1, B_2, \ldots$ is a sequence of independent, mean zero, integer valued random variables, which are bounded, i.e. $P\{|B_i| \leq M\} = 1$ for some M.
(a) Let $X_0 = 0$ and $X_n = B_1 + \ldots + B_n$ for $n \geq 0$. Show that X is recurrent.
(b) Suppose $Y_0 = 0$ and $Y_{n+1} = Y_n + B_n + L_n$, where $L_n = (-(Y_n + B_n))_+$. The process Y is a reflected version of X. Show that Y is recurrent.

6.22 Positive recurrence of reflected random walk with negative drift Suppose $B_1, B_2, \ldots$ is a sequence of independent, integer valued random variables, each with mean $\overline{B} < 0$ and second moment $\overline{B^2} < +\infty$. Suppose $X_0 = 0$ and $X_{n+1} = X_n + B_n + L_n$, where $L_n = (-(X_n + B_n))_+$. Show that X is positive recurrent, and give an upper bound on the mean under the equilibrium distribution, $\overline{X}$. (Note, it is not assumed that the Bs are bounded.)

6.23 Routeing with two arrival streams (a) Generalize Example 6.12 to the scenario shown,

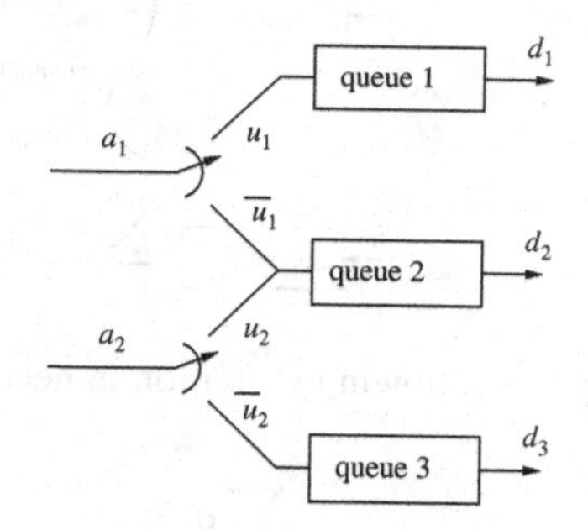

where $a_i, d_j \in (0, 1)$ for $1 \leq i \leq 2$ and $1 \leq j \leq 3$. In particular, determine conditions on a_1 and a_2 that insure there is a choice of $u = (u_1, u_2)$ which makes the system positive recurrent. Under those conditions, find an upper bound on $\overline{X}_1 + \overline{X}_2 + \overline{X}_3$, and select u to minimize the bound.
(b) Generalize Example 6.13 to the scenario shown. In particular, can you find a version of route-to-shorter routeing so that the bound found in part (a) still holds?

6.24 An inadequacy of a linear potential function Consider the system of Example 6.13 (a discrete-time model, using the route-to-shorter policy, with ties broken in favor of queue 1, so $u = I_{\{x_1 \leq x_2\}}$):

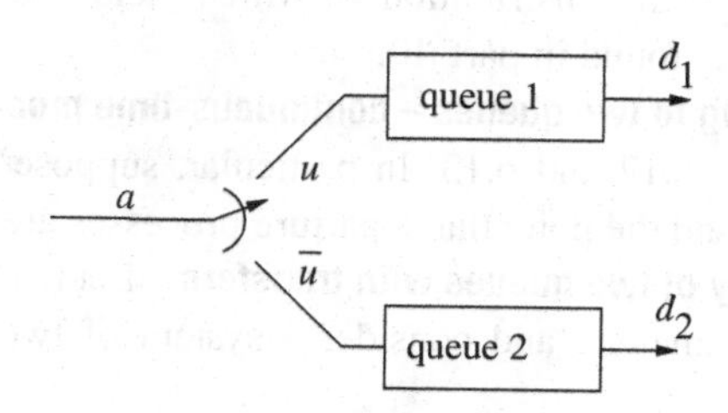

Assume $a = 0.7$ and $d_1 = d_2 = 0.4$. The system is positive recurrent. Explain why the function $V(x) = x_1 + x_2$ does *not* satisfy the Foster–Lyapunov stability criteria for positive recurrence, for any choice of the constant b and the finite set C.

6.25 Allocation of service Prove the claim in Example 6.17 about the largest value of ϵ.

6.26 Opportunistic scheduling (Based on Tassiulas & Ephremides (1993)) Suppose N queues are in parallel, and suppose the arrivals to a queue i form an independent, identically distributed sequence, with the number of arrivals in a given slot having mean $a_i > 0$ and finite second moment K_i. Let $S(t)$ for each t be a subset of $E = \{1, \ldots, N\}$ and $t \geq 0$. The random sets $S(t) : t \geq 0$ are assumed to be independent with common distribution w. The interpretation is that there is a single server, and in slot i, it can serve one packet from one of the queues in $S(t)$. For example, the queues might be in the base station of a wireless network with packets queued for N mobile users, and $S(t)$ denotes the set of mobile users that have working channels for time slot $[t, t+1)$. See the illustration:

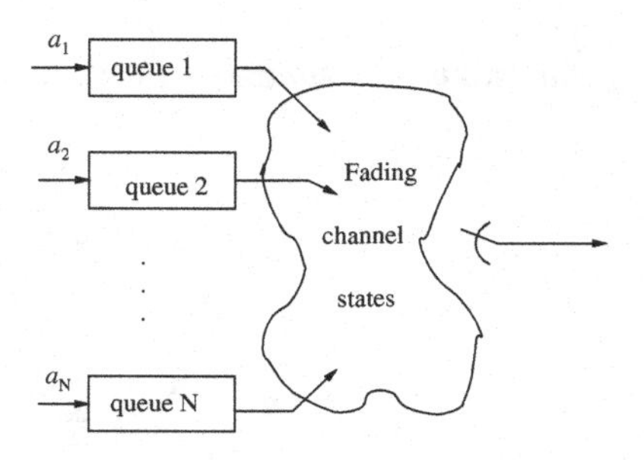

(a) Explain why the following condition is necessary for stability: For all $s \subset E$ with $s \neq \emptyset$,

$$\sum_{i \in s} a_i < \sum_{B: B \cap s \neq \emptyset} w(B). \tag{6.35}$$

(b) Consider u of the form $u = (u(i,s) : i \in E, s \subset E)$, with $u(i,s) \geq 0$, $u(i,s) = 0$ if $i \notin s$, and $\sum_{i \in E} u(i,s) = I_{\{s \neq \emptyset\}}$. Suppose that given $S(t) = s$, the queue that is given a potential service opportunity has probability distribution $(u(i,s) : i \in E)$. Then the probability of a potential service at queue i is given by $\mu_i(u) = \sum_s u(i,s)w(s)$ for $i \in E$. Show that under the condition (6.35), for some $\epsilon > 0$, u can be selected so that $a_i + \epsilon \leq \mu_i(u)$ for $i \in E$. (Hint: Apply the min-cut, max-flow theorem to an appropriate graph.)

(c) Show that using the u found in part (b) the process is positive recurrent.

(d) Suggest a dynamic scheduling method which does not require knowledge of the arrival rates or the distribution w, which yields the same bound on the mean sum of queue lengths found in part (b).

6.27 Routeing to two queues – continuous-time model Give a continuous-time analog of Examples 6.12 and 6.13. In particular, suppose that the arrival process is Poisson with rate λ and the potential departure processes are Poisson with rates μ_1 and μ_2.

6.28 Stability of two queues with transfers Let $(\lambda_1, \lambda_2, \nu, \mu_1, \mu_2)$ be a vector of strictly positive parameters, and consider a system of two service stations with transfers as pictured.

Station i has Poisson arrivals at rate λ_i and an exponential type server, with rate μ_i. In addition, customers are transferred from station 1 to station 2 at rate $u\nu$, where u is a

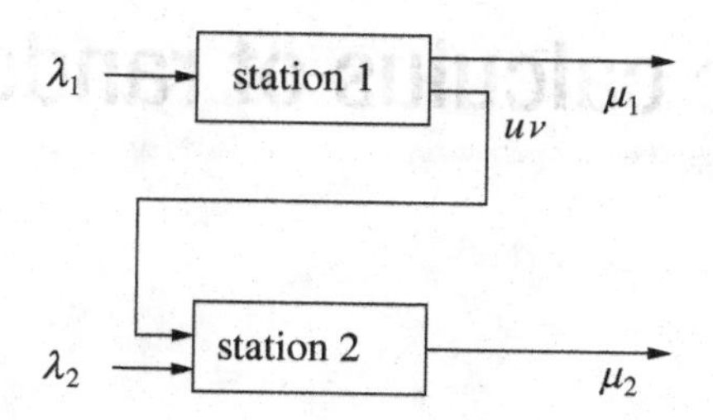

constant with $u \in U = [0,1]$. (Rather than applying dynamic programming here, we will apply the method of Foster–Lyapunov stability theory in continuous time.) The system is described by a continuous-time Markov process on $\mathbb{Z}_+^2$ with some transition rate matrix Q. (You don't need to write out Q.)

(a) Under what condition on $(\lambda_1, \lambda_2, \nu, \mu_1, \mu_2)$ is there a choice of the constant u such that the Markov process describing the system is positive recurrent?

(b) Let V be the quadratic Lyapunov function, $V(x_1, x_2) = \frac{x_1^2}{2} + \frac{x_2^2}{2}$. Compute the drift vector QV.

(c) Under the condition of part (a), and using the moment bound associated with the Foster–Lyapunov criteria, find an upper bound on the mean number in the system in equilibrium, $\overline{X}_1 + \overline{X}_2$. (The smaller the bound the better.)

6.29 Stability of a system with two queues and modulated server Consider two queues, queue 1 and queue 2, such that in each time slot, queue i receives a new packet with probability a_i, where $0 < a_1 < 1$ and $0 < a_2 < 1$. Suppose the server is described by a three state Markov process, with transition probabilities depending on a constant b, with $0 < b < \frac{1}{2}$, as shown.

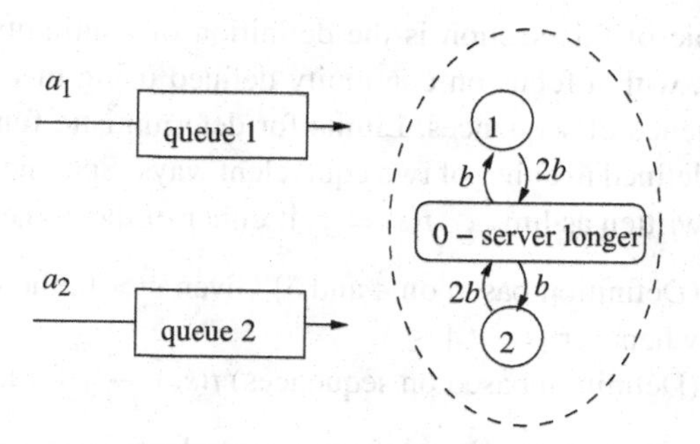

If the server process is in state i for $i \in \{1, 2\}$ at the beginning of a slot, then a potential service is given to station i. If the server process is in state 0 at the beginning of a slot, then a potential service is given to the longer queue (with ties broken in favor of queue 1). Then during the slot, the server state jumps with probability $2b$. (Note that a packet can arrive and depart in one time slot.) For what values of a_1 and a_2 is the process stable? Briefly explain your answer (but rigorous proof is not required).

7 Basic calculus of random processes

The calculus of deterministic functions revolves around continuous functions, derivatives, and integrals. These concepts all involve the notion of limits. See the appendix for a review of continuity, differentiation, and integration. In this chapter the same concepts are treated for random processes. We've seen four different senses in which a sequence of random variables can converge: almost surely (a.s.), in probability (p.), in mean square (m.s.), and in distribution (d.). Of these senses, we will use the mean square sense of convergence the most, and make use of the correlation version of the Cauchy criterion for m.s. convergence, and the associated facts that for m.s. convergence, the means of the limits are the limits of the means, and correlations of the limits are the limits of correlations (Proposition 2.11 and Corollaries 2.12 and 2.13). Ergodicity and the Karhunen–Loéve expansions are discussed as applications of integration of random processes.

7.1 Continuity of random processes

The topic of this section is the definition of continuity of a continuous-time random process, with a focus on continuity defined using m.s. convergence. Chapter 2 covers convergence of sequences. Limits for deterministic functions of a continuous variable can be defined in either of two equivalent ways. Specifically, a function f on $\mathbb{R}$ has a limit y at t_o, written as $\lim_{s \to t_o} f(s) = y$, if either of the two equivalent conditions is true:

(1) (Definition based on ϵ and δ) Given $\epsilon > 0$, there exists $\delta > 0$ so $| f(s) - y | \leq \epsilon$ whenever $|s - t_o| \leq \delta$.

(2) (Definition based on sequences) $f(s_n) \to y$ for any (s_n) such that $s_n \to t_o$.

Let us check that (1) and (2) are equivalent. Suppose (1) is true, and let (s_n) be such that $s_n \to t_o$. Let $\epsilon > 0$ and then let δ be as in condition (1). Since $s_n \to t_o$, it follows that there exists n_o so that $|s_n - t_o| \leq \delta$ for all $n \geq n_o$. But then $|f(s_n) - y| \leq \epsilon$ by the choice of δ. Thus, $f(s_n) \to y$. That is, (1) implies (2).

For the converse direction, it suffices to prove the contrapositive: if (1) is not true then (2) is not true. Suppose (1) is not true. Then there exists an $\epsilon > 0$ so that, for any $n \geq 1$, there exists a value s_n with $|s_n - t_o| \leq \frac{1}{n}$ such that $|f(s_n) - y| > \epsilon$. But then $s_n \to t_o$, and yet $f(s_n) \not\to y$, so (2) is false. That is, not (1) implies not (2). This completes the proof that (1) and (2) are equivalent.

Similarly, and by essentially the same reasons, convergence for a continuous-time random process can be defined using either ϵ and δ, or using sequences, at least for limits in the p., m.s., or d. senses. As we will see, the situation is slightly different for a.s. limits. Let $X = (X_t : t \in \mathbb{T})$ be a random process such that the index set $\mathbb{T}$ is equal to either all of $\mathbb{R}$, or an interval in $\mathbb{R}$, and fix $t_o \in \mathbb{T}$.

Definition 7.1 (Limits for continuous-time random processes.) The process $(X_t : t \in \mathbb{T})$ has limit Y at t_o :

(i) in the m.s. sense, written $\lim_{s\to t_o} X_s = Y$ *m.s.*, if for any $\epsilon > 0$, there exists $\delta > 0$ so that $E[(X_s - Y)^2] < \epsilon$ whenever $s \in \mathbb{T}$ and $|s - t_o| < \delta$. An equivalent condition is $X_{s_n} \stackrel{m.s.}{\to} Y$ as $n \to \infty$, whenever $s_n \to t_o$.

(ii) in probability, written $\lim_{s\to t_o} X_s = Y$ *p.*, if given any $\epsilon > 0$, there exists $\delta > 0$ so that $P\{|X_s - Y| \geq \epsilon\} < \epsilon$ whenever $s \in \mathbb{T}$ and $|s - t_o| < \delta$. An equivalent condition is $X_{s_n} \stackrel{p.}{\to} Y$ as $n \to \infty$, whenever $s_n \to t_o$.

(iii) in distribution, written $\lim_{s\to t_o} X_s = Y$ *d.*, if given any continuity point c of F_Y and any $\epsilon > 0$, there exists $\delta > 0$ so that $|F_{X,1}(c,s) - F_Y(c)| < \epsilon$ whenever $s \in \mathbb{T}$ and $|s - t_o| < \delta$. An equivalent condition is $X_{s_n} \stackrel{d.}{\to} Y$ as $n \to \infty$, whenever $s_n \to t_o$. (Recall that $F_{X,1}(c,s) = P\{X_s \leq c\}$.)

(iv) almost surely, written $\lim_{s\to t_o} X_s = Y$ *a.s.*, if there is an event F_{t_o} having probability one such that $F_{t_o} \subset \{\omega : \lim_{s\to t_o} X_s(\omega) = Y(\omega)\}$.[1]

The relationship among the above four types of convergence in continuous time is the same as the relationship among the four types of convergence of sequences, illustrated in Figure 2.8. That is, the following is true:

Proposition 7.2 *The following statements hold as $s \to t_o$ for a fixed t_o in $\mathbb{T}$: If either $X_s \stackrel{a.s.}{\to} Y$ or $X_s \stackrel{m.s.}{\to} Y$ then $X_s \stackrel{p.}{\to} Y$. If $X_s \stackrel{p.}{\to} Y$ then $X_s \stackrel{d.}{\to} Y$. Also, if there is a random variable Z with $E[Z^2] < \infty$ and $|X_t| \leq Z$ for all t, and if $X_s \stackrel{p.}{\to} Y$ then $X_s \stackrel{m.s.}{\to} Y$.*

Proof As indicated in Definition 7.1, the first three types of convergence are equivalent to convergence along sequences, in the corresponding senses. The fourth type of convergence, namely a.s. convergence as $s \to t_o$, implies convergence along sequences (Example 7.1 shows that the converse is not true). That is true because if (s_n) is a sequence converging to t_o,

$$\{\omega : \lim_{s\to t_o} X_t(\omega) = Y(\omega)\} \subset \{\omega : \lim_{n\to\infty} X_{s_n}(\omega) = Y(\omega)\}.$$

Therefore, if the first of these sets contains an event which has probability one, the second of these sets is an event which has probability one. The proposition then follows from the same relations for convergence of sequences. In particular, a.s. convergence

[1] This definition is complicated by the fact that the set $\{\omega : \lim_{s\to t_o} X_s(\omega) = Y(\omega)\}$ involves uncountably many random variables, and it is not necessarily an event. There is a way to simplify the definition as follows, but it requires an extra assumption. A probability space $(\Omega, \mathcal{F}, P)$ is *complete*, if whenever N is an event having probability zero, all subsets of N are events. If $(\Omega, \mathcal{F}, P)$ is complete, the definition of $\lim_{s\to t_o} X_s = Y$ *a.s.* is equivalent to the requirement that $\{\omega : \lim_{s\to t_o} X_s(\omega) = Y(\omega)\}$ be an event and have probability one.

for continuous time implies a.s. convergence along sequences (as just shown), which implies convergence in p. along sequences, which is the same as convergence in probability. The other implications of the proposition follow directly from the same implications for sequences, and the fact the first three definitions of convergence for continuous time have a form based on sequences. □

The following example shows that a.s. convergence as $s \to t_o$ is strictly stronger than a.s. convergence along sequences.

Example 7.1 Let U be uniformly distributed on the interval $[0, 1]$. Let $X_t = 1$ if $t - U$ is a rational number, and $X_t = 0$ otherwise. Each sample path of X takes values zero and one in any finite interval, so that X is not a.s. convergent at any t_o. However, for any fixed t, $P\{X_t = 0\} = 1$. Therefore, for any sequence s_n, since there are only countably many terms, $P\{X_{s_n} = 0 \text{ for all } n\} = 1$ so that $X_{s_n} \to 0$ *a.s.*

Definition 7.3 (Four types of continuity at a point for a random process) For each $t_o \in \mathbb{T}$ fixed, the random process $X = (X_t : t \in \mathbb{T})$ is *continuous at t_o* in any one of the four senses: m.s., p., a.s., or d., if $\lim_{s \to t_o} X_s = X_{t_o}$ in the corresponding sense.

The following is immediately implied by Proposition 7.2. It shows that for convergence of a random process at a single point, the relations illustrated in Figure 2.8 again hold.

Corollary 7.4 *If X is continuous at t_o in either the a.s. or m.s. sense, then X is continuous at t_o in probability. If X is continuous at t_o in probability, then X is continuous at t_o in distribution. Also, if there is a random variable Z with $E[Z^2] < \infty$ and $|X_t| \leq Z$ for all t, and if X is continuous at t_o in probability, then it is continuous at t_o in the m.s. sense.*

A deterministic function f on $\mathbb{R}$ is simply called continuous if it is continuous at all points. Since we have four senses of continuity at a point for a random process, this gives four types of continuity for random processes. Before stating them formally, we describe a fifth type of continuity of random processes, which is often used in applications. Recall that for a fixed $\omega \in \Omega$, the random process X gives a sample path, which is a function on $\mathbb{T}$. Continuity of a sample path is thus defined as it is for any deterministic function. The subset of Ω, $\{\omega : X_t(\omega) \text{ is a continuous function of } t\}$, or more concisely, $\{X_t \text{ is a continuous function of } t\}$, is the set of ω such that the sample path for ω is continuous. The fifth type of continuity requires that the sample paths be continuous, if a set of probability zero is ignored.

Definition 7.5 (Five types of continuity for a whole random process) A random process $X = (X_t : t \in \mathbb{T})$ is said to be

m.s. continuous if it is m.s. continuous at each t,
continuous in p. if it is continuous in p. at each t,

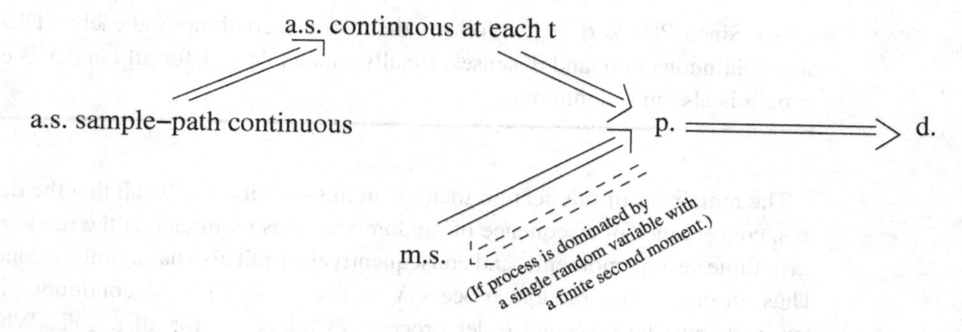

Figure 7.1 Relationships between five types of continuity of random processes.

continuous in d. if it is continuous in d. at each t,
a.s. continuous at each t, if it is a.s. continuous at each t[2]
a.s. sample-path continuous, if $F \subset \{X_t$ is continuous in $t\}$ for some event F with $P(F) = 1$.

The relationships between the five types of continuity for a whole random process are pictured in Figure 7.1 and summarized in the following proposition.

Proposition 7.6 *If a process is a.s. sample-path continuous it is a.s. continuous at each t. If a process is a.s. continuous at each t or m.s. continuous, it is continuous in p. If a process is continuous in p. it is continuous in d. Also, if there is a random variable Y with* $E[Y^2] < \infty$ *and* $|X_t| \le Y$ *for all t, and if X is continuous in p., then X is m.s. continuous.*

Proof Suppose X is a.s. sample-path continuous. Then for any $t_o \in \mathbb{T}$,

$$\{\omega : X_t(\omega) \text{ is continuous at all } t \in \mathbb{T}\} \subset \{\omega : X_t(\omega) \text{ is continuous at } t_o\}. \tag{7.1}$$

Since X is a.s. sample-path continuous, the set on the lefthand side of (7.1) contains an event F with $P(F) = 1$ and F is also a subset of the set on the righthand side of (7.1). Thus, X is a.s. continuous at t_o. Since t_o was an arbitrary element of $\mathbb{T}$, it follows that X is a.s. continuous at each t. The remaining implications of the proposition follow from Corollary 7.4. □

Example 7.2 (Shows a.s. sample-path continuity is strictly stronger than a.s. continuity at each t.) Let $X = (X_t : 0 \le t \le t)$ be given by $X_t = I_{\{t \ge U\}}$ for $0 \le t \le 1$, where U is uniformly distributed over [0, 1]. Thus, each sample path of X has a single upward jump of size one, at a random time U uniformly distributed over [0, 1]. So every sample path is discontinuous, and therefore X is not a.s. sample-path continuous. For any fixed t and ω, if $U(\omega) \ne t$ (i.e. if the jump of X is not exactly at time t) then $X_s(\omega) \to X_t(\omega)$ as

[2] We avoid using the terminology "a.s. continuous" for the whole random process, because such terminology could too easily be confused with a.s. sample-path continuous.

$s \to t$. Since $P\{U \neq t\} = 1$, it follows that X is a.s. continuous at each t. Therefore X is also continuous in p. and d. senses. Finally, since $|X_t| \leq 1$ for all t and X is continuous in p., it is also m.s. continuous.

The remainder of this section focuses on m.s. continuity. Recall that the definition of m.s. convergence of a sequence of random variables requires that the random variables have finite second moments, and consequently the limit also has a finite second moment. Thus, in order for a random process $X = (X_t : t \in \mathbb{T})$ to be continuous in the m.s. sense, it must be a second order process: $E[X_t^2] < \infty$ for all $t \in \mathbb{T}$. Whether X is m.s. continuous depends only on the correlation function R_X, as shown in the following proposition.

Proposition 7.7 *Suppose $(X_t : t \in \mathbb{T})$ is a second order process. The following are equivalent:*

(i) *R_X is continuous at all points of the form (t, t). (This condition involves R_X for points in and near the set of points of the form (t, t). It is stronger than requiring $R_X(t, t)$ to be continuous in t – see Example 7.3.)*
(ii) *X is m.s. continuous.*
(iii) *R_X is continuous over $\mathbb{T} \times \mathbb{T}$.*

If X is m.s. continuous, then the mean function, $\mu_X(t)$, is continuous. If X is wide sense stationary, the following are equivalent:

(i′) *$R_X(\tau)$ is continuous at $\tau = 0$.*
(ii′) *X is m.s. continuous.*
(iii′) *$R_X(\tau)$ is continuous over all of $\mathbb{R}$.*

Proof *((i) implies (ii))* Fix $t \in \mathbb{T}$ and suppose that R_X is continuous at the point (t, t). Then $R_X(s, s)$, $R_X(s, t)$, and $R_X(t, s)$ all converge to $R_X(t, t)$ as $s \to t$. Therefore, $\lim_{s\to t} E[(X_s - X_t)^2] = \lim_{s\to t}(R_X(s, s) - R_X(s, t) - R_X(t, s) + R_X(t, t)) = 0$. So X is m.s. continuous at t. Therefore if R_X is continuous at all points of the form $(t, t) \in \mathbb{T} \times \mathbb{T}$, then X is m.s. continuous at all $t \in \mathbb{T}$. Therefore (i) implies (ii).

((ii) implies (iii)) Suppose condition (ii) is true. Let $(s, t) \in \mathbb{T} \times \mathbb{T}$, and suppose $(s_n, t_n) \in \mathbb{T} \times \mathbb{T}$ for all $n \geq 1$ such that $\lim_{n\to\infty}(s_n, t_n) = (s, t)$. Therefore, $s_n \to s$ and $t_n \to t$ as $n \to \infty$. By condition (ii), it follows that $X_{s_n} \stackrel{m.s.}{\to} X_s$ and $X_{t_n} \stackrel{m.s.}{\to} X_t$ as $n \to \infty$. Since the limit of the correlations is the correlation of the limit for a pair of m.s. convergent sequences (Corollary 2.12) it follows that $R_X(s_n, t_n) \to R_X(s, t)$ as $n \to \infty$. Thus, R_X is continuous at (s, t), where (s, t) was an arbitrary point of $\mathbb{T} \times \mathbb{T}$. Therefore R_X is continuous over $\mathbb{T} \times \mathbb{T}$, proving that (ii) implies (iii).

Obviously (iii) implies (i), so the proof of the equivalence of (i)–(iii) is complete.

If X is m.s. continuous, then, by definition, for any $t \in \mathbb{T}$, $X_s \stackrel{m.s.}{\to} X_t$ as $s \to t$. It thus follows that $\mu_X(s) \to \mu_X(t)$, because the limit of the means is the mean of the limit, for a m.s. convergent sequence (Corollary 2.13). Thus, m.s. continuity of X implies that the deterministic mean function, μ_X, is continuous.

Finally, if X is WSS, then $R_X(s,t) = R_X(\tau)$ where $\tau = s - t$, and the three conditions (i)–(iii) become (*i′*)–(*iii′*), so the equivalence of (i)–(iii) implies the equivalence of (*i′*)–(*iii′*). □

Example 7.3 Let $X = (X_t : t \in \mathbb{R})$ be defined by $X_t = U$ for $t < 0$ and $X_t = V$ for $t \geq 0$, where U and V are independent random variables with mean zero and variance one. Let t_n be a sequence of strictly negative numbers converging to 0. Then $X_{t_n} = U$ for all n and $X_0 = V$. Since $P\{|U - V| \geq \epsilon\} \neq 0$ for ϵ small enough, X_{t_n} does not converge to X_0 in p. sense. So X is not continuous in probability at zero. It is thus not continuous in the m.s or a.s. sense at zero either. The only one of the five senses that the whole process could be continuous is continuous in distribution. The process X is continuous in distribution if and only if U and V have the same distribution. Finally, let us check the continuity properties of the autocorrelation function. The autocorrelation function is given by $R_X(s,t) = 1$ if either $s,t < 0$ or if $s,t \geq 0$, and $R_X(s,t) = 0$ otherwise. So R_X is not continuous at $(0,0)$, because $R(\frac{1}{n}, -\frac{1}{n}) = 0$ for all $n \geq 1$, so $R(\frac{1}{n}, -\frac{1}{n}) \not\to R_X(0,0) = 1$ as $n \to \infty$. However, it is true that $R_X(t,t) = 1$ for all t, so that $R_X(t,t)$ is a continuous function of t. This illustrates the fact that continuity of the function of two variables, $R_X(s,t)$, at a particular point of the form (t_o, t_o), is a stronger requirement than continuity of the function of one variable, $R_X(t,t)$, at $t = t_o$.

Example 7.4 Let $W = (W_t : t \geq 0)$ be a Brownian motion with parameter σ^2. Then $E[(W_t - W_s)^2] = \sigma^2|t-s| \to 0$ as $s \to t$. Therefore W is m.s. continuous. Another way to show W is m.s. continuous is to observe that the autocorrelation function, $R_W(s,t) = \sigma^2(s \wedge t)$, is continuous. Since W is m.s. continuous, it is also continuous in the p. and d. senses. As we stated in defining W, it is a.s. sample-path continuous, and therefore a.s. continuous at each $t \geq 0$, as well.

Example 7.5 Let $N = (N_t : t \geq 0)$ be a Poisson process with rate $\lambda > 0$. Then for fixed t, $E[(N_t - N_s)^2] = \lambda(t-s) + (\lambda(t-s))^2 \to 0$ as $s \to t$. Therefore N is m.s. continuous. As required, R_N, given by $R_N(s,t) = \lambda(s \wedge t) + \lambda^2 st$, is continuous. Since N is m.s. continuous, it is also continuous in the p. and d. senses. N is also a.s. continuous at any fixed t, because the probability of a jump at exactly time t is zero for any fixed t. However, N is not a.s. sample continuous. In fact, $P\{N \text{ is continuous on } [0,a]\} = e^{-\lambda a}$ and so $P\{N \text{ is continuous on } \mathbb{R}_+\} = 0$.

Definition 7.8 A random process $(X_t : t \in \mathbb{T})$, such that $\mathbb{T}$ is a bounded interval (open, closed, or mixed) in $\mathbb{R}$ with endpoints $a < b$, is *piecewise m.s. continuous*, if there exist $n \geq 1$ and $a = t_0 < t_1 < \ldots < t_n = b$, such that, for $1 \leq k \leq n$: X is m.s. continuous over (t_{k-1}, t_k) and has m.s. limits at the endpoints of (t_{k-1}, t_k).
More generally, if $\mathbb{T}$ is all of $\mathbb{R}$ or an interval in $\mathbb{R}$, X is piecewise m.s. continuous over $\mathbb{T}$ if it is piecewise m.s. continuous over every bounded subinterval of $\mathbb{T}$.

7.2 Mean square differentiation of random processes

Before considering the m.s. derivative of a random process, we review the definition of the derivative of a function (also see Appendix 11.4). Let the index set $\mathbb{T}$ be either all of $\mathbb{R}$ or an interval in $\mathbb{R}$. Suppose f is a deterministic function on $\mathbb{T}$. Recall that for a fixed t in $\mathbb{T}$, f is differentiable at t if $\lim_{s\to t}\frac{f(s)-f(t)}{s-t}$ exists and is finite, and if f is differentiable at t, the value of the limit is the derivative, $f'(t)$. The whole function f is called differentiable if it is differentiable at all t. The function f is called continuously differentiable if f is differentiable, and the derivative function f' is continuous.

In many applications of calculus, it is important that a function f be not only differentiable, but continuously differentiable. In much of the applied literature, when there is an assumption that a function is differentiable, it is understood that the function is continuously differentiable. For example, by the *fundamental theorem of calculus*,

$$f(b) - f(a) = \int_a^b f'(s)ds \tag{7.2}$$

holds if f is a continuously differentiable function with derivative f'. Example 11.6 shows that (7.2) might not hold if f is simply assumed to be differentiable.

Let $X = (X_t : t \in \mathbb{T})$ be a second order random process such that the index set $\mathbb{T}$ is equal to either all of $\mathbb{R}$ or an interval in $\mathbb{R}$. The following definition for m.s. derivatives is analogous to the definition of derivatives for deterministic functions.

Definition 7.9 For each t fixed, the random process $X = (X_t : t \in \mathbb{T})$ is *mean square (m.s.) differentiable at t* if the following limit exists:

$$\lim_{s\to t}\frac{X_s - X_t}{s-t}\ m.s.$$

The limit, if it exists, is the m.s. derivative of X at t, denoted by X'_t. The whole random process X is said to be *m.s. differentiable* if it is m.s. differentiable at each t, and it is said to be *m.s. continuously differentiable* if it is m.s. differentiable and the derivative process X' is m.s. continuous.

Let ∂_i denote the operation of taking the partial derivative with respect to the ith argument. For example, if $f(x,y) = x^2y^3$ then $\partial_2 f(x,y) = 3x^2y^2$ and $\partial_1\partial_2 f(x,y) = 6xy^2$. The partial derivative of a function is the same as the ordinary derivative with respect to one variable, with the other variables held fixed. We shall be applying ∂_1 and ∂_2 to an autocorrelation function $R_X = \{R_X(s,t) : (s,t) \in \mathbb{T}\times\mathbb{T}\}$, which is a function of two variables.

Proposition 7.10 *(a) (The derivative of the mean is the mean of the derivative) If X is m.s. differentiable, then the mean function μ_X is differentiable, and $\mu'_X(t) = \mu_{X'}(t)$ (i.e. the operations of (i) taking expectation, which basically involves integrating over ω, and (ii) differentiation with respect to t, can be done in either order).*

(b) *If X is m.s. differentiable, then $R_{X'X} = \partial_1 R_X$ and $R_{XX'} = \partial_2 R_X$, and the autocorrelation function of X' is given by $R_{X'} = \partial_1\partial_2 R_X = \partial_2\partial_1 R_X$. (In particular, the indicated partial derivatives exist.)*

(c) *X is m.s. differentiable at t if and only if the following limit exists and is finite:*

$$\lim_{s,\,s' \to t} \frac{R_X(s,s') - R_X(s,t) - R_X(t,s') + R_X(t,t)}{(s-t)(s'-t)}. \tag{7.3}$$

(Therefore, the whole process X is m.s. differentiable if and only if the limit in (7.3) exists and is finite for all $t \in \mathbb{T}$.)

(d) *X is m.s. continuously differentiable if and only if R_X, $\partial_2 R_X$, and $\partial_1\partial_2 R_X$ exist and are continuous. (By symmetry, if X is m.s. continuously differentiable, then also $\partial_1 R_X$ is continuous.)*

(e) *(Specialization of (d) for WSS case) Suppose X is WSS. Then X is m.s. continuously differentiable if and only if $R_X(\tau)$, $R_X'(\tau)$, and $R_X''(\tau)$ exist and are continuous functions of τ. If X is m.s. continuously differentiable then X and X' are jointly WSS, X' has mean zero (i.e. $\mu_{X'} = 0$) and autocorrelation function given by $R_{X'}(\tau) = -R_X''(\tau)$, and the cross correlation functions are given by $R_{X'X}(\tau) = R_X'(\tau)$ and $R_{XX'}(\tau) = -R_X'(\tau)$.*

(f) *(A necessary condition for m.s. differentiability) If X is WSS and m.s. differentiable, then $R_X'(0)$ exists and $R_X'(0) = 0$.*

(g) *If X is a m.s. differentiable Gaussian process, then X and its derivative process X' are jointly Gaussian.*

Proof (a) Suppose X is m.s. differentiable. Then for any t fixed,

$$\frac{X_s - X_t}{s-t} \xrightarrow{m.s.} X_t' \text{ as } s \to t.$$

It thus follows that

$$\frac{\mu_X(s) - \mu_X(t)}{s-t} \to \mu_{X'}(t) \text{ as } s \to t, \tag{7.4}$$

because the limit of the means is the mean of the limit, for a m.s. convergent sequence (Corollary 2.13). But (7.4) is just the definition of the statement that the derivative of μ_X at t is equal to $\mu_{X'}(t)$. That is, $\frac{d\mu_X}{dt}(t) = \mu_{X'}(t)$ for all t, or more concisely, $\mu_X{}' = \mu_{X'}$.

(b) Suppose X is m.s. differentiable. Since the limit of the correlations is the correlation of the limits for m.s. convergent sequences (Corollary 2.12), for $t, t' \in \mathbb{T}$,

$$\begin{aligned} R_{X'X}(t,t') &= \lim_{s\to t} E\left[\left(\frac{X(s)-X(t)}{s-t}\right) X(t')\right] \\ &= \lim_{s\to t} \frac{R_X(s,t') - R_X(t,t')}{s-t} = \partial_1 R_X(t,t'). \end{aligned}$$

Thus, $R_{X'X} = \partial_1 R_X$, and in particular, the partial derivative $\partial_1 R_X$ exists. Similarly, $R_{XX'} = \partial_2 R_X$. Also, by the same reasoning,

$$\begin{aligned} R_{X'}(t,t') &= \lim_{s'\to t'} E\left[X'(t)\left(\frac{X(s')-X(t')}{s'-t'}\right)\right] \\ &= \lim_{s'\to t'} \frac{R_{X'X}(t,s')-R_{X'X}(t,t')}{s'-t'} \\ &= \partial_2 R_{X'X}(t,t') = \partial_2\partial_1 R_X(t,t'), \end{aligned}$$

so that $R_{X'} = \partial_2\partial_1 R_X$. Similarly, $R_{X'} = \partial_1\partial_1 R_X$.

(c) By the correlation form of the Cauchy criterion (Proposition 2.11), X is m.s. differentiable at t if and only if the following limit exists and is finite:

$$\lim_{s,\,s'\to t} E\left[\left(\frac{X(s)-X(t)}{s-t}\right)\left(\frac{X(s')-X(t)}{s'-t}\right)\right]. \tag{7.5}$$

Multiplying out the terms in the numerator in (7.5) and using $E[X(s)X(s')] = R_X(s,s')$, $E[X(s)X(t)] = R_X(s,t)$, and so on, shows that (7.5) is equivalent to (7.3). So part (c) is proved.

(d) The numerator in (7.3) involves R_X evaluated at the four corners of the rectangle $[t,s]\times[t,s']$, shown in Figure 7.2. Suppose R_X, $\partial_2 R_X$, and $\partial_1\partial_2 R_X$ exist and are continuous functions. Then by the fundamental theorem of calculus,

$$\begin{aligned} &(R_X(s,s')-R_X(s,t)) - (R_X(t,s')-R_X(t,t)) \\ &\quad = \int_t^{s'} \partial_2 R_X(s,v)dv - \int_t^{s'} \partial_2 R_X(t,v)dv \\ &\quad = \int_t^{s'} [\partial_2 R_X(s,v) - \partial_2 R_X(t,v)]\,dv \\ &\quad = \int_t^{s'}\int_t^{s} \partial_1\partial_2 R_X(u,v)dudv. \end{aligned} \tag{7.6}$$

Therefore, the ratio in (7.3) is the average value of $\partial_1\partial_2 R_X$ over the rectangle $[t,s]\times[t,s']$. Since $\partial_1\partial_2 R_X$ is assumed to be continuous, the limit in (7.3) exists and it is equal to $\partial_1\partial_2 R_X(t,t)$. Therefore, by part (c) already proved, X is m.s. differentiable. By part

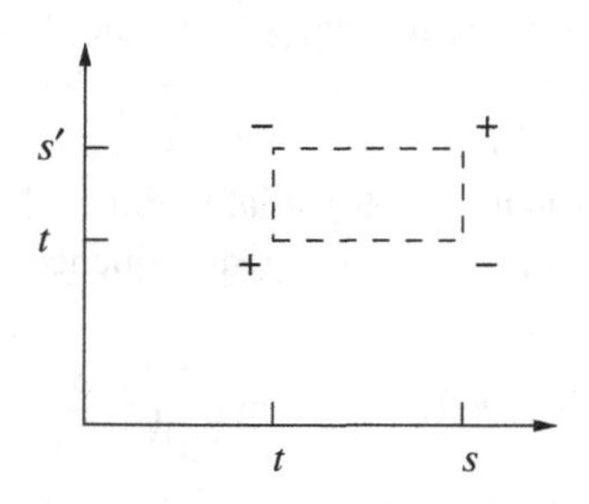

Figure 7.2 Sampling points of R_X.

(b), the autocorrelation function of X' is $\partial_1\partial_2 R_X$. Since this is assumed to be continuous, it follows that X' is m.s. continuous. Thus, X is m.s. continuously differentiable.

(e) If X is WSS, then $R_X(s-t) = R_X(\tau)$ where $\tau = s-t$. Suppose $R_X(\tau)$, $R'_X(\tau)$ and $R''_X(\tau)$ exist and are continuous functions of τ. Then

$$\partial_1 R_X(s,t) = R'_X(\tau) \quad \text{and} \quad \partial_2\partial_1 R_X(s,t) = -R''_X(\tau). \tag{7.7}$$

The minus sign in (7.7) appears because $R_X(s,t) = R_X(\tau)$ where $\tau = s-t$, and the derivative of τ with respect to t is -1. So, the hypotheses of part (d) hold, so that X is m.s. differentiable. Since X is WSS, its mean function μ_X is constant, which has derivative zero, so X' has mean zero. Also by part (c) and (7.7), $R_{X'X}(\tau) = R'_X(\tau)$ and $R_{X'X'} = -R_X''$. Similarly, $R_{XX'}(\tau) = -R'_X(\tau)$. Note that X and X' are each WSS and the cross correlation functions depend on τ alone, so X and X' are jointly WSS.

(f) If X is WSS then

$$E\left[\left(\frac{X(t)-X(0)}{t}\right)^2\right] = -\frac{2(R_X(t)-R_X(0))}{t^2}. \tag{7.8}$$

Therefore, if X is m.s. differentiable then the right side of (7.8) must converge to a finite limit as $t \to 0$, so in particular it is necessary that $(R_X(t) - R_X(0))/t \to 0$ as $t \to 0$. Therefore $R'_X(0) = 0$.

(g) The derivative process X' is obtained by taking linear combinations and m.s. limits of random variables in $X = (X_t; t \in \mathbb{T})$. Therefore, (g) follows from the fact that the joint Gaussian property is preserved under linear combinations and limits (Proposition 3.8(c)). □

Example 7.6 Let $f(t) = t^2\sin(1/t^2)$ for $t \neq 0$ and $f(0) = 0$ as in Example 11.6, and let $X = (X_t : t \in \mathbb{R})$ be the deterministic random process such that $X(t) = f(t)$ for all $t \in \mathbb{R}$. Since X is differentiable as an ordinary function, it is also m.s. differentiable, and its m.s. derivative X' is equal to f'. Since X', as a deterministic function, is not continuous at zero, it is also not continuous at zero in the m.s. sense. We have $R_X(s,t) = f(s)f(t)$ and $\partial_2 R_X(s,t) = f(s)f'(t)$, which is not continuous. So indeed the conditions of Proposition 7.10(d) do not hold, as required.

Example 7.7 A Brownian motion $W = (W_t : t \geq 0)$ is not m.s. differentiable. If it were, then for any fixed $t \geq 0$, $\frac{W(s)-W(t)}{s-t}$ would converge in the m.s. sense as $s \to t$ to a random variable with a finite second moment. For a m.s. convergent sequence, the second moments of the variables in the sequence converge to the second moment of the limit random variable, which is finite. But $W(s) - W(t)$ has mean zero and variance $\sigma^2|s-t|$, so that

$$\lim_{s\to t} E\left[\left(\frac{W(s)-W(t)}{s-t}\right)^2\right] = \lim_{s\to t}\frac{\sigma^2}{|s-t|} = +\infty. \tag{7.9}$$

Thus, W is not m.s. differentiable at any t. For another approach, we could appeal to Proposition 7.10 to deduce this result. The limit in (7.9) is the same as the limit in (7.5), but with s and s' restricted to be equal. Hence (7.5), or equivalently (7.3), is not a finite limit, implying that W is not differentiable at t.

Similarly, a Poisson process is not m.s. differentiable at any t. A WSS process X with $R_X(\tau) = e^{-\alpha|\tau|}$ is not m.s. differentiable because $R'_X(0)$ does not exist. A WSS process X with $R_X(\tau) = \frac{1}{1+\tau^2}$ is m.s. differentiable, and its derivative process X' is WSS with mean 0 and covariance function

$$R_{X'}(\tau) = -\left(\frac{1}{1+\tau^2}\right)'' = \frac{2-6\tau^2}{(1+\tau^2)^3}.$$

Proposition 7.11 *Suppose X is a m.s. differentiable random process and f is a differentiable function. Then the product $Xf = (X(t)f(t) : t \in \mathbb{R})$ is mean square differentiable and $(Xf)' = X'f + Xf'$.*

Proof:
Fix t. Then for each $s \neq t$,

$$\frac{X(s)f(s) - X(t)f(t)}{s-t} = \frac{(X(s)-X(t))f(s)}{s-t} + \frac{X(t)(f(s)-f(t))}{s-t}$$
$$\stackrel{m.s.}{\to} X'(t)f(t) + X(t)f'(t) \quad \text{as } s \to t.$$

Definition 7.12 A random process X on a bounded interval (open, closed, or mixed) with endpoints $a < b$ is *continuous and piecewise continuously differentiable in the m.s. sense*, if X is m.s. continuous over the interval, and if there exists $n \geq 1$ and $a = t_0 < t_1 < \ldots < t_n = b$, such that, for $1 \leq k \leq n$: X is m.s. continuously differentiable over (t_{k-1}, t_k) and X' has finite limits at the endpoints of (t_{k-1}, t_k).
More generally, if $\mathbb{T}$ is all of $\mathbb{R}$ or a subinterval of $\mathbb{R}$, then a random process $X = (X_t : t \in \mathbb{T})$ is continuous and piecewise continuously differentiable in the m.s. sense if its restriction to any bounded interval is continuous and piecewise continuously differentiable in the m.s. sense.

7.3 Integration of random processes

Let $X = (X_t : a \leq t \leq b)$ be a random process and let h be a function on a finite interval $[a, b]$. How shall we define the following integral?

$$\int_a^b X_t h(t)dt. \tag{7.10}$$

One approach is to note that for each fixed ω, $X_t(\omega)$ is a deterministic function of time, and so the integral can be defined as the integral of a deterministic function for each ω. We shall focus on another approach, namely mean square (m.s.) integration. An advantage of m.s. integration is that it relies much less on properties of sample paths of random processes.

As for integration of deterministic functions, the m.s. Riemann integrals are based on Riemann sums, defined as follows. Given:

- A partition of $(a, b]$ of the form $(t_0, t_1], (t_1, t_2], \ldots, (t_{n-1}, t_n]$, where $n \geq 0$ and $a = t_0 < t_1 \ldots < t_n = b$,
- A sampling point from each subinterval, $v_k \in (t_{k-1}, t_k]$, for $1 \leq k \leq n$,

the corresponding *Riemann sum* for Xh is defined by

$$\sum_{k=1}^{n} X_{v_k} h(v_k)(t_k - t_{k-1}).$$

The *norm of the partition* is defined to be $\max_k |t_k - t_{k-1}|$.

Definition 7.13 The *Riemann integral* $\int_a^b X_t h(t)dt$ is said to exist in the m.s. sense and its value is the random variable I if the following is true. Given any $\epsilon > 0$, there is a $\delta > 0$ so that $E[(\sum_{k=1}^n X_{v_k} h(v_k)(t_k - t_{k-1}) - I)^2] \leq \epsilon$ whenever the norm of the partition is less than or equal to δ. This definition is equivalent to the following condition, expressed using convergence of sequences. The m.s. Riemann integral exists and is equal to I, if for any sequence of partitions, specified by $((t_1^m, t_2^m, \ldots, t_{n_m}^m) : m \geq 1)$, with corresponding sampling points $((v_1^m, \ldots, v_{n_m}^m) : m \geq 1)$, such that the norm of the mth partition converges to zero as $m \to \infty$, the corresponding sequence of Riemann sums converges in the m.s. sense to I as $m \to \infty$. The process $X_t h(t)$ is said to be *m.s. Riemann integrable* over $(a, b]$ if the integral $\int_a^b X_t h(t)dt$ exists and is finite.

Next, suppose $X_t h(t)$ is defined over the whole real line. If $X_t h(t)$ is m.s. Riemann integrable over every bounded interval $[a, b]$, then the Riemann integral of $X_t h(t)$ over $\mathbb{R}$ is defined by

$$\int_{-\infty}^{\infty} X_t h(t)dt = \lim_{a,b\to\infty} \int_{-a}^{b} X_t h(t)dt \quad m.s.$$

provided that the indicated limits exist as a, b jointly converge to $+\infty$.

Whether an integral exists in the m.s. sense is determined by the autocorrelation function of the random process involved, as shown next. The condition involves Riemann integration of a deterministic function of two variables. As reviewed in Appendix 11.5, a two-dimensional Riemann integral over a bounded rectangle is defined as the limit of Riemann sums corresponding to a partition of the rectangle into subrectangles and choices of sampling points within the subrectangles. If the sampling points for the Riemann sums are required to be horizontally and vertically aligned, then we say the two-dimensional Riemann integral exists with aligned sampling.

Proposition 7.14 *The integral $\int_a^b X_t h(t)dt$ exists in the m.s. Riemann sense if and only if*

$$\int_a^b \int_a^b R_X(s,t)h(s)h(t)dsdt \tag{7.11}$$

exists as a two dimensional Riemann integral with aligned sampling. The m.s. integral exists, in particular, if X is m.s. piecewise continuous over $[a,b]$ and h is piecewise continuous over $[a,b]$.

Proof By definition, the m.s. integral of $X_t h(t)$ exists if and only if the Riemann sums converge in the m.s. sense for an arbitrary sequence of partitions and sampling points, such that the norms of the partitions converge to zero. So consider an arbitrary sequence of partitions of $(a,b]$ into intervals specified by the collection of endpoints, $((t_0^m, t_1^m, \ldots, t_{n^m}^m) : m \geq 1)$, with corresponding sampling point $v_k^m \in (t_{k-1}^m, t_k^m]$ for each m and $1 \leq k \leq n^m$, such that the norm of the mth partition converges to zero as $m \to \infty$. For each $m \geq 1$, let S_m denote the corresponding Riemann sum:

$$S_m = \sum_{k=1}^{n^m} X_{v_k^m} h(v_k^m)(t_k^m - t_{k-1}^m).$$

By the correlation form of the Cauchy criterion for m.s. convergence (Proposition 2.11), the sequence $(S_m : m \geq 1)$ converges in the m.s. sense if and only if $\lim_{m,m' \to \infty} E[S_m S_{m'}]$ exists and is finite. Note that

$$E[S_m S_{m'}] = \sum_{j=1}^{n^m} \sum_{k=1}^{n^{m'}} R_X(v_j^m, v_k^{m'})h(v_j^m)h(v_k^{m'})(t_j^m - t_{j-1}^m)(t_k^{m'} - t_{k-1}^{m'}), \tag{7.12}$$

and the righthand side of (7.12) is the Riemann sum for the integral (7.11), for the partition of $(a,b] \times (a,b]$ into rectangles of the form $(t_{j-1}^m, t_j^m] \times (t_{k-1}^{m'}, t_k^{m'}]$ and the sampling points $(v_j^m, v_k^{m'})$. Note that the mm' sampling points are aligned, in that they are determined by the $m + m'$ numbers $v_1^m, \ldots, v_{n^m}^m, v_1^{m'}, \ldots, v_{n^{m'}}^{m'}$. Moreover, any Riemann sum for the integral (7.11) with aligned sampling can arise in this way. Further, as $m, m' \to \infty$, the norm of this partition, which is the maximum length or width of any rectangle of the partition, converges to zero. Thus, the limit $\lim_{m,m' \to \infty} E[S_m S_{m'}]$ exists for any sequence of partitions and sampling points if and only if the integral (7.11) exists as a two-dimensional Riemann integral with aligned sampling.

Finally, if X is piecewise m.s. continuous over $[a,b]$ and h is piecewise continuous over $[a,b]$, then there is a partition of $[a,b]$ into intervals of the form $(s_{k-1}, s_k]$ such that X is m.s. continuous over (s_{k-1}, s_k) with m.s. limits at the endpoints, and h is continuous over (s_{k-1}, s_k) with finite limits at the endpoints. Therefore, $R_X(s,t)h(s)h(t)$ restricted to each rectangle of the form $(s_{j-1}, s_j) \times (s_{k-1}, s_k)$, is the restriction of a continuous function on $[s_{j-1}, s_j] \times [s_{k-1}, s_k]$. Thus $R_X(s,t)h(s)h(t)$ is Riemann integrable over $[a,b] \times [a,b]$. □

Proposition 7.15 *Suppose $X_t h(t)$ and $Y_t k(t)$ are both m.s. integrable over $[a,b]$. Then*

$$E\left[\int_a^b X_t h(t)dt\right] = \int_a^b \mu_X(t)h(t)dt, \tag{7.13}$$

$$E\left[\left(\int_a^b X_t h(t)dt\right)^2\right] = \int_a^b\int_a^b R_X(s,t)h(s)h(t)dsdt, \tag{7.14}$$

$$\mathrm{Var}\left(\int_a^b X_t h(t)dt\right) = \int_a^b\int_a^b C_X(s,t)h(s)h(t)dsdt, \tag{7.15}$$

$$E\left[\left(\int_a^b X_s h(s)ds\right)\left(\int_a^b Y_t k(t)dt\right)\right] = \int_a^b\int_a^b R_{XY}(s,t)h(s)k(t)dsdt, \tag{7.16}$$

$$\mathrm{Cov}\left(\int_a^b X_s h(s)ds, \int_a^b Y_t k(t)dt\right) = \int_a^b\int_a^b C_{XY}(s,t)h(s)k(t)dsdt, \tag{7.17}$$

$$\int_a^b X_t h(t) + Y_t k(t)dt = \int_a^b X_t h(t)dt + \int_a^b Y_t k(t))dt. \tag{7.18}$$

Proof Let (S_m) denote the sequence of Riemann sums appearing in the proof of Proposition 7.14. Since the mean of a m.s. convergent sequence of random variables is the limit of the means (Corollary 2.13),

$$\begin{aligned} E\left[\int_a^b X_t h(t)dt\right] &= \lim_{m\to\infty} E[S_m] \\ &= \lim_{m\to\infty} \sum_{k=1}^{n_m} \mu_X(v_k^m)h(v_k^m)(t_k^m - t_{k-1}^m). \end{aligned} \tag{7.19}$$

The righthand side of (7.19) is a limit of Riemann sums for $\int_a^b \mu_X(t)h(t)dt$. Since this limit exists and is equal to $E\left[\int_a^b X_t h(t)dt\right]$ for any sequence of partitions and sample points, it follows that $\int_a^b \mu_X(t)h(t)dt$ exists as a Riemann integral, and is equal to $E\left[\int_a^b X_t h(t)dt\right]$, so (7.13) is proved.

The second moment of the m.s. limit of $(S_m : m \geq 0)$ is $\lim_{m,m'\to\infty} E[S_m S_{m'}]$, by the correlation form of the Cauchy criterion for m.s. convergence (Proposition 2.11), which implies (7.14). It follows from (7.13) that

$$E\left[\left(\int_a^b X_t h(t)dt\right)\right]^2 = \int_a^b\int_a^b \mu_X(s)\mu_X(t)h(s)h(t)dsdt.$$

Subtracting each side of this from the corresponding side of (7.14) yields (7.15). The proofs of (7.16) and (7.17) are similar to the proofs of (7.14) and (7.15), and are left to the reader.

For any partition of $[a,b]$ and choice of sampling points, the Riemann sums for the three integrals appearing in (7.17) satisfy the corresponding additivity condition, implying (7.17). $\square$

The fundamental theorem of calculus, stated in Appendix 11.5, states that the increments of a continuous, piecewise continuously differentiable function are equal to integrals of the derivative of the function. The following is the generalization of the fundamental theorem of calculus to the m.s. calculus.

Theorem 7.16 *(Fundamental theorem of m.s. calculus) Let X be a m.s. continuously differentiable random process. Then for $a < b$,*

$$X_b - X_a = \int_a^b X'_t dt \quad \text{(m.s. Riemann integral)}. \tag{7.20}$$

More generally, if X is continuous and piecewise continuously differentiable, (11.4) holds with X'_t replaced by the righthand derivative, D_+X_t. (Note that $D_+X_t = X'_t$ whenever X'_t is defined.)

Proof The m.s. Riemann integral in (7.20) exists because X' is assumed to be m.s. continuous. Let $B = X_b - X_a - \int_a^b X'_t dt$, and let Y be an arbitrary random variable with a finite second moment. It suffices to show that $E[YB] = 0$, because a possible choice of Y is B itself. Let $\varphi(t) = E[YX_t]$. Then for $s \neq t$,

$$\frac{\varphi(s) - \varphi(t)}{s - t} = E\left[Y\left(\frac{X_s - X_t}{s - t}\right)\right].$$

Taking a limit as $s \to t$ and using the fact that the correlation of a limit is the limit of the correlations for m.s. convergent sequences, it follows that φ is differentiable and $\varphi'(t) = E[YX'_t]$. Since X' is m.s. continuous, it similarly follows that φ' is continuous.

Next, we use the fact that the integral in (7.20) is the m.s. limit of Riemann sums, with each Riemann sum corresponding to a partition of $(a, b]$ specified by some $n \geq 1$ and $a = t_0 < \ldots < t_n = b$ and sampling points $v_k \in (t_{k-1}, t_k]$ for $a \leq k \leq n$. Since the limit of the correlation is the correlation of the limit for m.s. convergence,

$$\begin{aligned} E\left[Y \int_a^b X'_t dt\right] &= \lim_{|t_k - t_{k-1}| \to 0} E\left[Y \sum_{k=1\breve{g}}^{n} X'_{v_k}(t_k - t_{k-1})\right] \\ &= \lim_{|t_k - t_{k-1}| \to 0} \sum_{k=1}^{n} \varphi'(v_k)(t_k - t_{k-1}) = \int_a^b \varphi'(t)dt. \end{aligned}$$

Therefore, $E[YB] = \varphi(b) - \varphi(a) - \int_a^b \varphi'(t)dt$, which is equal to zero by the fundamental theorem of calculus for deterministic continuously differentiable functions. This establishes (7.20) when X is m.s. continuously differentiable. If X is m.s. continuous and only piecewise continuously differentiable, we can use essentially the same proof, observing that φ is continuous and piecewise continuously differentiable, so that $E[YB] = \varphi(b) - \varphi(a) - \int_a^b \varphi'(t)dt = 0$ by the fundamental theorem of calculus for deterministic continuous, piecewise continuously differential functions. □

Proposition 7.17 *Suppose X is a Gaussian random process. Then X, together with all mean square derivatives of X that exist, and all m.s. Riemann integrals of X of the form $I(a, b) = \int_a^b X_t h(t)dt$ that exist, are jointly Gaussian.*

Proof The m.s. derivatives and integrals of X are obtained by taking m.s. limits of linear combinations of $X = (X_t; t \in \mathbb{T})$. Therefore, the proposition follows from the fact that the joint Gaussian property is preserved under linear combinations and limits (Proposition 3.8(c)). □

Theoretical exercise

Suppose $X = (X_t : t \geq 0)$ is a random process such that R_X is continuous. Let $Y_t = \int_0^t X_s ds$. Show that Y is m.s. differentiable, and $P\{Y_t' = X_t\} = 1$ for $t \geq 0$.

Example 7.8 Let $(W_t : t \geq 0)$ be a Brownian motion with $\sigma^2 = 1$, and let $X_t = \int_0^t W_s ds$ for $t \geq 0$. Let us find R_X and $P\{|X_t| \geq t\}$ for $t > 0$. Since $R_W(u, v) = u \wedge v$,

$$\begin{aligned} R_X(s,t) &= E\left[\int_0^s W_u du \int_0^t W_v dv\right] \\ &= \int_0^s \int_0^t (u \wedge v) dv du. \end{aligned}$$

To proceed, consider first the case $s \geq t$ and partition the region of integration into three parts as shown in Figure 7.3. The contributions from the two triangular subregions is the same, so

$$\begin{aligned} R_X(s,t) &= 2\int_0^t \int_0^u v dv du + \int_t^s \int_0^t v dv du \\ &= \frac{t^3}{3} + \frac{t^2(s-t)}{2} = \frac{t^2 s}{2} - \frac{t^3}{6}. \end{aligned}$$

Still assuming that $s \geq t$, this expression can be rewritten as

$$R_X(s,t) = \frac{st(s \wedge t)}{2} - \frac{(s \wedge t)^3}{6}. \tag{7.21}$$

Although we have found (7.21) only for $s \geq t$, both sides are symmetric in s and t. Thus (7.21) holds for all s, t.

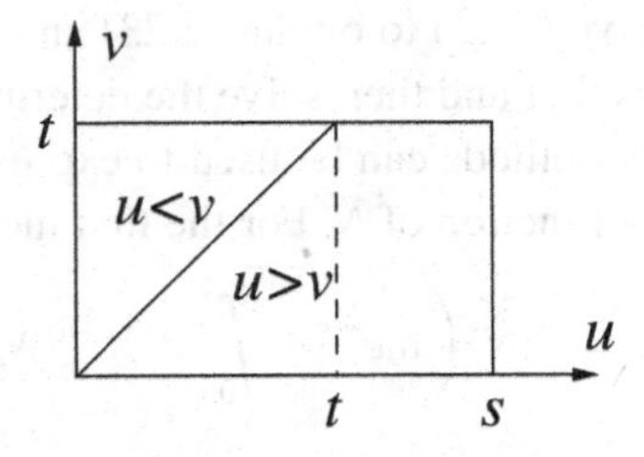

Figure 7.3 Partition of region of integration.

Since W is a Gaussian process, X is a Gaussian process. Also, $E[X_t] = 0$ (because W is mean zero) and $E[X_t^2] = R_X(t,t) = \frac{t^3}{3}$. Thus,

$$P\{|X_t| \geq t\} = 2P\left\{\frac{X_t}{\sqrt{\frac{t^3}{3}}} \geq \frac{t}{\sqrt{\frac{t^3}{3}}}\right\} = 2Q\left(\sqrt{\frac{3}{t}}\right).$$

Note that $P\{X_t| \geq t\} \to 1$ as $t \to +\infty$.

Example 7.9 Let $N = (N_t : t \geq 0)$ be a second order process with a continuous autocorrelation function R_N and let x_0 be a constant. Consider the problem of finding a m.s. differentiable random process $X = (X_t : t \geq 0)$ satisfying the linear differential equation

$$X_t' = -X_t + N_t, \qquad X_0 = x_0. \tag{7.22}$$

Guided by the case that N_t is a smooth nonrandom function, we write

$$X_t = x_0 e^{-t} + \int_0^t e^{-(t-v)} N_v dv \tag{7.23}$$

or

$$X_t = x_0 e^{-t} + e^{-t}\int_0^t e^v N_v dv. \tag{7.24}$$

Using Proposition 7.11, it is not difficult to check that (7.24) indeed gives the solution to (7.22).

Next, let us find the mean and autocovariance functions of X in terms of those of N. Taking the expectation on each side of (7.23) yields

$$\mu_X(t) = x_0 e^{-t} + \int_0^t e^{-(t-v)} \mu_N(v) dv. \tag{7.25}$$

A different way to derive (7.25) is to take expectations in (7.22) to yield the deterministic linear differential equation:

$$\mu_X'(t) = -\mu_X(t) + \mu_N(t), \qquad \mu_X(0) = x_0,$$

which can be solved to yield (7.25). To summarize, we have found two methods to start with the stochastic differential equation (7.23) to derive (7.25), thereby expressing the mean function of the solution X in terms of the mean function of the driving process N. The first is to solve (7.22) to obtain (7.23) and then take expectations, the second is to take expectations first and then solve the deterministic differential equation for μ_X.

The same two methods can be used to express the covariance function of X in terms of the covariance function of N. For the first method, we use (7.23) to obtain

$$\begin{aligned} C_X(s,t) &= \mathrm{Cov}\left(x_0 e^{-s} + \int_0^s e^{-(s-u)} N_u du,\ x_0 e^{-t} + \int_0^t e^{-(t-v)} N_v dv\right) \\ &= \int_0^s \int_0^t e^{-(s-u)} e^{-(t-v)} C_N(u,v) dv du. \end{aligned} \tag{7.26}$$

The second method is to derive deterministic differential equations. To begin, note that

$$\partial_1 C_X(s,t) = \mathrm{Cov}\,(X'_s, X_t) \quad = \quad \mathrm{Cov}\,(-X_s + N_s, X_t)$$

so

$$\partial_1 C_X(s,t) = -C_X(s,t) + C_{NX}(s,t). \tag{7.27}$$

For t fixed, this is a differential equation in s. Also, $C_X(0,t) = 0$. If somehow the cross covariance function C_{NX} is found, (7.27) and the boundary condition $C_X(0,t) = 0$ can be used to find C_X. So we turn next to finding a differential equation for C_{NX}:

$$\partial_2 C_{NX}(s,t) = \mathrm{Cov}(N_s, X'_t) \quad = \quad \mathrm{Cov}(N_s, -X_t + N_t)$$

so

$$\partial_2 C_{NX}(s,t) = -C_{NX}(s,t) + C_N(s,t). \tag{7.28}$$

For s fixed, this is a differential equation in t with initial condition $C_{NX}(s,0) = 0$. Solving (7.28) yields

$$C_{NX}(s,t) = \int_0^t e^{-(t-v)} C_N(s,v)dv. \tag{7.29}$$

Using (7.29) to replace C_{NX} in (7.27) and solving (7.27) yields (7.26).

7.4 Ergodicity

Let X be a stationary or WSS random process. Ergodicity generally means that certain time averages are asymptotically equal to certain statistical averages. For example, suppose $X = (X_t : t \in \mathbb{R})$ is WSS and m.s. continuous. The mean μ_X is defined as a statistical average: $\mu_X = E[X_t]$ for any $t \in \mathbb{R}$.

The time average of X over the interval $[0, t]$ is given by

$$\frac{1}{t}\int_0^t X_u du.$$

Of course, for t fixed, the time average is a random variable, and is typically not equal to the statistical average μ_X. The random process X is called *mean ergodic* (in the m.s. sense) if

$$\lim_{t\to\infty} \frac{1}{t}\int_0^t X_u du = \mu_X \quad m.s.$$

A discrete time WSS random process X is similarly called mean ergodic (in the m.s. sense) if

$$\lim_{n\to\infty} \frac{1}{n}\sum_{i=1}^{n} X_i = \mu_X \quad m.s. \tag{7.30}$$

For example, by the m.s. version of the law of large numbers, if $X = (X_n : n \in \mathbb{Z})$ is WSS with $C_X(n) = I_{\{n=0\}}$ (so that the X_is are uncorrelated) then (7.30) is true. For another example, if $C_X(n) = 1$ for all n, then X_0 has variance one and $P\{X_k = X_0\} = 1$ for all k (because equality holds in the Schwarz inequality: $C_X(n) \leq C_X(0)$). Then for all $n \geq 1$,

$$\frac{1}{n}\sum_{k=1}^{n} X_k = X_0.$$

Since X_0 has variance one, the process X is not ergodic if $C_X(n) = 1$ for all n. In general, whether X is m.s. ergodic in the m.s. sense is determined by the autocovariance function, C_X. The result is stated and proved next for continuous time, and the discrete-time version is true as well.

Proposition 7.18 *Let X be a real-valued, WSS, m.s. continuous random process. Then X is mean ergodic (in the m.s. sense) if and only if*

$$\lim_{t\to\infty} \frac{2}{t}\int_0^t \left(\frac{t-\tau}{t}\right) C_X(\tau)d\tau = 0. \tag{7.31}$$

Sufficient conditions are

(a) $\lim_{\tau\to\infty} C_X(\tau) = 0$. *(This condition is also necessary if* $\lim_{\tau\to\infty} C_X(\tau)$ *exists.)*
(b) $\int_{-\infty}^{\infty} |C_X(\tau)|d\tau < +\infty$.
(c) $\lim_{\tau\to\infty} R_X(\tau) = 0$.
(d) $\int_{-\infty}^{\infty} |R_X(\tau)|d\tau < +\infty$.

Proof By the definition of m.s. convergence, X is mean ergodic if and only if

$$\lim_{t\to\infty} E\left[\left(\frac{1}{t}\int_0^t X_u du - \mu_X\right)^2\right] = 0. \tag{7.32}$$

Since $E\left[\frac{1}{t}\int_0^t X_u du\right] = \frac{1}{t}\int_0^t \mu_X du = \mu_X$, (7.32) is equivalent to the condition $\text{Var}\left(\frac{1}{t}\int_0^t X_u du\right) \to 0$ as $t \to \infty$. By the properties of m.s. integrals,

$$\begin{aligned}
\text{Var}\left(\frac{1}{t}\int_0^t X_u du\right) &= \text{Cov}\left(\frac{1}{t}\int_0^t X_u du, \frac{1}{t}\int_0^t X_v dv\right) \\
&= \frac{1}{t^2}\int_0^t\int_0^t C_X(u-v)dudv \qquad (7.33) \\
&= \frac{1}{t^2}\int_0^t\int_{-v}^{t-v} C_X(\tau)d\tau dv \qquad (7.34) \\
&= \frac{1}{t^2}\int_0^t\int_0^{t-\tau} C_X(\tau)dvd\tau + \int_{-t}^0\int_{-\tau}^t C_X(\tau)dvd\tau \qquad (7.35) \\
&= \frac{1}{t}\int_{-t}^t \left(\frac{t-|\tau|}{t}\right) C_X(\tau)d\tau \\
&= \frac{2}{t}\int_0^t \left(\frac{t-\tau}{t}\right) C_X(\tau)d\tau,
\end{aligned}$$

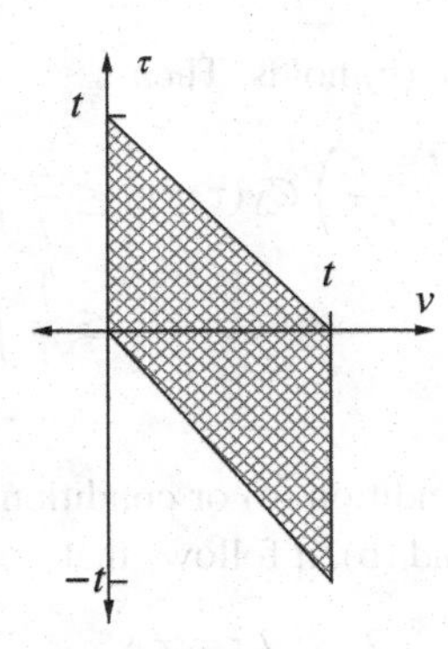

Figure 7.4 Region of integration for (7.34) and (7.35).

where for v fixed the variable $\tau = u - v$ was introduced, and we use the fact that in both (7.34) and (7.35), the pair (v, τ) ranges over the region pictured in Figure 7.4. This establishes the first statement of the proposition.

For the remainder of the proof, it is important to keep in mind that the integral in (7.33) is simply the average of $C_X(u - v)$ over the square $[0, t] \times [0, t]$. The function $C_X(u - v)$ is equal to $C_X(0)$ along the diagonal of the square, and the magnitude of the function is bounded by $C_X(0)$ everywhere in the square. Thus, if $C_X(u, v)$ is small for $u - v$ larger than some constant, if t is large the average of $C_X(u - v)$ over the square will be small. The integral in (7.31) is equivalent to the integral in (7.33), and both can be viewed as a weighted average of $C_X(\tau)$, with a triangular weighting function.

It remains to prove the assertions regarding (a)–(d). Suppose $C_X(\tau) \to c$ as $\tau \to \infty$. We claim the left side of (7.31) is equal to c. Indeed, given $\varepsilon > 0$ there exists $L > 0$ so that $|C_X(\tau) - c| \le \varepsilon$ whenever $\tau \ge L$. For $0 \le \tau \le L$ we can use the Schwarz inequality to bound $C_X(\tau)$, namely $|C_X(\tau)| \le C_X(0)$. Therefore for $t \ge L$,

$$
\begin{aligned}
\left| \frac{2}{t} \int_0^t \left(\frac{t-\tau}{t} \right) C_X(\tau) d\tau - c \right| &= \left| \frac{2}{t} \int_0^t \left(\frac{t-\tau}{t} \right) (C_X(\tau) - c)\, d\tau \right| \\
&\le \frac{2}{t} \int_0^t \left(\frac{t-\tau}{t} \right) |C_X(\tau) - c|\, d\tau \\
&\le \frac{2}{t} \int_0^L (C_X(0) + |c|)\, d\tau + \frac{2\varepsilon}{t} \int_L^t \frac{t-\tau}{t} d\tau \\
&\le \frac{2L\,(C_X(0) + |c|)}{t} + \frac{2\varepsilon}{L} \int_0^t \frac{t-\tau}{t} d\tau \\
&= \frac{2L\,(C_X(0) + |c|)}{t} + \varepsilon \\
&\le 2\varepsilon \text{ for } t \text{ large enough.}
\end{aligned}
$$

Thus the left side of (7.31) is equal to c, as claimed. Hence if $\lim_{\tau \to \infty} C_X(\tau) = c$, (7.31) holds if and only if $c = 0$. It remains to prove that (b), (c), and (d) each imply (7.31).

Suppose condition (b) holds. Then

$$\left|\frac{2}{t}\int_0^t\left(\frac{t-\tau}{t}\right)C_X(\tau)d\tau\right| \leq \frac{2}{t}\int_0^t |C_X(\tau)|d\tau$$
$$\leq \frac{1}{t}\int_{-\infty}^{\infty}|C_X(\tau)|d\tau \to 0 \text{ as } t\to\infty,$$

so that (7.31) holds.

Suppose either condition (c) or condition (d) holds. By the same arguments applied to C_X for parts (a) and (b), it follows that

$$\frac{2}{t}\int_0^t\left(\frac{t-\tau}{t}\right)R_X(\tau)d\tau \to 0 \text{ as } t\to\infty.$$

Since the integral in (7.31) is the variance of a random variable, it is nonnegative. Also, the integral is a weighted average of $C_X(\tau)$, and $C_X(\tau) = R_X(\tau) - \mu_X^2$. Therefore,

$$0 \leq \frac{2}{t}\int_0^t\left(\frac{t-\tau}{t}\right)C_X(\tau)dt$$
$$= -\mu_X^2 + \frac{2}{t}\int_0^t\left(\frac{t-\tau}{t}\right)R_X(\tau)d\tau \to -\mu_X^2 \text{ as } t\to\infty.$$

Thus, (7.31) holds, so that X is mean ergodic in the m.s. sense. In addition, we see that conditions (c) and (d) also each imply that $\mu_X = 0$. □

Example 7.10 Let f_c be a nonzero constant, let Θ be a random variable such that $\cos(\Theta)$, $\sin(\Theta)$, $\cos(2\Theta)$, and $\sin(2\Theta)$ have mean zero, and let A be a random variable independent of Θ such that $E[A^2] < +\infty$. Let $X = (X_t : t \in \mathbb{R})$ be defined by $X_t = A\cos(2\pi f_c t+\Theta)$. Then X is WSS with mean zero and $R_X(\tau) = C_X(\tau) = \frac{E[A^2]\cos(2\pi f_c\tau)}{2}$. Condition (7.31) is satisfied, so X is mean ergodic. Mean ergodicity can also be directly verified:

$$\left|\frac{1}{t}\int_0^t X_u du\right| = \left|\frac{A}{t}\int_0^t \cos(2\pi f_c u+\Theta)du\right|$$
$$= \left|\frac{A(\sin(2\pi f_c t+\Theta)-\sin(\Theta))}{2\pi f_c t}\right|$$
$$\leq \frac{|A|}{\pi f_c t} \to 0 \; m.s. \text{ as } t\to\infty.$$

Example 7.11 (Composite binary source) A student has two biased coins, each with a zero on one side and a one on the other. Whenever the first coin is flipped the outcome is a one with probability $\frac{3}{4}$. Whenever the second coin is flipped the outcome is a one with probability $\frac{1}{4}$. Consider a random process $(W_k : k \in \mathbb{Z})$ formed as follows. First, the student selects one of the coins, each coin being selected with equal probability. Then the selected coin is used to generate the W_ks – the other coin is not used at all.

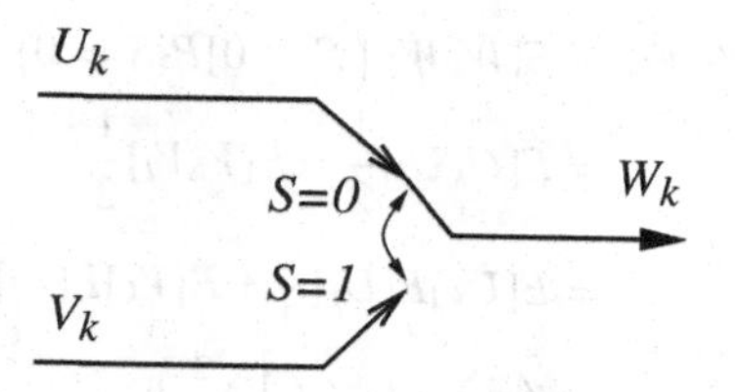

Figure 7.5 A composite binary source.

This scenario can be modeled as in Figure 7.5, using the following random variables:

- $(U_k : k \in \mathbb{Z})$ are independent $Be\left(\frac{3}{4}\right)$ random variables,
- $(V_k : k \in \mathbb{Z})$ are independent $Be\left(\frac{1}{4}\right)$ random variables,
- S is a $Be\left(\frac{1}{2}\right)$ random variable,
- The above random variables are all independent,
- $W_k = (1 - S)U_k + SV_k$.

The variable S can be thought of as a switch state. Value $S = 0$ corresponds to using the coin with probability of heads equal to $\frac{3}{4}$ for each flip.

Clearly W is stationary, and hence also WSS. Is W mean ergodic? One approach to answering this is the direct one. Clearly

$$\begin{aligned}\mu_W = E[W_k] &= E[W_k|S = 0]P\{S = 0\} + E[W_k \mid S = 1]P\{S = 1\} \\ &= \frac{3}{4} \cdot \frac{1}{2} + \frac{1}{4} \cdot \frac{1}{2} \;=\; \frac{1}{2}.\end{aligned}$$

So the question is whether

$$\frac{1}{n}\sum_{k=1}^{n} W_k \overset{?}{\to} \frac{1}{2} \quad m.s.$$

But by the strong law of large numbers

$$\begin{aligned}\frac{1}{n}\sum_{k=1}^{n} W_k &= \frac{1}{n}\sum_{k=1}^{n}((1 - S)U_k + SV_k) \\ &= (1 - S)\left(\frac{1}{n}\sum_{k=1}^{n} U_k\right) + S\left(\frac{1}{n}\sum_{k=1}^{n} V_k\right) \\ &\overset{m.s.}{\to} (1 - S)\frac{3}{4} + S\frac{1}{4} \;=\; \frac{3}{4} - \frac{S}{2}.\end{aligned}$$

Thus, the limit is a random variable, rather than the constant $\frac{1}{2}$. Intuitively, the process W has such strong memory due to the switch mechanism that even averaging over long time intervals does not diminish the randomness due to the switch.

Another way to show that W is not mean ergodic is to find the covariance function C_W and use the necessary and sufficient condition (7.31) for mean ergodicity. Note that for k fixed, $W_k^2 = W_k$ with probability one, so $E[W_k^2] = \frac{1}{2}$. If $k \neq l$, then

$$E[W_kW_l] = E[W_kW_l \mid S=0]P\{S=0\} + E[W_kW_l \mid S=1]P\{S=1\}$$
$$= E[U_kU_l]\frac{1}{2} + E[V_kV_l]\frac{1}{2}$$
$$= E[U_k]E[U_l]\frac{1}{2} + E[V_k]E[V_l]\frac{1}{2}$$
$$= \left(\frac{3}{4}\right)^2\frac{1}{2} + \left(\frac{1}{4}\right)^2\frac{1}{2} = \frac{5}{16}.$$

Therefore,

$$C_W(n) = \begin{cases} \frac{1}{4} & \text{if } n = 0 \\ \frac{1}{16} & \text{if } n \neq 0 \end{cases}.$$

Since $\lim_{n\to\infty} C_W(n)$ exists and is not zero, W is not mean ergodic.

In many applications, we are interested in averages of functions that depend on multiple random variables. We discuss this topic for a discrete time stationary random process, $(X_n : n \in \mathbb{Z})$. Let h be a bounded, Borel measurable function on $\mathbb{R}^k$ for some k. What time average would we expect to be a good approximation to the statistical average $E[h(X_1,\ldots,X_k)]$? A natural choice is

$$\frac{1}{n}\sum_{j=1}^{n} h(X_j, X_{j+1},\ldots,X_{j+k-1}).$$

We define a stationary random process $(X_n : n \in \mathbb{Z})$ to be ergodic if

$$\lim_{n\to\infty}\frac{1}{n}\sum_{j=1}^{n} h(X_j,\ldots,X_{j+k-1}) = E[h(X_1,\ldots,X_k)]$$

for every $k \geq 1$ and for every bounded Borel measurable function h on $\mathbb{R}^k$, where the limit is taken in any of the three senses a.s., p., or m.s.[3] An interpretation of the definition is that if X is ergodic then all of its finite dimensional distributions are determined as time averages.

As an example, suppose

$$h(x_1,x_2) = \begin{cases} 1 & \text{if } x_1 > 0 \geq x_2 \\ 0 & \text{else} \end{cases}.$$

Then $h(X_1,X_2)$ is one if the process (X_k) makes a "down crossing" of level 0 between times one and two. If X is ergodic then with probability 1,

$$\lim_{n\to\infty}\frac{1}{n}\begin{pmatrix}\text{number of down crossings}\\ \text{between times 1 and } n+1\end{pmatrix} = P\{X_1 > 0 \geq X_2\}. \tag{7.36}$$

[3] The mathematics literature uses a different definition of ergodicity for stationary processes, which is equivalent. There are also definitions of ergodicity that do not require stationarity.

Equation (7.36) relates quantities that are quite different in nature. The lefthand side of (7.36) is the long time-average downcrossing rate, whereas the righthand side of (7.36) involves only the joint statistics of two consecutive values of the process.

Ergodicity is a strong property. Two types of ergodic random processes are the following:

- a process $X = (X_k)$ such that the X_ks are iid,
- a stationary Gaussian random process X such that $\lim_{n\to\infty} R_X(n) = 0$ or, $\lim_{n\to\infty} C_X(n) = 0$.

7.5 Complexification, Part I

In some application areas, primarily in connection with spectral analysis as we shall see, complex valued random variables naturally arise. Vectors and matrices over $\mathbb{C}$ are reviewed in the Appendix. A complex random variable $X = U + jV$ can be thought of as essentially a two dimensional random variable with real coordinates U and V. Similarly, a random complex n-dimensional vector X can be written as $X = U + jV$, where U and V are each n-dimensional real vectors. As far as distributions are concerned, a random vector in n-dimensional complex space $\mathbb{C}^n$ is equivalent to a random vector with $2n$ real dimensions. For example, if the $2n$ real variables in U and V are jointly continuous, then X is a continuous type complex random vector and its density is given by a function $f_X(x)$ for $x \in \mathbb{C}^n$. The density f_X is related to the joint density of U and V by $f_X(u + jv) = f_{UV}(u, v)$ for all $u, v \in \mathbb{R}^n$.

As far as moments are concerned, all the second order analysis covered in the notes up to this point can be easily modified to hold for complex random variables, simply by inserting complex conjugates in appropriate places. To begin, if X and Y are complex random variables, we define their correlation by $E[XY^*]$ and similarly their covariance as $E[(X - E[X])(Y - E[Y])^*]$, where * is used to denote the operation on vectors or matrices of taking the transpose and then taking the complex conjugate of each coordinate. The Schwarz inequality becomes $|E[XY^*]| \le \sqrt{E[|X|^2]E[|Y|^2]}$ and its proof is essentially the same as for real-valued random variables. The cross correlation matrix for two complex random vectors X and Y is given by $E[XY^*]$, and similarly the cross covariance matrix is given by $\mathrm{Cov}(X, Y) = E[(X - E[X])(Y - E[Y])^*]$. As before, $\mathrm{Cov}(X) = \mathrm{Cov}(X, X)$. The various formulas for covariance still apply. For example, if A and C are complex matrices and b and d are complex vectors, then $\mathrm{Cov}(AX + b, CY + d) = A\mathrm{Cov}(X, Y)C^*$. Just as in the case of real-valued random variables, a matrix K is a valid covariance matrix (in other words, there exists some random vector X such that $K = \mathrm{Cov}(X)$) if and only if K is Hermitian symmetric and positive semidefinite.

Complex valued random variables X and Y with finite second moments are said to be *orthogonal* if $E[XY^*] = 0$, and with this definition the orthogonality principle holds for complex valued random variables. If X and Y are complex random vectors, then again $E[X|Y]$ is the MMSE estimator of X given Y, and the covariance matrix of the error vector is given by $\mathrm{Cov}(X) - \mathrm{Cov}(E[X|Y])$. The MMSE estimator for X of the form $AY + b$ (i.e. the best linear estimator of X based on Y) and the covariance of the corresponding error vector are given just as for vectors made of real random variables:

$$\hat{E}[X|Y] = E[X] + \text{Cov}(X, Y)\text{Cov}(Y)^{-1}(Y - E[Y]),$$
$$\text{Cov}(X - \hat{E}[X|Y]) = \text{Cov}(X) - \text{Cov}(X, Y)\text{Cov}(Y)^{-1}\text{Cov}(Y, X).$$

By definition, a sequence $X_1, X_2, \ldots$ of complex valued random variables converges in the m.s. sense to X if $E[|X_n|^2] < \infty$ for all n and if $\lim_{n\to\infty} E[|X_n - X|^2] = 0$. The various Cauchy criteria still hold with minor modification. A sequence with $E[|X_n|^2] < \infty$ for all n is a Cauchy sequence in the m.s. sense if $\lim_{m,n\to\infty} E[|X_n - X_m|^2] = 0$. As before, a sequence converges in the m.s. sense if and only if it is a Cauchy sequence. In addition, a sequence $X_1, X_2, \ldots$ of complex valued random variables with $E[|X_n|^2] < \infty$ for all n converges in the m.s. sense if and only if $\lim_{m,n\to\infty} E[X_m X_n^*]$ exists and is a finite constant c. If the m.s. limit exists, then the limiting random variable X satisfies $E[|X|^2] = c$.

Let $X = (X_t : t \in \mathbb{T})$ be a complex random process. Write $X_t = U_t + jV_t$ where U and V are each real-valued random processes. The process X is defined to be a second order process if $E[|X_t|^2] < \infty$ for all t. Since $|X_t|^2 = U_t^2 + V_t^2$ for each t, X being a second order process is equivalent to both U and V being second order processes. The correlation function of a second order complex random process X is defined by $R_X(s, t) = E[X_s X_t^*]$. The covariance function is given by $C_X(s, t) = \text{Cov}(X_s, X_t)$ where the definition of Cov for complex random variables is used. The definitions and results given for m.s. continuity, m.s. differentiation, and m.s. integration all carry over to the case of complex processes, because they are based on the use of the Cauchy criteria for m.s. convergence, which also carries over. For example, a complex valued random process is m.s. continuous if and only if its correlation function R_X is continuous. Similarly, the cross correlation function for two second order random processes X and Y is defined by $R_{XY}(s, t) = E[X_s Y_t^*]$. Note that $R_{XY}(s, t) = R_{YX}^*(t, s)$.

Let $X = (X_t : t \in \mathbb{T})$ be a complex random process such that $\mathbb{T}$ is either the real line or the set of integers, and write $X_t = U_t + jV_t$ where U and V are each real-valued random processes. By definition, X is *stationary* if and only if for any $t_1, \ldots, t_n \in \mathbb{T}$, the joint distribution of $(X_{t_1+s}, \ldots, X_{t_n+s})$ is the same for all $s \in \mathbb{T}$. Equivalently, X is stationary if and only if U and V are jointly stationary. The process X is defined to be WSS if X is a second order process such that $E[X_t]$ does not depend on t, and $R_X(s, t)$ is a function of $s - t$ alone. If X is WSS we use $R_X(\tau)$ to denote $R_X(s, t)$, where $\tau = s - t$. In a pair of complex-valued random processes, X and Y are defined to be jointly WSS if both X and Y are WSS and if the cross correlation function $R_{XY}(s, t)$ is a function of $s - t$. If X and Y are jointly WSS then $R_{XY}(-\tau) = R_{YX}^*(\tau)$.

In summary, everything discussed in this section regarding complex random variables, vectors, and processes can be considered a simple matter of notation. One simply needs to use $|X|^2$ instead of X^2, and to use a star "*" for Hermitian transpose in place of "T" for transpose. We shall begin using the notation at this point, and return to a discussion of the topic of complex valued random processes in

Section 8.6. In particular, we shall examine complex normal random vectors and their densities, and we shall see that there is somewhat more to complexification than just notation.

7.6 The Karhunen–Loève expansion

We have seen that under a change of coordinates, an n-dimensional random vector X is transformed into a vector $Y = U^*X$ such that the coordinates of Y are orthogonal random variables. Here U is the unitary matrix such that $E[XX^*] = U\Lambda U^*$. The columns of U are eigenvectors of the Hermitian symmetric matrix $E[XX^*]$ and the corresponding nonnegative eigenvalues of $E[XX^*]$ comprise the diagonal of the diagonal matrix Λ. The columns of U form an orthonormal basis for $\mathbb{C}^n$. The Karhunen–Loève expansion gives a similar change of coordinates for a random process on a finite interval, using an orthonormal basis of functions instead of an orthonormal basis of vectors.

Fix a finite interval $[a, b]$. The L^2 norm of a real or complex valued function f on the interval $[a, b]$ is defined by

$$||f|| = \sqrt{\int_a^b |f(t)|^2 dt}.$$

We write $L^2[a, b]$ for the set of all functions on $[a, b]$ which have finite L^2 norm. The inner product of two functions f and g in $L^2[a, b]$ is defined by

$$\langle f, g\rangle = \int_a^b f(t)g^*(t)dt.$$

The functions f and g are said to be orthogonal if $\langle f, g\rangle = 0$. Note that $||f|| = \sqrt{\langle f, f\rangle}$ and the Schwarz inequality holds: $|\langle f, g\rangle| \le ||f||\cdot||g||$. A finite or infinite set of functions (φ_n) in $L^2[a, b]$ is said to be an *orthonormal system* if the functions in the set are mutually orthogonal and have norm one, or in other words, $\langle \varphi_i, \varphi_j\rangle = I_{\{i=j\}}$ for all i and j.

In many applications it is useful to use representations of the form

$$f(t) = \sum_{n=1}^{N} c_n\varphi_n(t), \tag{7.37}$$

for some orthonormal system $\varphi_1, \ldots, \varphi_N$. In such a case, we think of $(c_1, \ldots, c_N)$ as the coordinates of f relative to the orthonormal system (φ_n), and we might write $f \leftrightarrow (c_1, \ldots, c_N)$. For example, transmitted signals in many digital communication systems have this form, where the coordinate vector $(c_1, , \ldots, c_N)$ represents a data symbol. The geometry of the space of all functions f of the form (7.37) for the fixed orthonormal system $\varphi_1, \ldots, \varphi_N$ is equivalent to the geometry of the coordinate vectors. For example, if g has a similar representation,

$$g(t) = \sum_{n=1}^{N} d_n\varphi_n(t),$$

or equivalently $g \leftrightarrow (d_1, \ldots, d_N)$, then $f+g \leftrightarrow (c_1, \ldots, c_N) + (d_1, \ldots, d_N)$ and

$$\begin{aligned}\langle f, g\rangle &= \int_a^b \left\{\sum_{m=1}^N c_m \varphi_m(t)\right\}\left\{\sum_{n=1}^N d_n^* \varphi_n^*(t)\right\} dt \\ &= \sum_{m=1}^N \sum_{n=1}^N c_m d_n^* \int_a^b \varphi_m(t)\varphi_n^*(t)dt \\ &= \sum_{m=1}^N \sum_{n=1}^N c_m d_n^* \langle \varphi_m, \varphi_n\rangle \\ &= \sum_{m=1}^N c_m d_m^*. \end{aligned} \tag{7.38}$$

That is, the inner product of the functions, $\langle f, g\rangle$, is equal to the inner product of their coordinate vectors. Note that for $1 \leq n \leq N$, $\varphi_n \leftrightarrow (0, \ldots, 0, 1, 0, \ldots, 0)$, such that the one is in the nth position. If $f \leftrightarrow (c_1, \ldots, c_N)$, then the nth coordinate of f is the inner product of f and φ_n :

$$\langle f, \varphi_n\rangle = \int_a^b \left(\sum_{m=1}^N c_m \varphi_m(t)\right) \varphi_n^*(t)dt = \sum_{m=1}^N c_m \langle \varphi_m, \varphi_n\rangle = c_n.$$

Another way to derive that $\langle f, \varphi_n\rangle = c_n$ is to note that $f \leftrightarrow (c_1, \ldots, c_N)$ and $\varphi_n \leftrightarrow (0, \ldots, 0, 1, 0, \ldots, 0)$, so $\langle f, \varphi_n\rangle$ is the inner product of $(c_1, \ldots, c_N)$ and $(0, \ldots, 0, 1, 0, \ldots, 0)$, or c_n. Thus, the coordinate vector for f is given by $f \leftrightarrow (\langle f, \varphi_1\rangle, \ldots, \langle f, \varphi_N\rangle)$.

The dimension of the space $L^2[a, b]$ is infinite, meaning that there are orthonormal systems $(\varphi_n : n \geq 1)$ with infinitely many functions. For such a system, a function f can have the representation

$$f(t) = \sum_{n=1}^{\infty} c_n \varphi_n(t). \tag{7.39}$$

In many instances encountered in practice, the sum (7.39) converges for each t, but in general what is meant is that the convergence is in the sense of the $L^2[a, b]$ norm:

$$\lim_{N\to\infty} \int_a^b \left| f(t) - \sum_{n=1}^N c_n \varphi_n(t)\right|^2 dt = 0,$$

or equivalently,

$$\lim_{N\to\infty} \left\| f - \sum_{n=1}^N c_n \varphi_n \right\| = 0.$$

The *span* of a set of functions $\varphi_1, \ldots, \varphi_N$ is the set of all functions of the form $a_1\varphi_1(t) + \ldots + a_N\varphi_N(t)$. If the functions $\varphi_1, \ldots, \varphi_N$ form an orthonormal system and if $f \in L^2[a, b]$, then the function $f^\sharp$ in the span of $\varphi_1, \ldots, \varphi_N$ that minimizes $||f - f^\sharp||$ is given by $f^\sharp(t) = \sum_{n=1}^N \langle f, \varphi_n\rangle \varphi_n(t)$. In fact, it is easy to check that $f - f^\sharp$ is orthogonal to φ_n for all n, implying that for any complex numbers $a_1, \ldots, a_N$,

$$||f - \sum_{n=1}^{N} a_n\varphi_n||^2 = ||f - f^\sharp||^2 + \sum_{n=1}^{N} |\langle f^\sharp, \varphi_n\rangle - a_n|^2.$$

Thus, the closest approximation is indeed given by $a_n = \langle f^\sharp, \varphi_n\rangle$. That is, $f^\sharp$ given by $f^\sharp(t) = \sum_{n=1}^{N} \langle f, \varphi_n\rangle\varphi_n(t)$ is the *projection* of f onto the span of the φs. Furthermore,

$$||f - f^\sharp||^2 = ||f||^2 - ||f^\sharp||^2 = ||f||^2 - \sum_{n=1}^{N} |\langle f, \varphi_n\rangle|^2. \tag{7.40}$$

The above reasoning is analogous to that in Proposition 3.5.

An orthonormal system (φ_n) is said to be an *orthonormal basis* for $L^2[a,b]$, if any $f \in L^2[a,b]$ can be represented as in (7.39). If (φ_n) is an orthonormal system then for any $f, g \in L^2[a,b]$, (7.38) still holds with N replaced by ∞ and is known as *Parseval's relation*:

$$\langle f, g\rangle = \sum_{n=1}^{\infty} \langle f, \varphi_n\rangle\langle g, \varphi_n\rangle^*.$$

In particular,

$$||f||^2 = \sum_{n=1}^{\infty} |\langle f, \varphi_n\rangle|^2.$$

A commonly used orthonormal basis is the following (with $[a,b] = [0,T]$ for some $T \geq 0$):

$$\varphi_1(t) = \frac{1}{\sqrt{T}}; \quad \varphi_{2k}(t) = \sqrt{\frac{2}{T}}\cos\left(\frac{2\pi kt}{T}\right),$$
$$\varphi_{2k+1}(t) = \sqrt{\frac{2}{T}}\sin\left(\frac{2\pi kt}{T}\right) \quad \text{for } k \geq 1. \tag{7.41}$$

What happens if f is replaced by a random process $X = (X_t : a \leq t \leq b)$? Suppose $(\varphi_n : 1 \leq n \leq N)$ is an orthonormal system consisting of continuous functions, with $N \leq \infty$. The system does not have to be a basis for $L^2[a,b]$, but if it is then there are infinitely many functions in the system. Suppose that X is m.s. continuous, or equivalently, that R_X is continuous as a function on $[a,b] \times [a,b]$. In particular, R_X is bounded. Then $E\left[\int_a^b |X_t|^2 dt\right] = \int_a^b R_X(t,t)dt < \infty$, so that $\int_a^b |X_t|^2 dt$ is finite with probability one. Suppose that X can be represented as

$$X_t = \sum_{n=1}^{N} C_n\varphi_n(t). \tag{7.42}$$

Such a representation exists if (φ_n) is a basis for $L^2[a,b]$, but some random processes have the form (7.42) even if N is finite or if N is infinite but the system is not a basis. The representation (7.42) reduces the description of the continuous-time random process to the description of the coefficients, (C_n). This representation of X is much easier to work with if the coordinate random variables are orthogonal.

Definition 7.19 A Karhunen–Loève (KL) expansion for a random process $X = (X_t : a \leq t \leq b)$ is a representation of the form (7.42) with $N \leq \infty$ such that:

(1) the functions (φ_n) are orthonormal: $\langle \varphi_m, \varphi_n \rangle = I_{\{m=n\}}$; and
(2) the coordinate random variables C_n are mutually orthogonal: $E[C_m C_n^*] = 0$.

Example 7.12 Let $X_t = A$ for $0 \leq t \leq T$, where A is a random variable with $0 < E[A^2] < \infty$. Then X has the form in (7.42) for $[a,b] = [0,T]$, $N = 1$, $C_1 = A\sqrt{T}$, and $\varphi_1(t) = \frac{I_{\{0 \leq t \leq T\}}}{\sqrt{T}}$. This is trivially a KL expansion, with only one term.

Example 7.13 Let $X_t = A\cos(2\pi t/T + \Theta)$ for $0 \leq t \leq T$, where A is a real-valued random variable with $0 < E[A^2] < \infty$, and Θ is a random variable uniformly distributed over the interval $[0, 2\pi]$ and independent of A. By the cosine angle addition formula, $X_t = A\cos(\Theta)\cos(2\pi t/T) - A\sin(\Theta)\sin(2\pi t/T)$. Then X has the form in (7.42) for $[a,b] = [0,T]$, $N = 2$,

$$\begin{aligned} C_1 &= A\sqrt{2T}\cos(\Theta), & C_2 &= -A\sqrt{2T}\sin(\Theta), \\ \varphi_1(t) &= \frac{\cos(2\pi t/T)}{\sqrt{2T}}, & \varphi_2(t) &= \frac{\sin(2\pi t/T)}{\sqrt{2T}}. \end{aligned}$$

In particular, φ_1 and φ_2 form an orthonormal system with $N = 2$ elements. To check whether this is a KL expansion, we see if $E[C_1 C_2^*] = 0$. Since $E[C_1 C_2^*] = -2TE[A^2]E[\cos(\Theta)\sin(\Theta)] = -TE[A^2]E[\sin(2\Theta)] = 0$, this is indeed a KL expansion, with two terms.

An important property of Karhunen–Loève (KL) expansions in practice is that they identify the most accurate finite dimensional approximations of a random process, as described in the following proposition. A random process $Z = (Z_t : a \leq t \leq b)$ is said to be *N-dimensional* if it has the form $Z_t = \sum_{n=1}^{N} B_n \psi_n(t)$ for some N random variables $B_1, \ldots, B_N$ and N functions $\psi_1, \ldots, \psi_N$.

Proposition 7.20 *Suppose X has a Karhunen–Loève (KL) expansion $X_t = \sum_{n=1}^{\infty} C_n \varphi_n(t)$ (see Definition 7.19). Let $\lambda_n = E[|C_n|^2]$ and suppose the terms are indexed so that $\lambda_1 \geq \lambda_2 \geq \ldots$. For any finite $N \geq 1$, the Nth partial sum, $X^{(N)}(t) = \sum_{n=1}^{N} C_n \varphi_n(t)$, is a choice for Z that minimizes $E[||X - Z||^2]$ over all N-dimensional random processes Z.*

Proof Suppose Z is a random linear combination of N functions, $\psi_1, \ldots, \psi_N$. Without loss of generality, assume that $\psi_1, \ldots, \psi_N$ is an orthonormal system. (If not, the Gram–Schmidt procedure could be applied to get an orthonormal system of N functions with the same span.) We first identify the optimal choice of random coefficients for the ψs fixed, and then consider the optimal choice of the ψs. For a given choice of ψs and a sample path of X, the L^2 norm $||X - Z||^2$ is minimized by projecting the sample path of X onto the span of the ψs, which means taking $Z_t = \sum_{j=1}^N \langle X, \psi_j\rangle \psi_j(t)$. That is, the sample path of Z has the form of $f^\sharp$ above, if f is the sample path of X. This determines the coefficients to be used for a given choice of ψs; it remains to determine the ψs. By (7.40), the (random) approximation error is

$$||X - Z||^2 = ||X||^2 - \sum_{j=1}^{N} |\langle X, \psi_j\rangle|^2.$$

Using the KL expansion for X yields

$$E[|\langle X, \psi_j\rangle|^2] = E\left[\left|\sum_{n=1}^{\infty} C_n\langle \varphi_n, \psi_j\rangle\right|^2\right] = \sum_{n=1}^{\infty} \lambda_n |\langle \varphi_n, \psi_j\rangle|^2.$$

Therefore,

$$E\left[||X - Z||^2\right] = E\left[||X||^2\right] - \sum_{n=1}^{\infty} \lambda_n b_n, \tag{7.43}$$

where $b_n = \sum_{j=1}^N |\langle \varphi_n, \psi_j\rangle|^2$. Note that (b_n) satisfies the constraints $0 \leq b_n \leq 1$, and $\sum_{n=1}^\infty b_n = N$. The righthand side of (7.43) is minimized over (b_n) subject to these constraints by taking $b_n = I_{\{1 \leq n \leq N\}}$. That can be achieved by taking $\psi_j = \varphi_j$ for $1 \leq j \leq N$, in which case $\langle X, \psi_j\rangle = C_j$, and Z becomes $X^{(N)}$. □

Proposition 7.21 *Suppose $X = (X_t : a \leq t \leq b)$ is m.s. continuous and (φ_n) is an orthonormal system of continuous functions. If (7.42) holds for some random variables (C_n), it is a KL expansion (i.e. the coordinate random variables are orthogonal) if and only if the φ_ns are eigenfunctions of R_X:*

$$R_X \varphi_n = \lambda_n \varphi_n, \tag{7.44}$$

where for $\varphi \in L^2[a, b]$, $R_X\varphi$ denotes the function $(R_X\varphi)(s) = \int_a^b R_X(s, t)\varphi(t)dt$. When (7.42) is a KL expansion, the eigenvalues are given by $\lambda_n = E[|C_n|^2]$.

Proof Suppose (7.42) holds. Then $C_n = \langle X, \varphi_n\rangle = \int_a^b X_t \varphi_n^*(t)dt$, so that

$$\begin{aligned}
E[C_m C_n^*] &= E\left[\langle X, \varphi_m\rangle \langle X, \varphi_n\rangle^*\right] \\
&= E\left[\left(\int_a^b X_s \varphi_m^*(s)ds\right)\left(\int_a^b X_t \varphi_n^*(t)dt\right)^*\right] \\
&= \int_a^b \int_a^b R_X(s, t)\varphi_m^*(s)\varphi_n(t)dsdt \\
&= \langle R_X \varphi_n, \varphi_m\rangle.
\end{aligned} \tag{7.45}$$

If the φ_ns are eigenfunctions of R_X, $E[C_m C_n^*] = \langle R_X\varphi_n, \varphi_m\rangle = \langle \lambda_n \varphi_n, \varphi_m\rangle = \lambda_n I_{\{m=n\}}$. In particular, $E[C_m C_n^*] = 0$ if $n \neq m$, so that (7.42) is a KL expansion. Also, taking $m = n$ yields $E[|C_n|^2] = \lambda_n$.

Conversely, suppose (7.42) is a KL expansion. Without loss of generality, suppose that the system (φ_n) is a basis of $L^2[a, b]$. (If it were not, it could be extended to a basis by augmenting it with functions from another basis and applying the Gram–Schmidt method of orthogonalizing.) Then for n fixed, $\langle R_X\varphi_n, \varphi_m\rangle = 0$ for all $m \neq n$. By the fact (φ_n) is a basis, the function $R_X\varphi_n$ has an expansion of the form (7.39), but all terms except possibly the nth are zero. Hence, $R_n\varphi_n = \lambda_n\varphi_n$ for some constant λ_n, so the eigenrelations (7.44) hold. Again, $E[|C_n|^2] = \lambda_n$ by the computation above. □

The following theorem is stated without proof.

Theorem 7.22 (Mercer's theorem) *If R_X is the autocorrelation function of a m.s. continuous random process $X = (X_t : a \leq t \leq b)$ (equivalently, if R_X is a continuous function on $[a, b] \times [a, b]$ that is positive semidefinite, i.e. $R_X(t_i, t_j)$ is a positive semidefinite matrix for any n and any $a \leq t_1 < t_2 < \ldots < t_n \leq b$), then there exists an orthonormal basis for $L^2[a, b]$, $(\varphi_n : n \geq 1)$, of continuous eigenfunctions and corresponding nonnegative eigenvalues $(\lambda_n : n \geq 1)$ for R_X, and R_X is given by the following series expansion:*

$$R_X(s, t) = \sum_{n=1}^{\infty} \lambda_n\varphi_n(s)\varphi_n^*(t). \tag{7.46}$$

The series converges uniformly in s, t, meaning that

$$\lim_{N\to\infty} \max_{s,\, t\in[a,b]} \left| R_X(s, t) - \sum_{n=1}^{N} \lambda_n\varphi_n(s)\varphi_n^*(t) \right| = 0.$$

Theorem 7.23 (Karhunen–Loève expansion) *If $X = (X_t : a \leq t \leq b)$ is a m.s. continuous random process it has a KL expansion,*

$$X_t = \sum_{n=1}^{\infty} \varphi_n(t)\langle X, \varphi_n\rangle,$$

and the series converges in the m.s. sense, uniformly over $t \in [a, b]$.

Proof Use the orthonormal basis (φ_n) guaranteed by Mercer's theorem. By (7.45), $E[\langle X, \varphi_m\rangle^* \langle X, \varphi_n\rangle] = \langle R_X\varphi_n, \varphi_m\rangle = \lambda_n I_{\{n=m\}}$. Also,

$$\begin{aligned} E[X_t\langle X, \varphi_n\rangle^*] &= E[X_t \int_a^b X_s^* \varphi_n(s)ds] \\ &= \int_a^b R_X(t, s)\varphi_n(s)ds = \lambda_n\varphi_n(t). \end{aligned}$$

These facts imply that for finite N,

$$E\left[\left| X_t - \sum_{n=1}^{N} \varphi_n(t)\langle X, \varphi_n\rangle \right|^2\right] = R_X(t, t) - \sum_{n=1}^{N} \lambda_n|\varphi_n(t)|^2, \tag{7.47}$$

which, since the series on the right side of (7.47) converges uniformly in t as $n \to \infty$, implies the stated convergence property for the representation of X. □

Remarks (1) The means of the coordinates of X in a KL expansion can be expressed using the mean function $\mu_X(t) = E[X_t]$ as follows:

$$E[\langle X, \varphi_n \rangle] = \int_a^b \mu_X(t)\varphi_n^*(t)dt = \langle \mu_X, \varphi_n \rangle.$$

Thus, the mean of the nth coordinate of X is the nth coordinate of the mean function of X.

(2) Symbolically, mimicking matrix notation, we can write the representation (7.46) of R_X as

$$R_X(s,t) = [\varphi_1(s)|\varphi_2(s)|\dots\ ] \begin{bmatrix} \lambda_1 & & & \\ & \lambda_2 & & \\ & & \lambda_3 & \\ & & & \ddots \end{bmatrix} \begin{bmatrix} \varphi_1^*(t) \\ \varphi_2^*(t) \\ \vdots \\ \\ \end{bmatrix}.$$

(3) If $f \in L^2[a,b]$ and $f(t)$ represents a voltage or current across a resistor, then the energy dissipated during the interval $[a,b]$ is, up to a multiplicative constant, given by

$$(\text{Energy of } f) = ||f||^2 = \int_a^b |f(t)|^2 dt = \sum_{n=1}^{\infty} |\langle f, \varphi_n \rangle|^2.$$

The mean total energy of $(X_t : a < t < b)$ is thus given by

$$\begin{aligned} E\left[\int_a^b |X_t|^2 dt\right] &= \int_a^b R_X(t,t)dt \\ &= \int_a^b \sum_{n=1}^{\infty} \lambda_n |\varphi_n(t)|^2 dt \\ &= \sum_{n=1}^{\infty} \lambda_n. \end{aligned}$$

(4) If $(X_t : a \le t \le b)$ is a real-valued mean zero Gaussian process and if the orthonormal basis functions are real valued, then the coordinates $\langle X, \varphi_n \rangle$ are uncorrelated, real-valued, jointly Gaussian random variables, and therefore are independent.

Example 7.14 Let $W = (W_t : t \ge 0)$ be a Brownian motion with parameter σ^2. Let us find the KL expansion of W over the interval $[0,T]$. Substituting $R_X(s,t) = \sigma^2(s \wedge t)$ into the eigenrelation (7.44) yields

$$\int_0^t \sigma^2 s\varphi_n(s)ds + \int_t^T \sigma^2 t\varphi_n(s)ds = \lambda_n \varphi_n(t). \tag{7.48}$$

Differentiating (7.48) with respect to t yields

$$\sigma^2 t\varphi_n(t) - \sigma^2 t\varphi_n(t) + \int_t^T \sigma^2 \varphi_n(s)ds = \lambda_n \varphi_n'(t), \tag{7.49}$$

and differentiating a second time yields that the eigenfunctions satisfy the differential equation $\lambda\varphi'' = -\sigma^2\varphi$. Also, setting $t = 0$ in (7.48) yields the boundary condition $\varphi_n(0) = 0$, and setting $t = T$ in (7.49) yields the boundary condition $\varphi_n'(T) = 0$. Solving yields that the eigenvalue and eigenfunction pairs for W are

$$\lambda_n = \frac{4\sigma^2 T^2}{(2n+1)^2\pi^2}, \quad \varphi_n(t) = \sqrt{\frac{2}{T}}\sin\left(\frac{(2n+1)\pi t}{2T}\right) \quad n \geq 0.$$

It can be shown that these functions form an orthonormal basis for $L^2[0, T]$.

Example 7.15 Let X be a white noise process. Such a process is not a random process as defined in these notes, but can be defined as a generalized process in the same way that a delta function can be defined as a generalized function. Generalized random processes, just like generalized functions, only make sense when multiplied by a suitable function and then integrated. For example, the delta function δ is defined by the requirement that for any function f that is continuous at $t = 0$,

$$\int_{-\infty}^{\infty} f(t)\delta(t)dt = f(0).$$

A white noise process X is such that integrals of the form $\int_{-\infty}^{\infty} f(t)X(t)dt$ exist for functions f with finite L^2 norm $||f||$. The integrals are random variables with finite second moments, mean zero and correlations given by

$$E\left[\left(\int_{-\infty}^{\infty} f(s)X_s ds\right)\left(\int_{-\infty}^{\infty} g(t)X_t dt\right)^*\right] = \sigma^2\int_{-\infty}^{\infty} f(t)g^*(t)dt.$$

In a formal or symbolic sense, this means that X is a WSS process with mean zero and autocorrelation function $R_X(s, t) = E[X_s X_t^*]$ given by $R_X(\tau) = \sigma^2\delta(\tau)$.

What would the KL expansion be for a white noise process over some fixed interval [a,b]? The eigenrelation (7.44) becomes simply $\sigma^2\varphi(t) = \lambda_n\varphi(t)$ for all t in the interval. Thus, all the eigenvalues of a white noise process are equal to σ^2, and any function φ with finite norm is an eigenfunction. Thus, if $(\varphi_n : n \geq 1)$ is an arbitrary orthonormal basis for $L^2[a, b]$, then the coordinates of the white noise process X, formally given by $X_n = \langle X, \varphi_n\rangle$, satisfy

$$E[X_n X_m^*] = \sigma^2 I_{\{n=m\}}. \tag{7.50}$$

This offers a reasonable interpretation of white noise. It is a generalized random process such that its coordinates $(X_n : n \geq 1)$ relative to an arbitrary orthonormal basis for a finite interval have mean zero and satisfy (7.50).

7.7 Periodic WSS random processes

Let $X = (X_t : t \in \mathbb{R})$ be a WSS random process and let T be a positive constant.

Proposition 7.24 *The following three conditions are equivalent:*

(a) $R_X(T) = R_X(0)$,

(b) $P\{X_{T+\tau} = X_\tau\} = 1$ *for all* $\tau \in \mathbb{R}$,
(c) $R_X(T+\tau) = R_X(\tau)$ *for all* $\tau \in \mathbb{R}$ *(i.e. periodic with period T).*

Proof Suppose (a) is true. Since $R_X(0)$ is real valued, so is $R_X(T)$, yielding

$$E[|X_{T+\tau} - X_\tau|^2] = E[X_{T+\tau}X^*_{T+\tau} - X_{T+\tau}X^*_\tau - X_\tau X^*_{T+\tau} + X_\tau X^*_\tau]$$
$$= R_X(0) - R_X(T) - R^*_X(T) + R_X(0) = 0.$$

Therefore, (a) implies (b). Next, suppose (b) is true and let $\tau \in \mathbb{R}$. Since two random variables that are equal with probability one have the same expectation, (b) implies that

$$R_X(T+\tau) = E[X_{T+\tau}X^*_0] = E[X_\tau X^*_0] = R_X(\tau).$$

Therefore (b) implies (c). Trivially (c) implies (a), so the equivalence of (a) through (c) is proved. □

Definition 7.25 We call X a periodic, WSS process of period T if X is WSS and any of the three equivalent properties (a), (b), or (c) of Proposition 7.24 holds.

Property (b) almost implies that the sample paths of X are periodic. However, for each τ it can be that $X_\tau \neq X_{\tau+T}$ on an event of probability zero, and since there are uncountably many real numbers τ, the sample paths need not be periodic. However, suppose (b) is true and define a process Y by $Y_t = X_{(t \bmod T)}$. (Recall that by definition, $(t \bmod T)$ is equal to $t + nT$, where n is selected so that $0 \leq t + nT < T$.) Then Y has periodic sample paths, and Y is a *version* of X, which by definition means that $P\{X_t = Y_t\} = 1$ for any $t \in \mathbb{R}$. Thus, the properties (a) through (c) are equivalent to the condition that X is WSS and there is a version of X with periodic sample paths of period T.

Suppose X is a m.s. continuous, periodic, WSS random process. Due to the periodicity of X, it is natural to consider the restriction of X to the interval $[0, T]$. The Karhunen–Loève expansion of X restricted to $[0, T]$ is described next. Let φ_n be the function on $[0, T]$ defined by

$$\varphi_n(t) = \frac{e^{2\pi jnt/T}}{\sqrt{T}}.$$

The functions $(\varphi_n : n \in \mathbb{Z})$ form an orthonormal basis for $L^2[0, T]$.[4] In addition, for any n fixed, both $R_X(\tau)$ and φ_n are periodic with period dividing T, so

$$\begin{aligned}\int_0^T R_X(s,t)\varphi_n(t)dt &= \int_0^T R_X(s-t)\varphi_n(t)dt \\ &= \int_{s-T}^{s} R_X(t)\varphi_n(s-t)dt \\ &= \int_0^T R_X(t)\varphi_n(s-t)dt \\ &= \frac{1}{\sqrt{T}}\int_0^T R_X(t)e^{2\pi jns/T}e^{-2\pi jnt/T}dt \\ &= \lambda_n\varphi_n(s),\end{aligned}$$

[4] Here it is more convenient to index the functions by the integers, rather than by the nonnegative integers. Sums of the form $\sum_{n=-\infty}^{\infty}$ should be interpreted as limits of $\sum_{n=-N}^{N}$ as $N \to \infty$.

where λ_n is given by

$$\lambda_n = \int_0^T R_X(t)e^{-2\pi jnt/T}dt = \sqrt{T}\langle R_X, \varphi_n\rangle. \tag{7.51}$$

Therefore φ_n is an eigenfunction of R_X with eigenvalue λ_n. The Karhunen–Loève expansion (5.20) of X over the interval $[0, T]$ can be written as

$$X_t = \sum_{n=-\infty}^{\infty} \hat{X}_n e^{2\pi jnt/T}, \tag{7.52}$$

where $\hat{X}_n$ is defined by

$$\hat{X}_n = \frac{1}{\sqrt{T}}\langle X, \varphi_n\rangle = \frac{1}{T}\int_0^T X_t e^{-2\pi jnt/T}dt.$$

Note that

$$E[\hat{X}_m \hat{X}_n^*] = \frac{1}{T}E[\langle X, \varphi_m\rangle\langle X, \varphi_n\rangle^*] = \frac{\lambda_n}{T}I_{\{m=n\}}.$$

Although the representation (7.52) has been derived only for $0 \le t \le T$, both sides of (7.52) are periodic with period T. Therefore, the representation (7.52) holds for all t. It is called the *spectral representation* of the periodic, WSS process X.

By (7.51), the series expansion (7.39) applied to the function R_X over the interval $[0, T]$ can be written as

$$\begin{aligned} R_X(t) &= \sum_{n=-\infty}^{\infty} \frac{\lambda_n}{T} e^{2\pi jnt/T} \\ &= \sum_{\omega} p_X(\omega)e^{j\omega t}, \end{aligned} \tag{7.53}$$

where p_X is the function on the real line $\mathbb{R} = (\omega : -\infty < \omega < \infty)$,[5] defined by

$$p_X(\omega) = \begin{cases} \lambda_n/T & \omega = \frac{2\pi n}{T} \text{ for some integer } n \\ 0 & \text{else} \end{cases}$$

and the sum in (7.53) is only over ω such that $p_X(\omega) \neq 0$. The function p_X is called the *power spectral mass function* of X. It is similar to a probability mass function, in that it is positive for at most a countable infinity of values. The value $p_X(\frac{2\pi n}{T})$ is equal to the power of the nth term in the representation (7.52):

$$E[|\hat{X}_n e^{2\pi jnt/T}|^2] = E[|\hat{X}_n|^2] = p_X\left(\frac{2\pi n}{T}\right),$$

and the total mass of p_X is the total power of X, $R_X(0) = E[|X_t|^2]$.

Periodicity is a rather restrictive assumption to place on a WSS process. In the next chapter we shall further investigate spectral properties of WSS processes. We shall see that many WSS random processes have a power spectral density. A given random

[5] The Greek letter ω is used here as it is traditionally used for frequency measured in radians per second. It is related to the frequency f measured in cycles per second by $\omega = 2\pi f$. Here ω is not the same as a typical element of the underlying space of all outcomes, Ω. The meaning of ω should be clear from the context.

variable might have a pmf or a pdf, and it definitely has a CDF. In the same way, a given WSS process might have a power spectral mass function or a power spectral density function, and it definitely has a cumulative power spectral distribution function. The periodic WSS processes of period T are precisely those WSS processes that have a power spectral mass function that is concentrated on the integer multiples of $\frac{2\pi}{T}$.

Problems

7.1 Calculus for a simple Gaussian random process Define $X = (X_t : t \in \mathbb{R})$ by $X_t = A + Bt + Ct^2$, where A, B, C are independent, $N(0, 1)$ random variables.
(a) Verify directly that X is m.s. differentiable.
(b) Express $P\left\{\int_0^1 X_s ds \geq 1\right\}$ in terms of Q, the standard normal complementary CDF.

7.2 Lack of sample path continuity of a Poisson process Let $N = (N_t : t \geq 0)$ be a Poisson process with rate $\lambda > 0$.
(a) Find the following two probabilities, explaining your reasoning: $P\{N$ is continuous over the interval $[0,T]\}$ for a fixed $T > 0$, and $P\{N$ is continuous over the interval $[0, \infty)\}$.
(b) Is N sample path continuous a.s.? Is N m.s. continuous?

7.3 Properties of a binary valued process Let $Y = (Y_t : t \geq 0)$ be given by $Y_t = (-1)^{N_t}$, where N is a Poisson process with rate $\lambda > 0$.
(a) Is Y a Markov process? If so, find the transition probability function $p_{i,j}(s, t)$ and the transition rate matrix Q.
(b) Is Y mean square continuous?
(c) Is Y mean square differentiable?
(d) Does $\lim_{T\to\infty} \frac{1}{T}\int_0^T y_t dt$ exist in the m.s. sense? If so, identify the limit.

7.4 Some statements related to the basic calculus of random processes
Classify each of the following statements as either true (meaning always holds) or false, and justify your answers.
(a) Let $X_t = Z$, where Z is a Gaussian random variable. Then $X = (X_t : t \in \mathbb{R})$ is mean ergodic in the m.s. sense.
(b) The function R_X defined by $R_X(\tau) = \begin{cases} \sigma^2 & |\tau| \leq 1 \\ 0 & \tau > 1 \end{cases}$ is a valid autocorrelation function.
(c) Suppose $X = (X_t : t \in \mathbb{R})$ is a mean zero stationary Gaussian random process, and suppose X is m.s. differentiable. Then for any fixed time t, X_t and X'_t are independent.

7.5 Differentiation of the square of a Gaussian random process (a) Show if random variables $(A_n : n \geq 0)$ are mean zero and jointly Gaussian and $\lim_{n\to\infty} A_n = A$ *m.s.*, then $\lim_{n\to\infty} A_n^2 = A^2$ *m.s.* (Hint: If A, B, C, and D are mean zero and jointly Gaussian, then $E[ABCD] = E[AB]E[CD] + E[AC]E[BD] + E[AD]E[BC]$.)
(b) Show that if random variables $(A_n, B_n : n \geq 0)$ are jointly Gaussian and $\lim_{n\to\infty} A_n = A$ *m.s.* and $\lim_{n\to\infty} B_n = B$ *m.s.* then $\lim_{n\to\infty} A_n B_n = AB$ *m.s.* (Hint: Use part (a) and the identity $ab = \frac{(a+b)^2 - a^2 - b^2}{2}$.)
(c) Let X be a mean zero, m.s. differentiable Gaussian random process, and let $Y_t = X_t^2$ for all t. Is Y m.s. differentiable? If so, justify your answer and express the derivative in terms of X_t and X'_t.

7.6 Continuity of a process passing through a nonlinearity Suppose X is a m.s. continuous random process and G is a bounded, continuous function on $\mathbb{R}$. Let $Y_t = G(X_t)$ for all $t \in \mathbb{R}$.
(a) Prove Y is m.s. continuous. (Hint: Use the connections between continuity in m.s. and p. senses. Also, a continuous function is uniformly continuous over any finite interval, so for any interval $[a, b]$ and $\epsilon > 0$, there is a $\delta > 0$ so that $|G(x) - G(x')| \leq \epsilon$ whenever $x, x' \in [a, b]$ with $|x - x'| \leq \delta$.)
(b) Give an example with G bounded but not continuous, such that Y is not m.s. continuous.
(c) Give an example with G continuous but not bounded, such that Y is not m.s. continuous.
7.7 Mean square differentiability of some random processes For each process described below, determine whether the process is m.s. differentiable in the m.s. sense. Justify your reasoning.
(a) $X_t = \int_0^t N_s ds$, where N is a Poisson random process with rate parameter one.
(b) Process Y, assumed to be a mean-zero Gaussian process with autocorrelation function $R_Y(s, t) = \begin{cases} 1 & \text{if } \lfloor s \rfloor = \lfloor t \rfloor \\ 0 & \text{else.} \end{cases}$. Here "$\lfloor x \rfloor$" denotes the greatest integer less than or equal to x.
(c) Process Z, defined by the series (which converges uniformly in the m.s. sense) $Z_t = \sum_{n=1}^{\infty} \frac{V_n \sin(nt)}{n^2}$, where the V_ns are independent, $N(0, 1)$ random variables.
7.8 Integral of OU process Suppose X is a stationary continuous-time Gaussian process with autocorrelation function $R_X(\tau) = Ae^{-|\tau|}$, and let $Y_t = \int_0^t X_u du$ for $t \geq 0$. (It follows that X has mean zero and is a Markov process. It is sometimes called the standard Ornstein–Uhlenbeck process, and it provides a model for the velocity of a particle moving in one dimension subject to random disturbances and friction, and thus Y would denote the position of the particle.)
(a) Find the mean and autocorrelation function of $(Y_t : t \geq 0)$.
(b) Find $g(t)$ for $t > 0$ so that $P\{|Y_t| \geq g(t)\} = 0.5$. (Hint: $Q(0.81) \approx 0.25$, where Q is the complementary CDF of the standard Gaussian distribution.)
(c) Find a function $f(\alpha)$ so that as $\alpha \to \infty$, the finite dimensional distributions of the process $Z_t \stackrel{\triangle}{=} f(\alpha) Y_{\alpha t}$ converge to the finite dimensional distributions of the standard Brownian motion process. (An interpretation is that $f(\alpha) X_{\alpha t}$ converges to white Gaussian noise.)
7.9 A two-state stationary Markov process Suppose X is a stationary Markov process with mean zero, state space $\{-1, 1\}$, and transition rate matrix $Q = \begin{pmatrix} -\alpha & \alpha \\ \alpha & -\alpha \end{pmatrix}$, where $\alpha \geq 0$. Note that $\alpha = 0$ is a possible case.
(a) Find the autocorrelation function, $R_X(\tau)$.
(b) For what value(s) of $\alpha \geq 0$ is X m.s. continuous?
(c) For what value(s) of $\alpha \geq 0$ is X m.s. continuously differentiable?
(d) For what value(s) of $\alpha \geq 0$ is X mean ergodic in the m.s. sense?
7.10 Cross correlation between a process and its m.s. derivative Suppose X is a m.s. differentiable random process. Show that $R_{X'X} = \partial_1 R_X$. (It follows, in particular, that $\partial_1 R_X$ exists.)

7.11 Fundamental theorem of calculus for m.s. calculus Suppose $X = (X_t : t \geq 0)$ is a m.s. continuous random process. Let Y be the process defined by $Y_t = \int_0^t X_u du$ for $t \geq 0$. Show that X is the m.s. derivative of Y. (It follows, in particular, that Y is m.s. differentiable.)

7.12 A windowed Poisson process Let $N = (N_t : t \geq 0)$ be a Poisson process with rate $\lambda > 0$, and let $X = (X_t : t \geq 0)$ be defined by $X_t = N_{t+1} - N_t$. Thus, X_t is the number of counts of N during the time window $(t, t+1]$.
(a) Sketch a typical sample path of N, and the corresponding sample path of X.
(b) Find the mean function $\mu_X(t)$ and covariance function $C_X(s,t)$ for $s, t \geq 0$. Express your answer in a simple form.
(c) Is X Markov? Why or why not?
(d) Is X mean-square continuous? Why or why not?
(e) Determine whether $\frac{1}{t}\int_0^t X_s ds$ converges in the mean square sense as $t \to \infty$.

7.13 An integral of white noise times an exponential Let $X_t = \int_0^t Z_u e^{-u} du$, for $t \geq 0$, where Z is white Gaussian noise with autocorrelation function $\delta(\tau)\sigma^2$, for some $\sigma^2 > 0$.
(a) Find the autocorrelation function, $R_X(s,t)$ for $s, t \geq 0$.
(b) Is X mean square differentiable? Justify your answer.
(c) Does X_t converge in the mean square sense as $t \to \infty$? Justify your answer.

7.14 A singular integral with a Brownian motion Consider $\int_0^1 \frac{w_t}{t} dt$, where w is a standard Brownian motion. Since $\text{Var}(\frac{w_t}{t}) = \frac{1}{t}$ diverges as $t \to 0$, we define the integral as $\lim_{\epsilon \to 0} \int_\epsilon^1 \frac{w_t}{t} dt$ *m.s.* if the limit exists.
(a) Does the limit exist? If so, what is the probability distribution of the limit?
(b) Similarly, we define $\int_1^\infty \frac{w_t}{t} dt$ to be $\lim_{T \to \infty} \int_1^T \frac{w_t}{t} dt$ *m.s.* if the limit exists. Does the limit exist? If so, what is the probability distribution of the limit?

7.15 An integrated Poisson process Let $N = (N_t : t \geq 0)$ denote a Poisson process with rate $\lambda > 0$, and let $Y_t = \int_0^t N_s ds$ for $s \geq 0$.
(a) Sketch a typical sample path of Y.
(b) Compute the mean function, $\mu_Y(t)$, for $t \geq 0$.
(c) Compute $\text{Var}(Y_t)$ for $t \geq 0$.
(d) Determine the value of the limit, $\lim_{t \to \infty} P\{Y_t < t\}$.

7.16 Recognizing m.s. properties Suppose X is a mean zero random process. For each choice of autocorrelation function shown, indicate which of the following properties X has: m.s. continuous, m.s. differentiable, m.s. integrable over finite length intervals, and mean ergodic in the the m.s. sense.
(a) X is WSS with $R_X(\tau) = (1 - |\tau|)_+$,
(b) X is WSS with $R_X(\tau) = 1 + (1 - |\tau|)_+$,
(c) X is WSS with $R_X(\tau) = \cos(20\pi\tau)\exp(-10|\tau|)$,
(d) $R_X(s,t) = \begin{cases} 1 & \text{if } \lfloor s \rfloor = \lfloor t \rfloor \\ 0 & \text{else} \end{cases}$, (not WSS, you do not need to check for mean ergodic property),
(e) $R_X(s,t) = \sqrt{s \wedge t}$ for $s, t \geq 0$. (not WSS, you do not need to check for mean ergodic property).

7.17 A random Taylor's approximation Suppose X is a mean zero WSS random process such that R_X is twice continuously differentiable. Guided by Taylor's approximation for

deterministic functions, we might propose the following estimator of X_t given X_0 and X'_0: $\widehat{X}_t = X_0 + tX'_0$.
(a) Express the covariance matrix for the vector $(X_0, X'_0, X_t)^T$ in terms of the function R_X and its derivatives.
(b) Express the mean square error $E[(X_t - \widehat{X}_t)^2]$ in terms of the function R_X and its derivatives.
(c) Express the optimal linear estimator $\widehat{E}[X_t|X_0, X'_0]$ in terms of X_0, X'_0, and the function R_X and its derivatives.
(d) (This part is optional – not required.) Compute and compare $\lim_{t\to 0}$ (mean square error)$/t^4$ for the two estimators, under the assumption that R_X is four times continuously differentiable.

7.18 A stationary Gaussian process Let $X = (X_t : t \in \mathbb{Z})$ be a real stationary Gaussian process with mean zero and $R_X(t) = \frac{1}{1+t^2}$. Answer the following unrelated questions.
(a) Is X a Markov process? Justify your answer.
(b) Find $E[X_3|X_0]$ and express $P\{|X_3 - E[X_3|X_0]| \geq 10\}$ in terms of Q, the standard Gaussian complementary cumulative distribution function.
(c) Find the autocorrelation function of X', the m.s. derivative of X.
(d) Describe the joint probability density of $(X_0, X'_0, X_1)^T$. You need not write it down in detail.

7.19 Integral of a Brownian bridge A standard Brownian bridge B can be defined by $B_t = W_t - tW_1$ for $0 \leq t \leq 1$, where W is a Brownian motion with parameter $\sigma^2 = 1$. A Brownian bridge is a mean zero, Gaussian random process which is a.s. sample path continuous, and has autocorrelation function $R_B(s,t) = s(1-t)$ for $0 \leq s \leq t \leq 1$.
(a) Why is the integral $X = \int_0^1 B_t dt$ well defined in the m.s. sense?
(b) Describe the joint distribution of the random variables X and W_1.

7.20 Correlation ergodicity of Gaussian processes (a) A WSS random process X is called correlation ergodic (in the m.s. sense) if for any constant h,

$$\lim_{t\to\infty} m.s. \frac{1}{t}\int_0^t X_{s+h}\overline{X_s}ds = E[X_{s+h}\overline{X_s}].$$

Suppose X is a mean zero, real-valued Gaussian process such that $R_X(\tau) \to 0$ as $|\tau| \to \infty$. Show that X is correlation ergodic. (Hints: Let $Y_t = X_{t+h}X_t$. Then correlation ergodicity of X is equivalent to mean ergodicity of Y. If A, B, C, and D are mean zero, jointly Gaussian random variables, then $E[ABCD] = E[AB]E[CD] + E[AC]E[BD] + E[AD]E[BC]$.)
(b) Give a simple example of a WSS random process that is mean ergodic in the m.s. sense but is not correlation ergodic in the m.s. sense.

7.21 A random process which changes at a random time Let $Y = (Y_t : t \in \mathbb{R})$ and $Z = (Z_t : t \in \mathbb{R})$ be stationary Gaussian Markov processes with mean zero and autocorrelation functions $R_Y(\tau) = R_Z(\tau) = e^{-|\tau|}$. Let U be a real-valued random variable and suppose Y, Z, and U, are mutually independent. Finally, let $X = (X_t : t \in \mathbb{R})$ be defined by

$$X_t = \begin{cases} Y_t & t < U \\ Z_t & t \geq U \end{cases}.$$

(a) Sketch a typical sample path of X.
(b) Find the first order distributions of X.
(c) Express the mean and autocorrelation function of X in terms of the CDF, F_U, of U.
(d) Under what condition on F_U is X m.s. continuous?
(e) Under what condition on F_U is X a Gaussian random process?

7.22 Gaussian review question Let $X = (X_t : t \in \mathbb{R})$ be a real-valued stationary Gaussian-Markov process with mean zero and autocorrelation function $C_X(\tau) = 9\exp(-|\tau|)$.
(a) A fourth degree polynomial of two variables is given by $p(x,y) = a+bx+cy+dxy+ex^2y+fxy^2+...$ such that all terms have the form cx^iy^j with $i+j \leq 4$. Suppose X_2 is to be estimated by an estimator of the form $p(X_0, X_1)$. Find the fourth degree polynomial p to minimize the MSE: $E[(X_2 - p(X_0, X_1))^2]$ and find the resulting MMSE. (Hint: Think! Very little computation is needed.)
(b) Find $P(X_2 \geq 4 | X_0 = \frac{1}{\pi}, X_1 = 3)$. You can express your answer using the Gaussian Q function $Q(c) = \int_c^\infty \frac{1}{\sqrt{2\pi}} e^{-u^2/2} du$. (Hint: Think! Very little computation is needed.)

7.23 First order differential equation driven by Gaussian white noise Let X be the solution of the ordinary differential equation $X' = -X + N$, with initial condition x_0, where $N = (N_t : t \geq 0)$ is a real-valued Gaussian white noise with $R_N(\tau) = \sigma^2\delta(\tau)$ for some constant $\sigma^2 > 0$. Although N is not an ordinary random process, we can interpret this as the condition that N is a Gaussian random process with mean $\mu_N = 0$ and correlation function $R_N(\tau) = \sigma^2\delta(\tau)$.
(a) Find the mean function $\mu_X(t)$ and covariance function $C_X(s,t)$.
(b) Verify that X is a Markov process by checking the necessary and sufficient condition: $C_X(r,s)C_X(s,t) = C_X(r,t)C_X(s,s)$ whenever $r < s < t$. (Note: The very definition of X also suggests that X is a Markov process, because if t is the "present time," the future of X depends only on X_t and the future of the white noise. The future of the white noise is independent of the past $(X_s : s \leq t)$. Thus, the present value X_t contains all the information from the past of X that is relevant to the future of X. This is the continuous-time analog of the discrete-time Kalman state equation.)
(c) Find the limits of $\mu_X(t)$ and $R_X(t+\tau, t)$ as $t \to \infty$. (Because these limits exist, X is said to be asymptotically WSS.)

7.24 KL expansion of a simple random process Let X be a WSS random process with mean zero and $R_X(\tau) = 100(\cos(10\pi\tau))^2 = 50 + 50\cos(20\pi\tau)$.
(a) Is X mean square differentiable? (Justify your answer.)
(b) Is X mean ergodic in the m.s. sense? (Justify your answer.)
(c) Describe a set of eigenfunctions and corresponding eigenvalues for the Karhunen–Loève expansion of $(X_t : 0 \leq t \leq 1)$.

7.25 KL expansion of a finite rank process Suppose $Z = (Z_t : 0 \leq t \leq T)$ has the form $Z_t = \sum_{n=1}^N X_n\xi_n(t)$ such that the functions $\xi_1, \ldots, \xi_N$ are orthonormal over $[0, T]$, and the vector $X = (X_1, ..., X_N)^T$ has a correlation matrix K with $\det(K) \neq 0$. The process Z is said to have *rank* N. Suppose K is not diagonal. Describe the Karhunen–Loève expansion of Z. That is, describe an orthornormal basis $(\varphi_n : n \geq 1)$, and eigenvalues for the KL expansion of X, in terms of the given functions (ξ_n) and correlation matrix K. Also, describe how the coordinates $\langle Z, \varphi_n \rangle$ are related to X.

7.26 KL expansion for derivative process Suppose that $X = (X_t : 0 \leq t \leq 1)$ is a m.s. continuously differentiable random process on the interval $[0, 1]$. Differentiating the KL expansion of X yields $X'(t) = \sum_n \langle X, \varphi_n \rangle \varphi_n'(t)$, which looks similar to a KL expansion for X', but it may be that the functions φ_n' are not orthonormal. For some cases it is not difficult to identify the KL expansion for X'. To explore this, let $(\varphi_n(t))$, $(\langle X, \varphi_n \rangle)$, and (λ_n) denote the eigenfunctions, coordinate random variables, and eigenvalues for the KL expansion of X over the interval $[0, 1]$. Let $(\psi_k(t))$, $(\langle X', \psi_k \rangle)$, and (μ_k) denote the corresponding quantities for X'. For each of the following choices of $(\varphi_n(t))$, express the eigenfunctions coordinate random variables, and eigenvalues for X' in terms of those for X:

(a) $\varphi_n(t) = e^{2\pi jnt}$, $n \in \mathbb{Z}$.
(b) $\varphi_1(t) = 1$, $\varphi_{2k}(t) = \sqrt{2}\cos(2\pi kt)$, and $\varphi_{2k+1}(t) = \sqrt{2}\sin(2\pi kt)$ for $k \geq 1$.
(c) $\varphi_n(t) = \sqrt{2}\sin(\frac{(2n+1)\pi t}{2})$, $n \geq 0$. (Hint: Sketch φ_n and φ_n' for $n = 1, 2, 3$.)
(d) $\varphi_1(t) = c_1(1 + \sqrt{3}t)$ and $\varphi_2(t) = c_2(1 - \sqrt{3}t)$. (Suppose $\lambda_n = 0$ for $n \notin \{1, 2\}$. The constants c_n should be selected so that $||\varphi_n|| = 1$ for $n = 1, 2$, but there is no need to calculate the constants for this problem.)

7.27 An infinitely differentiable process Let $X = (X_t : t \in \mathbb{R})$ be WSS with autocorrelation function $R_X(\tau) = e^{-\tau^2/2}$.

(a) Show that X is k-times differentiable in the m.s. sense, for all $k \geq 1$.
(b) Let $X^{(k)}$ denote the kth derivative process of X, for $k \geq 1$. Is $X^{(k)}$ mean ergodic in the m.s. sense for each k? Justify your answer.

7.28 KL expansion of a Brownian bridge Let B be a Gaussian random process on the interval $[0, 1]$ with $R_B(s, t) = (s \wedge t) - st$. Derive the eigen expansion of R_B guaranteed by Mercer's theorem, and describe the KL expansion of B. (Hint: Follow the method of Example 7.14.)

7.29 Periodicity of a random frequency sinusoidal process Suppose $X_t = A\exp(2\pi j\Phi t)$, where A and Φ are independent real-valued random variables such that $E[A^2] < \infty$.

(a) Under the additional assumption $P\{A > 0\} = 1$, under what conditions on the distributions of A and Φ is X a WSS periodic random process? (Here and in part (c), "periodic" means with a deterministic period.)
(b) Among the possibilities identified in part (a), under what additional conditions is X mean ergodic in the m.s. sense?
(c) Under the additional assumption $\mathrm{Var}(A) > 0$ (but not assuming $P\{A > 0\} = 1$), under what conditions on the distribution of A and Φ is X a WSS periodic random process?
(d) Among the possibilities identified in part (c), under what additional conditions is X mean ergodic in the m.s. sense?

7.30 Mean ergodicity of a periodic WSS random process Let X be a mean zero periodic WSS random process with period $T > 0$. Recall that X has a power spectral representation

$$X_t = \sum_{n \in \mathbb{Z}} \widehat{X}_n e^{2\pi jnt/T},$$

where the coefficients $\widehat{X}_n$ are orthogonal random variables. The power spectral mass function of X is the discrete mass function p_X supported on frequencies of the form $\frac{2\pi n}{T}$, such that $E[|\widehat{X}_n|^2] = p_X(\frac{2\pi n}{T})$. Under what conditions on p_X is the process X mean ergodic in the m.s. sense? Justify your answer.

7.31 Application of the KL expansion to estimation Let $X = (X_t : 0 \leq T)$ be a random process given by $X_t = AB\sin(\frac{\pi t}{T})$, where A and T are positive constants and B is an $N(0, 1)$ random variable. Think of X as an amplitude modulated random signal.

(a) What is the expected total energy of X?

(b) What are the mean and covariance functions of X?

(c) Describe the Karhunen–Loéve expansion of X. (Hint: Only one eigenvalue is nonzero, call it λ_1. What are λ_1, the corresponding eigenfunction φ_1, and the first coordinate $X_1 = \langle X, \varphi_1 \rangle$? You don't need to explicitly identify the other eigenfunctions $\varphi_2, \varphi_3, \ldots$ They can simply be taken to fill out an orthonormal basis.)

(d) Let $N = (X_t : 0 \leq t \leq T)$ be a real-valued Gaussian white noise process independent of X with $R_N(\tau) = \sigma^2\delta(\tau)$, and let $Y = X + N$. Think of Y as a noisy observation of X. The same basis functions used for X can be used for the Karhunen–Loève expansions of N and Y. Let $N_1 = \langle N, \varphi_1 \rangle$ and $Y_1 = \langle Y, \varphi_1 \rangle$. Note that $Y_1 = X_1 + N_1$. Find $E[B|Y_1]$ and the resulting mean square error. (Remark: The other coordinates $Y_2, Y_3, \ldots$ are independent of both X and Y_1, and are thus useless for the purpose of estimating B. Thus, $E[B|Y_1]$ is equal to $E[B|Y]$, the MMSE estimate of B given the entire observation process Y.)

7.32* An autocorrelation function or not? Let $R_X(s, t) = \cosh(a(|s - t| - 0.5))$ for $-0.5 \leq s, t \leq 0.5$ where a is a positive constant. Is R_X the autocorrelation function of a random process of the form $X = (X_t : -0.5 \leq t \leq 0.5)$? If not, explain why not. If so, give the Karhunen–Loève expansion for X.

7.33* On the conditions for m.s. differentiability

(a) Let $f(t) = \begin{cases} t^2\sin(1/t^2) & t \neq 0 \\ 0 & t = 0 \end{cases}$. Sketch f and show that f is differentiable over all of $\mathbb{R}$, and find the derivative function f'. Note that f' is not continuous, and $\int_{-1}^{1} f'(t)dt$ is not well defined, whereas this integral would equal $f(1) - f(-1)$ if f' were continuous.

(b) Let $X_t = Af(t)$, where A is a random variable with mean zero and variance one. Show that X is m.s. differentiable.

(c) Find R_X. Show that $\partial_1 R_X$ and $\partial_2\partial_1 R_X$ exist but are not continuous.

8 Random processes in linear systems and spectral analysis

Random processes can be passed through linear systems in much the same way as deterministic signals can. A time-invariant linear system is described in the time domain by an impulse response function, and in the frequency domain by the Fourier transform of the impulse response function. In a sense we shall see that Fourier transforms provide a diagonalization of WSS random processes, just as the Karhunen–Loève expansion allows for the diagonalization of a random process defined on a finite interval. While a m.s. continuous random process on a finite interval has a finite average energy, a WSS random process has a finite mean average energy per unit time, called the power.

Nearly all the definitions and results of this chapter can be carried through in either discrete time or continuous time. The set of frequencies relevant for continuous-time random processes is all of $\mathbb{R}$, while the set of frequencies relevant for discrete-time random processes is the interval $[-\pi, \pi]$. For ease of notation we shall primarily concentrate on continuous-time processes and systems in the first two sections, and give the corresponding definition for discrete time in the third section.

Representations of baseband random processes and narrowband random processes are discussed in Sections 8.4 and 8.5. Roughly speaking, baseband random processes are those which have power only in low frequencies. A baseband random process can be recovered from samples taken at a sampling frequency that is at least twice as large as the largest frequency component of the process. Thus, operations and statistical calculations for a continuous-time baseband process can be reduced to considerations for the discrete-time sampled process. Roughly speaking, narrowband random processes are those processes which have power only in a band (i.e. interval) of frequencies. A narrowband random process can be represented as a baseband random process that is modulated by a deterministic sinusoid. Complex random processes naturally arise as baseband equivalent processes for real-valued narrowband random processes. A related discussion of complex random processes is given in the last section of the chapter.

8.1 Basic definitions

The output $(Y_t : t \in \mathbb{R})$ of a linear system with *impulse response function* $h(s, t)$ and a random process input $(X_t : t \in \mathbb{R})$ is defined by

$$Y_s = \int_{-\infty}^{\infty} h(s, t) X_t dt \tag{8.1}$$

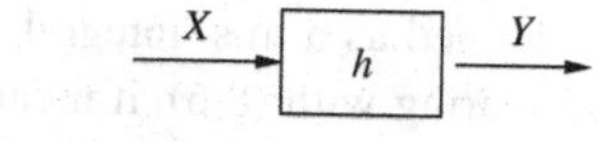

Figure 8.1 A linear system with input X, impulse response function h, and output Y.

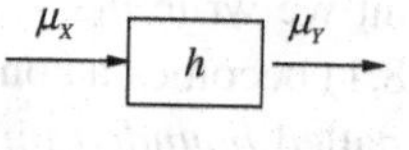

Figure 8.2 A linear system with input μ_X and impulse response function h.

see Figure 8.1. For example, the linear system could be a simple integrator from time zero, defined by

$$Y_s = \begin{cases} \int_0^s X_t dt & s \geq 0 \\ 0 & s < 0, \end{cases}$$

in which case the impulse response function is

$$h(s,t) = \begin{cases} 1 & s \geq t \geq 0 \\ 0 & \text{otherwise.} \end{cases}$$

The integral (8.1) defining the output Y will be interpreted in the m.s. sense. Thus, the integral defining Y_s for s fixed exists if and only if the following Riemann integral exists and is finite:

$$\int_{-\infty}^{\infty}\int_{-\infty}^{\infty} h^*(s,\tau)h(s,t)R_X(t,\tau)dtd\tau. \tag{8.2}$$

A sufficient condition for Y_s to be well defined is that R_X is a bounded continuous function, and $h(s,t)$ is continuous in t with $\int_{-\infty}^{\infty}|h(s,t)|dt < \infty$. The mean function of the output is given by

$$\mu_Y(s) = E\left[\int_{-\infty}^{\infty} h(s,t)X_t dt\right] = \int_{-\infty}^{\infty} h(s,t)\mu_X(t)dt. \tag{8.3}$$

As illustrated in Figure 8.2, the mean function of the output is the result of passing the mean function of the input through the linear system. The cross correlation function between the output and input processes is given by

$$\begin{aligned} R_{YX}(s,\tau) &= E\left[\int_{-\infty}^{\infty} h(s,t)X_t dt X_\tau^*\right] \\ &= \int_{-\infty}^{\infty} h(s,t)R_X(t,\tau)dt, \end{aligned} \tag{8.4}$$

and the correlation function of the output is given by

$$\begin{aligned} R_Y(s,u) &= E\left[Y_s\left(\int_{-\infty}^{\infty} h(u,\tau)X_\tau d\tau\right)^*\right] \\ &= \int_{-\infty}^{\infty} h^*(u,\tau)R_{YX}(s,\tau)d\tau \qquad (8.5) \\ &= \int_{-\infty}^{\infty}\int_{-\infty}^{\infty} h^*(u,\tau)h(s,t)R_X(t,\tau)dtd\tau. \qquad (8.6) \end{aligned}$$

Recall that Y_s is well defined as a m.s. integral if and only if the integral (8.2) is well defined and finite. Comparing with (8.6), it means that Y_s is well defined if and only if the right side of (8.6) with $u = s$ is well defined and gives a finite value for $E[|Y_s|^2]$.

The linear system is *time invariant* if $h(s,t)$ depends on s,t only through $s - t$. If the system is time invariant we write $h(s-t)$ instead of $h(s,t)$, and with this substitution the defining relation (8.1) becomes a convolution: $Y = h * X$.

A linear system is called *bounded input bounded output (bibo) stable* if the output is bounded whenever the input is bounded. When the system is time invariant, bibo stability is equivalent to the condition

$$\int_{-\infty}^{\infty} |h(\tau)| d\tau < \infty. \tag{8.7}$$

In particular, if (8.7) holds and if an input signal x satisfies $|x_s| < L$ for all s, then the output signal $y = x * h$ satisfies

$$|y(t)| \leq \int_{-\infty}^{\infty} |h(t-s)| L ds = L \int_{-\infty}^{\infty} |h(\tau)| d\tau$$

for all t. If X is a WSS random process then by the Schwarz inequality, R_X is bounded by $R_X(0)$. Thus, if X is WSS and m.s. continuous, and if the linear system is time-invariant and bibo stable, the integral in (8.2) exists and is bounded by

$$R_X(0) \int_{-\infty}^{\infty} \int_{-\infty}^{\infty} |h(s-\tau)||h(s-t)| dt d\tau = R_X(0) \left(\int_{-\infty}^{\infty} |h(\tau)| d\tau \right)^2 < \infty.$$

Thus, the output of a linear, time-invariant bibo stable system is well defined in the m.s. sense if the input is a stationary, m.s. continuous process.

A paragraph about convolutions is in order. It is useful to be able to recognize convolution integrals in disguise. If f and g are functions on $\mathbb{R}$, the convolution is the function $f * g$ defined by

$$f * g(t) = \int_{-\infty}^{\infty} f(s) g(t-s) ds,$$

or equivalently

$$f * g(t) = \int_{-\infty}^{\infty} f(t-s) g(s) ds,$$

or equivalently, for any real a and b

$$f * g(a+b) = \int_{-\infty}^{\infty} f(a+s) g(b-s) ds.$$

A simple change of variable shows that the above three expressions are equivalent. However, in order to immediately recognize a convolution, the salient feature is that the convolution is the integral of the product of f and g, with the arguments of both f and g ranging over $\mathbb{R}$ in such a way that the sum of the two arguments is held constant. The value of the constant is the value at which the convolution is being evaluated. Convolution is commutative: $f * g = g * f$ and associative: $(f * g) * k = f * (g * k)$ for three functions f, g, k. We simply write $f * g * k$ for $(f * g) * k$. The convolution $f * g * k$

is equal to a double integral of the product of f, g, and k, with the arguments of the three functions ranging over all triples in $\mathbb{R}^3$ with a constant sum. The value of the constant is the value at which the convolution is being evaluated. For example,

$$f * g * k(a+b+c) = \int_{-\infty}^{\infty}\int_{-\infty}^{\infty} f(a+s+t)g(b-s)k(c-t)dsdt.$$

Suppose that X is WSS and that the linear system is time invariant. Then (8.3) becomes

$$\mu_Y(s) = \int_{-\infty}^{\infty} h(s-t)\mu_X dt = \mu_X \int_{-\infty}^{\infty} h(t)dt.$$

Observe that $\mu_Y(s)$ does not depend on s. Equation (8.4) becomes

$$\begin{aligned} R_{YX}(s,\tau) &= \int_{-\infty}^{\infty} h(s-t)R_X(t-\tau)dt \\ &= h * R_X(s-\tau), \end{aligned} \tag{8.8}$$

which in particular means that $R_{YX}(s,\tau)$ is a function of $s-\tau$ alone. Equation (8.5) becomes

$$R_Y(s,u) = \int_{-\infty}^{\infty} h^*(u-\tau)R_{YX}(s-\tau)d\tau. \tag{8.9}$$

The right side of (8.9) looks nearly like a convolution, but as τ varies the sum of the two arguments is $u-\tau+s-\tau$, which is not constant as τ varies. To arrive at a true convolution, define the new function $\widetilde{h}$ by $\widetilde{h}(v) = h^*(-v)$. Using the definition of $\widetilde{h}$ and (8.8) in (8.9) yields

$$\begin{aligned} R_Y(s,u) &= \int_{-\infty}^{\infty} \widetilde{h}(\tau-u)(h * R_X)(s-\tau)d\tau \\ &= \widetilde{h} * (h * R_X)(s-u) = \widetilde{h} * h * R_X(s-u), \end{aligned}$$

which in particular means that $R_Y(s,u)$ is a function of $s-u$ alone.

To summarize, if X is WSS and if the linear system is time invariant, then X and Y are jointly WSS with

$$\mu_Y = \mu_X \int_{-\infty}^{\infty} h(t)dt, \qquad R_{YX} = h * R_X, \qquad R_Y = h * \widetilde{h} * R_X. \tag{8.10}$$

The convolution $\widetilde{h} * h$, equal to $h * \widetilde{h}$, can also be written as

$$\begin{aligned} h * \widetilde{h}(t) &= \int_{-\infty}^{\infty} h(s)\widetilde{h}(t-s)ds \\ &= \int_{-\infty}^{\infty} h(s)h^*(s-t)ds. \end{aligned} \tag{8.11}$$

The expression shows that $h * \widetilde{h}(t)$ is the correlation between h and h^* translated by t from the origin.

The equations derived in this section for the correlation functions R_X, R_{YX}, and R_Y also hold for the covariance functions C_X, C_{YX}, and C_Y. The derivations are the same except that covariances rather than correlations are computed. In particular, if X is WSS and the system is linear and time invariant, then $C_{YX} = h * C_X$ and $C_Y = h * \widetilde{h} * C_X$.

8.2 Fourier transforms, transfer functions, and power spectral densities

Fourier transforms convert convolutions into products, so this is a good point to begin using Fourier transforms. The Fourier transform of a function g mapping $\mathbb{R}$ to the complex numbers $\mathbb{C}$ is formally defined by

$$\widehat{g}(\omega) = \int_{-\infty}^{\infty} e^{-j\omega t} g(t) dt. \tag{8.12}$$

Some important properties of Fourier transforms are stated next:

Linearity: $\widehat{ag + bh} = a\widehat{g} + b\widehat{h}$,
Inversion: $g(t) = \int_{-\infty}^{\infty} e^{j\omega t}\widehat{g}(\omega)\frac{d\omega}{2\pi}$,
Convolution to multiplication: $\widehat{g * h} = \widehat{g}\widehat{h}$ and $\widehat{g} * \widehat{h} = 2\pi\widehat{gh}$,
Parseval's identity: $\int_{-\infty}^{\infty} g(t)h^*(t)dt = \int_{-\infty}^{\infty} \widehat{g}(\omega)\widehat{h}^*(\omega)\frac{d\omega}{2\pi}$,
Transform of time reversal: $\widehat{\widetilde{h}} = \widehat{h}^*$, where $\widetilde{h}(t) = h^*(-t)$,
Differentiation to multiplication by $j\omega$: $\widehat{\frac{dg}{dt}}(\omega) = (j\omega)\widehat{g}(\omega)$,
Pure sinusoid to delta function: For ω_o fixed: $\widehat{e^{j\omega_o t}}(\omega) = 2\pi\delta(\omega - \omega_o)$,
Delta function to pure sinusoid: For t_o fixed: $\widehat{\delta(t - t_o)}(\omega) = e^{-j\omega t_o}$.

The inversion formula above shows that a function g can be represented as an integral (basically a limiting form of linear combination) of sinusoidal functions of time $e^{j\omega t}$, and $\widehat{g}(\omega)$ is the coefficient in the representation for each ω. Parseval's identity applied with $g = h$ yields that the total energy of g (the square of the L^2 norm) can be computed in either the time or frequency domain: $||g||^2 = \int_{-\infty}^{\infty} |g(t)|^2 dt = \int_{-\infty}^{\infty} |\widehat{g}(\omega)|^2 \frac{d\omega}{2\pi}$. The factor 2π in the formulas can be attributed to the use of frequency ω in radians. If $\omega = 2\pi f$, then f is the frequency in Hertz (Hz) and $\frac{d\omega}{2\pi}$ is simply df.

The Fourier transform can be defined for a very large class of functions, including generalized functions such as delta functions. Here we won't attempt a systematic treatment, but will use Fourier transforms with impunity. In applications, one is often forced to determine in what senses the transform is well defined on a case-by-case basis. Two sufficient conditions for the Fourier transform of g to be well defined are mentioned in the remainder of this paragraph. The relation (8.12) defining a Fourier transform of g is well defined if, for example, g is a continuous function which is integrable: $\int_{-\infty}^{\infty} |g(t)|dt < \infty$, and in this case the dominated convergence theorem implies that $\widehat{g}$ is a continuous function. The Fourier transform can also be naturally defined whenever g has a finite L^2 norm, through the use of Parseval's identity. The idea is that if g has finite L^2 norm, then it is the limit in the L^2 norm of a sequence of functions g_n which are integrable. Owing to Parseval's identity, the Fourier transforms $\widehat{g}_n$ form a Cauchy sequence in the L^2 norm, and hence have a limit, which is defined to be $\widehat{g}$.

Return now to consideration of a linear time-invariant system with an impulse response function $h = (h(\tau) : \tau \in \mathbb{R})$. The Fourier transform of h is used so often that a special name and notation is used: it is called the *transfer function* and is denoted by $H(\omega)$.

The output signal $y = (y_t : t \in \mathbb{R})$ for an input signal $x = (x_t : t \in \mathbb{R})$ is given in the time domain by the convolution $y = x * h$. In the frequency domain this becomes $\widehat{y}(\omega) = H(\omega)\widehat{x}(\omega)$. For example, given $a < b$ let $H_{[a,b]}(\omega)$ be the ideal bandpass transfer function for frequency band $[a, b]$, defined by

$$H_{[a,b]}(\omega) = \begin{cases} 1 & a \leq \omega \leq b \\ 0 & \text{otherwise} \end{cases}. \tag{8.13}$$

If x is the input and y is the output of a linear system with transfer function $H_{[a,b]}$, then the relation $\widehat{y}(\omega) = H_{[a,b]}(\omega)\widehat{x}(\omega)$ shows that the frequency components of x in the frequency band $[a, b]$ pass through the filter unchanged, and the frequency components of x outside of the band are completely nulled. The total energy of the output function y can therefore be interpreted as the energy of x in the frequency band $[a, b]$. Therefore,

$$\text{Energy of } x \text{ in frequency interval } [a,b]$$
$$= ||y||^2 = \int_{-\infty}^{\infty} |H_{[a,b]}(\omega)|^2 |\widehat{x}(\omega)|^2 \frac{d\omega}{2\pi} = \int_a^b |\widehat{x}(\omega)|^2 \frac{d\omega}{2\pi}.$$

Consequently, it is appropriate to call $|\widehat{x}(\omega)|^2$ the energy spectral density of the deterministic signal x.

Given a WSS random process $X = (X_t : t \in \mathbb{R})$, the Fourier transform of its correlation function R_X is denoted by S_X. For reasons that we will soon see, the function S_X is called the *power spectral density* of X. Similarly, if Y and X are jointly WSS, then the Fourier transform of R_{YX} is denoted by S_{YX}, called the cross power spectral density function of Y and X. The Fourier transform of the time reverse complex conjugate function $\widetilde{h}$ is equal to H^*, so $|H(\omega)|^2$ is the Fourier transform of $h * \widetilde{h}$. With the above notation, the second moment relationships in (8.10) become:

$$S_{YX}(\omega) = H(\omega)S_X(\omega), \qquad S_Y(\omega) = |H(\omega)|^2 S_X(\omega).$$

Let us examine some of the properties of the power spectral density, S_X. If $\int_{-\infty}^{\infty} |R_X(t)| dt < \infty$ then S_X is well defined and is a continuous function. Because $R_{YX} = \widetilde{R}_{XY}$, it follows that $S_{YX} = S_{XY}^*$. In particular, taking $Y = X$ yields $R_X = \widetilde{R}_X$ and $S_X = S_X^*$, meaning that S_X is real valued.

The Fourier inversion formula applied to S_X yields $R_X(\tau) = \int_{-\infty}^{\infty} e^{j\omega\tau} S_X(\omega) \frac{d\omega}{2\pi}$. In particular,

$$E[|X_t|^2] = R_X(0) = \int_{-\infty}^{\infty} S_X(\omega) \frac{d\omega}{2\pi}. \tag{8.14}$$

The expectation $E[|X_t|^2]$ is called the power (or total power) of X, because if X_t is a voltage or current across a resistor, $|X_t|^2$ is the instantaneous rate of dissipation of heat energy. Therefore, (8.14) means that the total power of X is the integral of S_X over $\mathbb{R}$. This is the first hint that the name power spectral density for S_X is justified.

Let $a < b$ and let Y denote the output when the WSS process X is passed through the linear time-invariant system with transfer function $H_{[a,b]}$ defined by (8.13). The process Y represents the part of X in the frequency band $[a, b]$. By the relation $S_Y = |H_{[a,b]}|^2 S_X$

and the power relationship (8.14) applied to Y, we have

$$\text{Power of } X \text{ in frequency interval } [a,b]$$
$$= E[|Y_t|^2] = \int_{-\infty}^{\infty} S_Y(\omega)\frac{d\omega}{2\pi} = \int_a^b S_X(\omega)\frac{d\omega}{2\pi}. \tag{8.15}$$

Two observations can be made concerning (8.15). First, the integral of S_X over any interval $[a,b]$ is nonnegative. If S_X is continuous, this implies that S_X is nonnegative. Even if S_X is not continuous, we can conclude that S_X is nonnegative except possibly on a set of zero measure. The second observation is that (8.15) fully justifies the name "power spectral density of X" given to S_X.

Example 8.1 Suppose X is a WSS process and that Y is a moving average of X with averaging window duration T for some $T > 0$:

$$Y_t = \frac{1}{T}\int_{t-T}^{t} X_s ds.$$

Equivalently, Y is the output of the linear time-invariant system with input X and impulse response function h given by

$$h(\tau) = \begin{cases} \frac{1}{T} & 0 \le \tau \le T \\ 0 & \text{else} \end{cases}.$$

The output correlation function is given by $R_Y = h * \widetilde{h} * R_X$. Using (8.11) and referring to Figure 8.3 we find that $h * \widetilde{h}$ is a triangular shaped waveform:

$$h * \widetilde{h}(\tau) = \frac{1}{T}\left(1 - \frac{|\tau|}{T}\right)_+.$$

Similarly, $C_Y = h * \widetilde{h} * C_X$. Let us find in particular an expression for the variance of Y_t in terms of the function C_X:

$$\begin{aligned}\text{Var}(Y_t) = C_Y(0) &= \int_{-\infty}^{\infty} (h * \widetilde{h})(0-\tau)C_X(\tau)d\tau \\ &= \frac{1}{T}\int_{-T}^{T}\left(1 - \frac{|\tau|}{T}\right)C_X(\tau)d\tau. \end{aligned} \tag{8.16}$$

The expression in (8.16) arose earlier, in the section on mean ergodicity.

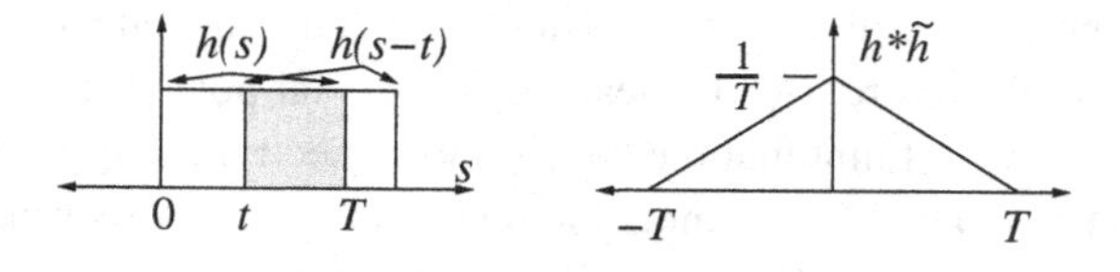

Figure 8.3 Convolution of two rectangle functions.

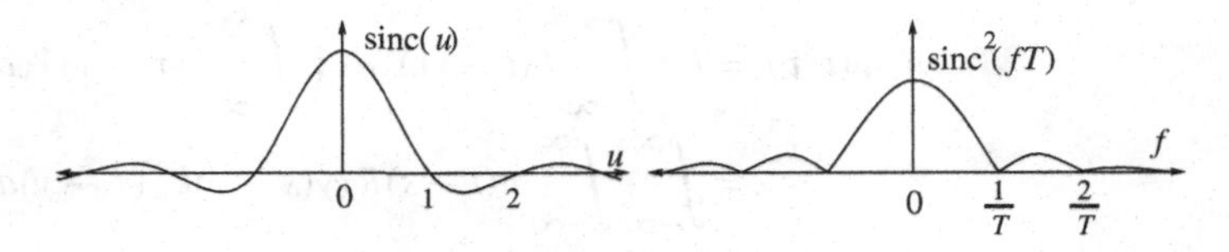

Figure 8.4 The sinc function and $|H(2\pi f)|^2 = |\text{sinc}(fT)|^2$.

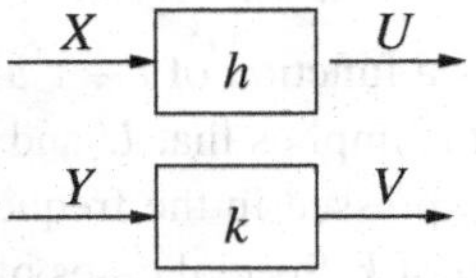

Figure 8.5 Parallel linear systems.

Let us see the effect of the linear system on the power spectral density of the input. Observe that

$$H(\omega) = \int_{-\infty}^{\infty} e^{-j\omega t} h(t)dt = \frac{1}{T}\left[\frac{e^{-j\omega T} - 1}{-j\omega}\right]$$
$$= \frac{2e^{-j\omega T/2}}{T\omega}\left[\frac{e^{j\omega T/2} - e^{-j\omega T/2}}{2j}\right]$$
$$= e^{-j\omega T/2}\left[\frac{\sin(\frac{\omega T}{2})}{\frac{\omega T}{2}}\right].$$

Equivalently, using the substitution $\omega = 2\pi f$,

$$H(2\pi f) = e^{-j\pi f T}\text{sinc}(fT)$$

where the *sinc function* is defined by

$$\text{sinc}(u) = \begin{cases} \frac{\sin(\pi u)}{\pi u} & u \neq 0 \\ 1 & u = 0 \end{cases}. \tag{8.17}$$

Therefore $|H(2\pi f)|^2 = |\text{sinc}(fT)|^2$, so that the output power spectral density is given by $S_Y(2\pi f) = S_X(2\pi f)|\text{sinc}(fT)|^2$, see Figure 8.4.

Example 8.2 Consider two linear time-invariant systems in parallel as shown in Figure 8.5. The first has input X, impulse response function h, and output U. The second has input Y, impulse response function k, and output V. Suppose that X and Y are jointly WSS. We can find R_{UV} as follows. The main trick is notational: to use enough different variables of integration so that none is used twice:

$$\begin{aligned}
R_{UV}(t,\tau) &= E\left[\int_{-\infty}^{\infty} h(t-s)X_s ds \left(\int_{-\infty}^{\infty} k(\tau-v)Y_v dv\right)^*\right] \\
&= \int_{-\infty}^{\infty}\int_{-\infty}^{\infty} h(t-s)R_{XY}(s-v)k^*(\tau-v)dsdv \\
&= \int_{-\infty}^{\infty} \{h * R_{XY}(t-v)\}k^*(\tau-v)dv \\
&= h * \widetilde{k} * R_{XY}(t-\tau).
\end{aligned}$$

Note that $R_{UV}(t,\tau)$ is a function of $t-\tau$ alone. Together with the fact that U and V are individually WSS, this implies that U and V are jointly WSS, and $R_{UV} = h * \widetilde{k} * R_{XY}$. The relationship is expressed in the frequency domain as $S_{UV} = HK^*S_{XY}$, where K is the Fourier transform of k. Special cases of this example include the case that $X = Y$ or $h = k$.

Example 8.3 Consider the circuit with a resistor and a capacitor shown in Figure 8.6. Take as the input signal the voltage difference on the left side, and as the output signal the voltage across the capacitor. Also, let q_t denote the charge on the upper side of the capacitor. Let us first identify the impulse response function by assuming a deterministic input x and a corresponding output y. The elementary equations for resistors and capacitors yield

$$\frac{dq}{dt} = \frac{1}{R}(x_t - y_t) \quad \text{and} \quad y_t = \frac{q_t}{C}.$$

Therefore

$$\frac{dy}{dt} = \frac{1}{RC}(x_t - y_t),$$

which in the frequency domain is

$$j\omega\widehat{y}(\omega) = \frac{1}{RC}(\widehat{x}(\omega) - \widehat{y}(\omega)),$$

so that $\widehat{y} = H\widehat{x}$ for the system transfer function H given by

$$H(\omega) = \frac{1}{1 + RCj\omega}.$$

Suppose, for example, that the input X is a real-valued, stationary Gaussian Markov process, so that its autocorrelation function has the form $R_X(\tau) = A^2e^{-\alpha|\tau|}$ for some constants A^2 and $\alpha > 0$. Then

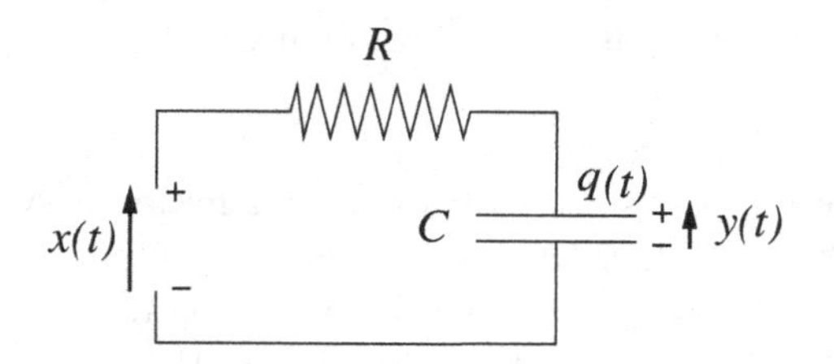

Figure 8.6 An RC circuit modeled as a linear system.

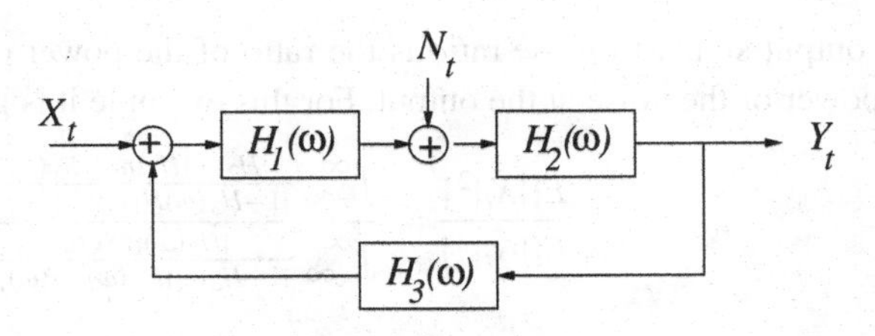

Figure 8.7 A feedback system.

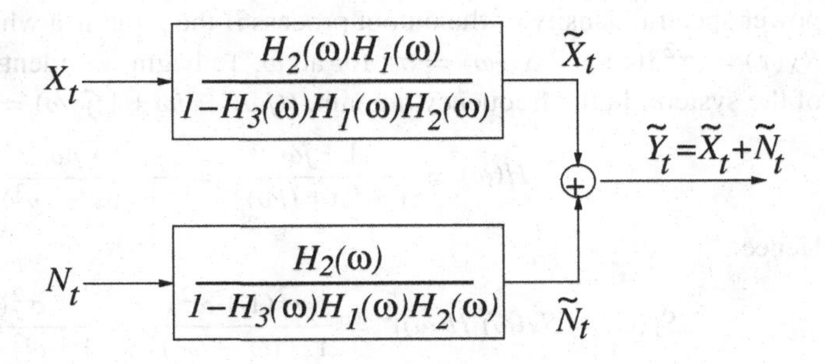

Figure 8.8 An equivalent representation.

$$S_X(\omega) = \frac{2A^2\alpha}{\omega^2 + \alpha^2}$$

and

$$S_Y(\omega) = S_X(\omega)|H(\omega)|^2 = \frac{2A^2\alpha}{(\omega^2 + \alpha^2)(1 + (RC\omega)^2)}.$$

Example 8.4 A random signal, modeled by the input random process X, is passed into a linear time-invariant system with feedback and with noise modeled by the random process N, as shown in Figure 8.7. The output is denoted by Y. Assume that X and N are jointly WSS and that the random variables comprising X are orthogonal to the random variables comprising N: $R_{XN} = 0$. Assume also, for the sake of system stability, that the magnitude of the gain around the loop satisfies $|H_3(\omega)H_1(\omega)H_2(\omega)| < 1$ for all ω such that $S_X(\omega) > 0$ or $S_N(\omega) > 0$. We shall express the output power spectral density S_Y in terms of the power spectral densities of X and N, and the three transfer functions H_1, H_2, and H_3. An expression for the signal-to-noise power ratio at the output will also be computed.

Under the assumed stability condition, the linear system can be written in the equivalent form shown in Figure 8.8. The process $\tilde{X}$ is the output due to the input signal X, and $\tilde{N}$ is the output due to the input noise N. The structure in Figure 8.8 is the same as considered in Example 8.2. Since $R_{XN} = 0$ it follows that $R_{\tilde{X}\tilde{N}} = 0$, so that $S_Y = S_{\tilde{X}} + S_{\tilde{N}}$. Consequently,

$$S_Y(\omega) = S_{\tilde{X}}(\omega) + S_{\tilde{N}}(\omega) = \frac{|H_2(\omega)^2| \left[|H_1(\omega)^2|S_X(\omega) + S_N(\omega)\right]}{|1 - H_3(\omega)H_1(\omega)H_2(\omega)|^2}.$$

The output signal-to-noise ratio is the ratio of the power of the signal at the output to the power of the noise at the output. For this example it is given by

$$\frac{E[|\tilde{X}_t|^2]}{E[|\tilde{N}_t|^2]} = \frac{\int_{-\infty}^{\infty} \frac{|H_2(\omega)H_1(\omega)|^2 S_X(\omega)}{|1-H_3(\omega)H_1(\omega)H_2(\omega)|^2}\frac{d\omega}{2\pi}}{\int_{-\infty}^{\infty} \frac{|H_2(\omega)|^2 S_N(\omega)}{|1-H_3(\omega)H_1(\omega)H_2(\omega)|^2}\frac{d\omega}{2\pi}}.$$

Example 8.5 Consider the linear time-invariant system defined as follows. For input signal x the output signal y is defined by $y''' + y' + y = x + x'$. We seek to find the power spectral density of the output process if the input is a white noise process X with $R_X(\tau) = \sigma^2\delta(\tau)$ and $S_X(\omega) = \sigma^2$ for all ω. To begin, we identify the transfer function of the system. In the frequency domain, $((j\omega)^3 + j\omega + 1)\widehat{y}(\omega) = (1 + j\omega)\widehat{x}(\omega)$, so that

$$H(\omega) = \frac{1 + j\omega}{1 + j\omega + (j\omega)^3} = \frac{1 + j\omega}{1 + j(\omega - \omega^3)}.$$

Hence,

$$S_Y(\omega) = S_X(\omega)|H(\omega)|^2 = \frac{\sigma^2(1 + \omega^2)}{1 + (\omega - \omega^3)^2} = \frac{\sigma^2(1 + \omega^2)}{1 + \omega^2 - 2\omega^4 + \omega^6}.$$

Observe that

$$\text{Output power} = \int_{-\infty}^{\infty} S_Y(\omega)\frac{d\omega}{2\pi} < \infty.$$

8.3 Discrete-time processes in linear systems

The basic definitions and use of Fourier transforms described above carry over naturally to discrete time. In particular, if the random process $X = (X_k : k \in \mathbb{Z})$ is the input of a linear, discrete-time system with impulse response function h, then the output Y is the random process given by

$$Y_k = \sum_{n=-\infty}^{\infty} h(k, n)X_n.$$

The equations in Section 8.1 can be modified to hold for discrete time simply by replacing integration over $\mathbb{R}$ by summation over $\mathbb{Z}$. In particular, if X is WSS and if the linear system is time invariant then (8.10) becomes

$$\mu_Y = \mu_X \sum_{n=-\infty}^{\infty} h(n), \qquad R_{YX} = h * R_X, \qquad R_Y = h * \widetilde{h} * R_X, \tag{8.18}$$

where the convolution in (8.18) is defined for functions g and h on $\mathbb{Z}$ by

$$g * h(n) = \sum_{k=-\infty}^{\infty} g(n - k)h(k).$$

Again, Fourier transforms can be used to convert convolution to multiplication. The Fourier transform of a function $g = (g(n) : n \in \mathbb{Z})$ is the function $\widehat{g}$ on $[-\pi, \pi]$ defined by

$$\widehat{g}(\omega) = \sum_{-\infty}^{\infty} e^{-j\omega n} g(n).$$

Some of the most basic properties are:

Linearity: $\widehat{ag + bh} = a\widehat{g} + b\widehat{h}$,
Inversion: $g(n) = \int_{-\pi}^{\pi} e^{j\omega n}\widehat{g}(\omega)\frac{d\omega}{2\pi}$,
Convolution to multiplication: $\widehat{g * h} = \widehat{g}\widehat{h}$ and $\widehat{g} * \widehat{h} = \frac{1}{2\pi}\widehat{gh}$,
Parseval's identity: $\sum_{n=-\infty}^{\infty} g(n)h^*(n) = \int_{-\pi}^{\pi} \widehat{g}(\omega)\widehat{h}^*(\omega)\frac{d\omega}{2\pi}$,
Transform of time reversal: $\widehat{\widetilde{h}} = \widehat{h}^*$, where $\widetilde{h}(t) = h(-t)^*$,
Pure sinusoid to delta function: For $\omega_o \in [-\pi, \pi]$ fixed: $\widehat{e^{j\omega_o n}}(\omega) = 2\pi\delta(\omega - \omega_o)$,
Delta function to pure sinusoid: For n_o fixed: $\widehat{I_{\{n=n_o\}}}(\omega) = e^{-j\omega n_o}$.

The inversion formula above shows that a function g on $\mathbb{Z}$ can be represented as an integral (basically a limiting form of linear combination) of sinusoidal functions of time $e^{j\omega n}$, and $\widehat{g}(\omega)$ is the coefficient in the representation for each ω. Parseval's identity applied with $g = h$ yields that the total energy of g (the square of the L^2 norm) can be computed in either the time or frequency domain: $||g||^2 = \sum_{n=-\infty}^{\infty} |g(n)|^2 = \int_{-\pi}^{\pi} |\widehat{g}(\omega)|^2 \frac{d\omega}{2\pi}$.

The Fourier transform and its inversion formula for discrete-time functions are equivalent to the Fourier series representation of functions in $L^2[-\pi, \pi]$ using the complete orthogonal basis $(e^{j\omega n} : n \in \mathbb{Z})$ for $L^2[-\pi, \pi]$, as discussed in connection with the Karhunen–Loève expansion. The functions in this basis all have norm 2π. Recall that when we considered the Karhunen–Loève expansion for a periodic WSS random process of period T, functions on a time interval were important and the power was distributed on the integers $\mathbb{Z}$ scaled by $\frac{1}{T}$. In this section, $\mathbb{Z}$ is considered to be the time domain and the power is distributed over an interval. That is, the role of $\mathbb{Z}$ and a finite interval are interchanged. The transforms used are essentially the same, but with j replaced by $-j$.

Given a linear time-invariant system in discrete time with an impulse response function $h = (h(\tau) : \tau \in \mathbb{Z})$, the Fourier transform of h is denoted by $H(\omega)$. The defining relation for the system in the time domain, $y = h * x$, becomes $\widehat{y}(\omega) = H(\omega)\widehat{x}(\omega)$ in the frequency domain. For $-\pi \leq a < b \leq \pi$,

$$\text{Energy of } x \text{ in frequency interval } [a, b] = \int_a^b |\widehat{x}(\omega)|^2 \frac{d\omega}{2\pi},$$

so it is appropriate to call $|\widehat{x}(\omega)|^2$ the energy spectral density of the deterministic, discrete-time signal x.

Given a WSS random process $X = (X_n : n \in \mathbb{Z})$, the Fourier transform of its correlation function R_X is denoted by S_X, and is called the power spectral density of X. Similarly, if Y and X are jointly WSS, then the Fourier transform of R_{YX} is denoted by S_{YX}, called the cross power spectral density function of Y and X. With the above notation, the second moment relationships in (8.18) become:

$$S_{YX}(\omega) = H(\omega)S_X(\omega), \qquad S_Y(\omega) = |H(\omega)|^2 S_X(\omega).$$

The Fourier inversion formula applied to S_X yields $R_X(n) = \int_{-\pi}^{\pi} e^{j\omega n} S_X(\omega) \frac{d\omega}{2\pi}$. In particular,

$$E[|X_n|^2] = R_X(0) = \int_{-\pi}^{\pi} S_X(\omega) \frac{d\omega}{2\pi}.$$

The expectation $E[|X_n|^2]$ is called the total power of X, and for $-\pi < a < b \le \pi$,

$$\text{power of } X \text{ in frequency interval } [a,b] = \int_a^b S_X(\omega) \frac{d\omega}{2\pi}.$$

8.4 Baseband random processes

Deterministic baseband signals are considered first. Let x be a continuous-time signal (i.e. a function on $\mathbb{R}$) such that its energy, $\int_{-\infty}^{\infty} |x(t)|^2 dt$, is finite. By the Fourier inversion formula, the signal x is an integral, which is essentially a sum, of sinusoidal functions of time, $e^{j\omega t}$. The weights are given by the Fourier transform $\widehat{x}(w)$. Let $f_o > 0$ and let $\omega_o = 2\pi f_o$. The signal x is called a *baseband signal,* with one-sided band limit f_o Hz, or equivalently ω_o radians/second, if $\widehat{x}(\omega) = 0$ for $|\omega| \ge \omega_o$. For such a signal, the Fourier inversion formula becomes

$$x(t) = \int_{-\omega_o}^{\omega_o} e^{j\omega t}\widehat{x}(\omega) \frac{d\omega}{2\pi}. \tag{8.19}$$

Equation (8.19) displays the baseband signal x as a linear combination of the functions $e^{j\omega t}$ indexed by $\omega \in [-\omega_o, \omega_o]$.

A celebrated theorem of Nyquist states that the baseband signal x is completely determined by its samples taken at sampling frequency $2f_o$. Specifically, define T by $\frac{1}{T} = 2f_o$. Then

$$x(t) = \sum_{n=-\infty}^{\infty} x(nT) \operatorname{sinc}\left(\frac{t-nT}{T}\right), \tag{8.20}$$

where the sinc function is defined by (8.17). Nyquist's equation (8.20) is indeed elegant. It obviously holds by inspection if $t = mT$ for some integer m, because for $t = mT$ the only nonzero term in the sum is the one indexed by $n = m$. The equation shows that the sinc function gives the correct interpolation of the narrowband signal x for times in between the integer multiples of T. We shall give a proof of (8.20) for deterministic signals, before considering its extension to random processes.

A proof of (8.20) goes as follows. Henceforth we will use ω_o more often than f_o, so it is worth remembering that $\omega_o T = \pi$. Taking $t = nT$ in (8.19) yields

$$\begin{aligned} x(nT) &= \int_{-\omega_o}^{\omega_o} e^{j\omega nT}\widehat{x}(\omega)\frac{d\omega}{2\pi} \\ &= \int_{-\omega_o}^{\omega_o} \widehat{x}(\omega)(e^{-j\omega nT})^* \frac{d\omega}{2\pi}. \end{aligned} \tag{8.21}$$

Equation (8.21) shows that $x(nT)$ is given by an inner product of $\hat{x}$ and $e^{-j\omega nT}$. The functions $e^{-j\omega nT}$, considered on the interval $-\omega_o < \omega < \omega_o$ and indexed by $n \in \mathbb{Z}$, form a complete orthogonal basis for $L^2[-\omega_o, \omega_o]$, and $\int_{-\omega_o}^{\omega_o} T\,|e^{-j\omega nT}|^2 \frac{d\omega}{2\pi} = 1$. Therefore, $\widehat{x}$ over the interval $[-\omega_o, \omega_o]$ has the following Fourier series representation:

$$\widehat{x}(\omega) = T \sum_{n=-\infty}^{\infty} e^{-j\omega nT} x(nT) \quad \omega \in [-\omega_o, \omega_o]. \tag{8.22}$$

Plugging (8.22) into (8.19) yields

$$x(t) = \sum_{n=-\infty}^{\infty} x(nT)T \int_{-\omega_o}^{\omega_o} e^{j\omega t} e^{-j\omega nT} \frac{d\omega}{2\pi}. \tag{8.23}$$

The integral in (8.23) can be simplified using

$$T \int_{-\omega_o}^{\omega_o} e^{j\omega\tau} \frac{d\omega}{2\pi} = \operatorname{sinc}\left(\frac{\tau}{T}\right), \tag{8.24}$$

with $\tau = t - nT$ to yield (8.20) as desired.

The sampling theorem extends naturally to WSS random processes. A WSS random process X with spectral density S_X is said to be a *baseband random process* with one-sided band limit ω_o if $S_X(\omega) = 0$ for $|\,\omega\,| \geq \omega_o$.

Proposition 8.1 *Suppose X is a WSS baseband random process with one-sided band limit ω_o and let $T = \pi/\omega_o$. Then for each $t \in \mathbb{R}$*

$$X_t = \sum_{n=-\infty}^{\infty} X_{nT} \operatorname{sinc}\left(\frac{t - nT}{T}\right) \quad m.s. \tag{8.25}$$

If B is the process of samples defined by $B_n = X_{nT}$, then the power spectral densities of B and X are related by

$$S_B(\omega) = \frac{1}{T} S_X\left(\frac{\omega}{T}\right) \quad \textit{for } |\,\omega\,| \leq \pi. \tag{8.26}$$

Proof Fix $t \in \mathbb{R}$. It must be shown that ϵ_N defined by the following expectation converges to zero as $N \to \infty$:

$$\varepsilon_N = E\left[\left|X_t - \sum_{n=-N}^{N} X_{nT} \operatorname{sinc}\left(\frac{t - nT}{t}\right)\right|^2\right].$$

When the square is expanded, terms of the form $E[X_aX_b^*]$ arise, where a and b take on the values t or nT for some n. But

$$E[X_aX_b^*] = R_X(a-b) = \int_{-\infty}^{\infty} e^{j\omega a}(e^{j\omega b})^* S_X(\omega)\frac{d\omega}{2\pi}.$$

Therefore, ε_N can be expressed as an integration over ω rather than as an expectation:

$$\varepsilon_N = \int_{-\infty}^{\infty} \left| e^{j\omega t} - \sum_{n=-N}^{N} e^{j\omega nT}\mathrm{sinc}\left(\frac{t-nT}{T}\right)\right|^2 S_X(\omega)\frac{d\omega}{2\pi}. \tag{8.27}$$

For t fixed, the function $(e^{j\omega t} : -\omega_o < \omega < \omega_o)$ has a Fourier series representation (use (8.24))

$$\begin{aligned} e^{j\omega t} &= T\sum_{-\infty}^{\infty} e^{j\omega nT}\int_{-\omega_o}^{\omega_o} e^{j\omega t}e^{-j\omega nT}\frac{d\omega}{2\pi} \\ &= \sum_{-\infty}^{\infty} e^{j\omega nT}\mathrm{sinc}\left(\frac{t-nT}{T}\right), \end{aligned}$$

so that the quantity inside the absolute value signs in (8.27) is the approximation error for the Nth partial Fourier series sum for $e^{j\omega t}$. Since $e^{j\omega t}$ is continuous in ω, a basic result in the theory of Fourier series yields that the Fourier approximation error is bounded by a single constant for all N and ω, and as $N \to \infty$ the Fourier approximation error converges to 0 uniformly on sets of the form $|\ \omega\ | \le \omega_o - \varepsilon$. Thus $\varepsilon_N \to 0$ as $N \to \infty$ by the dominated convergence theorem. The representation (8.25) is proved.

Clearly B is a WSS discrete-time random process with $\mu_B = \mu_X$ and

$$\begin{aligned} R_B(n) = R_X(nT) &= \int_{-\infty}^{\infty} e^{jnT\omega}S_X(\omega)\frac{d\omega}{2\pi} \\ &= \int_{-\omega_o}^{\omega_o} e^{jnT\omega}S_X(\omega)\frac{d\omega}{2\pi}, \end{aligned}$$

so, using a change of variable $\nu = T\omega$ and the fact $T = \frac{\pi}{\omega_o}$ yields

$$R_B(n) = \int_{-\pi}^{\pi} e^{jn\nu}\frac{1}{T}S_X\left(\frac{\nu}{T}\right)\frac{d\nu}{2\pi}. \tag{8.28}$$

But $S_B(\omega)$ is the unique function on $[-\pi, \pi]$ such that

$$R_B(n) = \int_{-\pi}^{\pi} e^{jn\omega}S_B(\omega)\frac{d\omega}{2\pi},$$

so (8.26) holds. The proof of Proposition 8.1 is complete. □

As a check on (8.26), we note that $B(0) = X(0)$, so the processes have the same total power. Thus, it must be that

$$\int_{-\pi}^{\pi} S_B(\omega)\frac{d\omega}{2\pi} = \int_{-\infty}^{\infty} S_X(\omega)\frac{d\omega}{2\pi}, \tag{8.29}$$

which is indeed consistent with (8.26).

Example 8.6 If $\mu_X = 0$ and the spectral density S_X of X is constant over the interval $[-\omega_o, \omega_o]$, then $\mu_B = 0$ and $S_B(\omega)$ is constant over the interval $[-\pi, \pi]$. Therefore $R_B(n) = C_B(n) = 0$ for $n \neq 0$, and the samples $(B(n))$ are mean zero, uncorrelated random variables.

Theoretical exercise

What does (8.26) become if X is *WSS* and has a power spectral density, but X is not a baseband signal?

8.5 Narrowband random processes

As noted in the previous section, a signal – modeled as either a deterministic finite energy signal or a WSS random process – can be reconstructed from samples taken at a sampling rate twice the highest frequency of the signal. For example, a typical voice signal may have highest frequency 5 kHz. If such a signal is multiplied by a signal with frequency 10^9 Hz, the highest frequency of the resulting product is about 200 000 times larger than that of the original signal. Naïve application of the sampling theorem would mean that the sampling rate would have to increase by the same factor. Fortunately, because the energy or power of such a modulated signal is concentrated in a narrow band, the signal is nearly as simple as the original baseband signal. The motivation of this section is to see how signals and random processes with narrow spectral ranges can be analyzed in terms of equivalent baseband signals. For example, the effects of filtering can be analyzed using baseband equivalent filters. As an application, an example at the end of the section is given which describes how a narrowband random process (to be defined) can be simulated using a sampling rate equal to twice the one-sided width of a frequency band of a signal, rather than twice the highest frequency of the signal.

Deterministic narrowband signals are considered first, and the development for random processes follows a similar approach. Let $\omega_c > \omega_o > 0$. A *narrowband signal* (relative to ω_o and ω_c) is a signal x such that $\widehat{x}(\omega) = 0$ unless ω is in the union of two intervals: the upper band, $(\omega_c - \omega_o, \omega_c + \omega_o)$, and the lower band, $(-\omega_c - \omega_o, -\omega_c + \omega_o)$. More compactly, $\widehat{x}(\omega) = 0$ if $|| \omega | - \omega_c| \geq \omega_o$.

A narrowband signal arises when a sinusoidal signal is modulated by a narrowband signal, as shown next. Let u and v be real-valued baseband signals, each with one-sided bandwidth less than ω_o, as defined at the beginning of the previous section. Define a signal x by

$$x(t) = u(t)\cos(\omega_c t) - v(t)\sin(\omega_c t). \tag{8.30}$$

Since $\cos(\omega_c t) = \frac{e^{j\omega_c t} + e^{-j\omega_c t}}{2}$ and $-\sin(\omega_c t) = \frac{je^{j\omega_c t} - je^{-j\omega_c t}}{2}$, (8.30) becomes

$$\widehat{x}(\omega) = \frac{1}{2}\left\{\widehat{u}(\omega - \omega_c) + \widehat{u}(\omega + \omega_c) + j\widehat{v}(\omega - \omega_c) - j\widehat{v}(\omega + \omega_c)\right\}. \tag{8.31}$$

Graphically, $\widehat{x}$ is obtained by sliding $\frac{1}{2}\widehat{u}$ to the right by ω_c, $\frac{1}{2}\widehat{u}$ to the left by ω_c, $\frac{j}{2}\widehat{v}$ to the right by ω_c, and $\frac{-j}{2}\widehat{v}$ to the left by ω_c, and then adding. Of course x is real valued by its definition. The reader is encouraged to verify from (8.31) that $\widehat{x}(\omega) = \widehat{x}^*(-\omega)$. Equation (8.31) shows that indeed x is a narrowband signal.

A convenient alternative expression for x is obtained by defining a complex valued baseband signal z by $z(t) = u(t) + jv(t)$. Then $x(t) = Re(z(t)e^{j\omega_c t})$. It is a good idea to keep in mind the case that ω_c is much larger than ω_o (written $\omega_c \gg \omega_o$). Then z varies slowly compared to the complex sinusoid $e^{j\omega_c t}$. In a small neighborhood of a fixed time t, x is approximately a sinusoid with frequency ω_c, peak amplitude $|z(t)|$, and phase given by the argument of $z(t)$. The signal z is called the complex envelope of x and $|z(t)|$ is called the real envelope of x.

So far we have shown that a real-valued narrowband signal x results from modulating sinusoidal functions by a pair of real-valued baseband signals, or equivalently, modulating a complex sinusoidal function by a complex-valued baseband signal. Does every real-valued narrowband signal have such a representation? The answer is yes, as we now show. Let x be a real-valued narrowband signal with finite energy. One attempt to obtain a baseband signal from x is to consider $e^{-j\omega_c t}x(t)$. This has Fourier transform $\widehat{x}(\omega + \omega_c)$, and the graph of this transform is obtained by sliding the graph of $\widehat{x}(\omega)$ to the left by ω_c. As desired, that shifts the portion of $\widehat{x}$ in the upper band to the baseband interval $(-\omega_o, \omega_o)$. However, the portion of $\widehat{x}$ in the lower band gets shifted to an interval centered about $-2\omega_c$, so that $e^{-j\omega_c t}x(t)$ is not a baseband signal.

An elegant solution to this problem is to use the Hilbert transform of x, denoted by $\check{x}$. By definition, $\check{x}(\omega)$ is the signal with Fourier transform $-j\text{sgn}(\omega)\widehat{x}(\omega)$, where

$$\text{sgn}(\omega) = \begin{cases} 1 & \omega > 0 \\ 0 & \omega = 0 \\ -1 & \omega < 0 \end{cases}.$$

Therefore $\check{x}$ can be viewed as the result of passing x through a linear, time-invariant system with transfer function $-j\text{sgn}(\omega)$ as pictured in Figure 8.9. Since this transfer function satisfies $H^*(\omega) = H(-\omega)$, the output signal $\check{x}$ is again real valued. In addition $|H(\omega)| = 1$ for all ω, except $\omega = 0$, so that the Fourier transforms of x and $\check{x}$ have the same magnitude for all nonzero ω. In particular, x and $\check{x}$ have equal energies.

Consider the Fourier transform of $x + j\check{x}$. It is equal to $2\widehat{x}(\omega)$ in the upper band and it is zero elsewhere. Thus, z defined by $z(t) = (x(t) + j\check{x}(t))e^{-j\omega_c t}$ is a baseband complex valued signal. Note that $x(t) = Re(x(t)) = Re(x(t) + j\check{x}(t))$, or equivalently

$$x(t) = Re\left(z(t)e^{j\omega_c t}\right). \tag{8.32}$$

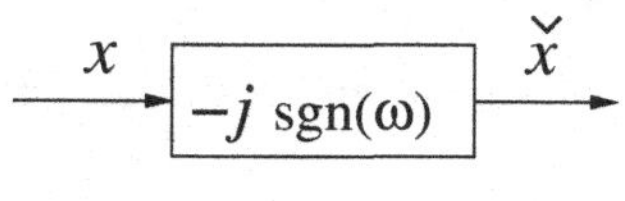

x → $-j\,\text{sgn}(\omega)$ → $\check{x}$

Figure 8.9 The Hilbert transform as a linear, time-invariant system.

If we let $u(t) = \mathrm{Re}(z(t))$ and $v(t) = \mathrm{Im}(z(t))$, then u and v are real-valued baseband signals such that $z(t) = u(t) + jv(t)$, and (8.32) becomes (8.30).

In summary, any finite energy real-valued narrowband signal x can be represented as (8.30) or (8.32), where $z(t) = u(t) + jv(t)$. The Fourier transform $\widehat{z}$ can be expressed in terms of $\widehat{x}$ by

$$\widehat{z}(\omega) = \begin{cases} 2\widehat{x}(\omega + \omega_c) & |\omega| \le \omega_o \\ 0 & \text{else} \end{cases}, \tag{8.33}$$

and $\widehat{u}$ is the Hermitian symmetric part of $\widehat{z}$ and $\widehat{v}$ is $-j$ times the Hermitian antisymmetric part of $\widehat{z}$:

$$\widehat{u}(\omega) = \frac{1}{2}\left(\widehat{z}(\omega) + \widehat{z}^*(-\omega)\right), \widehat{v}(\omega) = \frac{-j}{2}\left(\widehat{z}(\omega) - \widehat{z}^*(-\omega)\right).$$

In the other direction, $\widehat{x}$ can be expressed in terms of $\widehat{u}$ and $\widehat{v}$ by (8.31).

If x_1 and x_2 are each narrowband signals with corresponding complex envelope processes z_1 and z_2, then the convolution $x = x_1 * x_2$ is again a narrowband signal, and the corresponding complex envelope is $\frac{1}{2}z_1 * z_2$. To see this, note that the Fourier transform, $\widehat{z}$, of the complex envelope z for x is given by (8.33). Similar equations hold for $\widehat{z_i}$ in terms of $\widehat{x_i}$ for $i = 1, 2$. Using these equations and the fact $\widehat{x}(\omega) = \widehat{x_1}(\omega)\widehat{x_2}(\omega)$, it is readily seen that $\widehat{z}(\omega) = \frac{1}{2}\widehat{z_1}(\omega)\widehat{z_2}(\omega)$ for all ω, establishing the claim. Thus, the analysis of linear, time invariant filtering of narrowband signals can be carried out in the baseband equivalent setting.

A similar development is considered next for WSS random processes. Let U and V be jointly WSS real-valued baseband random processes, and let X be defined by

$$X_t = U_t \cos(\omega_c t) - V_t \sin(\omega_c t), \tag{8.34}$$

or equivalently, defining Z_t by $Z_t = U_t + jV_t$,

$$X_t = Re\left(Z_t e^{j\omega_c t}\right). \tag{8.35}$$

In some sort of generalized sense, we expect that X is a narrowband process. However, such an X need not even be WSS. Let us find the conditions on U and V that make X WSS. First, in order that $\mu_X(t)$ not depend on t, it must be that $\mu_U = \mu_V = 0$.

Using the notation $c_t = \cos(\omega_c t)$, $s_t = \sin(\omega_c t)$, and $\tau = a - b$,

$$R_X(a, b) = R_U(\tau)c_a c_b - R_{UV}(\tau)c_a s_b - R_{VU}(\tau)s_a c_b + R_V(\tau)s_a s_b.$$

Using the trigonometric identities such as $c_a c_b = (c_{a-b} + c_{a+b})/2$, this can be rewritten as

$$\begin{aligned} R_X(a, b) = &\left(\frac{R_U(\tau) + R_V(\tau)}{2}\right) c_{a-b} + \left(\frac{R_{UV}(\tau) - R_{VU}(\tau)}{2}\right) s_{a-b} \\ &+ \left(\frac{R_U(\tau) - R_V(\tau)}{2}\right) c_{a+b} - \left(\frac{R_{UV}(\tau) + R_{VU}(\tau)}{2}\right) s_{a+b}. \end{aligned}$$

Therefore, in order that $R_X(a, b)$ is a function of $a - b$, it must be that $R_U = R_V$ and $R_{UV} = -R_{VU}$. Since in general $R_{UV}(\tau) = R_{VU}(-\tau)$, the condition $R_{UV} = -R_{VU}$ means that R_{UV} is an odd function: $R_{UV}(\tau) = -R_{UV}(-\tau)$.

We summarize the results as a proposition.

Proposition 8.2 *Suppose X is given by (8.34) or (8.35), where U and V are jointly WSS. Then X is WSS if and only if U and V are mean zero with $R_U = R_V$ and $R_{UV} = -R_{VU}$. Equivalently, X is WSS if and only if $Z = U + jV$ is mean zero and $E[Z_a Z_b] = 0$ for all a, b. If X is WSS then*

$$R_X(\tau) = R_U(\tau)\cos(\omega_c\tau) + R_{UV}(\tau)\sin(\omega_c\tau),$$

$$S_X(\omega) = \frac{1}{2}[S_U(\omega-\omega_c) + S_U(\omega+\omega_c) - jS_{UV}(\omega-\omega_c) + jS_{UV}(\omega+\omega_c)],$$

and, with $R_Z(\tau)$ defined by $R_Z(a-b) = E[Z_a Z_b^*]$,

$$R_X(\tau) = \frac{1}{2}Re(R_Z(\tau)e^{j\omega_c\tau}).$$

The functions S_X, S_U, and S_V are nonnegative, even functions, and S_{UV} is a purely imaginary odd function (i.e. $S_{UV}(\omega) = Im(S_{UV}(\omega)) = -S_{UV}(-\omega)$).

Let X be any WSS real-valued random process with a spectral density S_X, and continue to let $\omega_c > \omega_o > 0$. Then X is defined to be a *narrowband random process* if $S_X(\omega) = 0$ whenever $|\,|\omega| - \omega_c\,| \geq \omega_o$. Equivalently, X is a narrowband random process if $R_X(t)$ is a narrowband function. We've seen how such a process can be obtained by modulating a pair of jointly WSS baseband random processes U and V. We show next that all narrowband random processes have such a representation.

To proceed as in the case of deterministic signals, we first wish to define the Hilbert transform of X, denoted by $\check{X}$. A slight concern about defining $\check{X}$ is that the function $-j\text{sgn}(\omega)$ does not have finite energy. However, we can replace this function by the function given by

$$H(\omega) = -j\text{sgn}(\omega)I_{|\omega|\leq\omega_o+\omega_c},$$

which has finite energy and it has a real-valued inverse transform h. Define $\check{X}$ as the output when X is passed through the linear system with impulse response h. Since X and h are real valued, the random process $\check{X}$ is also real valued. As in the deterministic case, define random processes Z, U, and V by $Z_t = (X_t + j\check{X}_t)e^{-j\omega_c t}$, $U_t = Re(Z_t)$, and $V_t = Im(Z_t)$.

Proposition 8.3 *Let X be a narrowband WSS random process, with spectral density S_X satisfying $S_X(\omega) = 0$ unless $\omega_c - \omega_o \leq |\omega| \leq \omega_c + \omega_o$, where $\omega_o < \omega_c$. Then $\mu_X = 0$ and the following representations hold:*

$$X_t = Re(Z_t e^{j\omega_c t}) = U_t\cos(\omega_c t) - V_t\sin(\omega_c t),$$

where $Z_t = U_t + jV_t$, and U and V are jointly WSS real-valued random processes with mean zero and

$$S_U(\omega) = S_V(\omega) = [S_X(\omega-\omega_c) + S_X(\omega+\omega_c)]\,I_{|\omega|\leq\omega_o} \tag{8.36}$$

and

$$S_{UV}(\omega) = j\,[S_X(\omega+\omega_c) - S_X(\omega-\omega_c)]\,I_{|\omega|\leq\omega_o}. \tag{8.37}$$

Equivalently,

$$R_U(\tau) = R_V(\tau) = R_X(\tau)\cos(\omega_c\tau) + \check{R}_X(\tau)\sin(\omega_c\tau) \tag{8.38}$$

and

$$R_{UV}(\tau) = R_X(\tau)\sin(\omega_c\tau) - \check{R}_X(\tau)\cos(\omega_c\tau). \tag{8.39}$$

Proof To show that $\mu_X = 0$, consider passing X through a linear, time-invariant system with transfer function $K(\omega) = 1$ if ω is in either the upper band or lower band, and $K(\omega) = 0$ otherwise. Then $\mu_Y = \mu_X \int_{-\infty}^{\infty} h(\tau)d\tau = \mu_X K(0) = 0$. Since $K(\omega) = 1$ for all ω such that $S_X(\omega) > 0$, it follows that $R_X = R_Y = R_{XY} = R_{YX}$. Therefore $E[|X_t - Y_t|^2] = 0$ so that X_t has the same mean as Y_t, namely zero, as claimed.

By the definitions of the processes Z, U, and V, using the notation $c_t = \cos(\omega_c t)$ and $s_t = \sin(\omega_c t)$, we have

$$U_t = X_t c_t + \check{X}_t s_t \quad V_t = -X_t s_t + \check{X}_t c_t.$$

The remainder of the proof consists of computing R_U, R_V, and R_{UV} as functions of two variables, because it is not yet clear that U and V are jointly WSS.

By the fact X is WSS and the definition of $\check{X}$, the processes X and $\check{X}$ are jointly WSS, and the various spectral densities are given by

$$S_{\check{X}X} = HS_X, \qquad S_{X\check{X}} = H^* S_X = -HS_X, \qquad S_{\check{X}} = |H|^2 S_X = S_X.$$

Therefore,

$$R_{\check{X}X} = \check{R}_X, \qquad R_{X\check{X}} = -\check{R}_X, \qquad R_{\check{X}} = R_X.$$

Thus, for real numbers a and b,

$$\begin{aligned} R_U(a,b) &= E\left[\left(X(a)c_a + \check{X}(a)s_a\right)\left(X(b)c_b + \check{X}(b)s_b\right)\right] \\ &= R_X(a-b)(c_a c_b + s_a s_b) + \check{R}_X(a-b)(s_a c_b - c_a s_b) \\ &= R_X(a-b)c_{a-b} + \check{R}_X(a-b)s_{a-b}. \end{aligned}$$

Thus, $R_U(a,b)$ is a function of $a - b$, and $R_U(\tau)$ is given by the right side of (8.38). The proof that R_V also satisfies (8.38), and the proof of (8.39) are similar. Finally, it is a simple matter to derive (8.36) and (8.37) from (8.38) and (8.39), respectively. □

Equations (8.36) and (8.37) have simple graphical interpretations, as illustrated in Figure 8.10. Equation (8.36) means that S_U and S_V are each equal to the sum of the upper lobe of S_X shifted to the left by ω_c and the lower lobe of S_X shifted to the right by ω_c. Similarly, equation (8.36) means that S_{UV} is equal to the sum of j times the upper lobe of S_X shifted to the left by ω_c and $-j$ times the lower lobe of S_X shifted to the right by ω_c. Equivalently, S_U and S_V are each twice the symmetric part of the upper lobe of S_X, and S_{UV} is j times the antisymmetric part of the upper lobe of S_X. Since R_{UV} is an odd function of τ, it follows that $R_{UV}(0) = 0$. Thus, for any fixed time t, U_t and V_t are uncorrelated. That does not imply that U_s and V_t are uncorrelated for all s and t, for the

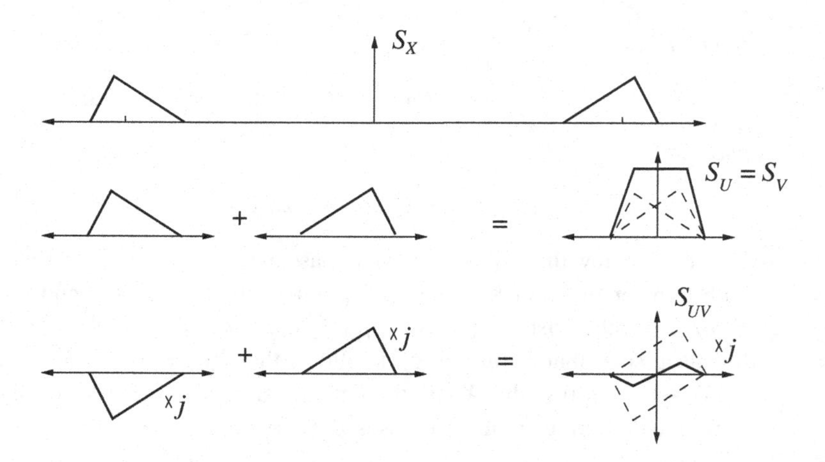

Figure 8.10 A narrowband power spectral density and associated baseband spectral densities.

cross correlation function R_{XY} is identically zero if and only if the upper lobe of S_X is symmetric about ω_c.

Example 8.7 (Baseband equivalent filtering of a random process) As noted above, filtering of narrowband deterministic signals can be described using equivalent baseband signals, namely the complex envelopes. The same is true for filtering of narrowband random processes. Suppose X is a narrowband WSS random process, suppose g is a finite energy narrowband signal, and suppose Y is the output process when X is filtered using impulse response function g. Then Y is also a WSS narrowband random process. Let Z denote the complex envelope of X, given in Proposition 8.3, and let z_g denote the complex envelope signal of g, meaning that z_g is the complex baseband signal such that $g(t) = Re(z_g(t)e^{j\omega_c t})$. It can be shown that the complex envelope process of Y is $\frac{1}{2}z_g * Z$.[1] Thus, the filtering of X by g is equivalent to the filtering of Z by $\frac{1}{2}z_g$.

Example 8.8 (Simulation of a narrowband random process) Let ω_o and ω_c be positive numbers with $0 < \omega_o < \omega_c$. Suppose S_X is a nonnegative function which is even (i.e. $S_X(\omega) = S_X(-\omega)$ for all ω) with $S_X(\omega) = 0$ if $||\omega| - \omega_c| \geq \omega_o$. We discuss briefly the problem of writing a computer simulation to generate a real-valued WSS random process X with power spectral density S_X.

By Proposition 8.2, it suffices to simulate baseband random processes U and V with the power spectral densities specified by (8.36) and cross power spectral density specified by (8.37). For increased tractability, we impose an additional assumption on S_X, namely that the upper lobe of S_X is symmetric about ω_c. This assumption is equivalent

[1] An elegant proof of this fact is based on spectral representation theory for WSS random processes, covered for example in (Doob 1953). The basic idea is to define the Fourier transform of a WSS random process, which, like white noise, is a generalized random process. Then essentially the same method we described for filtering of deterministic narrowband signals works.

to the assumption that S_{UV} vanishes, and therefore that the processes U and V are uncorrelated with each other. Thus, the processes U and V can be generated independently.

In turn, the processes U and V can be simulated by first generating sequences of random variables U_{nT} and V_{nT} for sampling frequency $\frac{1}{T} = 2f_o = \frac{\omega_o}{\pi}$. A discrete-time random process with power spectral density S_U can be generated by passing a discrete-time white noise sequence with unit variance through a discrete-time linear time-invariant system with real-valued impulse response function such that the transfer function H satisfies $S_U = |H|^2$. For example, taking $H(\omega) = \sqrt{S_U(\omega)}$ works, though it might not be the most well behaved linear system. (The problem of finding a transfer function H with additional properties such that $S_U = |H|^2$ is called the problem of spectral factorization, which we shall return to in the next chapter.) The samples V_{kT} can be generated similarly.

For a specific example, suppose that (using kHz for kilohertz, or thousands of Hertz)

$$S_X(2\pi f) = \begin{cases} 1 & 9000 \text{ kHz} < |f| < 9020 \text{ kHz} \\ 0 & \text{else} \end{cases}. \tag{8.40}$$

Note that the parameters ω_o and ω_c are not uniquely determined by S_X. They must simply be positive numbers with $\omega_o < \omega_c$ such that

$$(9000 \text{ kHz}, 9020 \text{ kHz}) \subset (f_c - f_o, f_c + f_o).$$

However, only the choice $f_c = 9010$ kHz makes the upper lobe of S_X symmetric around f_c. Therefore we take $f_c = 9010$ kHz. We take the minimum allowable value for f_o, namely $f_o = 10$ kHz. For this choice, (8.36) yields

$$S_U(2\pi f) = S_V(2\pi f) = \begin{cases} 2 & |f| < 10 \text{ kHz} \\ 0 & \text{else} \end{cases} \tag{8.41}$$

and (8.37) yields $S_{UV}(2\pi f) = 0$ for all f. The processes U and V are continuous-time baseband random processes with one-sided bandwidth limit 10 kHz. To simulate these processes it is therefore enough to generate samples of them with sampling period $T = 0.5 \times 10^{-4}$, and then use the Nyquist sampling representation described in Section 8.4. The processes of samples will, according to (8.26), have power spectral density equal to 4×10^4 over the interval $[-\pi, \pi]$. Consequently, the samples can be taken to be uncorrelated with $E[|A_k|^2] = E[|B_k|^2] = 4 \times 10^4$. For example, these variables can be taken to be independent real Gaussian random variables. Putting the steps together, we find the following representation for X:

$$X_t = \cos(\omega_c t)\left(\sum_{n=-\infty}^{\infty} A_n \text{sinc}\left(\frac{t-nT}{T}\right)\right) - \sin(\omega_c t)\left(\sum_{n=-\infty}^{\infty} B_n \text{sinc}\left(\frac{t-nT}{T}\right)\right).$$

8.6 Complexification, Part II

A complex random variable Z is said to be *circularly symmetric* if Z has the same distribution as $e^{j\theta}Z$ for every real value of θ. If Z has a pdf f_Z, circular symmetry of

Z means that $f_Z(z)$ is invariant under rotations about zero, or, equivalently, $f_Z(z)$ depends on z only through $|z|$. A collection of random variables $(Z_i : i \in I)$ is said to be *jointly circularly symmetric* if for every real value of θ, the collection $(Z_i : i \in I)$ has the same finite dimensional distributions as the collection $(Z_i e^{j\theta} : i \in I)$. Note that if $(Z_i : i \in I)$ is jointly circularly symmetric, and if $(Y_j : j \in J)$ is another collection of random variables such that each Y_j is a linear combination of Z_is (with no constants added in) then the collection $(Y_j : j \in J)$ is also jointly circularly symmetric.

Recall that a complex random vector Z, expressed in terms of real random vectors U and V as $Z = U + jV$, has mean $EZ = EU + jEV$ and covariance matrix $\mathrm{Cov}(Z) = E[(Z - EZ)(Z - EZ)^*]$. The *pseudo-covariance matrix* of Z is defined by $\mathrm{Cov}^p(Z) = E[(Z - EZ)(Z - EZ)^T]$, and it differs from the covariance of Z in that a transpose, rather than a Hermitian transpose, is involved. Note that $\mathrm{Cov}(Z)$ and $\mathrm{Cov}^p(Z)$ are readily expressed in terms of $\mathrm{Cov}(U)$, $\mathrm{Cov}(V)$, and $\mathrm{Cov}(U, V)$ as:

$$\mathrm{Cov}(Z) = \mathrm{Cov}(U) + \mathrm{Cov}(V) + j\left(\mathrm{Cov}(V, U) - \mathrm{Cov}(U, V)\right),$$
$$\mathrm{Cov}^p(Z) = \mathrm{Cov}(U) - \mathrm{Cov}(V) + j\left(\mathrm{Cov}(V, U) + \mathrm{Cov}(U, V)\right),$$

where $\mathrm{Cov}(V, U) = \mathrm{Cov}(U, V)^T$. Conversely,

$$\mathrm{Cov}(U) = \mathrm{Re}\left(\mathrm{Cov}(Z) + \mathrm{Cov}^p(Z)\right)/2, \quad \mathrm{Cov}(V) = \mathrm{Re}\left(\mathrm{Cov}(Z) - \mathrm{Cov}^p(Z)\right)/2,$$

and

$$\mathrm{Cov}(U, V) = \mathrm{Im}\left(-\mathrm{Cov}(Z) + \mathrm{Cov}^p(Z)\right)/2.$$

The vector Z is defined to be Gaussian if the random vectors U and V are jointly Gaussian.

Suppose that Z is a complex Gaussian random vector. Then its distribution is fully determined by its mean and the matrices $\mathrm{Cov}(U)$, $\mathrm{Cov}(V)$, and $\mathrm{Cov}(U, V)$, or equivalently by its mean and the matrices $\mathrm{Cov}(Z)$ and $\mathrm{Cov}^p(Z)$. Therefore, for a real value of θ, Z and $e^{j\theta}Z$ have the same distribution if and only if they have the same mean, covariance matrix, and pseudo-covariance matrix. Since $E[e^{j\theta}Z] = e^{j\theta}EZ$, $\mathrm{Cov}(e^{j\theta}Z) = \mathrm{Cov}(Z)$, and $\mathrm{Cov}^p(e^{j\theta}Z) = e^{j2\theta}\mathrm{Cov}^p(Z)$, Z and $e^{j\theta}Z$ have the same distribution if and only if $(e^{j\theta} - 1)EZ = 0$ and $(e^{j2\theta} - 1)\mathrm{Cov}^p(Z) = 0$. Hence, if θ is not a multiple of π, Z and $e^{j\theta}Z$ have the same distribution if and only if $EZ = 0$ and $\mathrm{Cov}^p(Z) = 0$. Consequently, a Gaussian random vector Z is circularly symmetric if and only if its mean vector and pseudo-covariance matrix are zero.

The joint density function of a circularly symmetric complex random vector Z with n complex dimensions and covariance matrix K, with $\det K \neq 0$, has the particularly elegant form:

$$f_Z(z) = \frac{\exp(-z^* K^{-1} z)}{\pi^n \det(K)}. \tag{8.42}$$

Equation (8.42) can be derived in the same way the density for Gaussian vectors with real components is derived. Namely, (8.42) is easy to verify if K is diagonal. If K is not diagonal, the Hermitian symmetric positive definite matrix K can be expressed as $K = U\Lambda U^*$, where U is a unitary matrix and Λ is a diagonal matrix with strictly positive diagonal entries. The random vector Y defined by $Y = U^*Z$ is Gaussian and circularly symmetric with covariance matrix Λ, and since $\det(\Lambda) = \det(K)$, it has pdf $f_Y(y) = \frac{\exp(-y^*\Lambda^{-1}y)}{\pi^n \det(K)}$. Since $|\det(U)| = 1, f_Z(z) = f_Y(U^*x)$, which yields (8.42).

Let us switch now to random processes. Let Z be a complex-valued random process and let U and V be the real-valued random processes such that $Z_t = U_t + jV_t$. Recall that Z is Gaussian if U and V are jointly Gaussian, and the covariance function of Z is defined by $C_Z(s,t) = \text{Cov}(Z_s, Z_t)$. The *pseudo-covariance function* of Z is defined by $C_Z^p(s,t) = \text{Cov}^p(Z_s, Z_t)$. As for covariance matrices of vectors, both C_Z and C_Z^p are needed to determine C_U, C_V, and C_{UV}.

Following the vast majority of the literature, we define Z to be *wide sense stationary* (WSS) if $\mu_Z(t)$ is constant and if $C_Z(s,t)$ (or $R_Z(s,t)$) is a function of $s-t$ alone. Some authors use a stronger definition of WSS, by defining Z to be WSS if either of the following two equivalent conditions is satisfied:

- $\mu_Z(t)$ is constant, and both $C_Z(s,t)$ and $C_Z^p(s,t)$ are functions of $s-t$,
- U and V are jointly WSS.

If Z is Gaussian then it is stationary if and only if it satisfies the stronger definition of WSS.

A complex random process $Z = (Z_t : t \in \mathbb{T})$ is called *circularly symmetric* if the random variables of the process, $(Z_t : t \in \mathbb{T})$, are jointly circularly symmetric. If Z is a complex Gaussian random process, it is circularly symmetric if and only if it has mean zero and $\text{Cov}_Z^p(s,t) = 0$ for all s, t. Proposition 8.3 shows that the baseband equivalent process Z for a Gaussian real-valued narrowband WSS random process X is circularly symmetric. Nearly all complex valued random processes in applications arise in this fashion. For circularly symmetric complex random processes, the definition of WSS we adopted, and the stronger definition mentioned in the previous paragraph, are equivalent. A circularly symmetric complex Gaussian random process is stationary if and only if it is WSS.

The interested reader can find more related to the material in this section in Neeser and Massey, "Proper complex random processes with applications to information theory," *IEEE Transactions on Information Theory*, **39**, (4), July 1993.

Problems

8.1 Baseband limiting Let X be a Gaussian random process with mean zero and autocorrelation function $R_X(\tau) = e^{-|\tau|}$.
(a) Find the numerical value of f_o in hertz so that 99% of the power of X is in the frequency band $[-f_o, f_o]$.
(b) Let Y be the output when X is passed through an ideal lowpass filter with cutoff frequency f_o (i.e. the transfer function is $H(2\pi f) = I_{[-f_o \le f \le f_o]}$). Are the random processes Y and $X - Y$ independent? Justify your answer.
(c) Find $P\{|X_t - Y_t| \ge 0.1\}$.

8.2 A second order stochastic differential equation Suppose X is a WSS m.s. continuous random process and Y is a WSS solution to the second order differential equation $Y'' + Y' + Y = X$.

(a) Express S_Y in terms of S_X.
(b) Suppose the power of X is one. What is the maximum possible power of Y, and for what choice of X is the maximum achieved?
(c) How small can the power of Y be, and for what choice of X (with power one) is the power of Y very small?

8.3 On filtering a WSS random process Suppose Y is the output of a linear time-invariant system with WSS input X, impulse response function h, and transfer function H. Indicate whether the following statements are true or false. Justify your answers.
(a) If $|H(\omega)| \leq 1$ for all ω then the power of Y is less than or equal to the power of X.
(b) If X is periodic (in addition to being WSS) then Y is WSS and periodic.
(c) If X has mean zero and strictly positive total power, and if $||h||^2 > 0$, then the output power is strictly positive.

8.4 On the cross spectral density Suppose X and Y are jointly WSS such that the power spectral densities S_X, S_Y, and S_{XY} are continuous. Show that for each ω, $|S_{XY}(\omega)|^2 \leq S_X(\omega)S_Y(\omega)$. Hint: Fix ω_o, let $\epsilon > 0$, and let J^ϵ denote the interval of length ϵ centered at ω_o. Consider passing both X and Y through a linear time-invariant system with transfer function $H^\epsilon(\omega) = I_{J^\epsilon}(\omega)$. Apply the Schwarz inequality to the output processes sampled at a fixed time, and let $\epsilon \to 0$.

8.5 Modulating and filtering a stationary process Let $X = (X_t : t \in Z)$ be a discrete-time mean-zero stationary random process with power $E[X_0^2] = 1$. Let Y be the stationary discrete-time random process obtained from X by modulation as follows:

$$Y_t = X_t \cos(80\pi t + \Theta),$$

where Θ is independent of X and is uniformly distributed over $[0, 2\pi]$. Let Z be the stationary discrete-time random process obtained from Y by the linear equations

$$Z_{t+1} = (1 - a)Z_t + aY_{t+1}$$

for all t, where a is a constant with $0 < a < 1$.
(a) Why is the random process Y stationary?
(b) Express the autocorrelation function of Y, $R_Y(\tau) = E[Y_\tau Y_0]$, in terms of the autocorrelation function of X. Similarly, express the power spectral density of Y, $S_Y(\omega)$, in terms of the power spectral density of X, $S_X(\omega)$.
(c) Find and sketch the transfer function $H(\omega)$ for the linear system describing the mapping from Y to Z.
(d) Can the power of Z be arbitrarily large (depending on a)? Explain your answer.
(e) Describe an input X satisfying the assumptions above so that the power of Z is at least 0.5, for any value of a with $0 < a < 1$.

8.6 Filtering a Gauss Markov process Let $X = (X_t : -\infty < t < +\infty)$ be a stationary Gauss Markov process with mean zero and autocorrelation function $R_X(\tau) = \exp(-|\tau|)$. Define a random process $Y = (Y_t : t \in \mathbb{R})$ by the differential equation $\dot{Y}_t = X_t - Y_t$.
(a) Find the cross correlation function R_{XY}. Are X and Y jointly stationary?
(b) Find $E[Y_5|X_5 = 3]$. What is the approximate numerical value?

(c) Is Y a Gaussian random process? Justify your answer.
(d) Is Y a Markov process? Justify your answer.

8.7 Slight smoothing Suppose Y is the output of the linear time-invariant system with input X and impulse response function h, such that X is WSS with $R_X(\tau) = \exp(-|\tau|)$, and $h(\tau) = \frac{1}{a}I_{\{|\tau|\leq\frac{a}{2}\}}$ for $a > 0$. If a is small, then h approximates the delta function $\delta(\tau)$, and consequently $Y_t \approx X_t$. This problem explores the accuracy of the approximation.
(a) Find $R_{YX}(0)$ and show $R_{YX}(0) = 1 - \frac{a}{4} + o(a)$ as $a \to 0$. (Hint: Use the power series expansion of e^u.)
(b) Find $R_Y(0)$ and show $R_Y(0) = 1 - \frac{a}{3} + o(a)$ as $a \to 0$.
(c) Show $E[|X_t - Y_t|^2] = \frac{a}{6} + o(a)$ as $a \to 0$.

8.8 A stationary two-state Markov process Let $X = (X_k : k \in \mathbb{Z})$ be a stationary Markov process with state space $\mathcal{S} = \{1, -1\}$ and one-step transition probability matrix

$$P = \begin{pmatrix} 1-p & p \\ p & 1-p \end{pmatrix},$$

where $0 < p < 1$. Find the mean, correlation function and power spectral density function of X. Hint: For nonnegative integers k:

$$P^k = \begin{pmatrix} \frac{1}{2} & \frac{1}{2} \\ \frac{1}{2} & \frac{1}{2} \end{pmatrix} + (1-2p)^k \begin{pmatrix} \frac{1}{2} & -\frac{1}{2} \\ -\frac{1}{2} & \frac{1}{2} \end{pmatrix}.$$

8.9 A stationary two-state Markov process in continuous time Let $X = (X_t : t \in \mathbb{R})$ be a stationary Markov process with state space $\mathcal{S} = \{1, -1\}$ and Q matrix

$$Q = \begin{pmatrix} -\alpha & \alpha \\ \alpha & -\alpha \end{pmatrix},$$

where $\alpha > 0$. Find the mean, correlation function and power spectral density function of X. Hint: Recall from the example in the chapter on Markov processes that for $s < t$, the matrix of transition probabilities $p_{ij}(s,t)$ is given by $H(\tau)$, where $\tau = t - s$ and

$$H(\tau) = \begin{pmatrix} \frac{1+e^{-2\alpha\tau}}{2} & \frac{1-e^{-2\alpha\tau}}{2} \\ \frac{1-e^{-2\alpha\tau}}{2} & \frac{1+e^{-2\alpha\tau}}{2} \end{pmatrix}.$$

8.10 A linear estimation problem Suppose X and Y are possibly complex valued jointly WSS processes with known autocorrelation functions, cross correlation function, and associated spectral densities. Suppose Y is passed through a linear time-invariant system with impulse response function h and transfer function H, and let Z be the output. The mean square error of estimating X_t by Z_t is $E[|X_t - Z_t|^2]$.
(a) Express the mean square error in terms of R_X, R_Y, R_{XY}, and h.
(b) Express the mean square error in terms of S_X, S_Y, S_{XY}, and H.
(c) Using your answer to part (b), find the choice of H that minimizes the mean square error. (Hint: Try working out the problem first assuming the processes are real valued. For the complex case, note that for $\sigma^2 > 0$ and complex numbers z and z_o, $\sigma^2|z|^2 - 2\mathrm{Re}(z^* z_o)$ is equal to $|\sigma z - \frac{z_o}{\sigma}|^2 - \frac{|z_o|^2}{\sigma^2}$, which is minimized with respect to z by $z = \frac{z_o}{\sigma^2}$.)

8.11 Linear time invariant, uncorrelated scattering channel A signal transmitted through a scattering environment can propagate over many different paths on its way to a receiver. The channel gains along distinct paths are often modeled as uncorrelated. The paths may differ in length, causing a delay spread. Let $h = (h_u : u \in \mathbb{Z})$ consist of uncorrelated, possibly complex valued random variables with mean zero and $E[|h_u|^2] = g_u$. Assume that $G = \sum_u g_u < \infty$. The variable h_u is the random complex gain for delay u, and $g = (g_u : u \in \mathbb{Z})$ is the energy gain delay mass function with total gain G. Given a deterministic signal x, the channel output is the random signal Y defined by $Y_i = \sum_{u=-\infty}^{\infty} h_u x_{i-u}$.
(a) Determine the mean and autocorrelation function for Y in terms of x and g.
(b) Express the average total energy of Y: $E[\sum_i Y_i^2]$, in terms of x and g.
(c) Suppose instead that the input is a WSS random process X with autocorrelation function R_X. The input X is assumed to be independent of the channel h. Express the mean and autocorrelation function of the output Y in terms of R_X and g. Is Y WSS?
(d) Since the impulse response function h is random, so is its Fourier transform $H = (H(\omega) : -\pi \leq \omega \leq \pi)$. Express the autocorrelation function of the random process H in terms of g.

8.12 The accuracy of approximate differentiation Let X be a WSS baseband random process with power spectral density S_X, and let ω_o be the one-sided band limit of X. The process X is m.s. differentiable and X' can be viewed as the output of a time-invariant linear system with transfer function $H(\omega) = j\omega$.
(a) What is the power spectral density of X'?
(b) Let $Y_t = \frac{X_{t+a} - X_{t-a}}{2a}$, for some $a > 0$. We can also view $Y = (Y_t : t \in \mathbb{R})$ as the output of a time-invariant linear system, with input X. Find the impulse response function k and transfer function K of the linear system. Show that $K(\omega) \to j\omega$ as $a \to 0$.
(c) Let $D_t = X'_t - Y_t$. Find the power spectral density of D.
(d) Find a value of a, depending only on ω_o, so that $E[|D_t|^2] \leq (0.01)E[|X'_t|]^2$. In other words, for such a, the m.s. error of approximating X'_t by Y_t is less than one percent of $E[|X'_t|^2]$. You can use the fact that $0 \leq 1 - \frac{\sin(u)}{u} \leq \frac{u^2}{6}$ for all real u. (Hint: Find a so that $S_D(\omega) \leq (0.01)S_{X'}(\omega)$ for $|\omega| \leq \omega_o$.)

8.13 Some linear transformations of some random processes Let $U = (U_n : n \in \mathbb{Z})$ be a random process such that the U_n are independent, identically distributed, with $E[U_n] = \mu$ and $\mathrm{Var}(U_n) = \sigma^2$, where $\mu \neq 0$ and $\sigma^2 > 0$. Please keep in mind that $\mu \neq 0$. Define $X = (X_n : n \in \mathbb{Z})$ by $X_n = \sum_{k=0}^{\infty} U_{n-k} a^k$, for a constant a with $0 < a < 1$.
(a) Is X stationary? Find the mean function μ_X and autocovariance function C_X for X.
(b) Is X a Markov process ? (Hint: X is not necessarily Gaussian. Does X have a state representation driven by U?)
(c) Is X mean ergodic in the m.s. sense?
Let U be as before, and let $Y = (Y_n : n \in \mathbb{Z})$ be defined by $Y_n = \sum_{k=0}^{\infty} U_{n-k} A^k$, where A is a random variable distributed on the interval $(0, 0.5)$ (the exact distribution is not specified), and A is independent of the random process U.
(d) Is Y stationary? Find the mean function μ_Y and autocovariance function C_Y for Y. (Your answer may include expectations involving A.)

(e) Is Y a Markov process? (Give a brief explanation.)
(f) Is Y mean ergodic in the m.s. sense?

8.14 Filtering Poisson white noise A Poisson process $N = (N_t : t \geq 0)$ has independent increments. The derivative of N, written N', does not exist as an ordinary random process, but it does exist as a generalized random process. Graphically, picture N' as a superposition of delta functions, one at each arrival time of the Poisson process. As a generalized random process, N' is stationary with mean and autocovariance functions given by $E[N'_t] = \lambda$, and $C_{N'}(s,t) = \lambda\delta(s-t)$, respectively, because, when integrated, these functions give the correct values for the mean and covariance of N: $E[N_t] = \int_0^t \lambda ds$ and $C_N(s,t) = \int_0^s \int_0^t \lambda\delta(u-v)dvdu$. The random process N' can be extended to be defined for negative times by augmenting the original random process N by another rate λ Poisson process for negative times. Then N' can be viewed as a stationary random process, and its integral over intervals gives rise to a process $N(a,b]$ as described in Problem 4.19. (The process $N' - \lambda$ is a white noise process, in that it is a generalized random process which is stationary, mean zero, and has autocorrelation function $\lambda\delta(\tau)$. Both N' and $N' - \lambda$ are called Poisson shot noise processes. One application for such processes is modeling noise in small electronic devices, in which effects of single electrons can be registered. For the remainder of this problem, N' is used instead of the mean zero version.) Let X be the output when N' is passed through a linear time-invariant filter with an impulse response function h, such that $\int_{-\infty}^{\infty} |h(t)|dt$ is finite. (Remark: In the special case that $h(t) = I_{\{0 \leq t < 1\}}$, X is the $M/D/\infty$ process of Problem 4.19.)
(a) Find the mean and covariance functions of X.
(b) Consider the special case $h(t) = e^{-t}I_{\{t \geq 0\}}$. Explain why X is a Markov process in this case. (Hint: What is the behavior of X between the arrival times of the Poisson process? What does X do at the arrival times?)

8.15 A linear system with a feedback loop The system with input X and output Y involves feedback with the loop transfer function shown.

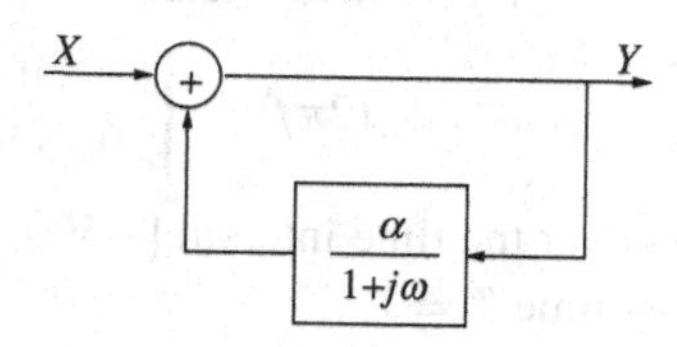

(a) Find the transfer function, K, of the system.
(b) Find the corresponding impulse response function.
(c) The power of Y divided by the power of X, depends on the power spectral density, S_X. Find the supremum of this ratio, over all choices of S_X, and describe what choice of S_X achieves this supremum.

8.16 Linear and nonlinear reconstruction from samples Suppose that $X_t = \sum_{n=-\infty}^{\infty} g(t-n-U)B_n$, where the B_ns are independent with mean zero and variance $\sigma^2 > 0$, g is a function with finite energy $\int |g(t)|^2 dt$ and Fourier transform $G(\omega)$, U is a random variable which is independent of B and uniformly distributed on the interval $[0,1]$. The process X is a typical model for a digital baseband signal, where the B_ns are random data symbols.

(a) Show that X is WSS, with mean zero and $R_X(t) = \sigma^2 g * \widetilde{g}(t)$.
(b) Under what conditions on G and T can the sampling theorem be used to recover
from its samples of the form $(X(nT) : n \in \mathbb{Z})$?
(c) Consider the particular case $g(t) = (1-|t|)_+$ and $T = 0.5$. Although this falls outsid
the conditions found in part (b), show that by using nonlinear operations, the proces
X can be recovered from its samples of the form $(X(nT) : n \in \mathbb{Z})$. (Hint: Consider
sample path of X.)

8.17 Sampling a cubed Gaussian process Let $X = (X_t : t \in \mathbb{R})$ be a baseband mea
zero stationary real Gaussian random process with one-sided band limit f_o Hz. Thus
$X_t = \sum_{n=-\infty}^{\infty} X_{nT}\text{sinc}\left(\frac{t-nT}{T}\right)$ where $\frac{1}{T} = 2f_o$. Let $Y_t = X_t^3$ for each t.
(a) Is Y stationary? Express R_Y in terms of R_X, and express S_Y in terms of S_X and/o
R_X. (Hint: If A, B are jointly Gaussian and mean zero, $\text{Cov}(A^3, B^3) = 6\text{Cov}(A, B)^3 +$
$9E[A^2]E[B^2]\text{Cov}(A, B)$.)
(b) At what rate $\frac{1}{T'}$ should Y be sampled so $Y_t = \sum_{n=-\infty}^{\infty} Y_{nT'}\ \text{sinc}\left(\frac{t-nT'}{T'}\right)$?
(c) Can Y be recovered with fewer samples than in part (b)? Explain.

8.18 An approximation of white noise White noise in continuous time can be approx
imated by a piecewise constant process as follows. Let T be a small positive constant
A_T be a positive scaling constant depending on T, and $(B_k : k \in \mathbb{Z})$ be a discrete-tim
white noise process with $R_B(k) = \sigma^2 I_{\{k=0\}}$. Define $(N_t : t \in \mathbb{R})$ by $N_t = A_T B_k$ fo
$t \in [kT, (k+1)T)$.
(a) Sketch a typical sample path of N and express $E[|\int_0^1 N_s ds|^2]$ in terms of A_T, T anc
σ^2. For simplicity assume $T = \frac{1}{K}$ for some large integer K.
(b) What choice of A_T makes the expectation found in part (a) equal to σ^2? Thi
choice makes N a good approximation to a continuous-time white noise process wit
autocorrelation function $\sigma^2\delta(\tau)$.
(c) What happens to the expectation found in (a) as $T \to 0$ if $A_T = 1$ for all T?

8.19 Simulating a baseband random process Suppose a real-valued Gaussian baseband process $X = (X_t : t \in \mathbb{R})$ with mean zero and power spectral density

$$S_X(2\pi f) = \begin{cases} 1 & \text{if } |f| \le 0.5 \\ 0 & \text{else} \end{cases}$$

is to be simulated over the time interval $[-500, 500]$ through use of the sampling theorem with sampling time $T = 1$.
(a) What is the joint distribution of the samples, $X_n : n \in Z$?
(b) Of course a computer cannot generate infinitely many random variables in a finite
amount of time. Therefore, consider approximating X by $X^{(N)}$ defined by

$$X_t^{(N)} = \sum_{n=-N}^{N} X_n \text{sinc}(t-n).$$

Find a condition on N to guarantee $E[(X_t - X_t^{(N)})^2] \le 0.01$ for $t \in [-500, 500]$. (Hint:
Use $|\text{sinc}(\tau)| \le \frac{1}{\pi|\tau|}$ and bound the series by an integral. Your choice of N should not
depend on t because the same N should work for all t in the interval $[-500, 500]$.)

8.20 Synthesizing a random process with specified spectral density This problem
deals with Monte Carlo simulation of a Gaussian stationary random process with a

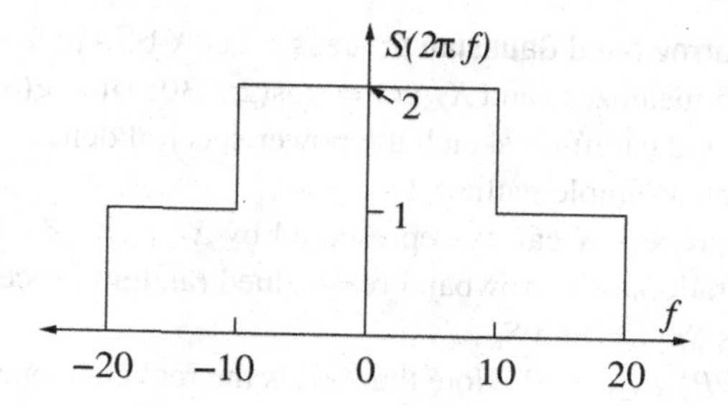

specified power spectral density. Give a representation of a random process X with the power spectral density S_X shown, using independent, $N(0, 1)$ random variables, and linear operations such as linear filtering and addition, as in the Nyquist sampling theorem representation of baseband processes. You do not need to address the fact that in practice, a truncation to a finite sum would be used to approximately simulate the process over a finite time interval, but do try to minimize the number of $N(0, 1)$ variables you use per unit time of simulation. Identify explicitly any functions you use, and also identify how many $N(0, 1)$ random variables you use per unit of time simulated.

8.21 Filtering to maximize signal to noise ratio Let X and N be continuous time, WSS random processes. Suppose $S_X(\omega) = |\omega| I_{\{|\omega|\leq\omega_o\}}$, and $S_N(\omega) = \sigma^2$ for all ω. Suppose also that X and N are uncorrelated with each other. Think of X as a signal, and N as noise. Suppose $X + N$ is passed through a linear time-invariant filter with transfer function H, which you are to specify. Let $\widetilde{X}$ denote the output signal and $\widetilde{N}$ denote the output noise. What choice of H, subject to the constraints (i) $|H(\omega)| \leq 1$ for all ω, and (ii) (power of $\widetilde{X}$) $\geq$ (power of X)/2, minimizes the power of $\widetilde{N}$?

8.22 Finding the envelope of a deterministic signal (a) Find the complex envelope $z(t)$ and real envelope $|z(t)|$ of $x(t) = \cos(2\pi(1000)t) + \cos(2\pi(1001)t)$, using the carrier frequency $f_c = 1000.5$ Hz. Simplify your answer.
(b) Repeat (a), using $f_c = 995$ Hz. (Hint: The real envelope should be the same as found in (a).)
(c) Explain why, in general, the real envelope of a narrowband signal does not depend on which frequency f_c is used to represent the signal (as long as f_c is chosen so that the upper band of the signal is contained in an interval $[f_c - a, f_c + a]$ with $a << f_c$).

8.23 Sampling a signal or process that is not band limited (a) Fix $T > 0$ and $\omega_o = \pi/T$. Given a finite energy signal x, let x^o be the band-limited signal with Fourier transform $\widehat{x^o}(\omega) = I_{\{|\omega|\leq\omega_o\}} \sum_{n=-\infty}^{\infty} \widehat{x}(\omega + 2n\omega_o)$. Show that $x(nT) = x^o(nT)$ for all integers n.
(b) Explain why $x^o(t) = \sum_{n=-\infty}^{\infty} x(nT)\text{sinc}\left(\frac{t-nT}{T}\right)$.
(c) Let X be a mean zero WSS random process, and R_X^o be the autocorrelation function for $S_X^o(\omega)$ defined by $S_X^o(\omega) = I_{\{|\omega|\leq\omega_o\}} \sum_{n=-\infty}^{\infty} S_X(\omega + 2n\omega_o)$. Show that $R_X(nT) = R_X^o(nT)$ for all integers n.
(d) Explain why Y defined by $Y_t = \sum_{n=-\infty}^{\infty} X_{nT}\text{sinc}\left(\frac{t-nT}{T}\right)$ is WSS with autocorrelation function R_X^o.
(e) Find S_X^o when $S_X(\omega) = \exp(-\alpha|\omega|)$ for $\omega \in \mathbb{R}$.

8.24 A narrowband Gaussian process Let X be a real-valued stationary Gaussian process with mean zero and $R_X(\tau) = \cos(2\pi(30\tau))(\mathrm{sinc}(6\tau))^2$.
(a) Find and carefully sketch the power spectral density of X.
(b) Sketch a sample path of X.
(c) The process X can be represented by $X_t = Re(Z_t e^{2\pi j30t})$, where $Z_t = U_t + jV_t$ for jointly stationary narrowband real-valued random processes U and V. Find the spectral densities S_U, S_V, and S_{UV}.
(d) Find $P\{|Z_{33}| > 5\}$. Note that $|Z_t|$ is the real envelope process of X.
8.25 Another narrowband Gaussian process Suppose a real-valued Gaussian random process $R = (R_t : t \in \mathbb{R})$ with mean 2 and power spectral density $S_R(2\pi f) = e^{-|f|/10^4}$ is fed through a linear time-invariant system with transfer function

$$H(2\pi f) = \begin{cases} 0.1 & 5000 \leq |f| \leq 6000 \\ 0 & \text{else} \end{cases}.$$

(a) Find the mean and power spectral density of the output process $(X_t : t \in \mathbb{R})$.
(b) Find $P\{X_{25} > 6\}$.
(c) The random process X is a narrowband random process. Find the power spectral densities S_U and S_V, and the cross spectral density S_{UV} of jointly WSS baseband random processes U and V so that

$$X_t = U_t \cos(2\pi f_c t) - V_t \sin(2\pi f_c t),$$

using $f_c = 5500$.
(d) Repeat part (c) with $f_c = 5000$.
8.26 Another narrowband Gaussian process (version 2) Suppose a real-valued Gaussian white noise process N (we assume white noise has mean zero) with power spectral density $S_N(2\pi f) \equiv \frac{N_o}{2}$ for $f \in \mathbb{R}$ is fed through a linear time-invariant system with transfer function H specified as follows, where f represents the frequency in gigahertz (GHz) and a gigahertz is 10^9 cycles per second:

$$H(2\pi f) = \begin{cases} 1 & 19.10 \leq |f| \leq 19.11 \\ \sqrt{\frac{19.12-|f|}{0.01}} & 19.11 \leq |f| \leq 19.12 \\ 0 & \text{else} \end{cases}.$$

(a) Find the mean and power spectral density of the output process $X = (X_t : t \in \mathbb{R})$.
(b) Express $P\{X_{25} > 2\}$ in terms of N_o and the standard normal complementary CDF function Q.
(c) The random process X is a narrowband random process. Find and sketch the power spectral densities S_U and S_V, and the cross spectral density S_{UV} of jointly WSS baseband random processes U and V so that

$$X_t = U_t \cos(2\pi f_c t) - V_t \sin(2\pi f_c t),$$

using $f_c = 19.11$ GHz.
(d) The complex envelope process is given by $Z = U + jV$ and the real envelope process is given by $|Z|$. Specify the distributions of Z_t and $|Z_t|$ for t fixed.

8.27 Declaring the center frequency for a given random process Let $a > 0$ and let g be a nonnegative function on $\mathbb{R}$ which is zero outside of the interval $[a, 2a]$. Suppose X is a narrowband WSS random process with power spectral density function $S_X(\omega) = g(|\omega|)$, or equivalently, $S_X(\omega) = g(\omega) + g(-\omega)$. The process X can thus be viewed as a narrowband signal for carrier frequency ω_c, for any choice of ω_c in the interval $[a, 2a]$. Let U and V be the baseband random processes in the usual complex envelope representation: $X_t = Re((U_t + jV_t)e^{j\omega_c t})$.
(a) Express S_U and S_{UV} in terms of g and ω_c.
(b) Describe which choice of ω_c minimizes $\int_{-\infty}^{\infty} |S_{UV}(\omega)|^2 \frac{d\omega}{d\pi}$. (Note: If g is symmetric around some frequency ν, then $\omega_c = \nu$. But what is the answer otherwise?)
8.28* Cyclostationary random processes A random process $X = (X_t : t \in \mathbb{R})$ is said to be cyclostationary with period T, if whenever s is an integer multiple of T, X has the same finite dimensional distributions as $(X_{t+s} : t \in \mathbb{R})$. This property is weaker than stationarity, because stationarity requires equality of finite dimensional distributions for all real values of s.
(a) What properties of the mean function μ_X and autocorrelation function R_X does any second order cyclostationary process possess? A process with these properties is called a wide sense cyclostationary process.
(b) Suppose X is cyclostationary and that U is a random variable independent of X that is uniformly distributed on the interval $[0, T]$. Let $Y = (Y_t : t \in \mathbb{R})$ be the random process defined by $Y_t = X_{t+U}$. Argue that Y is stationary, and express the mean and autocorrelation function of Y in terms of the mean function and autocorrelation function of X. Although X is not necessarily WSS, it is reasonable to define the power spectral density of X to equal the power spectral density of Y.
(c) Suppose B is a stationary discrete-time random process and that g is a deterministic function. Let X be defined by

$$X_t = \sum_{n=-\infty}^{\infty} g(t - nT)B_n.$$

Show that X is a cyclostationary random process. Find the mean function and autocorrelation function of X in terms of g, T, and the mean and autocorrelation function of B. If your answer is complicated, identify special cases which make the answer nice.
(d) Suppose Y is defined as in part (b) for the specific X defined in part (c). Express the mean μ_Y, autocorrelation function R_Y, and power spectral density S_Y in terms of g, T, μ_B, and S_B.
8.29* Zero crossing rate of a stationary Gaussian process Consider a stationary Gaussian process X with $S_X(2\pi f) = |f| - 50$ for $50 \leq |f| \leq 60$, and $S_X(2\pi f) = 0$ otherwise. Assume the process has continuous sample paths (it can be shown that such a version exists). A zero crossing from above is said to occur at time t if $X(t) = 0$ and $X(s) > 0$ for all s in an interval of the form $[t - \epsilon, t)$ for some $\epsilon > 0$. Determine the mean rate of zero crossings from above for X. If you can find an analytical solution, great. Alternatively, you can estimate the rate (aim for three significant digits) by Monte Carlo simulation of the random process.

9 Wiener filtering

Wiener filtering is a framework for minimum mean square error (MMSE) linear estimation in the context of wide sense stationary random processes. The estimators are obtained by passing the observed processes through linear filters. Equations that the optimal filters must satisfy follow readily from the orthogonality principle. In the case of noncausal estimation, the equations are relatively easy to solve in the frequency domain. For causal estimation, where the theory of Wiener filtering overlaps Kalman filtering, the optimality equations are of Wiener–Hopf type, and can be solved by the method of spectral factorization of rational spectral densities, as shown in Section 9.4. Section 9.5 explains the connection between the Wiener–Hopf equations in discrete time and the linear innovations approach used for deriving the Kalman filtering equations. The WSS assumptions of Wiener filtering are restrictive and not needed for Kalman filtering, but if the processes involved are WSS, then Wiener filtering gives insight about signals and noise that complements the Kalman filtering approach.

9.1 Return of the orthogonality principle

Consider the problem of estimating a random process X at some fixed time t given observation of a random process Y over an interval $[a, b]$. Suppose both X and Y are mean zero second order random processes and that the minimum mean square error is to be minimized. Let $\widehat{X}_t$ denote the best linear estimator of X_t based on the observations $(Y_s : a \leq s \leq b)$. In other words, define

$$\mathcal{V}_o = \{c_1 Y_{s_1} + \ldots + c_n Y_{s_n} : n \geq 1, s_1, \ldots, s_n \in [a, b], c_1, \ldots, c_n \in \mathbb{R}\}.$$

and let $\mathcal{V}$ be the m.s. closure of $\mathcal{V}$, which includes $\mathcal{V}_o$ and any random variable that is the m.s. limit of a sequence of random variables in $\mathcal{V}_o$. Then $\widehat{X}_t$ is the random variable in $\mathcal{V}$ that minimizes the mean square error, $E[|X_t - \widehat{X}_t|^2]$. By the orthogonality principle, the estimator $\widehat{X}_t$ exists and it is unique in the sense that any two solutions to the estimation problem are equal with probability one.

Perhaps the most useful part of the orthogonality principle is that a random variable W is equal to $\widehat{X}_t$ if and only if (i) $W \in \mathcal{V}$, and (ii) $(X_t - W) \perp Z$ for all $Z \in \mathcal{V}$. Equivalently, W is equal to $\widehat{X}_t$ if and only if (i) $W \in \mathcal{V}$, and (ii) $(X_t - W) \perp Y_u$ for all $u \in [a, b]$. Furthermore, the minimum mean square error (i.e. the error for the optimal estimator $\widehat{X}_t$) is given by $E[|X_t|^2] - E[|\widehat{X}_t|^2]$.

Note that m.s. integrals of the form $\int_a^b h(t,s)Y_s ds$ are in $\mathcal{V}$, because m.s. integrals are m.s. limits of finite linear combinations of the random variables of Y. Typically the set $\mathcal{V}$ is larger than the set of all m.s. integrals of Y. For example, if u is a fixed time in $[a,b]$ then $Y_u \in \mathcal{V}$. In addition, if Y is m.s. differentiable, then Y'_u is also in $\mathcal{V}$. Typically neither Y_u nor Y'_u can be expressed as a m.s. integral of $(Y_s : s \in \mathbb{R})$. However, Y_u can be obtained as an integral of the process Y multiplied by a delta function, though the integration has to be taken in a generalized sense.

The integral $\int_a^b h(t,s)Y_s ds$ is the linear MMSE estimator if and only if

$$X_t - \int_a^b h(t,s)Y_s ds \perp Y_u \quad \text{for } u \in [a,b],$$

or equivalently

$$E\left[\left(X_t - \int_a^b h(t,s)Y_s ds\right) Y_u^*\right] = 0 \quad \text{for } u \in [a,b],$$

or equivalently

$$R_{XY}(t,u) = \int_a^b h(t,s)R_Y(s,u)ds \quad \text{for } u \in [a,b].$$

Suppose now that the observation interval is the whole real line $\mathbb{R}$ and X and Y are jointly WSS. Then for t and v fixed, the problem of estimating X_t from $(Y_s : s \in \mathbb{R})$ is the same as estimating X_{t+v} from $(Y_{s+v} : s \in \mathbb{R})$. Therefore, if $h(t,s)$ for t fixed is the optimal function to use for estimating X_t from $(Y_s : s \in \mathbb{R})$, then it is also the optimal function to use for estimating X_{t+v} from $(Y_{s+v} : s \in \mathbb{R})$. Therefore, $h(t,s) = h(t+v, s+v)$, so that $h(t,s)$ is a function of $t-s$ alone, meaning that the optimal impulse response function h corresponds to a time-invariant system. Thus, we seek to find an optimal estimator of the form $\hat{X}_t = \int_{-\infty}^{\infty} h(t-s)Y_s ds$. The optimality condition becomes

$$X_t - \int_{-\infty}^{\infty} h(t-s)Y_s ds \perp Y_u \quad \text{for } u \in \mathbb{R},$$

which is equivalent to the condition

$$R_{XY}(t-u) = \int_{-\infty}^{\infty} h(t-s)R_Y(s-u)ds \quad \text{for } u \in \mathbb{R},$$

or $R_{XY} = h * R_Y$. In the frequency domain the condition is $S_{XY}(\omega) = H(\omega)S_Y(\omega)$ for all ω. Consequently, the optimal filter H is given by

$$H(\omega) = \frac{S_{XY}(\omega)}{S_Y(\omega)},$$

and the corresponding minimum mean square error is given by

$$E[|X_t - \widehat{X}_t|^2] = E[|X_t|^2] - E[|\widehat{X}_t|^2] = \int_{-\infty}^{\infty} \left(S_X(\omega) - \frac{|S_{XY}(\omega)|^2}{S_Y(\omega)}\right) \frac{d\omega}{2\pi}.$$

Example 9.1 Consider estimating a random process from observation of the random process plus noise, as shown in Figure 9.1. Assume that X and N are jointly WSS with mean zero. Suppose X and N have known autocorrelation functions and suppose that $R_{XN} \equiv 0$, so the variables of the process X are uncorrelated with the variables of the process N. The observed process is given by $Y = X+N$. Then $S_{XY} = S_X$ and $S_Y = S_X+S_N$, so the optimal filter is given by

$$H(\omega) = \frac{S_{XY}(\omega)}{S_Y(\omega)} = \frac{S_X(\omega)}{S_X(\omega) + S_N(\omega)}.$$

The associated minimum mean square error is given by

$$\begin{aligned} E[|X_t - \widehat{X}_t|^2] &= \int_{-\infty}^{\infty} \left(S_X(\omega) - \frac{S_X(\omega)^2}{S_X(\omega) + S_N(\omega)} \right) \frac{d\omega}{2\pi} \\ &= \int_{-\infty}^{\infty} \frac{S_X(\omega)S_N(\omega)}{S_X(\omega) + S_N(\omega)} \frac{d\omega}{2\pi}. \end{aligned}$$

Example 9.2 This example is a continuation of the previous example, for a particular choice of power spectral densities. Suppose that the signal process X is WSS with mean zero and power spectral density $S_X(\omega) = \frac{1}{1+\omega^2}$, suppose the noise process N is WSS with mean zero and power spectral density $\frac{4}{4+\omega^2}$, and suppose $S_{XN} \equiv 0$. Equivalently, $R_X(\tau) = \frac{e^{-|\tau|}}{2}$, $R_N(\tau) = e^{-2|\tau|}$, and $R_{XN} \equiv 0$. We seek the optimal linear estimator of X_t given $(Y_s : s \in \mathbb{R})$, where $Y = X + N$. Seeking an estimator of the form

$$\widehat{X}_t = \int_{-\infty}^{\infty} h(t-s)Y_s ds,$$

we find from the previous example that the transform H of h is given by

$$H(\omega) = \frac{S_X(\omega)}{S_X(\omega) + S_N(\omega)} = \frac{\frac{1}{1+\omega^2}}{\frac{1}{1+\omega^2} + \frac{4}{4+\omega^2}} = \frac{4+\omega^2}{8+5\omega^2}.$$

We will find h by finding the inverse transform of H. First, note that

$$\frac{4+\omega^2}{8+5\omega^2} = \frac{\frac{8}{5}+\omega^2}{8+5\omega^2} + \frac{\frac{12}{5}}{8+5\omega^2} = \frac{1}{5} + \frac{\frac{12}{5}}{8+5\omega^2}.$$

We know that $\frac{1}{5}\delta(t) \leftrightarrow \frac{1}{5}$. Also, for any $\alpha > 0$,

$$e^{-\alpha|t|} \leftrightarrow \frac{2\alpha}{\omega^2 + \alpha^2}, \tag{9.1}$$

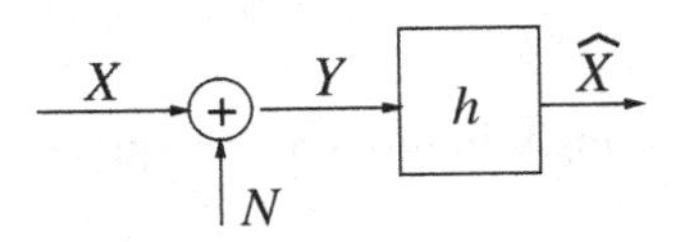

Figure 9.1 An estimator of a signal from signal plus noise, as the output of a linear filter.

so

$$\frac{1}{8+5\omega^2} = \frac{\frac{1}{5}}{\frac{8}{5}+\omega^2} = \left(\frac{1}{5\cdot 2}\sqrt{\frac{5}{8}}\right)\frac{2\sqrt{\frac{8}{5}}}{(\frac{8}{5}+\omega^2)} \leftrightarrow \left(\frac{1}{4\sqrt{10}}\right)e^{-\sqrt{\frac{8}{5}}|t|}.$$

Therefore the optimal filter is given in the time domain by

$$h(t) = \frac{1}{5}\delta(t) + \left(\frac{3}{5\sqrt{10}}\right)e^{-\sqrt{\frac{8}{5}}|t|}.$$

The associated minimum mean square error is given by (one way to do the integration is to use the fact that if $k \leftrightarrow K$ then $\int_{-\infty}^{\infty} K(\omega)\frac{d\omega}{2\pi} = k(0)$):

$$E[|X_t - \widehat{X}_t|^2] = \int_{-\infty}^{\infty} \frac{S_X(\omega)S_N(\omega)}{S_X(\omega)+S_N(\omega)}\frac{d\omega}{2\pi} = \int_{-\infty}^{\infty} \frac{4}{8+5\omega^2}\frac{d\omega}{2\pi} = \frac{1}{\sqrt{10}}.$$

In an example later in this chapter we will return to the same random processes, but seek the best linear estimator of X_t given $(Y_s : s \le t)$.

9.2 The causal Wiener filtering problem

A linear system is causal if the value of the output at any given time does not depend on the future of the input. That is to say that the impulse response function satisfies $h(t,s) = 0$ for $s > t$. In the case of a linear, time-invariant system, causality means that the impulse response function satisfies $h(\tau) = 0$ for $\tau < 0$. Suppose X and Y are mean zero and jointly WSS. In this section we will consider estimates of X given Y obtained by passing Y through a causal linear time-invariant system. For convenience in applications, a fixed parameter T is introduced. Let $\widehat{X}_{t+T|t}$ be the minimum mean square error linear estimate of X_{t+T} given $(Y_s : s \le t)$. Note that if Y is the same process as X and $T > 0$, then we are addressing the problem of predicting X_{t+T} from $(X_s : s \le t)$.

An estimator of the form $\int_{-\infty}^{\infty} h(t-s)Y_s ds$ is sought such that h corresponds to a causal system. Once again, the orthogonality principle implies that the estimator is optimal if and only if it satisfies

$$X_{t+T} - \int_{-\infty}^{\infty} h(t-s)Y_s ds \perp Y_u \quad \text{for } u \le t,$$

which is equivalent to the condition

$$R_{XY}(t+T-u) = \int_{-\infty}^{\infty} h(t-s)R_Y(s-u)ds \quad \text{for } u \le t$$

or $R_{XY}(t+T-u) = h * R_Y(t-u)$. Setting $\tau = t-u$ and combining this optimality condition with the constraint that h is a causal function, the problem is to find an impulse response function h satisfying:

$$R_{XY}(\tau + T) = h * R_Y(\tau) \quad \text{for } \tau \geq 0, \tag{9.2}$$

$$h(v) = 0 \quad \text{for } v < 0. \tag{9.3}$$

Equations (9.2) and (9.3) are called the Wiener–Hopf equations. We shall show how to solve them in the case where the power spectral densities are rational functions by using the method of spectral factorization. The next section describes some of the tools needed for the solution.

9.3 Causal functions and spectral factorization

A function h on $\mathbb{R}$ is said to be causal if $h(\tau) = 0$ for $\tau < 0$, and it is said to be anticausal if $h(\tau) = 0$ for $\tau > 0$. Any function h on $\mathbb{R}$ can be expressed as the sum of a causal function and an anticausal function as follows. Simply let $u(t) = I_{\{t \geq 0\}}$ and notice that $h(t)$ is the sum of the causal function $u(t)h(t)$ and the anticausal function $(1 - u(t))h(t)$. More compactly, we have the representation $h = uh + (1 - u)h$.

A transfer function H is said to be of *positive type* if the corresponding impulse response function h is causal, and H is said to be of *negative type* if the corresponding impulse response function is anticausal. Any transfer function can be written as the sum of a positive type transfer function and a negative type transfer function. Indeed, suppose H is the Fourier transform of an impulse response function h. Define $[H]_+$ to be the Fourier transform of uh and $[H]_-$ to be the Fourier transform of $(1-u)h$. Then $[H]_+$ is called the positive part of H and $[H]_-$ is called the negative part of H. The following properties hold:

- $H = [H]_+ + [H]_-$ (because $h = uh + (1 - u)h$),
- $[H]_+ = H$ if and only if H is positive type,
- $[H]_- = 0$ if and only if H is positive type,
- $[[H]_+]_- = 0$ for any H,
- $[[H]_+]_+ = [H]_+$ and $[[H]_-]_- = [H]_-$,
- $[H + G]_+ = [H]_+ + [G]_+$ and $[H + G]_- = [H]_- + [G]_-$.

Note that uh is the causal function that is closest to h in the L^2 norm. That is, uh is the projection of h onto the space of causal functions. Indeed, if k is any causal function, then

$$\begin{aligned}\int_{-\infty}^{\infty} |h(t) - k(t)|^2 dt &= \int_{-\infty}^{0} |h(t)|^2 dt + \int_{0}^{\infty} |h(t) - k(t)|^2 dt \\ &\geq \int_{-\infty}^{0} |h(t)|^2 dt,\end{aligned} \tag{9.4}$$

and equality holds in (9.4) if and only if $k = uh$ (except possibly on a set of measure zero). By Parseval's relation, it follows that $[H]_+$ is the positive type function that is closest to H in the L^2 norm. Equivalently, $[H]_+$ is the projection of H onto the space of positive type functions. Similarly, $[H]_-$ is the projection of H onto the space of negative type functions. Up to this point in this book, Fourier transforms have been defined for real values of ω only. However, for the purposes of factorization to be covered later, it is useful to consider the analytic continuation of the Fourier transforms to larger sets in $\mathbb{C}$. We use the same notation $H(\omega)$ for the function H defined for real values of ω only, and its continuation defined for complex ω. The following examples illustrate the use of the projections $[\ ]_+$ and $[\ ]_-$, and consideration of transforms for complex ω.

Example 9.3 Let $g(t) = e^{-\alpha|t|}$ for a constant $\alpha > 0$. The functions g, ug and $(1-u)g$ are pictured in Figure 9.2. The corresponding transforms are given by:

$$[G]_+(\omega) = \int_0^{\infty} e^{-\alpha t} e^{-j\omega t} dt = \frac{1}{j\omega + \alpha},$$

$$[G]_-(\omega) = \int_{-\infty}^{0} e^{\alpha t} e^{-j\omega t} dt = \frac{1}{-j\omega + \alpha},$$

$$G(\omega) = [G]_+(\omega) + [G]_-(\omega) = \frac{2\alpha}{\omega^2 + \alpha^2}.$$

Note that $[G]_+$ has a pole at $\omega = j\alpha$, so that the imaginary part of the pole of $[G]_+$ is positive. Equivalently, the pole of $[G]_+$ is in the upper half plane.

More generally, suppose that $G(\omega)$ has the representation

$$G(\omega) = \sum_{n=1}^{N_1} \frac{\gamma_n}{j\omega + \alpha_n} + \sum_{n=N_1+1}^{N} \frac{\gamma_n}{-j\omega + \alpha_n},$$

where $\mathrm{Re}(\alpha_n) > 0$ for all n. Then

$$[G]_+(\omega) = \sum_{n=1}^{N_1} \frac{\gamma_n}{j\omega + \alpha_n}, \qquad [G]_-(\omega) = \sum_{n=N_1+1}^{N} \frac{\gamma_n}{-j\omega + \alpha_n}.$$

Example 9.4 Let G be given by

$$G(\omega) = \frac{1 - \omega^2}{(j\omega + 1)(j\omega + 3)(j\omega - 2)}.$$

Note that G has only three simple poles. The numerator of G has no factors in common with the denominator, and the degree of the numerator is smaller than the degree of

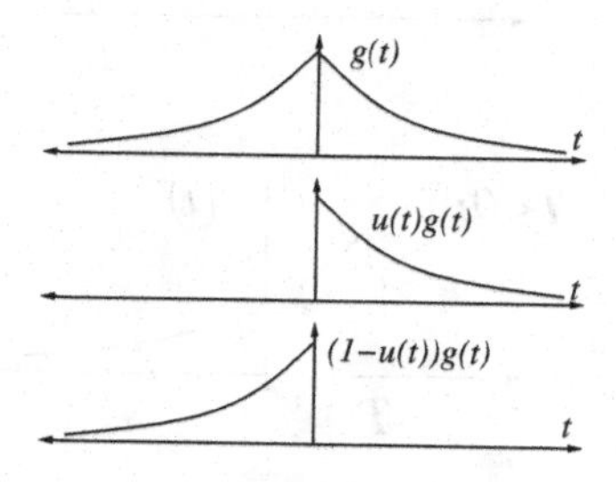

Figure 9.2 Decomposition of a two-sided exponential function.

the denominator. By the theory of partial fraction expansions in complex analysis, it therefore follows that G can be written as

$$G(\omega) = \frac{\gamma_1}{j\omega + 1} + \frac{\gamma_2}{j\omega + 3} + \frac{\gamma_3}{j\omega - 2}.$$

In order to identify γ_1, for example, multiply both expressions for G by $(j\omega + 1)$ and then let $j\omega = -1$. The other constants are found similarly. Thus

$$\gamma_1 = \left.\frac{1 - \omega^2}{(j\omega + 3)(j\omega - 2)}\right|_{j\omega=-1} = \frac{1 + (-1)^2}{(-1 + 3)(-1 - 2)} = -\frac{1}{3},$$

$$\gamma_2 = \left.\frac{1 - \omega^2}{(j\omega + 1)(j\omega - 2)}\right|_{j\omega=-3} = \frac{1 + 3^2}{(-3 + 1)(-3 - 2)} = 1,$$

$$\gamma_3 = \left.\frac{1 - \omega^2}{(j\omega + 1)(j\omega + 3)}\right|_{j\omega=2} = \frac{1 + 2^2}{(2 + 1)(2 + 3)} = \frac{1}{3}.$$

Consequently,

$$[G]_+(\omega) = -\frac{1}{3(j\omega + 1)} + \frac{1}{j\omega + 3} \quad \text{and} \quad [G]_-(\omega) = \frac{1}{3(j\omega - 2)}.$$

Example 9.5 Suppose that $G(\omega) = \frac{e^{-j\omega T}}{(j\omega+\alpha)}$. Multiplication by $e^{-j\omega T}$ in the frequency domain represents a shift by T in the time domain, so that

$$g(t) = \begin{cases} e^{-\alpha(t-T)} & t \geq T \\ 0 & t < T \end{cases},$$

as pictured in Figure 9.3. Consider two cases. First, if $T \geq 0$, then g is causal, G is positive type, and therefore $[G]_+ = G$ and $[G]_- = 0$. Second, if $T \leq 0$ then

$$g(t)u(t) = \begin{cases} e^{\alpha T} e^{-\alpha t} & t \geq 0 \\ 0 & t < 0 \end{cases},$$

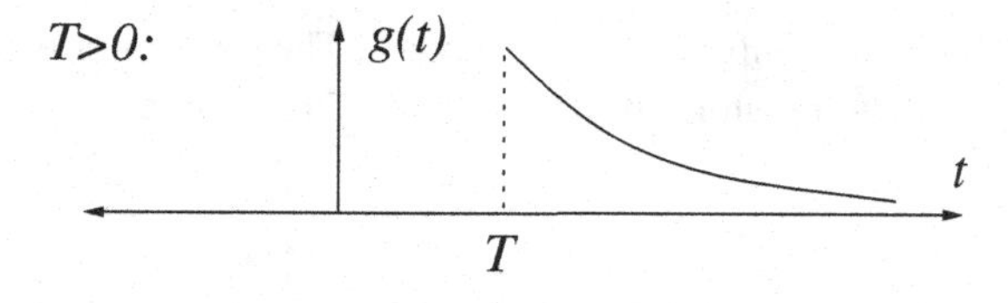

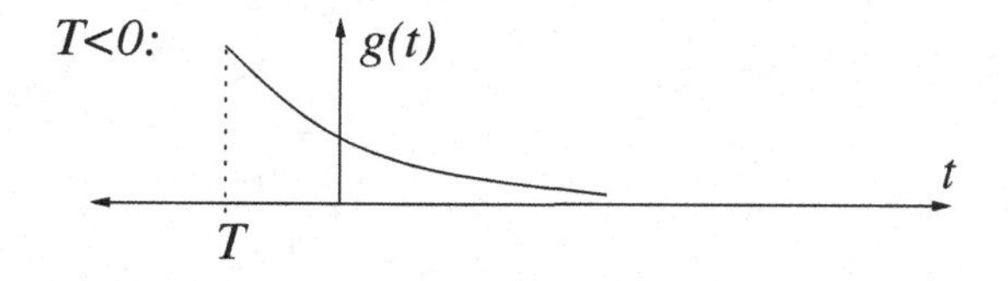

Figure 9.3 Exponential function shifted by T.

so that $[G]_+(\omega) = \frac{e^{\alpha T}}{j\omega+\alpha}$ and $[G]_-(\omega) = G(\omega) - [G]_+(\omega) = \frac{e^{-j\omega T}-e^{\alpha T}}{(j\omega+\alpha)}$. We can also find $[G]_-$ by computing the transform of $(1-u(t))g(t)$ (still assuming that $T \le 0$):

$$[G]_-(\omega) = \int_T^0 e^{\alpha(T-t)} e^{-j\omega t} dt = \left. \frac{e^{\alpha T-(\alpha+j\omega)t}}{-(\alpha+j\omega)} \right|_{t=T}^{0} = \frac{e^{-j\omega T} - e^{\alpha T}}{(j\omega+\alpha)}.$$

Example 9.6 Suppose H is the transfer function for impulse response function h. Let us unravel the notation and express

$$\int_{-\infty}^{\infty} \left| \left[e^{j\omega T} H(\omega) \right]_+ \right|^2 \frac{d\omega}{2\pi}$$

in terms of h and T. (Note that the factor $e^{j\omega T}$ is used, rather than $e^{-j\omega T}$ as in the previous example.) Multiplication by $e^{j\omega T}$ in the frequency domain corresponds to shifting by $-T$ in the time domain, so that

$$e^{j\omega T} H(\omega) \;\leftrightarrow\; h(t+T)$$

and thus

$$\left[e^{j\omega T} H(\omega) \right]_+ \;\leftrightarrow\; u(t) h(t+T).$$

Applying Parseval's identity, the definition of u, and a change of variables yields

$$\begin{aligned} \int_{-\infty}^{\infty} \left| \left[e^{j\omega T} H(\omega) \right]_+ \right|^2 \frac{d\omega}{2\pi} &= \int_{-\infty}^{\infty} |u(t) h(t+T)|^2 dt \\ &= \int_0^{\infty} |h(t+T)|^2 dt \\ &= \int_T^{\infty} |h(t)|^2 dt. \end{aligned}$$

The integral decreases from the energy of h to zero as T ranges from $-\infty$ to ∞.

Example 9.7 Suppose $[H]_- = [K]_- = 0$. Let us find $[HK]_-$. As usual, let h denote the inverse transform of H, and k denote the inverse transform of K. The supposition implies that h and k are both causal functions. Therefore the convolution $h * k$ is also a causal function. Since HK is the transform of $h * k$, it follows that HK is a positive type function. Equivalently, $[HK]_- = 0$.

The decomposition $H = [H]_+ + [H]_-$ is an additive one. Next we turn to multiplicative decomposition, concentrating on rational functions. A function H is said to be rational if it can be written as the ratio of two polynomials. Since polynomials can be factored over the complex numbers, a rational function H can be expressed in the form

$$H(\omega) = \gamma \frac{(j\omega+\beta_1)(j\omega+\beta_2)\ldots(j\omega+\beta_K)}{(j\omega+\alpha_1)(j\omega+\alpha_2)\ldots(j\omega+\alpha_N)}$$

for complex constants $\gamma, \alpha_1, \ldots, \alpha_N, \beta_1, \ldots, \beta_K$. Without loss of generality, we assume that $\{\alpha_i\} \cap \{\beta_j\} = \emptyset$. We also assume that the real parts of the constants $\alpha_1, \ldots, \alpha_N, \beta_1, \ldots, \beta_K$ are nonzero. The function H is positive type if and only if $\mathrm{Re}(\alpha_i) > 0$ for all i, or equivalently, if and only if all the poles of $H(\omega)$ are in the upper half plane $\mathrm{Im}(\omega) > 0$.

A positive type function H is said to have *minimum phase* if $\mathrm{Re}(\beta_i) > 0$ for all i. Thus a positive type function H is minimum phase if and only if $1/H$ is also positive type.

Suppose that S_Y is the power spectral density of a WSS random process and that S_Y is a rational function. The function S_Y, being nonnegative, is also real-valued, so $S_Y = S_Y^*$. Thus, if the denominator of S_Y has a factor of the form $j\omega+\alpha$ then the denominator must also have a factor of the form $-j\omega + \alpha^*$. Similarly, if the numerator of S_Y has a factor of the form $j\omega + \beta$ then the numerator must also have a factor of the form $-j\omega + \beta^*$.

Example 9.8 The function S_Y given by

$$S_Y(\omega) = \frac{8 + 5\omega^2}{(1 + \omega^2)(4 + \omega^2)}$$

can be factorized as

$$S_Y(\omega) = \underbrace{\sqrt{5}\frac{(j\omega + \sqrt{\frac{8}{5}})}{(j\omega + 2)(j\omega + 1)}}_{S_Y^+(\omega)} \underbrace{\sqrt{5}\frac{(-j\omega + \sqrt{\frac{8}{5}})}{(-j\omega + 2)(-j\omega + 1)}}_{S_Y^-(\omega)}, \tag{9.5}$$

where S_Y^+ is a positive type, minimum phase function and S_Y^- is a negative type function with $S_Y^- = (S_Y^+)^*$.

Note that the operators $[\ ]_+$ and $[\ ]_-$ give us an additive decomposition of a function H into the sum of a positive type and a negative type function, whereas spectral factorization has to do with products. At least formally, the factorization can be accomplished by taking a logarithm, doing an additive decomposition, and then exponentiating:

$$S_X(\omega) = \underbrace{\exp([\ln S_X(\omega)]_+)}_{S_X^+(\omega)} \underbrace{\exp([\ln S_X(\omega)]_-)}_{S_X^-(\omega)}. \tag{9.6}$$

Note that if $h \leftrightarrow H$ then, formally,

$$1 + h + \frac{h * h}{2!} + \frac{h * h * h}{3!} \cdots \leftrightarrow \exp(H) = 1 + H + \frac{H^2}{2!} + \frac{H^2}{3!} \cdots$$

so that if H is positive type, then $\exp(H)$ is also positive type. Thus, the factor S_X^+ in (9.6) is indeed a positive type function, and the factor S_X^- is a negative type function. Use of (9.6) is called the cepstrum method. Unfortunately, there is a host of problems, both numerical and analytical, in using the method, so that it will not be used further in this book.

9.4 Solution of the causal Wiener filtering problem for rational power spectral densities

In this section we show how to solve the causal Wiener filtering problem in the case where the spectral densities are rational functions. The Wiener–Hopf equations (9.2) and (9.3) can be formulated in the frequency domain as follows: find a positive type transfer function H such that

$$\left[e^{j\omega T}S_{XY} - HS_Y\right]_+ = 0. \tag{9.7}$$

Suppose S_Y is factored as $S_Y = S_Y^+ S_Y^-$ such that S_Y^+ is a minimum phase, positive type transfer function and $S_Y^- = (S_Y^+)^*$. Then S_Y^- and $\frac{1}{S_Y^-}$ are negative type functions. Since the product of two negative type functions is again negative type, (9.7) is equivalent to the equation obtained by multiplying the quantity within square brackets in (9.7) by $\frac{1}{S_Y^-}$, yielding the equivalent problem: find a positive type transfer function H such that

$$\left[\frac{e^{j\omega T}S_{XY}}{S_Y^-} - HS_Y^+\right]_+ = 0. \tag{9.8}$$

The function HS_Y^+, being the product of two positive type functions, is itself positive type. Thus (9.8) becomes

$$\left[\frac{e^{j\omega T}S_{XY}}{S_Y^-}\right]_+ - HS_Y^+ = 0.$$

Solving for H yields that the optimal transfer function is given by

$$H = \frac{1}{S_Y^+}\left[\frac{e^{j\omega T}S_{XY}}{S_Y^-}\right]_+. \tag{9.9}$$

The orthogonality principle yields that the mean square error satisfies

$$\begin{aligned} E[|X_{t+T} - \widehat{X}_{t+T|t}|^2] &= E[|X_{t+T}|^2] - E[|\widehat{X}_{t+T|t}|^2] \\ &= R_X(0) - \int_{-\infty}^{\infty} |H(\omega)|^2 S_Y(\omega)\frac{d\omega}{2\pi} \\ &= R_X(0) - \int_{-\infty}^{\infty} \left|\left[\frac{e^{j\omega T}S_{XY}}{S_Y^-}\right]_+\right|^2 \frac{d\omega}{2\pi}, \end{aligned} \tag{9.10}$$

where we use the fact that $|S_Y^+|^2 = S_Y$.

Another expression for the MMSE, which involves the optimal filter h, is the following:

$$\begin{aligned} \text{MMSE} &= E[(X_{t+T} - \widehat{X}_{t+T|t})(X_{t+T} - \widehat{X}_{t+T|t})^*] \\ &= E[(X_{t+T} - \widehat{X}_{t+T|t})X_{t+T}^*] = R_X(0) - R_{\widehat{X}X}(t, t+T) \\ &= R_X(0) - \int_{-\infty}^{\infty} h(s)R_{XY}^*(s+T)ds. \end{aligned}$$

Exercise Evaluate the limit as $T \to -\infty$ and the limit as $T \to \infty$ in (9.10).

Example 9.9 This example involves the same model as in Example 9.2, but here a causal estimator is sought. The observed random process is $Y = X + N$, where X is WSS with mean zero and power spectral density $S_X(\omega) = \frac{1}{1+\omega^2}$, N is WSS with mean zero and power spectral density $S_N(\omega) = \frac{4}{4+\omega^2}$, and $S_{XN} = 0$. We seek the optimal causal linear estimator of X_t given $(Y_s : s \le t)$. The power spectral density of Y is given by

$$S_Y(\omega) = S_X(\omega) + S_N(\omega) = \frac{8 + 5\omega^2}{(1+\omega^2)(4+\omega^2)}$$

and its spectral factorization is given by (9.5), yielding S_Y^+ and S_Y^-. Since $R_{XN} = 0$ it follows that

$$S_{XY}(\omega) = S_X(\omega) = \frac{1}{(j\omega+1)(-j\omega+1)}.$$

Therefore

$$\frac{S_{XY}(\omega)}{S_Y^-(\omega)} = \frac{(-j\omega+2)}{\sqrt{5}(j\omega+1)(-j\omega+\sqrt{\frac{8}{5}})} = \frac{\gamma_1}{j\omega+1} + \frac{\gamma_2}{-j\omega+\sqrt{\frac{8}{5}}},$$

where

$$\gamma_1 = \left.\frac{-j\omega+2}{\sqrt{5}(-j\omega+\sqrt{\frac{8}{5}})}\right|_{j\omega=-1} = \frac{3}{\sqrt{5}+\sqrt{8}},$$

$$\gamma_2 = \left.\frac{-j\omega+2}{\sqrt{5}(j\omega+1)}\right|_{j\omega=\sqrt{\frac{8}{5}}} = \frac{-\sqrt{\frac{8}{5}}+2}{\sqrt{5}+\sqrt{8}}.$$

Therefore

$$\left[\frac{S_{XY}(\omega)}{S_Y^-(\omega)}\right]_+ = \frac{\gamma_1}{j\omega+1} \tag{9.11}$$

and thus

$$H(\omega) = \frac{\gamma_1(j\omega+2)}{\sqrt{5}(j\omega+\sqrt{\frac{8}{5}})} = \frac{3}{5+2\sqrt{10}}\left(1 + \frac{2-\sqrt{\frac{8}{5}}}{j\omega+\sqrt{\frac{8}{5}}}\right),$$

so that the optimal causal filter is

$$h(t) = \frac{3}{5+2\sqrt{10}}\left(\delta(t) + (2-\sqrt{\frac{8}{5}})u(t)e^{-t\sqrt{\frac{8}{5}}}\right).$$

Finally, by (9.10) with $T = 0$, (9.11), and (9.1), the minimum mean square error is given by

$$E[|X_t - \widehat{X}_t|^2] = R_X(0) - \int_{-\infty}^{\infty} \frac{\gamma_1^2}{1+\omega^2}\frac{d\omega}{2\pi} = \frac{1}{2} - \frac{\gamma_1^2}{2} \approx 0.3246,$$

which is slightly larger than $\frac{1}{\sqrt{10}} \approx 0.3162$, the MMSE found for the best noncausal estimator in Example 9.2, and slightly smaller than $\frac{1}{3}$, the MMSE for the best "instantaneous" estimator of X_t given Y_t, which is $\frac{X_t}{3}$.

Example 9.10 A special case of the causal filtering problem formulated above is when the observed process Y is equal to X itself. This leads to the pure prediction problem. Let X be a WSS mean zero random process and let $T > 0$. Then the optimal linear predictor of X_{t+T} given $(X_s : s \leq t)$ corresponds to a linear time-invariant system with transfer function H given by (because $S_{XY} = S_X$, $S_Y = S_X$, $S_Y^+ = S_X^+$, and $S_Y^- = S_X^-$):

$$H = \frac{1}{S_X^+}\left[S_X^+ e^{j\omega T}\right]_+. \tag{9.12}$$

To be more specific, suppose $S_X(\omega) = \frac{1}{\omega^4+4}$. Observe that $\omega^4 + 4 = (\omega^2 + 2j)(\omega^2 - 2j)$. Since $2j = (1+j)^2$, we have that $(\omega^2 + 2j) = (\omega + 1 + j)(\omega - 1 - j)$. Factoring the term $(\omega^2 - 2j)$ in a similar way, and rearranging terms as needed, yields that the factorization of S_X is given by

$$S_X(\omega) = \underbrace{\frac{1}{(j\omega + (1+j))(j\omega + (1-j))}}_{S_X^+(\omega)} \underbrace{\frac{1}{(-j\omega + (1+j))(-j\omega + (1-j))}}_{S_X^-(\omega)},$$

so that

$$\begin{aligned} S_X^+(\omega) &= \frac{1}{(j\omega + (1+j))(j\omega + (1-j))} \\ &= \frac{\gamma_1}{j\omega + (1+j)} + \frac{\gamma_2}{j\omega + (1-j)}, \end{aligned}$$

where

$$\gamma_1 = \left.\frac{1}{j\omega + (1-j)}\right|_{j\omega = -(1+j)} = \frac{j}{2},$$

$$\gamma_2 = \left.\frac{1}{j\omega + (1+j)}\right|_{j\omega = -1+j} = \frac{-j}{2},$$

yielding that the inverse Fourier transform of S_X^+ is given by

$$S_X^+ \leftrightarrow \frac{j}{2}e^{-(1+j)t}u(t) - \frac{j}{2}e^{-(1-j)t}u(t).$$

Hence

$$S_X^+(\omega)e^{j\omega T} \leftrightarrow \begin{cases} \frac{j}{2}e^{-(1+j)(t+T)} - \frac{j}{2}e^{-(1-j)(t+T)} & t \geq -T \\ 0 & \text{else} \end{cases}$$

so that

$$\left[S_X^+(\omega)e^{j\omega T}\right]_+ = \frac{je^{-(1+j)T}}{2(j\omega + (1+j))} - \frac{je^{-(1-j)T}}{2(j\omega + (1-j))}.$$

The formula (9.12) for the optimal transfer function yields

$$\begin{aligned} H(\omega) &= \frac{je^{-(1+j)T}(j\omega + (1-j))}{2} - \frac{je^{-(1-j)T}(j\omega + (1+j))}{2} \\ &= e^{-T}\left[\frac{e^{jT}(1+j) - e^{-jT}(1-j)}{2j} + \frac{j\omega(e^{jT} - e^{-jT})}{2j}\right] \\ &= e^{-T}\left[\cos(T) + \sin(T) + j\omega\sin(T)\right], \end{aligned}$$

so that the optimal predictor for this example is given by

$$\widehat{X}_{t+T|t} = X_t e^{-T}(\cos(T) + \sin(T)) + X_t' e^{-T}\sin(T).$$

9.5 Discrete-time Wiener filtering

Causal Wiener filtering for discrete-time random processes can be handled in much the same way that it is handled for continuous-time random processes. An alternative approach can be based on the use of whitening filters and linear innovations sequences. Both of these approaches are discussed in this section, but first the topic of spectral factorization for discrete-time processes is discussed.

Spectral factorization for discrete-time processes naturally involves z-transforms. The z-transform of a function $(h_k : k \in \mathbb{Z})$ is given by

$$\mathcal{H}(z) = \sum_{k=-\infty}^{\infty} h(k)z^{-k}$$

for $z \in \mathbb{C}$. Setting $z = e^{j\omega}$ yields the Fourier transform: $H(\omega) = \mathcal{H}(e^{j\omega})$ for $0 \le \omega \le 2\pi$. Thus, the z-transform $\mathcal{H}$ restricted to the unit circle in $\mathbb{C}$ is equivalent to the Fourier transform H on $[0, 2\pi]$, and $\mathcal{H}(z)$ for other $z \in \mathbb{C}$ is an analytic continuation of its values on the unit circle.

Let $\widetilde{h}(k) = h^*(-k)$ as before. Then the z-transform of $\widetilde{h}$ is related to the z-transform $\mathcal{H}$ of h as follows:

$$\begin{aligned} \sum_{k=-\infty}^{\infty} \widetilde{h}(k)z^{-k} &= \sum_{k=-\infty}^{\infty} h^*(-k)z^{-k} = \sum_{l=-\infty}^{\infty} h^*(l)z^{l} \\ &= \left(\sum_{l=-\infty}^{\infty} h(l)(1/z^*)^{-l}\right)^* = \mathcal{H}^*(1/z^*). \end{aligned}$$

The impulse response function h is called causal if $h(k) = 0$ for $k < 0$, and the z-transform $\mathcal{H}$ is said to be positive type if h is causal. Note that if $\mathcal{H}$ is positive type, then $\lim_{|z|\to\infty} \mathcal{H}(z) = h(0)$. The projection $[\mathcal{H}]_+$ is defined as it was for Fourier transforms – it is the z transform of the function $u(k)h(k)$, where $u(k) = I_{\{k\ge 0\}}$. (We will not need to define or use $[\ ]_-$ for discrete-time functions.)

If X is a discrete-time WSS random process with correlation function R_X, then the z-transform of R_X is denoted by $\mathcal{S}_X$. Similarly, if X and Y are jointly WSS then the z-transform of R_{XY} is denoted by $\mathcal{S}_{XY}$. Recall that if Y is the output random process when X is passed through a linear time-invariant system with impulse response function h, then X and Y are jointly WSS and

$$R_{YX} = h * R_X, \quad R_{XY} = \widetilde{h} * R_X, \quad R_Y = h * \widetilde{h} * R_X,$$

which in the z-transform domain becomes:

$$\mathcal{S}_{YX}(z) = \mathcal{H}(z)\mathcal{S}_X(z), \quad \mathcal{S}_{XY}(z) = \mathcal{H}^*(1/z^*)\mathcal{S}_X(z), \quad \mathcal{S}_Y(z) = \mathcal{H}(z)\mathcal{H}^*(1/z^*)\mathcal{S}_X(z).$$

Example 9.11 Suppose Y is the output when white noise W with $R_W(k) = I_{\{k=0\}}$ is passed through a linear time-invariant system with impulse response function $h(k) = \rho^k I_{\{k\geq 0\}}$, where ρ is a complex constant with $|\rho| < 1$. Let us find $\mathcal{H}$, $\mathcal{S}_Y$, and R_Y. To begin,

$$\mathcal{H}(z) = \sum_{k=0}^{\infty}(\rho/z)^k = \frac{1}{1-\rho/z}$$

and the z-transform of $\widetilde{h}$ is $\frac{1}{1-\rho^* z}$. Note that the z-transform for h converges absolutely for $|z| > |\rho|$, whereas the z-transform for $\widetilde{h}$ converges absolutely for $|z| < 1/|\rho|$. Then

$$\mathcal{S}_Y(z) = \mathcal{H}(z)\mathcal{H}^*(1/z^*)\mathcal{S}_X(z) = \frac{1}{(1-\rho/z)(1-\rho^* z)}.$$

The autocorrelation function R_Y can be found using $R_Y = h * \widetilde{h} * R_W$ or by inverting the z-transform $\mathcal{S}_Y$. Taking the latter approach, factor out z and use the method of partial fraction expansion to obtain

$$\begin{aligned}
\mathcal{S}_Y(z) &= \frac{z}{(z-\rho)(1-\rho^* z)} \\
&= z\left(\frac{1}{(1-|\rho|^2)(z-\rho)} + \frac{1}{((1/\rho^*)-\rho)(1-\rho^* z)}\right) \\
&= \frac{1}{(1-|\rho|^2)}\left(\frac{1}{1-\rho/z} + \frac{z\rho^*}{1-\rho^* z}\right),
\end{aligned}$$

which is the z-transform of

$$R_Y(k) = \begin{cases} \frac{\rho^k}{1-|\rho|^2} & k \geq 0 \\ \frac{(\rho^*)^{-k}}{1-|\rho|^2} & k < 0 \end{cases}.$$

The z-transform $\mathcal{S}_Y$ of R_Y converges absolutely for $|\rho| < z < 1/|\rho|$.

Suppose $\mathcal{H}(z)$ is a rational function of z, meaning it is a ratio of two polynomials of z with complex coefficients. We assume that the numerator and denominator have no zeros in common, and that neither has a root on the unit circle. The function $\mathcal{H}$ is positive type (the z-transform of a causal function) if its poles (the zeros of its denominator polynomial) are inside the unit circle in the complex plane. If $\mathcal{H}$ is positive type and if its zeros are also inside the unit circle, then h and $\mathcal{H}$ are said to be minimum phase functions (in the time domain and z-transform domain, respectively). A positive-type minimum phase function $\mathcal{H}$ has the property that both $\mathcal{H}$ and its inverse $1/\mathcal{H}$ are causal functions. Two linear time-invariant systems in series, one with transfer function $\mathcal{H}$ and one with transfer function $1/\mathcal{H}$, passes all signals. Thus if $\mathcal{H}$ is positive type and minimum phase, we say that $\mathcal{H}$ is causal and causally invertible.

Assume that $\mathcal{S}_Y$ corresponds to a WSS random process Y and that $\mathcal{S}_Y$ is a rational function with no poles or zeros on the unit circle in the complex plane. We shall investigate the symmetries of $\mathcal{S}_Y$, with an eye towards its factorization. First,

$$R_Y = \widetilde{R}_Y \quad \text{so that} \quad \mathcal{S}_Y(z) = \mathcal{S}_Y^*(1/z^*). \tag{9.13}$$

Therefore, if z_0 is a pole of $\mathcal{S}_Y$ with $z_0 \neq 0$, then $1/z_0^*$ is also a pole. Similarly, if z_0 is a zero of $\mathcal{S}_Y$ with $z_0 \neq 0$, then $1/z_0^*$ is also a zero of $\mathcal{S}_Y$. These observations imply that $\mathcal{S}_Y$ can be uniquely factored as

$$\mathcal{S}_Y(z) = \mathcal{S}_Y^+(z)\mathcal{S}_Y^-(z)$$

such that for some constant $\beta > 0$:

- $\mathcal{S}_Y^+$ is a minimum phase, positive type z-transform,
- $\mathcal{S}_Y^-(z) = (\mathcal{S}_Y^+(1/z^*))^*$,
- $\lim_{|z|\to\infty} \mathcal{S}_Y^+(z) = \beta$.

There is an additional symmetry if R_Y is real valued:

$$\mathcal{S}_Y(z) = \sum_{k=-\infty}^{\infty} R_Y(k)z^{-k} = \sum_{k=-\infty}^{\infty} (R_Y(k)(z^*)^{-k})^* = \mathcal{S}_Y^*(z^*) \quad \text{(for real } R_Y). \tag{9.14}$$

Therefore, if R_Y is real and if z_0 is a nonzero pole of $\mathcal{S}_Y$, then z_0^* is also a pole. Combining (9.13) and (9.14) yields that if R_Y is real then the real-valued nonzero poles of $\mathcal{S}_Y$ come in pairs: z_0 and $1/z_0$, and the other nonzero poles of $\mathcal{S}_Y$ come in quadruples: $z_0, z_0^*, 1/z_0$ and $1/z_0^*$. A similar statement concerning the zeros of $\mathcal{S}_Y$ also holds true. Some example factorizations are as follows (where $|\rho| < 1$ and $\beta > 0$):

$$\mathcal{S}_Y(z) = \underbrace{\frac{\beta}{1-\rho/z}}_{\mathcal{S}_Y^+(z)}\ \underbrace{\frac{\beta}{1-\rho^* z}}_{\mathcal{S}_Y^-(z)},$$

$$\mathcal{S}_Y(z) = \underbrace{\frac{\beta(1-.8/z)}{(1-.6/z)(1-.7/z)}}_{\mathcal{S}_Y^+(z)}\ \underbrace{\frac{\beta(1-.8z)}{(1-.6z)(1-.7z)}}_{\mathcal{S}_Y^-(z)},$$

$$\mathcal{S}_Y(z) = \underbrace{\frac{\beta}{(1-\rho/z)(1-\rho^*/z)}}_{\mathcal{S}_Y^+(z)}\ \underbrace{\frac{\beta}{(1-\rho z)(1-\rho^* z)}}_{\mathcal{S}_Y^-(z)}.$$

An important application of spectral factorization is the generation of a discrete-time WSS random process with a specified correlation function R_Y. The idea is to start with a discrete-time white noise process W with $R_W(k) = I_{\{k=0\}}$, or equivalently, $\mathcal{S}_W(z) \equiv 1$, and then pass it through an appropriate linear, time-invariant system. The appropriate filter is given by taking $\mathcal{H}(z) = \mathcal{S}_Y^+(z)$, for then the spectral density of the output is indeed given by

$$\mathcal{H}(z)\mathcal{H}^*(1/z^*)\mathcal{S}_W(z) = \mathcal{S}_Y^+(z)\mathcal{S}_Y^-(z) = \mathcal{S}_Y(z).$$

The spectral factorization can be used to solve the causal filtering problem in discrete time. Arguing just as in the continuous time case, we find that if X and Y are jointly WSS random processes, then the best estimator of X_{n+T} given $(Y_k : k \le n)$ having the form

$$\widehat{X}_{n+T|n} = \sum_{k=-\infty}^{\infty} Y_k h(n-k)$$

for a causal function h is the function h satisfying the Wiener–Hopf equations (9.2) and (9.3), and the z transform of the optimal h is given by

$$\mathcal{H} = \frac{1}{\mathcal{S}_Y^+}\left[\frac{z^T\mathcal{S}_{XY}}{\mathcal{S}_Y^-}\right]_+. \tag{9.15}$$

Finally, an alternative derivation of (9.15) is given, based on the use of a whitening filter. The idea is the same as the idea of linear innovations sequence considered in Chapter 3. The first step is to note that the causal estimation problem is particularly simple if the observed process is white noise. Indeed, if the observed process Y is white noise with $R_Y(k) = I_{\{k=0\}}$ then for each $k \ge 0$ the choice of $h(k)$ is simply made to minimize the mean square error when X_{n+T} is estimated by the single term $h(k)Y_{n-k}$. This gives $h(k) = R_{XY}(T+k)I_{\{k\ge 0\}}$. Another way to get the same result is to solve the Wiener–Hopf equations (9.2) and (9.3) in discrete time where $R_Y(k) = I_{\{k=0\}}$. In general, of course, the observed process Y is not white, but the idea is to replace Y by an equivalent observed process Z that is white.

Let Z be the result of passing Y through a filter with transfer function $\mathcal{G}(z) = 1/S^+(z)$. Since $S^+(z)$ is a minimum phase function, $\mathcal{G}$ is a positive type function and the system is causal. Thus, any random variable in the m.s. closure of the linear span of $(Z_k : k \le n)$ is also in the m.s. closure of the linear span of $(Y_k : k \le n)$. Conversely, since Y can be recovered from Z by passing Z through the causal linear time-invariant system with transfer function $S^+(z)$, any random variable in the m.s. closure of the linear span of $(Y_k : k \le n)$ is also in the m.s. closure of the linear span of $(Z_k : k \le n)$. Hence, the optimal causal linear estimator of X_{n+T} based on $(Y_k : k \le n)$ is equal to the optimal causal linear estimator of X_{n+T} based on $(Z_k : k \le n)$. By the previous paragraph, such an estimator is obtained by passing Z through the linear time-invariant system with impulse response function $R_{XZ}(T+k)I_{\{k\ge 0\}}$, which has z transform $[z^T\mathcal{S}_{XZ}]_+$, see Figure 9.4.

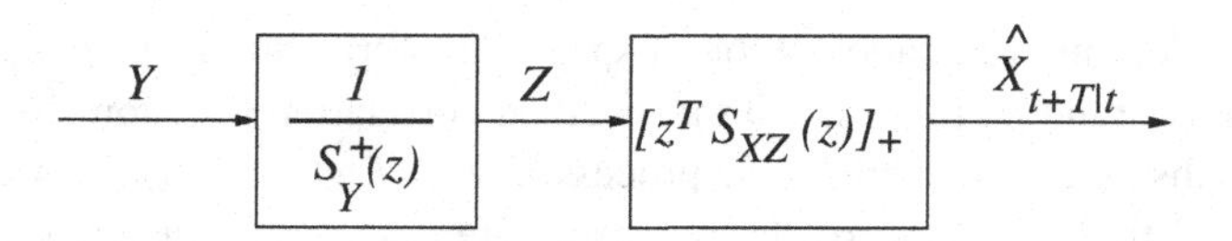

Figure 9.4 Optimal filtering based on whitening first.

The transfer function for two linear, time-invariant systems in series is the product of their z-transforms. In addition,

$$\mathcal{S}_{XZ}(z) = \mathcal{G}^*(1/z^*)\mathcal{S}_{XY}(z) = \frac{\mathcal{S}_{XY}(z)}{\mathcal{S}_Y^-(z)}.$$

Hence, the series system shown in Figure 9.4 is indeed equivalent to passing Y through the linear time-invariant system with $\mathcal{H}(z)$ given by (9.15).

Example 9.12 Suppose that X and N are discrete-time mean zero WSS random processes such that $R_{XN} = 0$. Suppose $\mathcal{S}_X(z) = \frac{1}{(1-\rho/z)(1-\rho z)}$ where $0 < \rho < 1$, and suppose that N is a discrete-time white noise with $\mathcal{S}_N(z) \equiv \sigma^2$ and $R_N(k) = \sigma^2 I_{\{k=0\}}$. Let the observed process Y be given by $Y = X+N$. Let us find the minimum mean square error linear estimator of X_n based on $(Y_k : k \leq n)$. We begin by factorizing $\mathcal{S}_Y$:

$$\begin{aligned}\mathcal{S}_Y(z) = \mathcal{S}_X(z) + \mathcal{S}_N(z) &= \frac{z}{(z-\rho)(1-\rho z)} + \sigma^2 \\ &= \frac{-\sigma^2\rho\left\{z^2 - (\frac{1+\rho^2}{\rho} + \frac{1}{\sigma^2\rho})z + 1\right\}}{(z-\rho)(1-\rho z)}.\end{aligned}$$

The quadratic expression in braces can be expressed as $(z - z_0)(z - 1/z_0)$, where z_0 is the smaller root of the expression in braces, yielding the factorization

$$\mathcal{S}_Y(z) = \underbrace{\frac{\beta(1-z_0/z)}{(1-\rho/z)}}_{\mathcal{S}_Y^+(z)} \underbrace{\frac{\beta(1-z_0 z)}{(1-\rho z)}}_{\mathcal{S}_Y^-(z)} \quad \text{where} \quad \beta^2 = \frac{\sigma^2\rho}{z_0}.$$

Using the fact $\mathcal{S}_{XY} = \mathcal{S}_X$, and appealing to a partial fraction expansion yields

$$\begin{aligned}\frac{\mathcal{S}_{XY}(z)}{\mathcal{S}_Y^-(z)} &= \frac{1}{\beta(1-\rho/z)(1-z_0 z)} \\ &= \frac{1}{\beta(1-\rho/z)(1-z_0\rho)} + \frac{z}{\beta((1/z_0)-\rho)(1-z_0 z)}.\end{aligned} \tag{9.16}$$

The first term in (9.16) is positive type, and the second is the z-transform of a function that is supported on the negative integers. Thus, the first term is equal to $\left[\frac{\mathcal{S}_{XY}}{\mathcal{S}_Y^-}\right]_+$. Finally, dividing by $\mathcal{S}_Y^+$ yields that the z-transform of the optimal filter is given by

$$\mathcal{H}(z) = \frac{1}{\beta^2(1-z_0\rho)(1-z_0/z)},$$

or in the time domain

$$h(n) = \frac{z_0^n I_{\{n \geq 0\}}}{\beta^2(1 - z_0\rho)}.$$

Problems

9.1 A quadratic predictor Suppose X is a mean zero, stationary discrete-time random process and that n is an integer with $n \geq 1$. Consider estimating X_{n+1} by a nonlinear one-step predictor of the form

$$\widehat{X}_{n+1} = h_0 + \sum_{k=1}^{n} h_1(k)X_k + \sum_{j=1}^{n}\sum_{k=1}^{j} h_2(j,k)X_jX_k.$$

(a) Find equations in term of the moments (second and higher, if needed) of X for the triple (h_0, h_1, h_2) to minimize the one-step prediction error: $E[(X_{n+1} - \widehat{X}_{n+1})^2]$.
(b) Explain how your answer to part (a) simplifies if X is a Gaussian random process.

9.2 A smoothing problem Suppose X and Y are mean zero, second order random processes in continuous time. Suppose the MMSE estimator of X_5 is to be found based on observation of $(Y_u : u \in [0,3] \cup [7,10])$. Assuming the estimator takes the form of an integral, derive the optimality conditions that must be satisfied by the kernel function (the function that Y is multiplied by before integrating). Use the orthogonality principle.

9.3 A simple prediction problem suppose X is a Gaussian stationary process with $R_X(\tau) = e^{-|\tau|}$ and mean zero. Suppose X_T is to be estimated given $(X_t : t \leq 0)$ where T is a fixed positive constant, and the mean square error is to be minimized. Without loss of generality, suppose the estimator has the form $\widehat{X}_T = \int_0^T g(t)X_t dt$ for some (possibly generalized) function g.
(a) Using the orthogonality principle, find equations that characterize g.
(b) Identify the solution g. (Hint: Does X have any special properties?)

9.4 A standard noncausal estimation problem (a) Derive the Fourier transform of the function $g(t) = \exp(-\alpha|t|)$.
(b) Find $\int_{-\infty}^{\infty} \frac{1}{a+b\omega^2}\frac{d\omega}{2\pi}$ for $a, b > 0$. (Hint: Use the result of part (a) and the fact, which follows from the inverse Fourier transform, that $\int_{-\infty}^{\infty} \widehat{g}(\omega)\frac{d\omega}{2\pi} = g(0) = 1$.)
(c) Suppose $Y = X + N$, where X and N are each WSS random processes with mean zero, and X and N are uncorrelated with each other. The observed process is $Y = X + N$. Suppose $R_X(\tau) = \exp(-\alpha|\tau|)$ and $R_N = \sigma^2\delta(\tau)$, so that N is a white noise process with two-sided power spectral density σ^2. Identify the transfer function H and impulse response function h of the filter for producing $\widehat{X}_t = \widehat{E}[X_t|Y]$, the MMSE estimator of X_t given $Y = (Y_s : s \in \mathbb{R})$.
(d) Find the resulting MMSE for the estimator you found in part (c). Check that the limits of your answer as $\sigma \to 0$ or $\sigma \to \infty$ make sense.
(e) Let $D_t = X_t - \widehat{X}_t$. Find the cross covariance function $C_{D,Y}$.

9.5 A simple, noncausal estimation problem Let $X = (X_t : t \in \mathbb{R})$ be a real-valued, stationary Gaussian process with mean zero and autocorrelation function

$R_X(t) = A^2\text{sinc}(f_o t)$, where A and f_o are positive constants. Let $N = (N_t : t \in \mathbb{R})$ be a real-valued Gaussian white noise process with $R_N(\tau) = \sigma^2\delta(\tau)$, which is independent of X. Define the random process $Y = (Y_t : t \in \mathbb{R})$ by $Y_t = X_t + N_t$. Let $\widehat{X}_t = \int_{-\infty}^{\infty} h(t-s)Y_s ds$, where the impulse response function h, which can be noncausal, is chosen to minimize $E[D_t^2]$ for each t, where $D_t = X_t - \widehat{X}_t$.
(a) Find h.
(b) Identify the probability distribution of D_t, for t fixed.
(c) Identify the conditional distribution of D_t given Y_t, for t fixed.
(d) Identify the autocorrelation function, R_D, of the error process D, and the cross correlation function, R_{DY}.

9.6 Interpolating a Gauss Markov process Let X be a real-valued, mean zero stationary Gaussian process with $R_X(\tau) = e^{-|\tau|}$. Let $a > 0$. Suppose thet X_0 is estimated by $\widehat{X}_0 = c_1 X_{-a} + c_2 X_a$ where the constants c_1 and c_2 are chosen to minimize the mean square error (MSE).
(a) Use the orthogonality principle to find c_1, c_2, and the resulting minimum MSE $E[(X_0 - \widehat{X}_0)^2]$. (Your answers should depend only on a.)
(b) Use the orthogonality principle again to show that $\widehat{X}_0$ as defined above is the minimum MSE estimator of X_0 given $(X_s : |s| \geq a)$. (This implies that X has a two-sided Markov property.)

9.7 Estimation of a filtered narrowband random process in noise Suppose X is a mean zero real-valued stationary Gaussian random process with the spectral density shown.

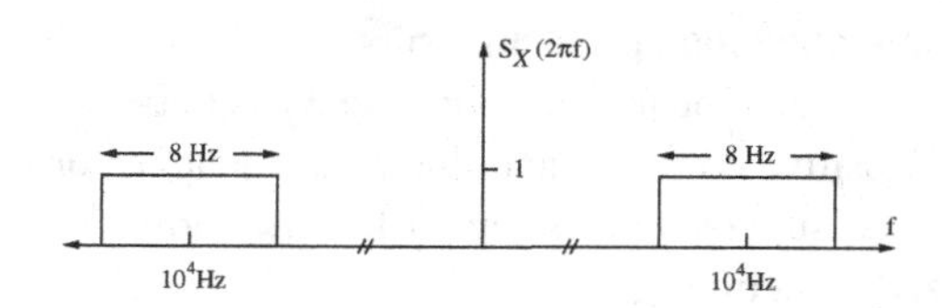

(a) Explain how X can be simulated on a computer using a pseudo-random number generator that generates standard normal random variables. Try to use the minimum number per unit time. How many normal random variables does your construction require per simulated unit time?
(b) Suppose X is passed through a linear time-invariant system with approximate transfer function $H(2\pi f) = 10^7/(10^7 + f^2)$. Find an approximate numerical value for the power of the output.
(c) Let $Z_t = X_t + W_t$ where W is a Gaussian white noise random process, independent of X, with $R_W(\tau) = \delta(\tau)$. Find h to minimize the mean square error $E[(X_t - \widehat{X}_t)^2]$, where $\widehat{X} = h * Z$.
(d) Find the mean square error for the estimator of part (c).

9.8 Proportional noise Suppose X and N are second order, mean zero random processes such that $R_{XN} \equiv 0$, and let $Y = X + N$. Suppose the correlation functions R_X and R_N are known, and that $R_N = \gamma^2 R_X$ for some nonnegative constant γ^2. Consider the problem of estimating X_t using a linear estimator based on $(Y_u : a \leq u \leq b)$, where a, b, and t are given times with $a < b$.

(a) Use the orthogonality principle to show that if $t \in [a,b]$, then the optimal estimator is given by $\widehat{X}_t = \kappa Y_t$ for some constant κ, and identify the constant κ and the corresponding MSE.
(b) Suppose in addition that X and N are WSS and that X_{t+T} is to be estimated from $(Y_s : s \le t)$. Show how the equation for the optimal causal filter reduces to your answer to part (a) when $T \le 0$.
(c) Continue under the assumptions of part (b), except consider $T > 0$. How is the optimal filter for estimating X_{t+T} from $(Y_s : s \le t)$ related to the problem of predicting X_{t+T} from $(X_s : s \le t)$?

9.9 Predicting the future of a simple WSS process Let X be a mean zero, WSS random process with power spectral density $S_X(\omega) = \frac{1}{\omega^4+13\omega^2+36}$.
(a) Find the positive type, minimum phase rational function S_X^+ such that $S_X(\omega) = |S_X^+(\omega)|^2$.
(b) Let T be a fixed known constant with $T \ge 0$. Find $\widehat{X}_{t+T|t}$, the MMSE linear estimator of X_{t+T} given $(X_s : s \le t)$. Be as explicit as possible. (Hint: Check that your answer is correct in the cases $T = 0$ and $T \to \infty$.)
(c) Find the MSE for the optimal estimator of part (b).

9.10 Short answer filtering questions (a) Prove or disprove: If H is a positive type function then so is H^2.
(b) Prove or disprove: Suppose X and Y are jointly WSS, mean zero random processes with continuous spectral densities such that $S_X(2\pi f) = 0$ unless $|f| \in$ [9012 MHz, 9015 MHz] and $S_Y(2\pi f) = 0$ unless $|f| \in$ [9022 MHz, 9025 MHz]. Then the best linear estimate of X_0 given $(Y_t : t \in \mathbb{R})$ is 0.
(c) Let $H(2\pi f) = \text{sinc}(f)$. Find $[H]_+$.

9.11 On the MSE for causal estimation Recall that if X and Y are jointly WSS and have power spectral densities, and if S_Y is rational with a spectral factorization, then the mean square error for linear estimation of X_{t+T} using $(Y_s : s \le t)$ is given by

$$(\text{MSE}) = R_X(0) - \int_{-\infty}^{\infty} \left| \left[\frac{e^{j\omega T} S_{XY}}{S_Y^-} \right]_+ \right|^2 \frac{d\omega}{2\pi}.$$

Evaluate and interpret the limits of this expression as $T \to -\infty$ and as $T \to \infty$.

9.12 A singular estimation problem Let $X_t = Ae^{j2\pi f_o t}$, where $f_o > 0$ and A is a mean zero complex valued random variable with $E[A^2] = 0$ and $E[|A|^2] = \sigma_A^2$. Let N be a white noise process with $R_N(\tau) = \sigma_N^2\delta(\tau)$. Let $Y_t = X_t + N_t$. Let $\widehat{X}$ denote the output process when Y is filtered using the impulse response function $h(\tau) = \alpha e^{-(\alpha - j2\pi f_o)t} I_{\{t\ge 0\}}$.
(a) Verify that X is a WSS periodic process, and find its power spectral density (the power spectral density only exists as a generalized function, i.e. there is a delta function in it).
(b) Give a simple expression for the output of the linear system when the input is X.
(c) Find the mean square error, $E[|X_t - \hat{X}_t|^2]$. How should the parameter α be chosen to approximately minimize the MSE?

9.13 Filtering a WSS signal plus noise Suppose X and N are jointly WSS, mean zero, continuous-time random processes with $R_{XN} \equiv 0$. The processes are the inputs to a system with the block diagram shown, for some transfer functions $K_1(\omega)$ and $K_2(\omega)$:

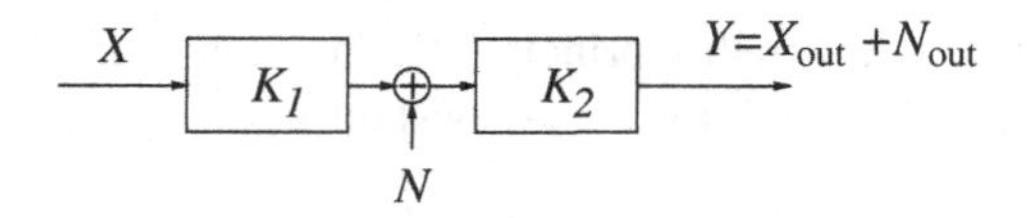

Suppose that for every value of ω, $K_i(\omega) \neq 0$ for $i = 1$ and $i = 2$. Because the two subsystems are linear, we can view the output process Y as the sum of two processes X_{out}, due to the input X, plus N_{out}, due to the input N. Your answers to the first four parts should be expressed in terms of K_1, K_2, and the power spectral densities S_X and S_N.

(a) What is the power spectral density S_Y?

(b) Find the signal-to-noise ratio at the output (the power of X_{out} divided by the power of N_{out}).

(c) Suppose Y is passed into a linear system with transfer function H, designed so that the output at time t is $\widehat{X}_t$, the best linear estimator of X_t given $(Y_s : s \in \mathbb{R})$. Find H.

(d) Find the resulting minimum mean square error.

(e) The correct answer to part (d) (the minimum MSE) does not depend on the filter K_2. Why?

9.14 A prediction problem Let X be a mean zero WSS random process with correlation function $R_X(\tau) = e^{-|\tau|}$. Using the Wiener filtering equations, find the optimal linear MMSE estimator (i.e. predictor) of X_{t+T} based on $(X_s : s \leq t)$ for a constant $T > 0$. Explain why your answer takes such a simple form.

9.15 Properties of a particular Gaussian process Let X be a stationary Gaussian continuous-time process with $\mu_X = 0$, $R_X(\tau) = (1 + |\tau|)e^{-|\tau|}$, and $S_X(\omega) = (2/(1+\omega^2))^2$. Answer the following questions, being sure to provide justification.

(a) Is X mean ergodic in the m.s. sense?

(b) Is X a Markov process?

(c) Is X differentiable in the m.s. sense?

(d) Find the causal, minimum phase filter h (or its transform H) such that if white noise with autocorrelation function $\delta(\tau)$ is filtered using h then the output autocorrelation function is R_X.

(e) Express X as the solution of a stochastic differential equation driven by white noise.

9.16 Spectral decomposition and factorization (a) Let x be the signal with Fourier transform given by $\widehat{x}(2\pi f) = \left[\text{sinc}(100f)e^{j2\pi fT}\right]_+$. Find the energy of x for all real values of the constant T.

(b) Find the spectral factorization of $S(\omega) = \frac{1}{\omega^4+16\omega^2+100}$. (Hint: $1+3j$ is a pole of S.)

9.17 A continuous-time Wiener filtering problem Suppose (X_t) and (N_t) are uncorrelated, mean zero processes with $R_X(t) = \exp(-2|t|)$ and $S_N(\omega) \equiv N_o/2$ for a positive constant N_o. Suppose that $Y_t = X_t + N_t$.

(a) Find the optimal (noncausal) filter for estimating X_t from the observations $(Y_s : -\infty < s < +\infty)$ and find the resulting mean square error. Comment on how the MMSE depends on N_o.

(b) Find the optimal causal filter with lead time T, that is, the Wiener filter for estimating X_{t+T} given $(Y_s : -\infty < s \leq t)$, and find the corresponding MMSE. For simplicity assume $T \geq 0$. Comment on the limiting value of the MMSE as $T \to \infty$, as $N_o \to \infty$, or as $N_o \to 0$.

9.18 Estimation of a random signal, using the KL expansion Suppose that X is a m.s. continuous, mean zero process over an interval $[a,b]$, and suppose N is a white noise process, with $R_{XN} \equiv 0$ and $R_N(s,t) = \sigma^2\delta(s-t)$. Let $(\varphi_k : k \geq 1)$ be a complete orthonormal basis for $L^2[a,b]$ consisting of eigenfunctions of R_X, and let $(\lambda_k : k \geq 1)$ denote the corresponding eigenvalues. Suppose that $Y = (Y_t : a \leq t \leq b)$ is observed.
(a) Fix an index i. Express the MMSE estimator of (X, φ_i) given Y in terms of the coordinates, $(Y, \varphi_1), (Y, \varphi_2), \ldots$ of Y, and find the corresponding mean square error.
(b) Now suppose f is a function in $L^2[a,b]$. Express the MMSE estimator of (X,f) given Y in terms of the coordinates $((f, \varphi_j) : j \geq 1)$ of f, the coordinates of Y, the λs, and σ. Also, find the mean square error.

9.19 Noiseless prediction of a baseband random process Fix positive constants T and ω_o, suppose $X = (X_t : t \in \mathbb{R})$ is a baseband random process with one-sided frequency limit ω_o, and let $H^{(n)}(\omega) = \sum_{k=0}^{n} \frac{(j\omega T)^k}{k!}$, which is a partial sum of the power series of $e^{j\omega T}$. Let $\widehat{X}^{(n)}_{t+T|t}$ denote the output at time t when X is passed through the linear time-invariant system with transfer function $H^{(n)}$. As the notation suggests, $\widehat{X}^{(n)}_{t+T|t}$ is an estimator (not necessarily optimal) of X_{t+T} given $(X_s : s \leq t)$.
(a) Describe $\widehat{X}^{(n)}_{t+T|t}$ in terms of X in the time domain. Verify that the linear system is causal.
(b) Show that $\lim_{n\to\infty} a_n = 0$, where $a_n = \max_{|\omega| \leq \omega_o} |e^{j\omega T} - H^{(n)}(\omega)|$. (This means that the power series converges uniformly for $\omega \in [-\omega_o, \omega_o]$.)
(c) Show that the mean square error can be made arbitrarily small by taking n sufficiently large. In other words, show that $\lim_{n\to\infty} E[|X_{t+T} - \widehat{X}^{(n)}_{t+T|t}|^2] = 0$.
(d) Thus, the future of a narrowband random process X can be predicted perfectly from its past. What is wrong with the following argument for general WSS processes? If X is an arbitrary WSS random process, we could first use a bank of (infinitely many) narrowband filters to split X into an equivalent set of narrowband random processes (call them "subprocesses") which sum to X. By the above, we can perfectly predict the future of each of the subprocesses from its past. So adding together the predictions would yield a perfect prediction of X from its past.

9.20 Linear innovations and spectral factorization Suppose X is a discrete-time WSS random process with mean zero. Suppose that the z-transform version of its power spectral density has the factorization: $\mathcal{S}_X(z) = \mathcal{S}_X^+(z)\mathcal{S}_X^-(z)$ such that $\mathcal{S}_X^+(z)$ is a minimum phase, positive type function, $\mathcal{S}_X^-(z) = (\mathcal{S}_X^+(1/z^*))^*$, and $\lim_{|z|\to\infty} \mathcal{S}_X^+(z) = \beta$ for some $\beta > 0$. The linear innovations sequence of X is the sequence $\widetilde{X}$ such that $\widetilde{X}_k = X_k - \widehat{X}_{k|k-1}$, where $\widehat{X}_{k|k-1}$ is the MMSE predictor of X_k given $(X_l : l \leq k-1)$. Note that there is no constant multiplying X_k in the definition of $\widetilde{X}_k$. You should use $\mathcal{S}_X^+(z)$, $\mathcal{S}_X^-(z)$, and/or β in giving your answers.
(a) Show that $\widetilde{X}$ can be obtained by passing X through a linear time-invariant filter, and identify the corresponding value of $\mathcal{H}$.
(b) Identify the mean square prediction error, $E[|X_k - \widehat{X}_{k|k-1}|^2]$.

9.21 A singular nonlinear estimation problem Suppose X is a standard Brownian motion with parameter $\sigma^2 = 1$ and suppose N is a Poisson random process with

rate $\lambda = 10$, which is independent of X. Let $Y = (Y_t : t \geq 0)$ be defined by $Y_t = X_t + N_t$.
(a) Find the optimal estimator of X_1 among the estimators that are linear functions of $(Y_t : 0 \leq t \leq 1)$ and the constants, and find the corresponding mean square error. Your estimator can include a constant plus a linear combination, or limits of linear combinations, of $Y_t : 0 \leq t \leq 1$. (Hint: There is a closely related problem elsewhere in this problem set.)
(b) Find the optimal possibly nonlinear estimator of X_1 given $(Y_t : 0 \leq t \leq 1)$, and find the corresponding mean square error. (Hint: No computation is needed. Draw sample paths of the processes.)

9.22 A discrete-time Wiener filtering problem Extend the discrete-time Wiener filtering problem considered in Section 9.5 to incorporate a lead time T. Assume T to be integer valued. Identify the optimal filter in both the z-transform domain and in the time domain. (Hint: Treat the case $T \leq 0$ separately. You need not identify the covariance of error.)

9.23 Causal estimation of a channel input process Let $X = (X_t : t \in \mathbb{R})$ and $N = (N_t : t \in \mathbb{R})$ denote WSS random processes with $R_X(\tau) = \frac{3}{2}e^{-|\tau|}$ and $R_N(\tau) = \delta(\tau)$. Think of X as an input signal and N as noise, and suppose X and N are orthogonal to each other. Let k denote the impulse response function given by $k(\tau) = 2e^{-3\tau}I_{\{\tau \geq 0\}}$, and suppose an output process Y is generated according to the block diagram shown:

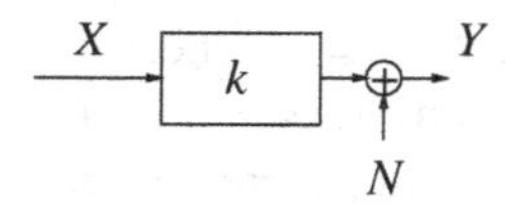

That is, $Y = X * k + N$. Suppose X_t is to be estimated by passing Y through a *causal* filter with impulse response function h, and transfer function H. Find the choice of H and h in order to minimize the mean square error.

9.24 Estimation given a strongly correlated process Suppose g and k are minimum phase causal functions in discrete time, with $g(0) = k(0) = 1$, and z-transforms $\mathcal{G}$ and $\mathcal{K}$. Let $W = (W_k : k \in \mathbb{Z})$ be a mean zero WSS process with $S_W(\omega) \equiv 1$, let $X_n = \sum_{i=-\infty}^{\infty} g(n-i)W_i$ and $Y_n = \sum_{i=-\infty}^{\infty} k(n-i)W_i$.
(a) Express R_X, R_Y, R_{XY}, $\mathcal{S}_X$, $\mathcal{S}_Y$, and $\mathcal{S}_{XY}$ in terms of g, k, $\mathcal{G}$, $\mathcal{K}$.
(b) Find h so that $\widehat{X}_{n|n} = \sum_{i=-\infty}^{\infty} Y_i h(n-i)$ is the MMSE linear estimator of X_n given $(Y_i : i \leq n)$.
(c) Find the resulting mean square error. Give an intuitive reason for your answer.

9.25 Estimation of a process with raised cosine spectrum Let $Y = X + N$, where X and N are independent, mean zero, WSS random processes with

$$S_X(\omega) = \frac{(1 + \cos(\frac{\pi\omega}{\omega_o}))}{2} I_{\{|\omega| \leq \omega_o\}} \quad \text{and} \quad S_N(\omega) = \frac{N_o}{2},$$

where $N_o > 0$ and $\omega_o > 0$.
(a) Find the transfer function H for the filter such that if the input process is Y, the output process at time t, $\widehat{X}_t$, is the optimal linear estimator of X_t based on $(Y_s : s \in \mathbb{R})$.
(b) Express the mean square error, $\sigma_e^2 = E[(\widehat{X}_t - X_t)^2]$, as an integral in the frequency domain. (You need not carry out the integration.)

(c) Describe the limits of your answers to (a) and (b) as $N_o \to 0$.
(c) Describe the limits of your answers to (a) and (b) as $N_o \to \infty$.

9.26 Linear and nonlinear filtering Let $Z = (Z_t : t \in \mathbb{R})$ be a stationary Markov process with state space $\mathcal{S} = \{3, 1, -1, -3\}$ and generator matrix $Q = (q_{i,j})$ with $q_{i,j} = \lambda$ if $i \neq j$ and $q_{i,i} = -3\lambda$, for $i,j \in \mathcal{S}$. Let $Y = (Y_t : t \in \mathbb{R})$ be a random process defined by $Y_t = Z_t + N_t$, where N is a white Gaussian noise process with $R_N(\tau) = \sigma^2\delta(\tau)$, for some $\sigma^2 > 0$.

(a) Find the stationary distribution π, the transition probabilities $p_{i,j}(\tau)$, the mean μ_Z, and autocorrelation function R_Z for Z.

(b) Find the transfer function H, so that if $\widehat{Z}$ is the output of the linear system with transfer function H, then $\widehat{Z}_t = \widehat{E}[Z_t|Y]$. Express the mean square error, $E[(Z_t - \widehat{Z}_t)^2]$ in terms of λ and σ^2.

(c) For t fixed, find a nonlinear function $\widehat{Z}_t^{(NL)}$ of Y such that $E[(Z_t - \widehat{Z}_t^{(NL)})^2]$ is strictly smaller than the MSE found in part (b). (You don't need to compute the MSE of your estimator.)

(d) Derive an estimation procedure using the fact that (Z, Y) is a continuous-time version of the hidden Markov model. Specifically, let $\Delta > 0$ be small and let $t_0 = K\Delta$ for some large integer K. Let $\widetilde{Y}_k = \int_{(k-1)\Delta}^{k\Delta} Y_t dt$ and $\widetilde{Z}_k = Z_{k\Delta}$. Then $(\widetilde{Z}_k, \widetilde{Y}_k : 1 \leq k \leq K)$ is approximately a hidden Markov model with observation space $\mathbb{R}$ instead of a finite observation space. Identify the (approximate) parameter (π, A, B) of this Markov model (note that $b_{i,y}$ for i fixed should be a pdf as a function of y). (Using this model, the forward-backward algorithm could be used to approximately compute the conditional pmf of X at a fixed time given Y, which becomes asymptotically exact as $\Delta \to 0$. An alternative to this approach is to simply start with a discrete-time model. Another alternative is to derive a continuous-time version of the forward-backward algorithm.)

9.27* Resolution of Wiener and Kalman filtering Consider the state and observation models:

$$X_n = FX_{n-1} + W_n,$$

$$Y_n = H^T X_n + V_n,$$

where $(W_n : -\infty < n < +\infty)$ and $(V_n : -\infty < n < +\infty)$ are independent vector-valued random sequences of independent, identically distributed mean zero random variables. Let Σ_W and Σ_V denote the respective covariance matrices of W_n and V_n. (F, H, and the covariance matrices must satisfy a stability condition. Can you find it?)

(a) What are the autocorrelation function R_X and crosscorrelation function R_{XY}?

(b) Use the orthogonality principle to derive conditions for the causal filter h that minimizes $E[\| X_{n+1} - \sum_{j=0}^{\infty} h(j)Y_{n-j} \|^2]$ (i.e. derive the basic equations for the Wiener-Hopf method).

(c) Write down and solve the equations for the Kalman predictor in steady state to derive an expression for h, and verify that it satisfies the orthogonality conditions.

10 Martingales

This chapter builds on the brief introduction to martingales given in Chapter 4, to give a glimpse of how martingales can be used to obtain bounds and prove convergence in many contexts, such as for estimation and control algorithms in a random environment. On one hand, the notion of a martingale is weak enough to include processes arising in applications involving estimation and control, and on the other hand, the notion is strong enough that important tools for handling sums of independent random variables, such as the law of large numbers, the central limit theorem, and large deviation estimates extend to martingales.

Two other topics in this book are closely related to martingales. The first is the use of linear innovations sequences discussed in Chapter 3. As explained in Example 10. below, martingale difference sequences arise as innovations sequences when the linearity constraint on predictors, imposed for linear innovations sequences, is dropped. The other topic in this book closely related to martingales is the Foster–Lyapunov theory for Markov processes, discussed in Chapter 6. A central feature of the Foster–Lyapunov theory is the drift of a function of a Markov process: $E[V(X_{t+1}) - V(X_t)|X_t = x]$. If this drift were zero then $V(X_t)$ would be a martingale. The assumptions used in the Foster-Lyapunov theory allow for a controlled difference from the martingale assumption. In a sense, martingale theory is what is left when the linearity and Markov assumptions are both dropped.

The chapter is organized as follows. The definition of a martingale involves conditional expectations, so to give the general definition of a martingale we first revisit the definition of conditional expectation in Section 10.1. The standard definition of martingales, in which σ-algebras are used to represent information, is given in Section 10.2. Section 10.3 explains how the Chernoff bound, central to large deviations theory, readily extends to sequences that are not independent. Section 10.4 discusses the use of stopping times together with martingales, for proving bounds for dynamical systems.

10.1 Conditional expectation revisited

The general definition of a martingale requires the general definition of conditional expectation. We begin by reviewing the definitions we have given so far. In Chapter 1 we reviewed the following elementary definition of $E[X|Y]$. If X and Y are both discrete random variables, and $\{u_1, u_2, \ldots\}$ denotes the set of possible values of X, then

$$E[X|Y=i] = \sum_j u_j P(X=u_j|Y=i),$$

which is well defined if $P\{Y=i\} > 0$ and either the sum restricted to $j : u_j > 0$ or to $j : u_j < 0$ is convergent. That is, $E[X|Y=i]$ is the mean of the conditional pmf of X given $Y=i$. Note that $g(i) = E[X|Y=i]$ is a function of i, and we let $E[X|Y]$ be the random variable defined by $E[X|Y] = g(Y)$. Similarly, if X and Y have a joint pdf, $E[X|Y=y] = \int x f_{X|Y}(x|y)dx = g(y)$ and $E[X|Y] = g(Y)$.

Chapter 3 shows $E[X|Y]$ can be defined whenever $E[X^2] < \infty$, even if X and Y are neither discrete random variables nor have a joint pdf. The definition is based on a projection, characterized by the orthogonality principle. Specifically, if $E[X^2] < \infty$, then $E[X|Y]$ is the unique random variable such that:

- it has the form $g(Y)$ for some (Borel measurable) g such that $E[(g(Y)^2] < \infty$, and
- $E[(X - E[X|Y])f(Y)] = 0$ for all (Borel measurable) functions f such that $E[(f(Y))^2] < \infty$.

That is, $E[X|Y]$ is an unconstrained estimator based on Y, such that the error $X - E[X|Y]$ is orthogonal to all functions of Y with finite second moments. By the orthogonality principle, $E[X|Y]$ exists and is unique, if differences on a set of probability zero are ignored. This second definition of $E[X|Y]$ is more general than the elementary definition, because it doesn't require X and Y to be discrete or to have a joint pdf, but it is less general because it requires $E[X^2] < \infty$.

The definition of $E[X|Y]$ given next generalizes the previously given definition in two ways. First, the definition applies as long as $E[|X|] < \infty$, which is a weaker requirement than $E[X^2] < \infty$. Second, the definition is based on having information represented by a σ-algebra, rather than by a random variable. Recall that, by definition, a σ-algebra $\mathcal{D}$ for a set Ω is a set of subsets of Ω such that:

(a) $\Omega \in \mathcal{D}$,

(b) if $A \in \mathcal{D}$ then $A^c \in \mathcal{D}$,

(c) if $A, B \in \mathcal{D}$ then $A \cup B \in \mathcal{D}$, and more generally, if $A_1, A_2, \ldots$ is such that $A_i \in \mathcal{D}$ for $i \geq 1$, then $\cup_{i=1}^{\infty} A_i \in \mathcal{D}$.

In particular, the set of events, $\mathcal{F}$, in a probability space $(\Omega, \mathcal{F}, P)$, is required to be a σ-algebra. The original motivation for introducing $\mathcal{F}$ in this context was a technical one, related to the impossibility of extending P to be defined on all subsets of Ω, for important examples such as $\Omega = [0,1]$ and $P((a,b)) = b - a$ for all intervals (a,b). However, σ-algebras are also useful for modeling the information available to an observer. We call $\mathcal{D}$ a sub-σ-algebra of $\mathcal{F}$ if $\mathcal{D}$ is a σ-algebra such that $\mathcal{D} \subset \mathcal{F}$. In applications, to say that the information available to an observer is modeled by a sub-σ-algebra $\mathcal{D}$, means that for any event $A \in \mathcal{D}$, the observer will learn whether A occurs, i.e. whether the selected value of ω is in A. A random variable Z is said to be *$\mathcal{D}$-measurable* if $\{Z \leq c\} \subset \mathcal{D}$ for all c. By definition, random variables are functions on Ω that are $\mathcal{F}$-measurable. The smaller the σ-algebra $\mathcal{D}$ is, the fewer the set of $\mathcal{D}$-measurable random variables. In practice, sub-σ-algebras are usually generated by collections of random variables:

Definition 10.1 The σ-algebra *generated* by a set of random variables $(Y_i : i \in I)$, denoted by $\sigma(Y_i : i \in I)$, is the smallest σ-algebra containing all sets of the form $\{Y_i \leq c\}$.[1] The σ-algebra generated by a single random variable Y is denoted by $\sigma(Y)$ and sometimes as $\mathcal{F}^Y$.

An equivalent definition would be that $\sigma(Y_i : i \in I)$ is the smallest σ-algebra such that each Y_i is measurable with respect to it.

As explained above, a sub-σ-algebra of $\mathcal{F}$ characterizes the knowledge an observer will gain when the probability experiment modeled by the probability space $(\Omega, \mathcal{F}, P)$ is conducted. In Chapter 3, using all estimators of the form $g(Y)$ corresponds to modeling an observer that learns the value of Y. It means that even before the experiment is conducted, we know the observer will learn the value of $Y(\omega)$ once ω is selected. An equivalent condition would be to allow any estimator that is a $\sigma(Y)$-measurable random variable. That is, as shown in Problem 9.8, if Y and Z are random variables on the same probability space, then $Z = g(Y)$ for some Borel measurable function g if and only if Z is $\sigma(Y)$ measurable. Using sub-σ-algebras is closer to the heart of modeling what an observer will learn about the outcome than using random variables for the modeling. For example, two different random variables, such as Y and Y^3, can generate the same sub-σ-algebra.

Example 10.1 (The trivial σ-algebra) Let $(\Omega, \mathcal{F}, P)$ be a probability space. Suppose X is a random variable such that, for some constant c_o, $X(\omega) = c_o$ for all $\omega \in \Omega$. Then X is measurable with respect to the trivial σ-algebra $\mathcal{D}$ defined by $\mathcal{D} = \{\emptyset, \Omega\}$. That is, constant random variables are $\{\emptyset, \Omega\}$-measurable.

Conversely, suppose Y is a $\{\emptyset, \Omega\}$-measurable random variable. Select an arbitrary $\omega_o \in \Omega$ and let $c_o = Y(\omega_o)$. On one hand, $\{\omega : Y(\omega) \leq c_o\}$ cannot be empty because it contains ω_o, so $\{\omega : Y(\omega) \leq c_o\} = \Omega$. On the other hand, $\{\omega : Y(\omega) \leq c\}$ doesn't contain ω_o for $c < c_o$, so $\{\omega : Y(\omega) \leq c\} = \emptyset$ for $c < c_o$. Therefore, $Y(\omega) = c_o$ for all ω. That is, $\{\emptyset, \Omega\}$-measurable random variables are constant.

Definition 10.2 If X is a random variable on $(\Omega, \mathcal{F}, P)$ with finite mean and $\mathcal{D}$ is a sub-σ-algebra of $\mathcal{F}$, the *conditional expectation of X given $\mathcal{D}$*, $E[X|\mathcal{D}]$, is the unique (two versions equal with probability one are considered to be the same) random variable on $(\Omega, \mathcal{F}, P)$ such that

(i) $E[X|\mathcal{D}]$ is $\mathcal{D}$-measurable,

(ii) $E[(X - E[X|\mathcal{D}])I_A] = 0$ for all $A \in \mathcal{D}$. (Here I_A is the indicator function of A.)

We remark that a possible choice of A in property (ii) of the definition is $A = \Omega$, so $E[X|\mathcal{D}]$ should satisfy $E[X - E[X|\mathcal{D}]] = 0$, or equivalently, since $E[X]$ is assumed to be

[1] The smallest one exists – it is equal to the intersection of all σ-algebras which contain all sets of the form $\{Y_i \leq c\}$.

finite, $E[X] = E[E[X|\mathcal{D}]]$. In particular, an implication of the definition is that $E[X|\mathcal{D}]$ also has a finite mean.

Proposition 10.3 *Definition 10.2 is well posed. Specifically, there exists a random variable satisfying conditions (i) and (ii), and it is unique.*

Proof (*Uniqueness*) Suppose U and V are each $\mathcal{D}$-measurable random variables such that $E[(X-U)I_A] = 0$ and $E[(X-V)I_A] = 0$ for all $A \in \mathcal{D}$. It follows that $E[(U-V)I_A] = E[(X-V)I_A] - E[(X-U)I_A] = 0$ for any $A \in \mathcal{D}$. A possible choice of A is $\{U > V\}$, so $E[(U-V)I_{\{U>V\}}] = 0$. Since $(U-V)I_{\{U>V\}}$ is nonnegative and is strictly positive on the event $\{U > V\}$, it must be that $P\{U > V\} = 0$. Similarly, $P\{U < V\} = 0$. So $P\{U = V\} = 1$.

(*Existence*) Existence is first proved assuming $P\{X \geq 0\} = 1$. Let $L^2(\mathcal{D})$ be the space of $\mathcal{D}$-measurable random variables with finite second moments. Then $L^2(\mathcal{D})$ is a closed, linear subspace of $L^2(\Omega, \mathcal{F}, P)$, so the orthogonality principle can be applied. For any $n \geq 0$, the random variable $X \wedge n$ is bounded and thus has a finite second moment. Let $\widehat{X}_n$ be the projection of $X \wedge n$ onto $L^2(\mathcal{D})$. Then by the orthogonality principle, $X \wedge n - \widehat{X}_n$ is orthogonal to any random variable in $L^2(\mathcal{D})$. In particular, $X \wedge n - \widehat{X}_n$ is orthogonal to I_A for any $A \in \mathcal{D}$. Therefore, $E[(X \wedge n - \widehat{X}_n)I_A] = 0$ for all $A \in \mathcal{D}$. Equivalently,

$$E[(X \wedge n)I_A] = E[\widehat{X}_n I_A]. \tag{10.1}$$

The next step is to take a limit as $n \to \infty$. Since $E[(X \wedge n)I_A]$ is nondecreasing in n for $A \in \mathcal{D}$, the same is true of $E[\widehat{X}_n I_A]$. Thus, for $n \geq 0$, $E[(\widehat{X}_{n+1} - \widehat{X}_n)I_A] \geq 0$ for any $A \in \mathcal{D}$. Taking $A = \{\widehat{X}_{n+1} - \widehat{X}_n < 0\}$ implies that $P\{\widehat{X}_{n+1} \geq \widehat{X}_n\} = 1$. Therefore, the sequence $(\widehat{X}_n)$ converges a.s., and we denote the limit by $\widehat{X}_\infty$. We show that $\widehat{X}_\infty$ satisfies the two properties, (i) and (ii), required of $E[X|\mathcal{D}]$. First, $\widehat{X}_\infty$ is $\mathcal{D}$-measurable because it is the limit of a sequence of $\mathcal{D}$-measurable random variables. Secondly, for any $A \in \mathcal{D}$, the sequences of random variables $(X \wedge n)I_A$ and $\widehat{X}_n I_A$ are a.s. nondecreasing and nonnegative, so by the monotone convergence theorem (Theorem 11.14) and (10.1):

$$E[XI_A] = \lim_{n\to\infty} E[(X \wedge n)I_A] = \lim_{n\to\infty} E[\widehat{X}_n I_A] = E[\widehat{X}_\infty I_A].$$

So property (ii), $E[(X - \widehat{X}_\infty)I_A] = 0$, is also satisfied. Existence is proved in the case $P\{X \geq 0\} = 1$.

For the general case, X can be represented as $X = X_+ - X_-$, where X_+ and X_- are nonnegative with finite means. By the case already proved, $E[X_+|\mathcal{D}]$ and $E[X_-|\mathcal{D}]$ exist, and, of course, they satisfy conditions (i) and (ii) in Definition 10.2. Therefore, with $E[X|\mathcal{D}] = E[X_+|\mathcal{D}] - E[X_-|\mathcal{D}]$, it is a simple matter to check that $E[X|\mathcal{D}]$ also satisfies conditions (i) and (ii), as required. □

Proposition 10.4 *Let X and Y be random variables on $(\Omega, \mathcal{F}, P)$ and let $\mathcal{A}$ and $\mathcal{D}$ be sub-σ-algebras of $\mathcal{F}$.*

1. *(Consistency with definition based on projection) If $E[X^2] < \infty$ and $\mathcal{V} = \{g(Y) : g$ is Borel measurable such that $E[g(Y)^2] < \infty\}$, then $E[X|Y]$, defined as the MMSE projection of X onto $\mathcal{V}$ (also written as $\Pi_{\mathcal{V}}(X)$) is equal to $E[X|\sigma(Y)]$.*

2. *(Linearity) If $E[X]$ and $E[Y]$ are finite, $aE[X|\mathcal{D}] + bE[Y|\mathcal{D}] = E[aX + bY|\mathcal{D}]$.*
3. *(Tower property) If $E[X]$ is finite and $\mathcal{A} \subset \mathcal{D} \subset \mathcal{F}$, $E[E[X|\mathcal{D}]|\mathcal{A}] = E[X|\mathcal{A}]$. (In particular, $E[E[X|\mathcal{D}]] = E[X]$.)*
4. *(Positivity preserving) If $E[X]$ is finite and $X \geq 0$ a.s. then $E[X|\mathcal{D}] \geq 0$ a.s.*
5. *(L1 contraction property) $E[|E[X|\mathcal{D}]|] \leq E[|X|]$.*
6. *(L1 continuity) If $E[X_n]$ is finite for all n and $E[|X_n - X_\infty|] \to 0$, then $E[|E[X_n|\mathcal{D}] - E[X_\infty|\mathcal{D}]|] \to 0$.*
7. *(Pull out property) If X is $\mathcal{D}$-measurable and $E[XY]$ and $E[Y]$ are finite, then $E[XY|\mathcal{D}] = XE[Y|\mathcal{D}]$.*

Proof *(Consistency with definition based on projection)* Suppose X and $\mathcal{V}$ are as in part 1. Then, by definition, $E[X|Y] \in \mathcal{V}$ and $E[(X - E[X|Y])Z] = 0$ for any $Z \in \mathcal{V}$. As mentioned above, a random variable has the form $g(Y)$ if and only if it is $\sigma(Y)$-measurable. In particular, $\mathcal{V}$ is simply the set of $\sigma(Y)$-measurable random variables Z such that $E[Z^2] < \infty$. Thus, $E[X|Y]$ is $\sigma(Y)$ measurable, and $E[(X - E[X|Y])Z] = 0$ for any $\sigma(Y)$-measurable random variable Z such that $E[Z^2] < \infty$. As a special case $E[(X - E[X|Y])I_A] = 0$ for any $A \in \sigma(Y)$. Thus, $E[X|Y]$ satisfies conditions (i) and (ii) in Definition 10.2 of $E[X|\sigma(Y)]$. So $E[X|Y] = E[X|\sigma(Y)]$.

(Linearity property) (This is similar to the proof of linearity for projections, Proposition 3.3.) It suffices to check that the linear combination $aE[X|\mathcal{D}] + bE[Y|\mathcal{D}]$ satisfies the two conditions that define $E[aX + bY|\mathcal{D}]$. First, $E[X|\mathcal{D}]$ and $E[Y|\mathcal{D}]$ are both $\mathcal{D}$-measurable, so their linear combination is also $\mathcal{D}$-measurable. Secondly, if $D \in \mathcal{D}$, then $E[(X - E[X|\mathcal{D}])I_A] = E[(Y - E[Y|\mathcal{D}])I_A] = 0$, from which it follows that

$$
\begin{aligned}
&E[(aX + bY - E[aX + bY|\mathcal{D}])\, I_A] \\
&\quad = aE[(X - E[X|\mathcal{D}])I_A] + bE[(Y - E[Y|\mathcal{D}])I_A] = 0.
\end{aligned}
$$

Therefore, $aE[X|\mathcal{D}] + bE[Y|\mathcal{D}] = E[aX + bY|\mathcal{D}]$.

(Tower property) (This is similar to the proof of Proposition 3.4, about projections onto nested subspaces.) It suffices to check that $E[E[X|\mathcal{D}]|\mathcal{A}]$ satisfies the two conditions that define $E[X|\mathcal{A}]$. First, $E[E[X|\mathcal{D}]|\mathcal{A}]$ itself is a conditional expectation given $\mathcal{A}$, so it is $\mathcal{A}$ measurable. Second, let $A \in \mathcal{A}$. Now $X - E[E[X|\mathcal{D}]|\mathcal{A}] = (X - E[X|\mathcal{D}]) + (E[X|\mathcal{D}] - E[E[X|\mathcal{D}]|\mathcal{A}])$, and (use the fact $A \in \mathcal{D}$) $E[(X - E[X|\mathcal{D}])I_A]$ and $E[(E[X|\mathcal{D}] - E[E[X|\mathcal{D}]|\mathcal{A}])I_A] = 0$. Adding these last two equations yields $E[(X - E[E[X|\mathcal{D}]|\mathcal{A}])I_A] = 0$. Therefore, $E[E[X|\mathcal{D}]|\mathcal{A}] = E[X|\mathcal{A}]$.

(Positivity preserving) Suppose $E[X]$ is finite and $X \geq 0$ a.s. Let A be the event given by $A = \{E[X|\mathcal{D}] < 0\}$. Observe that $A \in \mathcal{D}$ because $E[X|\mathcal{D}]$ is $\mathcal{D}$-measurable. So $E[E[X|\mathcal{D}]I_A] = E[XI_A] \geq 0$, while $P\{E[X|\mathcal{D}]I_A \leq 0\} = 1$. Hence, $P\{E[X|\mathcal{D}]I_A = 0\} = 1$, which is to say that $E[X|\mathcal{D}] \geq 0$ a.s.

(L1 contraction property) (This property is a special case of the conditional version of Jensen's inequality, established in Problem 10.9. Here a different proof is given.) $X = X_+ - X_-$, where X_+ is the positive part of X and X_- is the negative part of X, given by $X_+ = X \vee 0$ and $X_- = (-X) \vee 0$. Since X is assumed to have a finite mean, the same is true of $X_\pm$. Moreover, $E[E[X_\pm|\mathcal{D}]] = E[X_\pm]$, and by the linearity property, $E[X|\mathcal{D}] = E[X_+|\mathcal{D}] - E[X_-|\mathcal{D}]$. By the positivity preserving property, $E[X_+|\mathcal{D}]$ and

$E[X_-|\mathcal{D}]$ are both nonnegative a.s., so $E[X_+|\mathcal{D}]+E[X_-|\mathcal{D}] \geq |E[X_+|\mathcal{D}]-E[X_-|\mathcal{D}]|$ a.s. (The inequality is strict for ω such that both $E[X_+|\mathcal{D}]$ and $E[X_-|\mathcal{D}]$ are strictly positive.) Therefore,

$$\begin{aligned} E[|X|] &= E[X_+] + E[X_-] \\ &= E[E[X_+|\mathcal{D}] + E[X_-|\mathcal{D}]] \\ &\geq E[|E[X_+|\mathcal{D}] - E[X_-|\mathcal{D}]|] \\ &= E[|E[X|\mathcal{D}]|], \end{aligned}$$

and the $L1$ contraction property is proved.

(L1 continuity) Since for any n, $|X_\infty| \leq |X_n| + |X_n - X_\infty|$, the hypotheses imply that X_∞ has a finite mean. By linearity and the $L1$ contraction property, $E[|E[X_n|\mathcal{D}] - E[X_\infty|\mathcal{D}]|] = E[|E[X_n - X_\infty|\mathcal{D}]|] \leq E[|E[X_n - X_\infty|]|]$, which implies the $L1$ continuity property.

(Pull out property) The pull out property will be proved first under the added assumption that X and Y are nonnegative random variables. Clearly $XE[Y|\mathcal{D}]$ is $\mathcal{D}$ measurable. Let $A \in \mathcal{D}$. It remains to show that

$$E[XYI_A] = E[XE[Y|\mathcal{D}]I_A]. \tag{10.2}$$

If X has form I_D for $D \in \mathcal{D}$ then (10.2) becomes $E[YI_{A\cap D}] = E[E[Y|\mathcal{D}]I_{A\cap D}]$, which holds by the definition of $E[Y|\mathcal{D}]$ and the fact $A \cap D \in \mathcal{D}$. Equation (10.2) is thus also true if X is a finite linear combination of random variables of the form I_D, that is, if X is a simple $\mathcal{D}$-measurable random variable. We now bring in the fact X is the a.s. limit of a nondecreasing sequence of nonnegative simple random variables X_n. Then (10.2) holds for X replaced by X_n:

$$E[X_nYI_A] = E[X_nE[Y|\mathcal{D}]I_A]. \tag{10.3}$$

Also, X_nYI_A is a nondecreasing sequence converging to XYI_A a.s., and $X_nE[Y|\mathcal{D}]I_A$ is a nondecreasing sequence converging to $XE[Y|\mathcal{D}]I_A$ a.s. By the monotone convergence theorem, taking $n \to \infty$ on both sides of (10.3) yields (10.2). This proves the pull out property under the added assumption that X and Y are nonnegative.

In the general case, $X = X_+ - X_-$, where $X_+ = X \vee 0$ and $X_- = (-X) \vee 0$, and similarly $Y = Y_+ - Y_-$. The hypotheses imply $E[X_\pm Y_\pm]$ and $E[Y_\pm]$ are finite so that $E[X_\pm Y_\pm|\mathcal{D}] = X_\pm E[Y_\pm|\mathcal{D}]$, and therefore

$$E[X_\pm Y_\pm I_A] = E[X_\pm E[Y_\pm|\mathcal{D}]I_A], \tag{10.4}$$

where in these equations the sign on both appearances of X should be the same, and the sign on both appearances of Y should be the same. The left side of (10.2) can be expressed as a linear combination of terms of the form $E[X_\pm Y_\pm I_A]$:

$$E[XYI_A] = E[X_+Y_+I_A] - E[X_+Y_-I_A] - E[X_-Y_+I_A] + E[X_-Y_-I_A].$$

Similarly, the right side of (10.2) can be expressed as a linear combination of terms of the form $E[X_\pm E[Y_\pm|\mathcal{D}]I_A]$. Therefore, (10.2) follows from (10.4). □

10.2 Martingales with respect to filtrations

A *filtration* of a σ-algebra $\mathcal{F}$ is a sequence of sub-σ-algebras $\boldsymbol{\mathcal{F}} = (\mathcal{F}_n : n \geq 0)$ of $\mathcal{F}$, such that $\mathcal{F}_n \subset \mathcal{F}_{n+1}$ for $n \geq 0$. If $Y = (Y_n : n \geq 0)$ or $Y = (Y_n : n \geq 1)$ is a sequence of random variables on $(\Omega, \mathcal{F}, P)$, the filtration generated by Y, often written as $\boldsymbol{\mathcal{F}}^Y = (\mathcal{F}_n^Y : n \geq 0)$, is defined by letting $\mathcal{F}_n^Y = \sigma(Y_k : k \leq n)$. (If there is no variable Y_0 defined, we take $\mathcal{F}_0^Y$ to be the trivial σ-algebra, $\mathcal{F}_0^Y = \{\emptyset, \Omega\}$, representing no observations.)

In practice, a filtration represents a sequence of observations or measurements. If the filtration is generated by a random process, then the information available at time n represents observation of the random process up to time n.

A random process $(X_n : n \geq 0)$ is *adapted* to a filtration $\boldsymbol{\mathcal{F}}$ if X_n is $\mathcal{F}_n$ measurable for each $n \geq 0$.

Definition 10.5 Let $Y = (Y_n : n \geq 0)$ be a random process on some probability space with a filtration $\boldsymbol{\mathcal{F}}$. Then Y is a *martingale* with respect to $\boldsymbol{\mathcal{F}}$ if for all $n \geq 0$:

(i) Y_n is $\mathcal{F}_n$ measurable (i.e. the process Y is adapted to $\boldsymbol{\mathcal{F}}$),
(ii) $E[|Y_n|] < \infty$, (equivalently $E[Y_n]$ exists and is finite)
(iii) $E[Y_{n+1}|\mathcal{F}_n] = Y_n$ *a.s.*

Y is a *submartingale* relative to $\boldsymbol{\mathcal{F}}$ if (i) and (ii) hold and $E[Y_{n+1}|\mathcal{F}_n] \geq Y_n$ *a.s.*
Y is a *supermartingale* relative to $\boldsymbol{\mathcal{F}}$ if (i) and (ii) hold and $E[Y_{n+1}|\mathcal{F}_n] \leq Y_n$ *a.s.*

Some comments are in order. Note that if $Y = (Y_n : n \geq 0)$ is a martingale with respect to a filtration $\boldsymbol{\mathcal{F}} = (\mathcal{F}_n : n \geq 0)$, then Y is also a martingale with respect to the filtration generated by Y itself. Indeed, for each n, Y_n is $\mathcal{F}_n$ measurable, whereas $\mathcal{F}_n^Y$ is the smallest σ-algebra with respect to which Y_n is measurable, so $\mathcal{F}_n^Y \subset \mathcal{F}_n$. Therefore, the tower property of conditional expectation, the fact Y is a martingale with respect to $\boldsymbol{\mathcal{F}}$, and the fact Y_n is $\mathcal{F}_n^Y$ measurable, imply

$$E[Y_{n+1}|\mathcal{F}_n^Y] = E[E[Y_{n+1}|\mathcal{F}_n]|\mathcal{F}_n^Y] = E[Y_n|\mathcal{F}_n^Y] = Y_n.$$

Thus, in practice, if Y is said to be a martingale and no filtration $\mathcal{F}$ is specified, at least Y is a martingale with respect to the filtration it generates.

Note that if Y is a martingale with respect to a filtration $\boldsymbol{\mathcal{F}}$, then for any $n, k \geq 0$,

$$E[Y_{n+k+1}|\mathcal{F}_n] = E[E[Y_{n+k+1}|\mathcal{F}_{n+k}]|\mathcal{F}_n] = E[Y_{n+k}|\mathcal{F}_n].$$

Therefore, by induction on k for n fixed:

$$E[Y_{n+k}|\mathcal{F}_n] = Y_n, \tag{10.5}$$

for $n, k \geq 0$.

Example 10.2 Suppose $(U_i : i \geq 1)$ is a collection of independent random variables, each with mean zero. Let $S_0 = 0$ and for $n \geq 1$, $S_n = \sum_{i=1}^n U_i$. Let $\boldsymbol{\mathcal{F}} = (\mathcal{F}_n : n \geq 0)$ denote the filtration generated by S: $\mathcal{F}_n = \sigma(S_0, \ldots, S_n)$. Equivalently, $\mathcal{F}_0 = \{\emptyset, \Omega\}$ and $\mathcal{F}_n = \sigma(U_1, \ldots, U_n)$ for $n \geq 1$. Then $S = (S_n : n \geq 0)$ is a martingale with respect to $\boldsymbol{\mathcal{F}}$:

$$E[S_{n+1}|\mathcal{F}_n] = E[U_{n+1}|\mathcal{F}_n] + E[S_n|\mathcal{F}_n] = 0 + S_n = S_n.$$

Example 10.3 Suppose S and $\mathcal{F}$ are defined as in Example 10.2 in terms of a sequence of independent random variables U. Suppose in addition $\text{Var}(U_i) = \sigma^2$ for some finite σ^2. Let $M_n = S_n^2 - n\sigma^2$ for $n \geq 0$. Then $M = (M_n : n \geq 0)$ is a martingale relative to $\mathcal{F}$. Indeed, M is adapted to $\mathcal{F}$. Since $S_{n+1} = S_n + U_{n+1}$, $M_{n+1} = M_n + 2S_nU_{n+1} + U_{n+1}^2 - \sigma^2$, so

$$\begin{aligned} E[M_{n+1}|\mathcal{F}_n] &= E[M_n|\mathcal{F}_n] + 2S_nE[U_{n+1}|\mathcal{F}_n] + E[U_{n+1}^2 - \sigma^2|\mathcal{F}_n] \\ &= M_n + 2S_nE[U_{n+1}] + E[U_{n+1}^2 - \sigma^2] \\ &= M_n. \end{aligned}$$

Example 10.4 Suppose $X_1, X_2, \ldots$ is a sequence of independent, identically distributed random variables and θ is a number such that $E[e^{\theta X_1}] < \infty$. Let $S_0 = 0$ and let $S_n = X_1 + \ldots + X_n$ for $n \geq 1$. Then (M_n) defined by $M_n = e^{\theta S_n}/E[e^{\theta X_1}]^n$ for $n \geq 0$ is a martingale.

Example 10.5 (Galton–Watson branching process) A Galton–Watson branching process starts with an initial set of individuals, called the zeroth generation. For example, there may be just one individual in the zeroth generation. The $(n+1)$th generation is the set of all offspring of individuals in the nth generation. The number of offspring of each individual has the same distribution as a given discrete random variable X, with the numbers of offspring of different individuals being mutually independent. Let G_n denote the number of individuals in the nth generation of a branching process. Suppose $a > 0$ satisfies $E[a^X] = 1$ and $E[a^{G_0}] < \infty$. Then, a^{G_n} is a martingale.

Example 10.6 (Doob martingale) Let $M_n = E[\Phi|\mathcal{F}_n]$ for $n \geq 0$, where Φ is a random variable with finite mean, and $\mathcal{F} = (\mathcal{F}_n : n \geq 0)$ is a filtration. By the tower property of conditional expectation, $M = (M_n : n \geq 0)$ is a martingale with respect to $\mathcal{F}$.

Definition 10.6 A *martingale difference sequence* $(D_n : n \geq 1)$ *relative to a filtration* $\mathcal{F} = (\mathcal{F}_n : n \geq 0)$ is a sequence of random variables $(D_n : n \geq 1)$ such that:

(i) $(D_n : n \geq 1)$ is adapted to $\mathcal{F}$ (i.e. D_n is $\mathcal{F}_n$ measurable for each $n \geq 1$),

(ii) $E[|D_n|] < \infty$ for $n \geq 1$,

(iii) $E[D_{n+1}|\mathcal{F}_n] = 0$ a.s. for all $n \geq 0$.

Equivalently, $(D_n : n \geq 1)$ has the form $D_n = M_n - M_{n-1}$ for $n \geq 1$, for some $(M_n : n \geq 0)$ which is a martingale with respect to $\mathcal{F}$.

Definition 10.7 A random process $(H_n : n \geq 1)$ is said to be *predictable with respect to a filtration* $\mathcal{F} = (\mathcal{F}_n : n \geq 0)$ if H_n is $\mathcal{F}_{n-1}$ measurable for all $n \geq 1$. (Sometimes this is called *one-step predictable*, because $\mathcal{F}_n$ determines H one step ahead.)

Example 10.7 (Nonlinear innovations process, a.k.a. Doob decomposition) Suppose $Y = (Y_n : n \geq 1)$ is a sequence of random variables with finite means that is adapted to a filtration $\mathcal{F}$. Let $H_n = E[Y_n|\mathcal{F}_{n-1}]$ for $n \geq 0$. Then $H = (H_n : n \geq 1)$ is a predictable process and $D = (D_n : n \geq 1)$, defined by $D_n = Y_n - H_n$, is a martingale difference sequence with respect to $\mathcal{F}$. The sequence (D_n) is the nonlinear innovations sequence for Y, and $Y_n = H_n + D_n$ for all $n \geq 1$. In summary, any such process Y is the sum of a predictable process H and a martingale difference sequence D. Moreover, for given Y and $\mathcal{F}$, this decomposition is unique up to events of measure zero, because a predictable martingale difference sequence is almost surely identically zero.

As described in Chapter 3 in connection with Kalman filtering, if Y is a second order random process, the linear innovations sequence associated with Y is the sequence $(\widetilde{Y}_n)$ of orthogonal random variables defined using linear MMSE estimators, rather than conditional expectations: $\widetilde{Y}_n = Y_n - \widehat{E}[Y_n|Y_1, \cdots, Y_{n-1}]$. For Gaussian random processes and $\mathcal{F}_n = \sigma(Y_1, \ldots, Y_n)$, the nonlinear and linear innovations sequences, (D_n) and $(\widetilde{Y}_n)$, coincide.

Example 10.8 Let $(D_n : n \geq 1)$ be a martingale difference sequence and $(H_k : k \geq 1)$ be a bounded predictable process, both relative to a filtration $\mathcal{F} = (\mathcal{F}_n : n \geq 0)$. We claim that the new process $\widetilde{D} = (\widetilde{D}_n : n \geq 1)$ defined by $\widetilde{D}_n = H_n D_n$ is also a martingale difference sequence with respect to $\mathcal{F}$. Indeed, it is adapted, has finite means, and

$$E[H_{n+1}D_{n+1}|\mathcal{F}_n] = H_{n+1}E[D_{n+1}|\mathcal{F}_n] = 0,$$

where we pulled out the $\mathcal{F}_n$ measurable random variable H_{n+1} from the conditional expectation given $\mathcal{F}_n$. An interpretation is that D_n is the net gain to a gambler if one dollar is bet on the outcome of a fair game in round n, and so $H_n D_n$ is the net gain if H_n dollars are bet in round n. The requirement that $(H_k : k \geq 1)$ be predictable means that the gambler must decide how much to bet in round n based only on information available at the end of round $n - 1$. It would be an unfair advantage if the gambler already knew D_n when deciding how much money to bet in round n.

If the initial reserves of the gambler were some constant M_0, then the reserves of the gambler after n rounds would be given by:

$$M_n = M_0 + \sum_{k=1}^{n} H_k D_k.$$

Then $(M_n : n \geq 0)$ is a martingale with respect to $\mathcal{F}$. The random variables $H_k D_k$, $1 \leq k \leq n$ are orthogonal. Also, $E[(H_k D_k)^2] = E[E[(H_k D_k)^2|\mathcal{F}_{k-1}]] = E[H_k^2 E[D_k^2|\mathcal{F}_{k-1}]]$. Therefore,

$$E[(M_n - M_0)^2] = \sum_{k=1}^{n} E[H_k^2 E[D_k^2|\mathcal{F}_{k-1}]].$$

10.3 Azuma–Hoeffding inequality

One of the simplest inequalities for martingales is the Azuma–Hoeffding inequality. It is proven in this section, and applications to prove concentration inequalities for some combinatorial problems are given.[2]

Lemma 10.8 *Suppose D is a mean zero random variable with $P\{|D-b| \leq d\} = 1$ for some constant b. Then for any $\alpha \in \mathbb{R}$, $E[e^{\alpha D}] \leq e^{(\alpha d)^2/2}$.*

Proof Since D has mean zero and lies in the interval $[b-d, b+d]$ with probability one, the interval must contain zero, so $|b| \leq d$. To avoid trivial cases we assume $|b| < d$. Since $e^{\alpha x}$ is convex in x, the value of $e^{\alpha x}$ for $x \in [b-d, b+d]$ is bounded above by the linear function that is equal to $e^{\alpha x}$ at the endpoints, $x = b \pm d$, of the interval:

$$e^{\alpha x} \leq \frac{x-b+d}{2d} e^{\alpha(b+d)} + \frac{b+d-x}{2d} e^{\alpha(b-d)}. \tag{10.6}$$

Since D lies in that interval with probability one, (10.6) remains true if x is replaced by the random variable D. Taking expectations on both sides and using $E[D] = 0$ yields

$$E[e^{\alpha D}] \leq \frac{d-b}{2d} e^{\alpha(b+d)} + \frac{b+d}{2d} e^{\alpha(b-d)}. \tag{10.7}$$

The proof is completed by showing that the right side of (10.7) is less than or equal to $e^{(\alpha d)^2/2}$ for any $|b| < d$. Letting $u = \alpha d$ and $\theta = b/d$, the inequality to be proved becomes $f(u) \leq e^{u^2/2}$, for $u \in \mathbb{R}$ and $|\theta| < 1$, where

$$f(u) = \ln\left(\frac{(1-\theta)e^{u(1+\theta)} + (1+\theta)e^{u(-1+\theta)}}{2}\right).$$

Taylor's formula implies that $f(u) = f(0) + f'(0)u + \frac{f''(v)u^2}{2}$ for some v in the interval with endpoints 0 and u. Elementary, but somewhat tedious, calculations show that

$$f'(u) = \frac{(1-\theta^2)(e^u - e^{-u})}{(1-\theta)e^u + (1+\theta)e^{-u}}$$

and

$$f''(u) = \frac{4(1-\theta^2)}{[(1-\theta)e^u + (1+\theta)e^{-u}]^2} = \frac{1}{\cosh^2(u+\beta)},$$

where $\beta = \frac{1}{2}\ln(\frac{1-\theta}{1+\theta})$. Note that $f(0) = f'(0) = 0$, and $f''(u) \leq 1$ for all $u \in \mathbb{R}$. Therefore, $f(u) \leq u^2/2$ for all $u \in \mathbb{R}$, as was to be shown. □

Proposition 10.9 *(Azuma–Hoeffding inequality with centering) Suppose $(Y_n : n \geq 0)$ is a martingale and $(B_n : n \geq 1)$ is a predictable process, both with respect to a filtration*

[2] See McDiarmid survey paper (McDiarmid 1989).

$\mathcal{F} = (\mathcal{F}_n : n \geq 0)$, *such that* $P\{|Y_{n+1} - B_{n+1}| \leq d_n\} = 1$ *for all* $n \geq 0$. *Then*

$$P\{|Y_n - Y_0| \geq \lambda\} \leq 2\exp\left(-\frac{\lambda^2}{2\sum_{i=1}^n d_i^2}\right).$$

Proof Let $n \geq 0$. The idea is to write $Y_n = Y_n - Y_{n-1} + Y_{n-1}$, to use the tower propert of conditional expectation, and to apply Lemma 10.8 to the random variable $Y_n - Y_{n-}$ for $d = d_n$. This yields:

$$\begin{aligned} E\left[e^{\alpha(Y_n - Y_0)}\right] &= E\left[E\left[e^{\alpha(Y_n - Y_{n-1} + Y_{n-1} - Y_0)}\middle|\mathcal{F}_{n-1}\right]\right] \\ &= E\left[e^{\alpha(Y_{n-1} - Y_0)}E\left[e^{\alpha(Y_n - Y_{n-1})}\middle|\mathcal{F}_{n-1}\right]\right] \\ &\leq E\left[e^{\alpha(Y_{n-1} - Y_0)}\right]e^{(\alpha d_n)^2/2}. \end{aligned}$$

Thus, by induction on n,

$$E\left[e^{\alpha(Y_n - Y_0)}\right] \leq e^{(\alpha^2/2)\sum_{i=1}^n d_i^2}.$$

The remainder of the proof is essentially the Chernoff inequality:

$$P\{Y_n - Y_0 \geq \lambda\} \leq E\left[e^{\alpha(Y_n - Y_0 - \lambda)}\right] \leq e^{(\alpha^2/2)\sum_{i=1}^n d_i^2 - \alpha\lambda}.$$

Finally, taking α to make this bound as tight as possible, i.e. $\alpha = \frac{\lambda}{\sum_{i=1}^n d_i^2}$, yields

$$P\{Y_n - Y_0 \geq \lambda\} \leq \exp\left(-\frac{\lambda^2}{2\sum_{i=1}^n d_i^2}\right).$$

Similarly, $P\{Y_n - Y_0 \leq -\lambda\}$ satisfies the same bound because the previous bound applie for (Y_n) replaced by $(-Y_n)$, yielding the proposition. □

Definition 10.10 A function f of n variables is said to satisfy the Lipschitz conditio with constant c if $|f(x_1, \ldots, x_n) - f(x_1, \ldots, x_{i-1}, y_i, x_{i+1}, \ldots, x_n)| \leq c$ for any $x_1, \ldots, x_n$ i, and y_i.[3]

Proposition 10.11 *(McDiarmid's inequality) Suppose* $F = f(X_1, \ldots, X_n)$, *where satisfies the Lipschitz condition with constant* c, *and* $X_1, \ldots, X_n$ *are independent randon variables. Then* $P\{|F - E[F]| \geq \lambda\} \leq 2\exp(-\frac{2\lambda^2}{nc^2})$.

Proof Let $(Z_k : 0 \leq k \leq n)$ denote the Doob martingale defined by $Z_k = E[F|\mathcal{F}_k^X]$ where, as usual, $\mathcal{F}_k^X = \sigma(X_k : 1 \leq k \leq n)$ is the filtration generated by (X_k). Note that $\mathcal{F}_0^X$ is the trivial σ-algebra $\{\emptyset, \Omega\}$, corresponding to no observations, so $Z_0 = E[F]$

[3] Equivalently, $|f(x) - f(y)| \leq c d_H(x, y)$, where $d_H(x, y)$ denotes the Hamming distance, which is the numbe of coordinates in which x and y differ. In other contexts, the Lipschitz condition is with respect to a differen distance metric, such as Euclidean distance.

Also, $Z_n = F$. In words, Z_k is the conditional expectation of F, given that the first k Xs are revealed.

For $0 \leq k \leq n-1$, let

$$g_k(x_1, \ldots, x_k, x_{k+1}) = E[f(x_1, \ldots, x_{k+1}, X_{k+2}, \ldots, X_n)].$$

Note that $Z_{k+1} = g_k(X_1, \ldots, X_{k+1})$. Since f satisfies the Lipschitz condition with constant c, the same is true of g_k. In particular, for $x_1, \ldots, x_k$ fixed, the set of possible values (i.e. range) of $g_k(x_1, \ldots, x_{k+1})$ as x_{k+1} varies, lies within some interval (depending on $x_1, \ldots, x_k$) with length at most c. Define $m_k(x_1, \cdots, x_k)$ to be the midpoint of the smallest such interval:

$$m_k(x_1, \ldots, x_k) = \frac{\sup_{x_{k+1}} g_k(x_1, \ldots, x_{k+1}) + \inf_{x_{k+1}} g_k(x_1, \ldots, x_{k+1})}{2}$$

and let $B_{k+1} = m_k(X_1, \ldots, X_k)$. Then B is a predictable process and $|Z_{k+1} - B_{k+1}| \leq \frac{c}{2}$ with probability one. Thus, the Azuma–Hoeffding inequality with centering can be applied with $d_i = \frac{c}{2}$ for all i, giving the desired result. □

Example 10.9 (Independent sets in an Erdős–Rényi graph) Let $V = \{v_1, \ldots, v_n\}$ be a finite set of cardinality $n \geq 1$. For each i, j with $1 \leq i < j \leq n$, suppose $Z_{i,j}$ is a Bernoulli random variable with parameter p, where $0 \leq p \leq 1$, such that the Zs are mutually independent. Let $G = (V, E)$ be a random graph, such that for $i < j$, there is an undirected edge between vertices v_i and v_j (i.e. v_i and v_j are neighbors) if and only if $Z_{i,j} = 1$. Equivalently, the set of edges is $E = \{\{i,j\} : i < j \text{ and } Z_{i,j} = 1\}$. (The graph G is called an *Erdős–Rényi random graph* with parameters n and p.) An *independent set* [4] in the graph is a set of vertices such that no two of the vertices in the set are neighbors. Let $\mathcal{I} = \mathcal{I}(G)$ denote the maximum of the cardinalities of all independent sets for G. Note that $\mathcal{I}$ is a random variable, because the graph is random. We shall apply McDiarmid's inequality to find a concentration bound for $\mathcal{I}(G)$. Note that $\mathcal{I}(G) = f((Z_{i,j} : 1 \leq i < j \leq n))$, for an appropriate function f. We could write a computer program for computing f, for example by cycling through all subsets of V, seeing which ones are independent sets, and reporting the largest cardinality of the independent sets. The running time for this algorithm is exponential in n. However, there is no need to be so explicit about how to compute f. Observe next that changing any one of the Zs would change $\mathcal{I}(G)$ by at most one. In particular, if there is an independent set in a graph, and if one edge is added to the graph, then at most one vertex would have to be removed from the independent set for the original graph to obtain an independent set for the new graph. Thus, f satisfies the Lipschitz condition with constant $c = 1$. Thus, by McDiarmid's inequality with $c = 1$ and m variables, where $m = n(n-1)/2$,

$$P\{|\mathcal{I} - E[\mathcal{I}]| \geq \lambda\} \leq 2\exp\left(-\frac{4\lambda^2}{n(n-1)}\right).$$

[4] The terminology "independent" here is not associated with statistical independence.

We next derive a tighter bound. For $1 \leq i \leq n$, let $X_i = (Z_{1,i}, Z_{2,i}, \ldots, Z_{i-1,i})$. In words, for each i, X_i determines which vertices with index less than i are neighbors of vertex v_i. Of course $\mathcal{I}$ is also determined by $X_1, \ldots, X_n$. Moreover, if any one of the Xs changes, $\mathcal{I}$ changes by at most one. That is, $\mathcal{I}$ can be expressed as a function of the n variables $X_1, \ldots, X_n$, such that the function satisfies the Lipschitz condition with constant $c = 1$. Therefore, by McDiarmid's inequality with $c = 1$ and n variables,[5]

$$P\{|\mathcal{I} - E[\mathcal{I}]| \geq \lambda\} \leq 2\exp\left(-\frac{2\lambda^2}{n}\right).$$

For example, if $\lambda = a\sqrt{n}$, we have

$$P\{|\mathcal{I} - E[\mathcal{I}]| \geq a\sqrt{n}\} \leq 2\exp(-2a^2)$$

whenever $n \geq 1$, $0 \leq p \leq 1$, and $a > 0$.

Note that McDiarmid's inequality, as illustrated in the above example, gives an upper bound on how spread out the distribution of a random variable is, without requiring specific knowledge about the mean of the random variable. Inequalities of this form are known as *concentration inequalities*. McDiarmid's inequality can similarly be applied to obtain concentration inequalities for many other numbers associated with random graphs, such as the size of a maximum matching (a matching is a set of edges, no two of which have a node in common), chromatic index (number of colors needed to color all edges so that all edges containing a single vertex are different colors), chromatic number (number of colors needed to color all vertices so that neighbors are different colors), minimum number of edges that need to be cut to break a graph into two equal size components, and so on.

10.4 Stopping times and the optional sampling theorem

Let $X = (X_k : k \geq 0)$ be a martingale with respect to $\boldsymbol{\mathcal{F}} = (\mathcal{F}_k : k \geq 0)$. Note that $E[X_{k+1}] = E[E[X_{k+1}|\mathcal{F}_k]] = E[X_k]$. So, by induction on n, $E[X_n] = E[X_0]$ for all $n \geq 0$.

A useful interpretation of a martingale $X = (X_k : k \geq 0)$ is that X_k is the reserve (amount of money on hand) that a gambler playing a fair game at each time step has after k time steps, if X_0 is the initial reserve. (If the gambler is allowed to go into debt, the reserve can be negative.) The condition $E[X_{k+1}|\mathcal{F}_k] = X_k$ means that, given the knowledge that is observable up to time k, the expected reserve after the next game is equal to the reserve at time k. The equality $E[X_n] = E[X_0]$ has the natural interpretation that the expected reserve of the gambler after n games have been played, is equal to the initial reserve X_0.

This section focuses on the following question. What happens if the gambler stops after a random number, T, of games? Is it true that $E[X_T] = E[X_0]$?

[5] Since X_n is degenerate, we could use $n - 1$ instead of n, but it makes little difference.

Example 10.10 Suppose $X_n = W_1 + \ldots + W_n$, such that the Ws are mutually independent and $P\{W_k = 1\} = P\{W_k = -1\} = 0.5$ for all k. Let T be the random time:

$$T = \begin{cases} 3 & \text{if } W_1 + W_2 + W_3 = 1 \\ 0 & \text{else} \end{cases}.$$

Then $X_T = 3$ with probability 3/8, and $X_T = 0$ otherwise. Hence, $E[X_T] = 3/8$.

Does Example 10.10 give a realistic strategy for a gambler to obtain a strictly positive expected payoff from a fair game? To implement the strategy, the gambler should stop gambling after T games. However, the event $\{T = 0\}$ depends on the outcomes W_1, W_2, and W_3. Thus, at time zero, the gambler is required to make a decision about whether to stop before any games are played based on the outcomes of the first three games. Unless the gambler can somehow predict the future, the gambler will be unable to implement the strategy of stopping play after T games.

Intuitively, a random time corresponds to an implementable stopping strategy if the gambler has enough information after n games to tell whether to play future games. That type of condition is captured by the notion of optional stopping time, defined as follows.

Definition 10.12 An *optional stopping time* T relative to $\mathcal{F} = (\mathcal{F}_k : k \geq 0)$ is a random variable with values in $\mathbb{Z}_+$ such that for any $n \geq 0$, $\{T \leq n\} \in \mathcal{F}_n$.

The intuitive interpretation of the condition $\{T \leq n\} \in \mathcal{F}_n$ is that the gambler should have enough information by time n to know whether to stop by time n. Since σ-algebras are closed under set complements, the condition in the definition of an optional stopping time is equivalent to requiring that, for any $n \geq 0$, $\{T > n\} \in \mathcal{F}_n$. This means that the gambler should have enough information by time n to know whether to continue gambling strictly beyond time n.

Example 10.11 Let $(X_n : n \geq 0)$ be a random process adapted to a filtration $\mathcal{F} = (\mathcal{F}_n : n \geq 0)$. Let A be some fixed (Borel measurable) subset of $\mathbb{R}$, and let $T = \min\{n \geq 0 : X_n \in A\}$. Then T is a stopping time relative to $\mathcal{F}$ because $\{T \leq n\} = \cup_{k=0}^{n}\{X_k \in A\} \in \mathcal{F}_n$.

Example 10.12 This example gives a highly risky strategy for a gambler betting variable amounts of money in a game of fair coin flips. The coin flips are modeled as independent Bernoulli random variables $W_1, W_2, \ldots$ with parameter $p = 0.5$. For each $n \geq 1$ the gambler bets some money at the beginning of the nth round, and if $W_n = 1$, the gambler wins back double the amount bet in that round, and if $W_n = 0$ the gambler loses the money bet in that round. Let X_n denote the reserve of the gambler after n rounds. For simplicity, we assume that the gambler can borrow money as needed, and that the initial reserve of the gambler is zero. So $X_0 = 0$.

Suppose the gambler adopts the following strategy. She bets 2^{n-1} units of money in the nth round and stops playing as soon as she wins one round. Let T be the number of

rounds the gambler plays. If she wins in the first round, i.e. if $T = 1$, then she woul
have had to borrow one unit of money in order to play the first round, and she has a ne
gain of one unit of money after playing. For $T \geq 2$, she loses $1+2+\ldots+2^{T-2} = 2^{T-1} -$
units of money in the first $T-1$ rounds, and she wins 2^{T-1} units of money in the Tt
round, so again she has a net gain of one unit of money. For this strategy, the number o
rounds, T, that the gambler plays has the geometric distribution with parameter $p = 0.5$
Thus, $E[T] = 2$. In particular, T is finite with probability one, and so $P\{X_T = 1\} =$
while $X_0 = 0$. The gambler always has a positive net gain of one unit of money, and sh
does not need to know the outcomes of the coin flips before they happen!

But don't run out and start playing this strategy, expecting to make money for sure
There is a catch – the amount borrowed can be very large. Indeed, let us compute th
expectation of B, the total amount borrowed before the final win. If $T = 1$ then $B =$
(only the one unit borrowed in the first round is counted). If $T = 2$ then $B = 3$ (the uni
lost in the first round, and two more borrowed in order to play the second round). I
general, $B = 2^T - 1$. Thus,

$$E[B] = E[2^T - 1] = \sum_{n=1}^{\infty}(2^n - 1)P\{T = n\} = \sum_{n=1}^{\infty}(2^n - 1)2^{-n} = \sum_{n=1}^{\infty}(1 - 2^{-n}) = +\infty$$

That is, the expected amount of money the gambler needs to borrow is infinite.

The following lemma shows that the positive expected gain resulting in Exam
ple 10.12 cannot happen at a fixed finite number of plays.

Lemma 10.13 *If X is a martingale and T is an optional stopping time relative t $(\Omega, \mathcal{F}, P)$, then $E[X_{T\wedge n}] = E[X_0]$ for any n.*

Proof Note that

$$X_{T\wedge(n+1)} - X_{T\wedge n} = \begin{cases} 0 & \text{if } T \leq n \\ X_{n+1} - X_n & \text{if } T > n \end{cases}$$
$$= (X_{n+1} - X_n)I_{\{T>n\}}.$$

Using this and the tower property of conditional expectation yields

$$\begin{aligned} E[X_{T\wedge(n+1)} - X_{T\wedge n}] &= E[E[(X_{n+1} - X_n)I_{\{T>n\}}|\mathcal{F}_n]] \\ &= E[E[X_{n+1} - X_n|\mathcal{F}_n]I_{\{T>n\}}] = 0 \end{aligned}$$

because $E[X_{n+1} - X_n|\mathcal{F}_n] = 0$. Therefore, $E[X_{T\wedge(n+1)}] = E[X_{T\wedge n}]$ for all $n \geq 0$. So by
induction on n, $E[X_{T\wedge n}] = E[X_0]$ for all $n \geq 0$. □

The following proposition follows immediately from Lemma 10.13. This propositio
and the corollaries following it represent a version of the *optional sampling theorem.*

Proposition 10.14 *If X is a martingale and T is an optional stopping time relative to $(\Omega, \mathcal{F}, P)$, then $E[X_0] = \lim_{n\to\infty} E[X_{T\wedge n}]$. In particular, if*

$$\lim_{n\to\infty} E[X_{T\wedge n}] = E[X_T] \tag{10.8}$$

then $E[X_T] = E[X_0]$.

By Proposition 10.14, $E[X_T] = E[X_0]$ holds true under any additional assumptions strong enough to imply (10.8). Note that $X_{T\wedge n} \stackrel{a.s.}{\to} X_T$ as $n \to \infty$, so (10.8) is simply requiring the convergence of the means to the mean of the limit, for an a.s. convergent sequence of random variables. There are several different sufficient conditions for this to happen, involving conditions on the martingale X, the stopping time T, or both. For example:

Corollary 10.15 *If X is a (discrete-time) martingale and T is an optional stopping time relative to $(\Omega, \mathcal{F}, P)$, and if T is bounded (so $P\{T \leq n_o\} = 1$ for some n_o) then $E[X_T] = E[X_0]$.*

Proof If $P\{T \leq n_o\} = 1$ then, for all $n \geq n_o$, $T \wedge n = T$ a.s, and so $E[X_{T\wedge n}] = E[X_T]$. Therefore, the corollary follows from Proposition 10.14. □

Corollary 10.16 *If X is a martingale and T is an optional stopping time relative to $(\Omega, \mathcal{F}, P)$, and if there is a random variable Y such that $|X_n| \leq Y$ a.s. for all n, and $E[Y] < \infty$, then $E[X_T] = E[X_0]$.*

Proof The assumptions imply $|X_{T\wedge n}| \leq Y$ a.s. for all n, so the dominated convergence theorem (Theorem 11.12 in the Appendix) implies (10.8). Thus the result follows from Proposition 10.14. □

Corollary 10.17 *Suppose $(X_n : n \geq 0)$ is a martingale relative to $(\Omega, \mathcal{F}, P)$ such that: (i) there is a constant c such that $E[|X_{n+1} - X_n| \,|\mathcal{F}_n] \leq c$ for $n \geq 0$, and (ii) T is stopping time such that $E[T] < \infty$. Then $E[X_T] = E[X_0]$.*

If, instead, $(X_n : n \geq 0)$ is a submartingale relative to $(\Omega, \mathcal{F}, P)$, satisfying (i) and (ii), then $E[X_T] \geq E[X_0]$.

Proof Suppose $(X_n : n \geq 0)$ is a martingale relative to $(\Omega, \mathcal{F}, P)$, satisfying (i) and (ii). We shall apply the dominated convergence theorem as in the proof of Corollary 10.16. Let Y be defined by

$$Y = |X_0| + |X_1 - X_0| + \ldots + |X_T - X_{T-1}|.$$

Clearly $|X_{T\wedge n}| \leq Y$ for all $n \geq 0$, so to apply the dominated convergence theorem it remains to show that $E[Y] < \infty$. But

$$\begin{aligned} E[Y] &= E[|X_0|] + E\left[\sum_{i=1}^{\infty} |X_i - X_{i-1}| I_{\{i \leq T\}}\right] \\ &= E[|X_0|] + E\left[\sum_{i=1}^{\infty} E\left[|X_i - X_{i-1}| I_{\{i \leq T\}} \mid \mathcal{F}_{i-1}\right]\right] \\ &= E[|X_0|] + E\left[\sum_{i=1}^{\infty} I_{\{i \leq T\}} E\left[|X_i - X_{i-1}| \mid \mathcal{F}_{i-1}\right]\right] \\ &= E[|X_0|] + cE[T] < \infty. \end{aligned}$$

The first statement of the corollary is proved. If instead X is a submartingale, then minor variation of Lemma 10.13 yields that $E[X_{T\wedge n}] \geq E[X_0]$. The proof for the first pa of the corollary, already given, shows that conditions (i) and (ii) imply that $E[X_{T\wedge n}] \to$ $E[X_T]$ as $n \to \infty$. Therefore, $E[X_T] \geq E[X_0]$. □

Martingale inequalities offer a way to provide upper and lower bounds on the comple tion times of algorithms. As an illustration, the following example shows how a lowe bound can be found for a particular game.

Example 10.13 Consider the following game. There is an urn, initially with r re marbles and b blue marbles. A player takes turns until the urn is empty, and the goa of the player is to minimize the expected number of turns required. At the beginning o each turn, the player can remove a set of marbles, and the set must be one of four types one red, one blue, one red and one blue, or two red and two blue. After removing the se of marbles, a fair coin is flipped. If tails appears, the turn is over. If heads appears, the some marbles are added back to the bag, according to Table 10.1. Our goal is to find lower bound on $E[T]$, where T is the number of turns needed by the player until the ur is empty. The bound should hold for any strategy the player adopts. Let X_n denote th total number of marbles in the urn after n turns. If the player elects to remove only on marble during a turn (either red or blue) then with probability one half, two marbles ar put back. Hence, for either set with one marble, the expected change in the total numbe of marbles in the urn is zero. If the player elects to remove two reds or two blues, the with probability one half, three marbles are put back into the urn. For these turns, th expected change in the number of marbles in the urn is -0.5. Hence, for any choice o u_n (representing the decision of the player for the $n+1$th turn),

$$E[X_{n+1}|X_n, u_n] \geq X_n - 0.5 \quad \text{on } \{T > n\}.$$

That is, the drift of X_n towards zero is at most 0.5 in magnitude, so we suspect tha no strategy can empty the urn in average time less than $(r+b)/0.5$. In fact, this resul is true, and it is now proved. Let $M_n = X_{n\wedge T} + \frac{n\wedge T}{2}$. By the observations above, M is a submartingale. Furthermore, $|M_{n+1} - M_n| \leq 2$. Either $E[T] = +\infty$ or $E[T] < \infty$ If $E[T] = +\infty$ then the inequality to be proved, $E[T] \geq 2(r+b)$, is trivially true, so suppose $E[T] < \infty$. Then by Corollary 10.17, $E[M_T] \geq E[M_0] = r+b$. Also, $M_T = \frac{T}{2}$ with probability one, so $E[T] \geq 2(r+b)$, as claimed.

Table 10.1 Rules of the marble game

Set removed	Set returned to bag on "heads"
one red	one red and one blue
one blue	one red and one blue
two reds	three blues
two blues	three reds

10.5 Notes

Material on Azuma–Hoeffding inequality and McDiarmid's method can be found in McDiarmid's tutorial article (McDiarmid 1989).

Problems

10.1 Two martingales associated with a simple branching process Let $G=(G_n : n \geq 0)$ denote the Galton–Watson branching process with random variable X denoting the number of offspring of a typical individual, as in Example 10.5.
(a) Identify the constant θ so that $\frac{G_n}{\theta^n}$ is a martingale.
(b) Let $\mathcal{E}$ denote the event of eventual extinction, and let $\alpha = P\{\mathcal{E}\}$. Show that $P(\mathcal{E}|G_0, \ldots, G_n) = \alpha^{G_n}$. Thus, $M_n = \alpha^{G_n}$ is a Doob martingale based on the random variable $\Phi = I_{\mathcal{E}}$.
(c) Using the fact $E[M_1] = E[M_0]$, find an equation for α. (It can be shown that α is the smallest positive solution to the equation, and $\alpha < 1$ if and only if $E[X] > 1$.)

10.2 A covering problem Consider a linear array of n cells. Suppose that m base stations are randomly placed among the cells, such that the locations of the base stations are independent and uniformly distributed among the n cell locations. Let r be a positive integer. Call a cell i covered if there is at least one base station at some cell j with $|i-j| \leq r-1$. Thus, each base station (except those near the edge of the array) covers $2r-1$ cells. Note that there can be more than one base station at a given cell, and interference between base stations is ignored.
(a) Let F denote the number of cells covered. Apply the method of bounded differences based on the Azuma–Hoeffding inequality to find an upper bound on $P\{|F-E[F]| \geq \gamma\}$.
(b) (This part is related to the coupon collector's problem, Problem 4.39, and may not have anything to do with martingales.) Rather than fixing the number of base stations, m, let X denote the number of base stations needed until all cells are covered. When $r=1$ we have seen that $P\{X \geq n \ln n + cn\} \to \exp(-e^{-c})$ (the coupon collector's problem). For general $r \geq 1$, find $g_1(r)$ and $g_2(r)$ so that for any $\epsilon > 0$, $P\{X \geq (g_2(r)+\epsilon) n \ln n\} \to 0$ and $P\{X \leq (g_1(r)-\epsilon) n \ln n\} \to 0$. (Ideally you can find a solution with $g_1 = g_2$, but if not, it would be nice if they are close.)

10.3 Doob decomposition (a) Show that a predictable martingale difference sequence is identically zero with probability one.
(b) Using your answer to part (a), show that the Doob decomposition described in Example 10.7 is unique up to events of probability zero.

10.4 Stopping time properties (a) Show that if S and T are stopping times for some filtration $\mathcal{F}$, then $S \wedge T$, $S \vee T$, and $S+T$ are also stopping times.
(b) Show that if $\mathcal{F}$ is a filtration and $X = (X_k : k \geq 0)$ is the random sequence defined by $X_k = I_{\{T \leq k\}}$ for some random time T with values in $\mathbb{Z}_+$, then T is a stopping time if and only if X is $\mathcal{F}$-adapted.
(c) If T is a stopping time for a filtration $\mathcal{F}$, then $\mathcal{F}_T$ is defined to be the set of events A such that $A \cap \{T \leq n\} \in \mathcal{F}_n$ for all n. (Or, for discrete time, the set of events A such that $A \cap \{T = n\} \in \mathcal{F}_n$ for all n.) Show that (i) $\mathcal{F}_T$ is a σ-algebra, (ii) T is $\mathcal{F}_T$ measurable, and (iii) if X is an adapted process then X_T is $\mathcal{F}_T$ measurable.

10.5 A stopped random walk Let $W_1, W_2, \ldots$ be a sequence of independent, identicall distributed mean zero random variables. with $P\{W_1 = 0\} \neq 1$. Let $S_0 = 0$ and $S_n =$ $W_1 + \ldots + W_n$ for $n \geq 1$. Fix a constant $c > 0$ and let $T = \min\{n \geq 0 : |S_n| \geq c\}$. Th goal of this problem is to show that $E[S_T] = 0$.
(a) Show that $E[S_T] = 0$ if there is a constant D so that $P\{|W_i| > D\} = 0$. (Hint: Invok a version of the optional stopping theorem.)
(b) In view of part (a), we need to address the case that the Ws are not bounded. Le

$$\widetilde{W}_n = \begin{cases} W_n & \text{if } |W_n| \leq 2c \\ a & \text{if } W_n > 2c \\ -b & \text{if } W_n < -2c \end{cases}$$ where the constants a and b are selected so that $a \geq 2c$

$b \geq 2c$, and $E[\widetilde{W}_n] = 0$. Let $\widetilde{S}_n = \widetilde{W}_1 + \ldots + \widetilde{W}_n$ for $n \geq 0$. Note that if $T < n$ an if $\widetilde{W}_n \neq W_n$, then $T = n$. Therefore, $T = \min\{n \geq 0 : |\widetilde{S}_n| \geq c\}$. Let $\widetilde{\sigma}^2 = \mathrm{Var}(\widetilde{W}_n$ and let $M_n = \widetilde{S}_n^2 - n\widetilde{\sigma}^2$. Show that M is a martingale. Hence, $E[M_{T \wedge n}] = 0$ for all n Conclude that $E[T] < \infty$.
(c) Show that $E[S_T] = 0$. (Hint: Use part (b) and invoke a version of the optiona stopping theorem.)

10.6 Bounding the value of a game Example 10.13 considers a game with marbles an shows that the mean completion time T satisfies $E[T] \geq 2(r+b)$ no matter what strateg is used. Suggest a strategy that approximately minimizes $E[T]$, and for that strategy, fin an upper bound on $E[T]$.

10.7 On the size of a maximum matching in a random bipartite graph
Given $1 \leq d < n$, let $U = \{u_1, \ldots, u_n\}$ and $V = \{v_1, \ldots, v_n\}$ be disjoint sets o cardinality n, and let G be a bipartite random graph with vertex set $U \cup V$, such tha if V_i denotes the set of neighbors of u_i, then $V_1, \ldots, V_n$ are independent, and each i uniformly distributed over the set of all $\binom{n}{d}$ subsets of V of cardinality d. A matchin for G is a subset of edges M such that no two edges in M have a common vertex. Let Z denote the maximum of the cardinalities of the matchings for G.
(a) Find bounds a and b, with $0 < a \leq b < n$, so that $a \leq E[Z] \leq b$.
(b) Give an upper bound on $P\{|Z - E[Z]| \geq \gamma\sqrt{n}\}$, for $\gamma > 0$, showing that for fixed d the distribution of Z is concentrated about its mean as $n \to \infty$.
(c) Suggest a greedy algorithm for finding a large cardinality matching.

10.8* On random variables of the form *g(Y)* Let Y and Z be random variables on th same probability space. The purpose of this problem is to establish $Z = g(Y)$ for som Borel measurable function g if and only if Z is $\sigma(Y)$ measurable.

("only if" part) Suppose $Z = g(Y)$ for a Borel measurable function g, and let $c \in \mathbb{R}$ It must be shown that $\{Z \leq c\} \in \sigma(Y)$. Since g is a Borel measurable function, by definition, $A = \{y : g(y) \leq c\}$ is a Borel subset of $\mathbb{R}$.
(a) Show $\{Z \leq c\} = \{Y \in A\}$.
(b) Using the definition of Borel sets, show $\{Y \in A\} \in \sigma(Y)$ for any Borel set A. The "only if" part follows.

("if" part) Suppose Z is $\sigma(Y)$ measurable. It must be shown Z has the form $g(Y)$ fo some Borel measurable function g.
(c) Prove this first in the special case Z has the form of an indicator function: $Z = I_B$ for some event B, which satisfies $B \in \sigma(Y)$. (Hint: Appeal to the definition of $\sigma(Y)$.)

(d) Prove the "if" part in general. (Hint: Z can be written as the supremum of a countable set of random variables, with each being a constant times an indicator function: $Z = \sup_n q_n I_{\{Z \le q_n\}}$, where $q_1, q_2, \ldots$ is an enumeration of the set of rational numbers.)

10.9* Regular conditional distributions Let X be a random variable on $(\Omega, \mathcal{F}, P)$ and let $\mathcal{D}$ be a sub-σ-algebra of $\mathcal{F}$. Conditional probabilities such as $P(X \le c|\mathcal{D})$ for a fixed constant c are defined by applying Definition 10.2 of conditional expectation to the indicator random variable, $I_{\{X \le c\}}$. This can sometimes have different versions, but any two such versions are equal with probability one. The idea of regular conditional distributions is to select a version of $P(X \le c|\mathcal{D})$ for every real number c so that, as a function of c for ω fixed, the result is a valid CDF (i.e. nondecreasing, right-continuous, with limit zero at $-\infty$ and limit one at $+\infty$). The difficulty is that there are uncountably many choices of c. Here is the definition. A *regular conditional CDF* of X given D, denoted by $F_{X|\mathcal{D}}(c|\omega)$, is a function of $(c, \omega) \in \mathbb{R} \times \Omega$ such that:

(1) for each ω fixed, as a function of c, $F_{X|\mathcal{D}}(c|\omega)$ is a valid CDF,

(2) for any $c \in \mathbb{R}$, as a function of ω, $F_{X|\mathcal{D}}(c|\omega)$ is a version of $P(X \le c|\mathcal{D})$.

The purpose of this problem is to prove the existence of a regular conditional CDF. For each rational number q, let $\Phi(q) = P(X \le q|\mathcal{D})$. That is, for each rational number q, pick $\Phi(q)$ to be one particular version of $P(X \le q|\mathcal{D})$. We sometimes write $\Phi(q, \omega)$ for the random variable $\Phi(q)$ to make explicit the dependence on ω. By the positivity preserving property of conditional expectations, $P\{0 \le \Phi(q) < \Phi(q') \le 1\} = 1$ if $q < q'$. Let $\{q_1, q_2, \ldots\}$ denote the set of rational numbers, listed in some order, and define the event G_1 by

$$G_1 = \cap_{n,m:q_n<q_m}\{0 \le \Phi(q_n) \le \Phi(q_m) \le 1\}.$$

Then $P(G_1) = 1$ because G_1 is the intersection of countably many events of probability one. The limits $\Phi(-\infty) \stackrel{\triangle}{=} \lim_{n\to-\infty} \Phi(n)$ and $\Phi(\infty) \stackrel{\triangle}{=} \lim_{n\to\infty} \Phi(n)$ both exist and take values in $[0, 1]$ for all $\omega \in G_1$ because bounded monotone sequences in the closed, bounded interval $[0, 1]$ have limits in the interval. Let $\Phi(-\infty) = 0$ and $\Phi(\infty) = 1$ for $\omega \in G_1^c$. Since $E[|\Phi(n) - 0|] = E[\Phi(n)] = P\{X \le n\} \to 0$ as $n \to -\infty$, it follows that $\Phi(n) \stackrel{p.}{\to} 0$ as $n \to -\infty$. Since the limit random variables for convergence in the p. and a.s. senses must be equal with probability one, $\Phi(-\infty) = 0$ with probability one. Likewise, $\Phi(\infty) = 1$ with probability one. Let $G_2 = G_1 \cap \{\Phi(-\infty) = 0\} \cap \{\Phi(\infty) = 1\}$. Then $P(G_2) = 1$.

Modify $\Phi(q, \omega)$ for $\omega \in G_2^c$ by letting $\Phi(q, \omega) = F_o(q)$ for $\omega \in G_2^c$ and all rational q, where F_o is an arbitrary, fixed CDF. Then for any $c \in \mathbb{R}$ and $\omega \in \Omega$, let

$$\Psi(c, \omega) = \inf_{q>c} \Phi(q, \omega).$$

Show that Ψ is a regular conditional CDF of X given $\mathcal{D}$. (Hint: To prove that $\Psi(c, \cdot)$ is a version of $P(X \le c|\mathcal{D})$ for any fixed $c \in \mathbb{R}$, use the definitions of $P(X \le c|\mathcal{D})$.)

10.10* An even more general definition of conditional expectation Let X be a random variable on $(\Omega, \mathcal{F}, P)$ and let $\mathcal{D}$ be a sub-σ-algebra of $\mathcal{F}$. Let $F_{X|\mathcal{D}}(c|\omega)$ be a regular conditional CDF of X given $\mathcal{D}$. Then for each ω, we can define $E[X|\mathcal{D}]$ at ω to equal the mean for the CDF $F_{X|\mathcal{D}}(\cdot|\omega)$, which is contained in the extended real lin $\mathbb{R} \cup \{-\infty, +\infty\}$. Symbolically: $E[X|\mathcal{D}](\omega) = \int_{\mathbb{R}} c F_{X|\mathcal{D}}(dc|\omega)$. Show, in the specia case $E[|X|] < \infty$, this definition is consistent with the one given previously. As a application, the following conditional version of Jensen's inequality holds: If φ is convex function on $\mathbb{R}$, then $E[\varphi(X)|\mathcal{D}] \geq \varphi(E[X|\mathcal{D}])$ a.s. The proof is given by applyin the ordinary Jensen's inequality for each ω fixed, for the regular conditional CDF of given $\mathcal{D}$ evaluated at ω.

11 Appendix

11.1 Some notation

The following notational conventions are used in this book.

$$A^c = \text{ complement of } A$$

$$AB = A \cap B$$

$$A \subset B \leftrightarrow \text{any element of } A \text{ is also an element of } B$$

$$A - B = AB^c$$

$$\bigcup_{i=1}^{\infty} A_i = \{a : a \in A_i \text{ for some } i\}$$

$$\bigcap_{i=1}^{\infty} A_i = \{a : a \in A_i \text{ for all } i\}$$

$$a \vee b \ = \ \max\{a, b\} = \begin{cases} a & \text{if} \quad a \geq b \\ b & \text{if} \quad a < b \end{cases}$$

$$a \wedge b = \min\{a, b\}$$

$$a_+ = a \vee 0 = \max\{a, 0\}$$

$$I_A(x) = \begin{cases} 1 & \text{if } x \in A \\ 0 & \text{else} \end{cases}$$

$$(a, b) = \{x : a < x < b\} \quad (a, b] = \{x : a < x \leq b\}$$

$$[a, b) = \{x : a \leq x < b\} \quad [a, b] = \{x : a \leq x \leq b\}$$

$$\mathbb{Z} - \text{set of integers}$$

$$\mathbb{Z}_+ - \text{set of nonnegative integers}$$

$$\mathbb{R} - \text{set of real numbers}$$

$$\mathbb{R}_+ - \text{set of nonnegative real numbers}$$

$$\mathbb{C} = \text{set of complex numbers}$$

$$A_1 \times \ldots \times A_n = \{(a_1, \ldots, a_n)^T : a_i \in A_i \text{ for } 1 \leq i \leq n\}$$

$$A^n = \underbrace{A \times \ldots \times A}_{n \text{ times}}$$

$$\lfloor t \rfloor = \text{greatest integer } n \text{ such that } n \leq t$$
$$\lceil t \rceil = \text{least integer } n \text{ such that } n \geq t$$

$$A \triangleq \text{expression} - \text{denotes that } A \text{ is defined by the expression}$$

All the trigonometric identities required in these notes can be easily derived from th
two identities:

$$\cos(a+b) = \cos(a)\cos(b) - \sin(a)\sin(b)$$
$$\sin(a+b) = \sin(a)\cos(b) + \cos(a)\sin(b)$$

and the facts $\cos(-a) = \cos(a)$ and $\sin(-b) = -\sin(b)$.

A set of numbers is *countably infinite* if the numbers in the set can be listed in sequence $x_i : i = 1, 2, \ldots$ For example, the set of rational numbers is countably infinit but the set of all real numbers in any interval of positive length is not countably infinit

11.2 Convergence of sequences of numbers

We begin with some basic definitions. Let $(x_n) = (x_1, x_2, \ldots)$ and $(y_n) = (y_1, y_2, \ldots)$ b sequences of numbers and let x be a number. By definition, x_n converges to x as $n \to \infty$ if for each $\epsilon > 0$ there is an integer n_ϵ so that $|\, x_n - x \,| < \epsilon$ for every $n \geq n_\epsilon$. We writ $\lim_{n\to\infty} x_n = x$ to denote that x_n converges to x.

Example 11.1 Let $x_n = \frac{2n+4}{n^2+1}$. Let us verify that $\lim_{n\to\infty} x_n = 0$. The inequality $|\, x_n \,| <$ ϵ holds if $2n + 4 \leq \epsilon(n^2 + 1)$. Therefore it holds if $2n + 4 \leq \epsilon n^2$. Therefore it hold if both $2n \leq \frac{\epsilon}{2} n^2$ and $4 \leq \frac{\epsilon}{2} n^2$. So if $n_\epsilon = \left\lceil \max\left\{ \frac{4}{\epsilon}, \sqrt{\frac{8}{\epsilon}} \right\} \right\rceil$ then $n \geq n_\epsilon$ implies tha $|\, x_n \,| < \epsilon$. So $\lim_{n\to\infty} x_n = 0$.

By definition, (x_n) converges to $+\infty$ as $n \to \infty$ if for every $K > 0$ there is an intege n_K so that $x_n \geq K$ for every $n \geq n_K$. Convergence to $-\infty$ is defined in a similar way. For example, $n^3 \to \infty$ as $n \to \infty$ and $n^3 - 2n^4 \to -\infty$ as $n \to \infty$.

Occasionally a two-dimensional array of numbers $(a_{m,n} : m \geq 1, n \geq 1)$ is considered. By definition, $a_{m,n}$ converges to a number a^* as m and n jointly go to infinity i for each $\epsilon > 0$ there is $n_\epsilon > 0$ so that $|\, a_{m,n} - a^* \,| < \epsilon$ for every $m, n \geq n_\epsilon$. We writ $\lim_{m,n\to\infty} a_{m,n} = a$ to denote that $a_{m,n}$ converges to a as m and n jointly go to infinity.

Theoretical exercises

1. Let $a_{m,n} = 1$ if $m = n$ and $a_{m,n} = 0$ if $m \neq n$. Show $\lim_{n\to\infty}(\lim_{m\to\infty} a_{m,n}) =$ $\lim_{m\to\infty}(\lim_{n\to\infty} a_{mn}) = 0$ but $\lim_{m,n\to\infty} a_{m,n}$ does not exist.

[1] Some authors reserve the word "convergence" for convergence to a finite limit. When we say a sequenc converges to $+\infty$ some would say the sequence diverges to $+\infty$.

2. Let $a_{m,n} = \frac{(-1)^{m+n}}{\min(m,n)}$. Show that $\lim_{m\to\infty} a_{m,n}$ does not exist for any n and $\lim_{n\to\infty} a_{m,n}$ does not exist for any m, but $\lim_{m,n\to\infty} a_{m,n} = 0$.
3. If $\lim_{m,n\to\infty} a_{mn} = a*$ and $\lim_{m\to\infty} a_{mn} = b_n$ for each n, show $\lim_{n\to\infty} b_n = a^*$.

The condition $\lim_{m,n\to\infty} a_{m,n} = a*$ can be expressed in terms of convergence of sequences depending on only one index (as can all the other limits discussed in these notes) as follows. Namely, $\lim_{m,n\to\infty} a_{m,n} = a*$ is equivalent to the following: $\lim_{k\to\infty} a_{m_k,n_k} = a^*$ whenever $((m_k, n_k) : k \geq 1)$ is a sequence of pairs of positive integers such that $m_k \to \infty$ and $n_k \to \infty$ as $k \to \infty$. The condition that the limit $\lim_{m,n\to\infty} a_{m,n}$ exists is equivalent to the condition that the limit $\lim_{k\to\infty} a_{m_k,n_k}$ exists whenever $((m_k, n_k) : k \geq 1)$ is a sequence of pairs of positive integers such that $m_k \to \infty$ and $n_k \to \infty$ as $k \to \infty$.[2]

A sequence $a_1, a_2, \ldots$ is said to be nondecreasing if $a_i \leq a_j$ for $i < j$. Similarly a function f on the real line is nondecreasing if $f(x) \leq f(y)$ whenever $x < y$. The sequence is called *strictly increasing* if $a_i < a_j$ for $i < j$ and the function is called strictly increasing if $f(x) < f(y)$ whenever $x < y$.[3] A strictly increasing or strictly decreasing sequence is said to be strictly monotone, and a nondecreasing or nonincreasing sequence is said to be monotone.

The sum of an infinite sequence is defined to be the limit of the partial sums. That is, by definition,

$$\sum_{k=1}^{\infty} y_k = x \text{ means that } \lim_{n\to\infty} \sum_{k=1}^{n} y_k = x.$$

Often we want to show that a sequence converges even if we don't explicitly know the value of the limit. A sequence (x_n) is bounded if there is a number L so that $| x_n | \leq L$ for all n. Any sequence that is bounded and monotone converges to a finite number.

Example 11.2 Consider the sum $\sum_{k=1}^{\infty} k^{-\alpha}$ for a constant $\alpha > 1$. For each n the nth partial sum can be bounded by comparison to an integral, based on the fact that for $k \geq 2$, the kth term of the sum is less than the integral of $x^{-\alpha}$ over the interval $[k-1, k]$:

$$\sum_{k=1}^{n} k^{-\alpha} \leq 1 + \int_1^n x^{-\alpha} dx = 1 + \frac{1 - n^{1-\alpha}}{(\alpha - 1)} \leq 1 + \frac{1}{\alpha - 1} = \frac{\alpha}{\alpha - 1}.$$

The partial sums are also monotone nondecreasing (in fact, strictly increasing). Therefore the sum $\sum_{k=1}^{\infty} k^{-\alpha}$ exists and is finite.

---

[2] We could add here the condition that the limit should be the same for all choices of sequences, but it is automatically true. If two sequences were to yield different limits of a_{m_k,n_k}, a third sequence could be constructed by interleaving the first two, and a_{m_k,n_k} would not be convergent for that sequence.

[3] We avoid simply saying "increasing," because for some authors it means strictly increasing and for other authors it means nondecreasing. While inelegant, our approach is safer.

A sequence (x_n) is a *Cauchy sequence* if $\lim_{m,n\to\infty} \mid x_m - x_n \mid = 0$. It is not har to show that if x_n converges to a finite limit x then (x_n) is a Cauchy sequence. Mor useful is the converse statement, called the *Cauchy criteria* for convergence, or th *completeness* property of $\mathbb{R}$: If (x_n) is a Cauchy sequence then x_n converges to a finit limit as $n \to \infty$.

Example 11.3 Suppose $(x_n : n \geq 1)$ is a sequence such that $\sum_{i=1}^{\infty} |x_{i+1} - x_i| < \infty$ The Cauchy criteria can be used to show that the sequence $(x_n : n \geq 1)$ is convergen Suppose $1 \leq m < n$. Then by the triangle inequality for absolute values:

$$|x_n - x_m| \leq \sum_{i=m}^{n-1} |x_{i+1} - x_i|$$

or, equivalently,

$$|x_n - x_m| \leq \left| \sum_{i=1}^{n-1} |x_{i+1} - x_i| - \sum_{i=1}^{m-1} |x_{i+1} - x_i| \right|. \quad (11.1$$

Inequality (11.1) also holds if $1 \leq n \leq m$. By definition of the sum, $\sum_{i=1}^{\infty} |x_{i+1} - x_i$ both sums on the right side of (11.1) converge to $\sum_{i=1}^{\infty} |x_{i+1} - x_i|$ as $m, n \to \infty$, so th right side of (11.1) converges to zero as $m, n \to \infty$. Thus, (x_n) is a Cauchy sequence and it is hence convergent.

Theoretical exercises

1. Show that if $\lim_{n\to\infty} x_n = x$ and $\lim_{n\to\infty} y_n = y$ then $\lim_{n\to\infty} x_n y_n = xy$.
2. Find the limits and prove convergence as $n \to \infty$ for the following sequences:
   (a) $x_n = \frac{\cos(n^2)}{n^2+1}$, (b) $y_n = \frac{n^2}{\log n}$, (c) $z_n = \sum_{k=2}^{n} \frac{1}{k \log k}$.

The *minimum* of a set of numbers, A, written $\min A$, is the smallest number in th set, if there is one. For example, $\min\{3, 5, 19, -2\} = -2$. Of course, $\min A$ is wel defined if A is finite (i.e. has finite cardinality). Some sets fail to have a minimum for example neither $\{1, 1/2, 1/3, 1/4, \ldots\}$ nor $\{0, -1, -2, \ldots\}$ have a smallest numbe The *infimum* of a set of numbers A, written $\inf A$, is the greatest lower bound for A If A is bounded below, then $\inf A = \max\{c : c \leq a \text{ for all } a \in A\}$. For example $\inf\{1, 1/2, 1/3, 1/4, \ldots\} = 0$. If there is no finite lower bound, the infimum is $-\infty$. Fo example, $\inf\{0, -1, -2, \ldots\} = -\infty$. By convention, the infimum of the empty set i $+\infty$. With these conventions, if $A \subset B$ then $\inf A \geq \inf B$. The infimum of any subset o $\mathbb{R}$ exists, and if $\min A$ exists, then $\min A = \inf A$, so the notion of infimum extends th notion of minimum to all subsets of $\mathbb{R}$.

Similarly, the *maximum* of a set of numbers A, written $\max A$, is the largest numbe in the set, if there is one. The *supremum* of a set of numbers A, written $\sup A$, is the leas upper bound for A. We have $\sup A = -\inf\{-a : a \in A\}$. In particular, $\sup A = +\infty$ if A is not bounded above, and $\sup \emptyset = -\infty$. The supremum of any subset of $\mathbb{R}$ exists, an if $\max A$ exists, then $\max A = \sup A$, so the notion of supremum extends the notion of maximum to all subsets of $\mathbb{R}$.

The notions of infimum and supremum of a set of numbers are useful because they exist for any set of numbers. There is a pair of related notions that generalizes the notion of limit. Not every sequence has a limit, but the following terminology is useful for describing the limiting behavior of a sequence, whether or not the sequence has a limit.

Definition 11.1 The *liminf* (also called *limit inferior*) of a sequence $(x_n : n \geq 1)$, is defined by

$$\liminf_{n\to\infty} x_n = \lim_{n\to\infty} [\inf\{x_k : k \geq n\}], \tag{11.2}$$

and the *limsup* (also called *limit superior*) is defined by

$$\limsup_{n\to\infty} x_n = \lim_{n\to\infty} \left[\sup\{x_k : k \geq n\}\right]. \tag{11.3}$$

The possible values of the liminf and limsup of a sequence are $\mathbb{R} \cup \{-\infty, +\infty\}$.

The limit on the right side of (11.2) exists because the infimum inside the square brackets is monotone nondecreasing in n. Similarly, the limit on the right side of (11.3) exists. So every sequence of numbers has a liminf and limsup.

Definition 11.2 A *subsequence* of a sequence $(x_n : n \geq 1)$ is a sequence of the form $(x_{k_i} : i \geq 1)$, where $k_1, k_2, \ldots$ is a strictly increasing sequence of integers. The set of *limit points* of a sequence is the set of all limits of convergent subsequences. The values $-\infty$ and $+\infty$ are possible limit points.

Example 11.4 Suppose $y_n = 121 - 25n^2$ for $n \leq 100$ and $y_n = 1/n$ for $n \geq 101$. The liminf and limsup of a sequence do not depend on any finite number of terms of the sequence, so the values of y_n for $n \leq 100$ are irrelevant. For all $n \geq 101$, $\inf\{x_k : k \geq n\} = \inf\{1/n, 1/(n+1), \ldots\} = 0$, which trivially converges to 0 as $n \to \infty$. So the liminf of (y_n) is zero. For all $n \geq 101$, $\sup\{x_k : k \geq n\} = \sup\{1/n, 1/(n+1), \ldots\} = \frac{1}{n}$, which converges also to 0 at $n \to \infty$. So the limsup of (y_n) is also zero. Zero is also the only limit point of (y_n).

Example 11.5 Consider the sequence of numbers $(2, -3/2, 4/3, -5/4, 6/5, \ldots)$, which we also write as $(x_n : n \geq 1)$ such that $x_n = \frac{(n+1)(-1)^{n+1}}{n}$. The maximum (and supremum) of the sequence is 2, and the minimum (and infimum) of the sequence is $-3/2$. But for large n, the sequence alternates between numbers near one and numbers near minus one. More precisely, the subsequence of odd numbered terms, $(x_{2i-1} : i \geq 1)$, converges to 1, and the subsequence of even numbered terms, $(x_{2i} : i \geq 1)$, has limit -1. Thus, both 1 and -1 are limit points of the sequence, and there are no other limit points. The overall sequence itself does not converge (i.e. does not have a limit) but $\liminf_{n\to\infty} x_n = -1$ and $\limsup_{n\to\infty} x_n = +1$.

Some simple facts about the limit, liminf, limsup, and limit points of a sequence are collected in the following proposition. The proof is left to the reader.

Proposition 11.3 *Let $(x_n : n \geq 1)$ denote a sequence of numbers.*

1. *The condition* $\liminf_{n\to\infty} x_n \geq x_\infty$ *is equivalent to the following: for any $\gamma < x_\infty$, $x_n \geq \gamma$ for all sufficiently large n.*
2. *The condition* $\limsup_{n\to\infty} x_n \leq x_\infty$ *is equivalent to the following: for any $\gamma > x_\infty$, $x_n \leq \gamma$ for all sufficiently large n.*
3. $\liminf_{n\to\infty} x_n \leq \limsup_{n\to\infty} x_n$.
4. $\lim_{n\to\infty} x_n$ *exists if and only if the liminf equals the limsup, and if the limit exists, then the limit, liminf, and limsup are equal.*
5. $\lim_{n\to\infty} x_n$ *exists if and only if the sequence has exactly one limit point, x^*, and if the limit exists, it is equal to that one limit point.*
6. *Both the liminf and limsup of the sequence are limit points. The liminf is the smallest limit point and the limsup is the largest limit point (keep in mind that $-\infty$ and $+\infty$ are possible values of the liminf, limsup, or a limit point).*

Theoretical exercises

1. Prove Proposition 11.3
2. Here's a more challenging one. Let r be an irrational constant, and let $x_n = nr - \lfloor nr \rfloor$ for $n \geq 1$. Show that every point in the interval $[0, 1]$ is a limit point of $(x_n : n \geq 1)$. (P. Bohl, W. Sierpinski, and H. Weyl independently proved a stronger result in 1909–1910: namely, the fraction of the first n values falling into a subinterval of $[0,1]$ converges to the length of the subinterval.)

11.3 Continuity of functions

Let f be a function on $\mathbb{R}^n$ for some n, and let $x_o \in \mathbb{R}^n$. The function has a limit y at x_o and such situation is denoted by $\lim_{x\to x_o} f(x) = y$, if the following is true. Given $\epsilon > 0$ there exists $\delta > 0$ so that $| f(x) - y | \leq \epsilon$ whenever $0 < \|x - x_o\| < \delta$. This convergence condition can also be expressed in terms of convergence of sequences, as follows. The condition $\lim_{x\to x_o} f(x) = y$ is equivalent to the condition $f(x_n) \to y$ for any sequence $x_1, x_2, \ldots$ from $\mathbb{R}^n - x_o$ such that $x_n \to x_o$.

The function f is said to be continuous at x_o, or equivalently, x_o is said to be a *continuity point* of f, if $\lim_{x\to x_o} f(x) = f(x_o)$. In terms of sequences, f is continuous at x_o if $f(x_n) \to f(x_o)$ whenever $x_1, x_2, \ldots$ is a sequence converging to x_o. The function f is simply said to be continuous if it is continuous at every point in $\mathbb{R}^n$.

Let $n = 1$, so consider a function f on $\mathbb{R}$, and let $x_o \in \mathbb{R}$. The function has a righthand limit y at x_o, and such situation is denoted by $f(x_o+) = y$ or $\lim_{x\searrow x_o} f(x) = y$, if the following is true. Given $\epsilon > 0$, there exists $\delta > 0$ so that $| f(x) - y | \leq \epsilon$ whenever $0 < x - x_o < \delta$. Equivalently, $f(x_o+) = y$ if $f(x_n) \to y$ for any sequence $x_1, x_2, \ldots$ from $(x_o, +\infty)$ such that $x_n \to x_o$. The lefthand limit $f(x_o-) = \lim_{x\nearrow x_o} f(x)$ is defined similarly. If f is monotone nondecreasing, then the lefthand and righthand limits exist, and $f(x_o-) \leq f(x_o) \leq f(x_o+)$ for all x_o.

A function f is called right-continuous at x_o if $f(x_o) = f(x_o+)$. A function f is simply called right-continuous if it is right-continuous at all points.

Definition 11.4 A function f on a bounded interval (open, closed, or mixed) with endpoints $a < b$ is *piecewise continuous*, if there exist $n \geq 1$ and $a = t_0 < t_1 < \ldots < t_n = b$, such that, for $1 \leq k \leq n$: f is continuous over (t_{k-1}, t_k) and has finite limits at the endpoints of (t_{k-1}, t_k).

More generally, if $\mathbb{T}$ is all of $\mathbb{R}$ or an interval in $\mathbb{R}$, f is piecewise continuous over $\mathbb{T}$ if it is piecewise continuous over every bounded subinterval of $\mathbb{T}$.

11.4 Derivatives of functions

Let f be a function on $\mathbb{R}$ and let $x_o \in \mathbb{R}$. Then f is *differentiable* at x_o if the following limit exists and is finite:

$$\lim_{x \to x_o} \frac{f(x) - f(x_o)}{x - x_o}.$$

The value of the limit is the derivative of f at x_o, written as $f'(x_o)$. In more detail, this condition that f is differentiable at x_o means there is a finite value $f'(x_o)$ so that, for any $\epsilon > 0$, there exists $\delta > 0$, so that

$$\left| \frac{f(x) - f(x_o)}{x - x_o} - f'(x_o) \right| \leq \delta$$

whenever $0 < |x - x_o| < \epsilon$. Alternatively, in terms of convergence of sequences, it means there is a finite value $f'(x_o)$ so that

$$\lim_{n \to \infty} \frac{f(x_n) - f(x_o)}{x_n - x_o} = f'(x_o)$$

whenever $(x_n : n \geq 1)$ is a sequence with values in $\mathbb{R} - \{x_o\}$ converging to x_o. The function f is *differentiable* if it is differentiable at all points.

The *righthand derivative* of f at a point x_o, denoted by $D_+f(x_o)$, is defined the same way as $f'(x_o)$, except the limit is taken using only x such that $x > x_o$. The extra condition $x > x_o$ is indicated by using a slanting arrow in the limit notation:

$$D_+f(x_0) = \lim_{x \searrow x_o} \frac{f(x) - f(x_o)}{x - x_o}.$$

Similarly, the lefthand derivative of f at x_o is $D_-f(x_0) = \lim_{x \nearrow x_o} \frac{f(x) - f(x_o)}{x - x_o}$.

Theoretical exercise

1. Suppose f is defined on an open interval containing x_o, then $f'(x_o)$ exists if and only if $D_+f(x_o) = D_-f(x_0)$. If $f'(x_o)$ exists then $D_+f(x_o) = D_-f(x_0) = f'(x_o)$.

We write f'' for the derivative of f'. For an integer $n \geq 0$ we write $f^{(n)}$ to denote the result of differentiating f n times.

Theorem 11.5 *(Mean value form of Taylor's theorem) Let f be a function on an interval (a, b) such that its nth derivative $f^{(n)}$ exists on (a, b). Then for $a < x, x_0 < b$,*

$$f(x) = \sum_{k=0}^{n-1} \frac{f^{(k)}(x_0)}{k!}(x - x_0)^k + \frac{f^{(n)}(y)(x - x_0)^n}{n!}$$

for some y between x and x_0.

Clearly differentiable functions are continuous. But they can still have rather odd properties, as indicated by the following example.

Example 11.6 Let $f(t) = t^2 \sin(1/t^2)$ for $t \neq 0$ and $f(0) = 0$. This function f is a classic example of a differentiable function with a derivative function that is not continuous. To check the derivative at zero, note that $|\frac{f(s)-f(0)}{s}| \leq |s| \to 0$ as $s \to 0$, so $f'(0) = 0$. The usual calculus can be used to compute $f'(t)$ for $t \neq 0$, yielding

$$f'(t) = \begin{cases} 2t\sin(\frac{1}{t^2}) - \frac{2\cos(\frac{1}{t^2})}{t} & t \neq 0 \\ 0 & t = 0 \end{cases}.$$

The derivative f' is not even close to being continuous at zero. As t approaches zero, the cosine term dominates, and f reaches both positive and negative values with arbitrarily large magnitude.

Even though the function f of Example 11.6 is differentiable, it does not satisfy the fundamental theorem of calculus (stated in the next section). One way to rule out the wild behavior of Example 11.6, is to assume that f is *continuously differentiable*, which means that f is differentiable and its derivative function is continuous. For some applications, it is useful to work with functions more general than continuously differentiable ones, but for which the fundamental theorem of calculus still holds. A possible approach is to use the following condition.

Definition 11.6 A function f on a bounded interval (open, closed, or mixed) with endpoints $a < b$ is *continuous and piecewise continuously differentiable*, if f is continuous over the interval, and if there exist $n \geq 1$ and $a = t_0 < t_1 < \ldots < t_n = b$, such that, for $1 \leq k \leq n$: f is continuously differentiable over (t_{k-1}, t_k) and f' has finite limits at the endpoints of (t_{k-1}, t_k).
More generally, if $\mathbb{T}$ is all of $\mathbb{R}$ or a subinterval of $\mathbb{R}$, then a function f on $\mathbb{T}$ is continuous and piecewise continuously differentiable if its restriction to any bounded interval is continuous and piecewise continuously differentiable.

Example 11.7 Two examples of continuous, piecewise continuously differentiable functions on $\mathbb{R}$ are: $f(t) = \min\{t^2, 1\}$ and $g(t) = |\sin(t)|$.

Example 11.8 The function given in Example 11.6 is not considered to be piecewise continuously differentiable because the derivative does not have finite limits at zero.

Theoretical exercises

1. Suppose f is a continuously differentiable function on an open bounded interval (a, b). Show that if f' has finite limits at the endpoints, then so does f.
2. Suppose f is a continuous function on a closed, bounded interval $[a, b]$ such that f' exists and is continuous on the open subinterval (a, b). Show that if the righthand limit of the derviative at a, $f'(a+) = \lim_{x\searrow a} f'(x)$, exists, then the righthand derivative at a, defined by

$$D_{+}f(a) = \lim_{x\searrow a} \frac{f(x) - f(a)}{x - a}$$

also exists, and the two limits are equal.

Let g be a function from $\mathbb{R}^n$ to $\mathbb{R}^m$. Thus for each vector $x \in \mathbb{R}^n$, $g(x)$ is an m vector. The derivative matrix of g at a point x, $\frac{\partial g}{\partial x}(x)$, is the $n \times m$ matrix with ijth entry $\frac{\partial g_i}{\partial x_j}(x)$. Sometimes for brevity we write $y = g(x)$ and think of y as a variable depending on x, and we write the derivative matrix as $\frac{\partial y}{\partial x}(x)$.

Theorem 11.7 *(Implicit function theorem) If $m = n$ and if $\frac{\partial y}{\partial x}$ is continuous in a neighborhood of x_0 and if $\frac{\partial y}{\partial x}(x_0)$ is nonsingular, then the inverse mapping $x = g^{-1}(y)$ is defined in a neighborhood of $y_0 = g(x_0)$ and*

$$\frac{\partial x}{\partial y}(y_0) = \left(\frac{\partial y}{\partial x}(x_0)\right)^{-1}.$$

11.5 Integration

11.5.1 Riemann integration

Let g be a bounded function on a bounded interval of the form $(a, b]$. Given:

- A partition of $(a, b]$ of the form $(t_0, t_1], (t_1, t_2], \ldots, (t_{n-1}, t_n]$, where $n \geq 0$ and $a = t_0 < t_1 \ldots < t_n = b$,
- A sampling point from each subinterval, $v_k \in (t_{k-1}, t_k]$, for $1 \leq k \leq n$,

the corresponding *Riemann sum* for g is defined by

$$\sum_{k=1}^{n} g(v_k)(t_k - t_{k-1}).$$

The *norm of the partition* is defined to be $\max_k |t_k - t_{k-1}|$. The *Riemann integral* $\int_a^b g(x)dx$ is said to exist and its value is I if the following is true. Given any $\epsilon > 0$, there is a $\delta > 0$ so that $|\sum_{k=1}^{n} g(v_k)(t_k - t_{k-1}) - I| \leq \epsilon$ whenever the norm of the partition is less than or equal to δ. This definition is equivalent to the following condition, expressed using convergence of sequences. The Riemann integral exists an is equal to I, if for any sequence of partitions, specified by $((t_1^m, t_2^m, \ldots, t_{n_m}^m) : m \geq 1)$, with corresponding sampling points $((v_1^m, \ldots, v_{n_m}^m) : m \geq 1)$, such that the norm of th mth partition converges to zero as $m \to \infty$, the corresponding sequence of Rieman sums converges to I as $m \to \infty$. The function g is said to be *Riemann integrable* ove $(a, b]$ if the integral $\int_a^b g(x)dx$ exists and is finite.

Next, suppose g is defined over the whole real line. If for every interval $(a, b]$, g i bounded over $[a, b]$ and Riemann integrable over $(a, b]$, then the Riemann integral of over $\mathbb{R}$ is defined by

$$\int_{-\infty}^{\infty} g(x)dx = \lim_{a,b\to\infty} \int_{-a}^{b} g(x)dx,$$

provided that the indicated limit exists as a, b jointly converge to $+\infty$. The values $+\infty$ or $-\infty$ are possible.

A function that is continuous, or just piecewise continuous, is Riemann integrabl over any bounded interval. Moreover, the following is true for Riemann integration:

Theorem 11.8 *(Fundamental theorem of calculus) Let f be a continuously differen tiable function on $\mathbb{R}$. Then for $a < b$,*

$$f(b) - f(a) = \int_a^b f'(x)dx. \tag{11.4}$$

More generally, if f is continuous and piecewise continuously differentiable, (11.4) *hold with $f'(x)$ replaced by the righthand derivative, $D_+f(x)$. (Note that $D_+f(x) = f'(x)$ whenever $f'(x)$ is defined.)*

We will have occasion to use Riemann integrals in two dimensions. Let g be bounded function on a bounded rectangle of the form $(a^1, b^1] \times (a^2, b^2]$. Given:

- A partition of $(a^1, b^1] \times (a^2, b^2]$ into $n^1 \times n^2$ rectangles of the form $(t_j^1, t_{j-1}^1] \times (t_k^2, t_{k-1}^2]$, where $n^i \geq 1$ and $a^i = t_0^i < t_1^i < \ldots < t_{n^i}^i = b^i$ for $i = 1, 2$,
- A sampling point (v_{jk}^1, v_{jk}^2) inside $(t_j^1, t_{j-1}^1] \times (t_k^2, t_{k-1}^2]$ for $1 \leq j \leq n^1$ and $1 \leq k \leq n^2$,

the corresponding Riemann sum for g is

$$\sum_{j=1}^{n^1} \sum_{k=1}^{n^2} g(v_{j,k}^1, v_{j,k}^2)(t_j^1 - t_{j-1}^1)(t_k^2 - t_{k-1}^2).$$

The norm of the partition is $\max_{i\in\{1,2\}} \max_k \mid t_k^i - t_{k-1}^i \mid$. As in the case of one dimension g is said to be Riemann integrable over $(a^1, b^1] \times (a^2, b^2]$, and $\int\int_{(a^1,b^1]\times(a^2,b^2]} g(x_1, x_2)dsdt = I$, if the value of the Riemann sum converges to I for any sequence of partitions and sampling point pairs, with the norms of the partitions converging to zero.

The above definition of a Riemann sum allows the $n^1 \times n^2$ sampling points to be selected arbitrarily from the $n^1 \times n^2$ rectangles. If, instead, the sampling points are restricted to have the form (v_j^1, v_k^2), for $n^1 + n^2$ numbers $v_1^1, \ldots, v_{n^1}^1, v_1^2, \ldots v_{n^2}^2$, we say the corresponding Riemann sum uses *aligned sampling.* We define a function g on $[a, b] \times [a, b]$ to be *Riemann integrable with aligned sampling* in the same way as we defined g to be Riemann integrable, except the family of Riemann sums used are the ones using aligned sampling. Since the set of sequences that must converge is more restricted for aligned sampling, a function g on $[a, b] \times [a, b]$ that is Riemann integrable is also Riemann integrable with aligned sampling.

Proposition 11.9 *A sufficient condition for g to be Riemann integrable (and hence Riemann integrable with aligned sampling) over $(a^1, b^1] \times (a^2, b^2]$ is that g be the restriction to $(a^1, b^1] \times (a^2, b^2]$ of a continuous function on $[a^1, b^1] \times [a^2, b^2]$. More generally, g is Riemann integrable over $(a^1, b^1] \times (a^2, b^2]$ if there is a partition of $(a^1, b^1] \times (a^2, b^2]$ into finitely many subrectangles of the form $(t_j^1, t_{j-1}^1] \times (t_k^2, t_{k-1}^2]$, such that g on $(t_j^1, t_{j-1}^1] \times (t_k^2, t_{k-1}^2]$ is the restriction to $(t_j^1, t_{j-1}^1] \times (t_k^2, t_{k-1}^2]$ of a continuous function on $[t_j^1, t_{j-1}^1] \times [t_k^2, t_{k-1}^2]$.*

Proposition 11.9 is a standard result in real analysis. Its proof uses the fact that continuous functions on bounded, closed sets are uniformly continuous, from which it follows that, for any $\epsilon > 0$, there is a $\delta > 0$ so that the Riemann sums for any two partitions with norm less than or equal to δ differ by at most ϵ. The Cauchy criterion for convergence of sequences of numbers is also used.

11.5.2 Lebesgue integration

Lebesgue integration of a random variable with respect to a probability measure (a.k.a. conditional expectation) is defined in Section 1.5 and is written as

$$E[X] = \int_\Omega X(\omega)P(d\omega).$$

The idea is to first define the expectation for simple random variables, then for nonnegative random variables, and then for general random variables by $E[X] = E[X_+] - E[X_-]$. The same approach can be used to define the Lebesgue integral

$$\int_{-\infty}^{\infty} g(\omega)d\omega$$

for Borel measurable functions g on $\mathbb{R}$. Such an integral is well defined if either $\int_{-\infty}^{\infty} g_+(\omega)d\omega < +\infty$ or $\int_{-\infty}^{\infty} g_-(\omega)d\omega < +\infty$.

11.5.3 Riemann–Stieltjes integration

Let g be a bounded function on a closed interval $[a, b]$ and let F be a nondecreasing function on $[a, b]$. The Riemann–Stieltjes integral

$$\int_a^b g(x)dF(x) \qquad \text{(Riemann–Stieltjes)}$$

is defined the same way as the Riemann integral, except that the Riemann sums ar changed to

$$\sum_{k=1}^{n} g(v_k)(F(t_k) - F(t_{k-1})).$$

Extension of the integral over the whole real line is done as it is for Riemann integratior An alternative definition of $\int_{-\infty}^{\infty} g(x)dF(x)$, preferred in the context of these notes, i given next.

11.5.4 Lebesgue–Stieltjes integration

Let F be a CDF. As seen in Section 1.3, there is a corresponding probability measur $\tilde{P}$ on the Borel subsets of $\mathbb{R}$. Given a Borel measurable function g on $\mathbb{R}$, the Lebesgue Stieltjes integral of g with respect to F is defined to be the Lebesgue integral of g wit respect to $\tilde{P}$:

$$\text{(Lebesgue–Stieltjes)} \quad \int_{-\infty}^{\infty} g(x)dF(x) = \int_{-\infty}^{\infty} g(x)\tilde{P}(dx) \quad \text{(Lebesgue).}$$

The same notation $\int_{-\infty}^{\infty} g(x)dF(x)$ is used for both Riemann–Stieltjes (RS) an Lebesgue–Stieltjes (LS) integration. If g is continuous and the LS integral is finite then the integrals agree. In particular, $\int_{-\infty}^{\infty} xdF(x)$ is identical as either an LS or R integral. However, for equivalence of the integrals

$$\int_{\Omega} g(X(\omega))P(d\omega) \text{ and } \int_{-\infty}^{\infty} g(x)dF(x),$$

even for continuous functions g, it is essential that the integral on the right be understoo as an LS integral. Hence, in these notes, only the LS interpretation is used, and R integration is not needed.

If F has a corresponding pdf f, then

$$\text{(Lebesgue–Stieltjes)} \quad \int_{-\infty}^{\infty} g(x)dF(x) = \int_{-\infty}^{\infty} g(x)f(x)dx \quad \text{(Lebesgue),}$$

for any Borel measurable function g.

11.6 On convergence of the mean

Suppose $(X_n : n \geq 1)$ is a sequence of random variables such that $X_n \xrightarrow{p.} X_\infty$, fo some random variable X_∞. The theorems in this section address the question of whethe $E[X_n] \to E[X_\infty]$. The hypothesis $X_n \xrightarrow{p.} X_\infty$ means that for any $\epsilon > 0$ and $\delta > 0$ $P\{|X_n - X_\infty| \leq \epsilon\} \geq 1 - \delta$. Thus, the event that X_n is close to X_∞ has probability clos to one. But the mean of X_n can differ greatly from the mean of X if, in the unlikely even that $|X_n - X_\infty|$ is not small, it is very, very large.

Example 11.9 Suppose U is a random variable with a finite mean, and suppose $A_1, A_2, \ldots$ is a sequence of events, each with positive probability, but such that $P(A_n) \to 0$, and let $b_1, b_2, \ldots$ be a sequence of nonzero numbers. Let $X_n = U + b_n I_{A_n}$ for $n \geq 1$. Then for any $\epsilon > 0$, $P\{|X_n - U| \geq \epsilon\} \leq P\{X_n \neq U\} = P(A_n) \to 0$ as $n \to \infty$, so $X_n \xrightarrow{p.} U$. However, $E[X_n] = E[U] + b_n P(A_n)$. Thus, if the b_n have very large magnitude, the mean $E[X_n]$ can be far larger or far smaller than $E[U]$, for all large n.

The simplest way to rule out the very, very large values of $|X_n - X_\infty|$ is to require the sequence (X_n) to be bounded. That would rule out using constants b_n with arbitrarily large magnitudes in Example 11.9. The following result is a good start – it is generalized to yield the dominated convergence theorem further below.

Theorem 11.10 *(Bounded convergence theorem) Let $X_1, X_2, \ldots$ be a sequence of random variables such that for some finite L, $P\{|X_n| \leq L\} = 1$ for all $n \geq 1$, and such that $X_n \xrightarrow{p.} X$ as $n \to \infty$. Then $E[X_n] \to E[X]$.*

Proof For $\epsilon > 0$, $P\{|X| \geq L + \epsilon\} \leq P\{|X - X_n| \geq \epsilon\} \to 0$, so $P\{|X| \geq L + \epsilon\} = 0$. Since ϵ was arbitrary, $P\{|X| \leq L\} = 1$. Therefore, $P\{|X - X_n| \leq 2L\} = 1$ for all $n \geq 1$. Again let $\epsilon > 0$. Then

$$|X - X_n| \leq \epsilon + 2L I_{\{|X - X_n| \geq \epsilon\}}, \tag{11.5}$$

so that $|E[X] - E[X_n]| = |E[X - X_n]| \leq E[|X - X_n|] \leq \{\epsilon + 2LP|X - X_n| \geq \epsilon\}$. By the hypotheses, $P\{|X - X_n| \geq \epsilon\} \to 0$ as $n \to \infty$. Thus, for n large enough, $|E[X] - E[X_n]| < 2\epsilon$. Since ϵ is arbitrary, $E[X_n] \to E[X]$. □

Equation (11.5) is central to the proof just given. It bounds the difference $|X - X_n|$ by ϵ on the event $\{|X - X_n| < \epsilon\}$, which has probability close to one for n large, and on the complement of this event, the difference $|X - X_n|$ is still bounded so that its contribution is small for n large enough.

The following lemma, used to establish the dominated convergence theorem, is similar to the bounded convergence theorem, but the variables are assumed to be bounded only on one side: specifically, the random variables are restricted to be greater than or equal to zero. The result is that $E[X_n]$ for large n can still be much larger than $E[X_\infty]$, but cannot be much smaller. The restriction to nonnegative X_ns would rule out using negative constants b_n with arbitrarily large magnitudes in Example 11.9. The statement of the lemma uses "liminf," which is defined in Appendix 11.2.

Lemma 11.11 *(Fatou's lemma) Suppose (X_n) is a sequence of nonnegative random variables such that $X_n \xrightarrow{p.} X_\infty$. Then $\liminf_{n\to\infty} E[X_n] \geq E[X_\infty]$. (Equivalently, for any $\gamma < E[X_\infty]$, $E[X_n] \geq \gamma$ for all sufficiently large n.)*

Proof We shall prove the equivalent form of the conclusion given in the lemma, so let γ be any constant with $\gamma < E[X_\infty]$. By the definition of $E[X_\infty]$, there is a simple random variable Z with $Z \leq X_\infty$ such that $E[Z] \geq \gamma$. Since $Z = X_\infty \wedge Z$,

$$|X_n \wedge Z - Z| = |X_n \wedge Z - X_\infty \wedge Z| \le |X_n - X_\infty| \xrightarrow{p.} 0,$$

so $X_n \wedge Z \xrightarrow{p.} Z$. By the bounded convergence theorem, $\lim_{n\to\infty} E[X_n \wedge Z] = E[Z] > \gamma$
Since $E[X_n] \ge E[X_n \wedge Z]$, it follows that $E[X_n] \ge \gamma$ for all sufficiently large n. □

Theorem 11.12 *(Dominated convergence theorem) If $X_1, X_2, \ldots$ is a sequence of random variables and X_∞ and Y are random variables such that the following three conditions hold:*

(i) $X_n \xrightarrow{p.} X_\infty$ *as* $n \to \infty$,
(ii) $P\{|X_n| \le Y\} = 1$ *for all* n,
(iii) $E[Y] < +\infty$,

then $E[X_n] \to E[X_\infty]$.

Proof The hypotheses imply that $(X_n + Y : n \ge 1)$ is a sequence of nonnegative random variables which converges in probability to $X_\infty + Y$. So Fatou's lemma implies that $\liminf_{n\to\infty} E[X_n + Y] \ge E[X_\infty + Y]$, or equivalently, subtracting $E[Y]$ from both sides, $\liminf_{n\to\infty} E[X_n] \ge E[X_\infty]$. Similarly, since $(-X_n + Y : n \ge 1)$ is a sequence of nonnegative random variables which converges in probability to $-X_\infty + Y$, Fatou's lemma implies that $\liminf_{n\to\infty} E[-X_n + Y] \ge E[-X_\infty + Y]$, or equivalently, $\limsup_{n\to\infty} E[X_n] \le E[X_\infty]$. Summarizing,

$$\limsup_{n\to\infty} E[X_n] \le E[X_\infty] \le \liminf_{n\to\infty} E[X_n].$$

In general, the liminf of a sequence is less than or equal to the limsup, and if the liminf is equal to the limsup, then the limit exists and is equal to both the liminf and limsup. Thus, $E[X_n] \to E[X_\infty]$. □

Corollary 11.13 *(A consequence of integrability) If Z has a finite mean, then given any $\epsilon > 0$, there exists a $\delta > 0$ so that if $P(A) < \delta$, then $|E[ZI_A]| \le \epsilon$.*

Proof If not, there would exist a sequence of events A_n with $P(A_n) \to 0$ with $|E[ZI_{A_n}]| \ge \epsilon$. But $ZI_{A_n} \xrightarrow{p.} 0$, and ZI_{A_n} is dominated by the integrable random variable $|Z|$ for all n, so the dominated convergence theorem implies that $E[ZI_{A_n}] \to 0$, which would result in a contradiction. □

The following theorem is based on a different way to control the difference between $E[X_n]$ for large n and $E[X_\infty]$. Rather than a domination condition, it is assumed that the sequence is monotone in n.

Theorem 11.14 *(Monotone convergence theorem) Let $X_1, X_2, \ldots$ be a sequence of random variables such that $E[X_1] > -\infty$ and such that $X_1(\omega) \le X_2(\omega) \le \ldots$ Then the limit X_∞ given by $X_\infty(\omega) = \lim_{n\to\infty} X_n(\omega)$ for all ω is an extended random variable (with possible value ∞) and $E[X_n] \to E[X_\infty]$ as $n \to \infty$.*

Proof By adding $\min\{0, -X_1\}$ to all the random variables involved if necessary, we can assume without loss of generality that $X_1, X_2, \ldots$, and therefore also X, are nonnegative. Recall that $E[X]$ is equal to the supremum of the expectation of simple random variables that are less than or equal to X. So let γ be any number such that $\gamma < E[X]$. Then, there is a simple random variable $\widetilde{X}$ less than or equal to X with $E[\widetilde{X}] \geq \gamma$. The simple random variable $\widetilde{X}$ takes only finitely many possible values. Let L be the largest. Then $\widetilde{X} \leq X \wedge L$, so that $E[X \wedge L] > \gamma$. By the bounded convergence theorem, $E[X_n \wedge L] \to E[X \wedge L]$. Therefore, $E[X_n \wedge L] > \gamma$ for all large enough n. Since $E[X_n \wedge L] \leq E[X_n] \leq E[X]$, it follows that $\gamma < E[X_n] \leq E[X]$ for all large enough n. Since γ is an arbitrary constant with $\gamma < E[X]$, the desired conclusion, $E[X_n] \to E[X]$, follows. □

11.7 Matrices

An $m \times n$ *matrix* over the reals $\mathbb{R}$ has the form

$$A = \begin{pmatrix} a_{11} & a_{12} & \cdots & a_{1n} \\ a_{21} & a_{22} & \cdots & a_{2n} \\ \vdots & \vdots & & \vdots \\ a_{m1} & a_{m2} & \cdots & a_{mn} \end{pmatrix},$$

where $a_{ij} \in \mathbb{R}$ for all i, j. This matrix has m rows and n columns. A matrix over the complex numbers $\mathbb{C}$ has the same form, with $a_{ij} \in \mathbb{C}$ for all i, j. The *transpose* of an $m \times n$ matrix $A = (a_{ij})$ is the $n \times m$ matrix $A^T = (a_{ji})$. For example

$$\begin{pmatrix} 1 & 0 & 3 \\ 2 & 1 & 1 \end{pmatrix}^T = \begin{pmatrix} 1 & 2 \\ 0 & 1 \\ 3 & 1 \end{pmatrix}.$$

The matrix A is *symmetric* if $A = A^T$. Symmetry requires that the matrix A be square: $m = n$. The *diagonal* of a matrix comprises the entries of the form a_{ii}. A square matrix A is called *diagonal* if the entries off the diagonal are zero. The $n \times n$ *identity matrix* is the $n \times n$ diagonal matrix with ones on the diagonal. We write I to denote an identity matrix of some dimension n.

If A is an $m \times k$ matrix and B is a $k \times n$ matrix, then the *product* AB is the $m \times n$ matrix with ijth element $\sum_{l=1}^{k} a_{il} b_{lj}$. A *vector* x is an $m \times 1$ matrix, where m is the dimension of the vector. Thus, vectors are written in column form:

$$x = \begin{pmatrix} x_1 \\ x_2 \\ \vdots \\ x_m \end{pmatrix}.$$

The set of all dimension m vectors over $\mathbb{R}$ is the m dimensional Euclidean space $\mathbb{R}^m$. The *inner product* of two vectors x and y of the same dimension m is the number $x^T y$, equal

to $\sum_{i=1}^{m} x_i y_i$. The vectors x and y are *orthogonal* if $x^T y = 0$. The *Euclidean length* c
norm of a vector x is given by $\|x\| = (x^T x)^{\frac{1}{2}}$. A set of vectors $\varphi_1, \ldots, \varphi_n$ is *orthonorma*
if the vectors are orthogonal to each other and $\|\varphi_i\| = 1$ for all i.

A set of vectors $v_1, \ldots, v_n$ in $\mathbb{R}^m$ is said to *span* $\mathbb{R}^m$ if any vector in $\mathbb{R}^m$ can b
expressed as a linear combination $\alpha_1 v_1 + \alpha_2 v_2 + \ldots + \alpha_n v_n$ for some $\alpha_1, \ldots, \alpha_n \in \mathbb{R}$
An orthonormal set of vectors $\varphi_1, \ldots, \varphi_n$ in $\mathbb{R}^m$ spans $\mathbb{R}^m$ if and only if $n = m$. A
orthonormal basis for $\mathbb{R}^m$ is an orthonormal set of m vectors in $\mathbb{R}^m$. An orthonorma
basis $\varphi_1, \ldots, \varphi_m$ corresponds to a coordinate system for $\mathbb{R}^m$. Given a vector v in $\mathbb{R}^m$
the coordinates of v relative to $\varphi_1, \ldots, \varphi_m$ are given by $\alpha_i = \varphi_i^T v$. The coordinate
$\alpha_1, \ldots, \alpha_m$ are the unique numbers such that $v = \alpha_1 \varphi_1 + \ldots + \alpha_m \varphi_m$.

A square matrix U is called *orthonormal* if any of the following three equivalen
conditions is satisfied:

1. $U^T U = I$,
2. $U U^T = I$,
3. the columns of U form an orthonormal basis.

Given an $m \times m$ orthonormal matrix U and a vector $v \in \mathbb{R}^m$, the coordinates of v relativ
to U are given by the vector $U^T v$. Given a square matrix A, a vector φ is an *eigenvecto*
of A and λ is an *eigenvalue* of A if the eigen relation $A\varphi = \lambda\varphi$ is satisfied.

A *permutation* π of numbers $1, \ldots, m$ is a one-to-one mapping of $\{1, 2, \ldots, m\}$ ont
itself. That is $(\pi(1), \ldots, \pi(m))$ is a reordering of $(1, 2, \ldots, m)$. Any permutation i
either even or odd. A permutation is *even* if it can be obtained by an even numbe
of transpositions of two elements. Otherwise a permutation is *odd*. We write

$$(-1)^\pi = \begin{cases} 1 & \text{if } \pi \text{ is even} \\ -1 & \text{if } \pi \text{ is odd} \end{cases}.$$

The *determinant* of a square matrix A, written $\det(A)$, is defined by

$$\det(A) = \sum_{\pi} (-1)^\pi \prod_{i=1}^{m} a_{i\pi(i)}.$$

The absolute value of the determinant of a matrix A is denoted by $|A|$, so $|A| =$
$|\det(A)|$.

Some important properties of determinants are the following. Let A and B be $m \times m$
matrices.

1. If B is obtained from A by multiplication of a row or column of A by a scala
constant c, then $\det(B) = c \det(A)$.
2. If $\mathcal{U}$ is a subset of $\mathbb{R}^m$ and $\mathcal{V}$ is the image of $\mathcal{U}$ under the linear transformatio
determined by A:

$$\mathcal{V} = \{Ax : x \in \mathcal{U}\},$$

then

$$(\text{the volume of } \mathcal{U}) = |A| \times (\text{the volume of } \mathcal{V}).$$

3. $\det(AB) = \det(A)\det(B)$.
4. $\det(A) = \det(A^T)$.
5. $|U| = 1$ if U is orthonormal.
6. The columns of A span $\mathbb{R}^n$ if and only if $\det(A) \neq 0$.
7. The equation $p(\lambda) = \det(\lambda I - A)$ defines a polynomial p of degree m called the *characteristic polynomial* of A.
8. The zeros $\lambda_1, \lambda_2, \ldots, \lambda_m$ of the characteristic polynomial of A, repeated according to multiplicity, are the *eigenvalues* of A, and $\det(A) = \prod_{i=1}^n \lambda_i$. The eigenvalues can be complex valued with nonzero imaginary parts.

If K is a symmetric $m \times m$ matrix, then the eigenvalues $\lambda_1, \lambda_2, \ldots, \lambda_m$, are real valued (not necessarily distinct) and there exists an orthonormal basis consisting of the corresponding eigenvectors $\varphi_1, \varphi_2, \ldots, \varphi_m$. Let U be the orthonormal matrix with columns $\varphi_1, \ldots, \varphi_m$ and let Λ be the diagonal matrix with diagonal entries given by the eigenvalues

$$\Lambda = \begin{pmatrix} \lambda_1 & & & \\ & \lambda_2 & & \\ & & \ddots & \\ & & & \lambda_m \end{pmatrix}.$$

Then the relations among the eigenvalues and eigenvectors can be written as $KU = U\Lambda$. Therefore $K = U\Lambda U^T$ and $\Lambda = U^T KU$. A symmetric $m \times m$ matrix A is *positive semidefinite* if $\alpha^T A\alpha \geq 0$ for all m-dimensional vectors α. A symmetric matrix is positive semidefinite if and only if its eigenvalues are nonnegative.

The remainder of this section deals with matrices over $\mathbb{C}$. The *Hermitian transpose* of a matrix A is the matrix A^*, obtained from A^T by taking the complex conjugate of each element of A^T. For example,

$$\begin{pmatrix} 1 & 0 & 3+2j \\ 2 & j & 1 \end{pmatrix}^* = \begin{pmatrix} 1 & 2 \\ 0 & -j \\ 3-2j & 1 \end{pmatrix}.$$

The set of all dimension m vectors over $\mathbb{C}$ is the m-complex dimensional space $\mathbb{C}^m$. The *inner product* of two vectors x and y of the same dimension m is the complex number y^*x, equal to $\sum_{i=1}^m x_i y_i^*$. The vectors x and y are *orthogonal* if $x^*y = 0$. The length or norm of a vector x is given by $\|x\| = (x^*x)^{\frac{1}{2}}$. A set of vectors $\varphi_1, \ldots, \varphi_n$ is *orthonormal* if the vectors are orthogonal to each other and $\|\varphi_i\| = 1$ for all i.

A set of vectors $v_1, \ldots, v_n$ in $\mathbb{C}^m$ is said to *span* $\mathbb{C}^m$ if any vector in $\mathbb{C}^m$ can be expressed as a linear combination $\alpha_1 v_1 + \alpha_2 v_2 + \ldots + \alpha_n v_n$ for some $\alpha_1, \ldots, \alpha_n \in \mathbb{C}$. An orthonormal set of vectors $\varphi_1, \ldots, \varphi_n$ in $\mathbb{C}^m$ spans $\mathbb{C}^m$ if and only if $n = m$. An *orthonormal basis* for $\mathbb{C}^m$ is an orthonormal set of m vectors in $\mathbb{C}^m$. An orthonormal basis $\varphi_1, \ldots, \varphi_m$ corresponds to a coordinate system for $\mathbb{C}^m$. Given a vector v in $\mathbb{R}^m$, the coordinates of v relative to $\varphi_1, \ldots, \varphi_m$ are given by $\alpha_i = \varphi_i^* v$. The coordinates $\alpha_1, \ldots, \alpha_m$ are the unique numbers such that $v = \alpha_1\varphi_1 + \ldots + \alpha_m\varphi_m$.

A square matrix U over $\mathbb{C}$ is called *unitary* (rather than orthonormal) if any of th following three equivalent conditions is satisfied:

1. $U^*U = I$,
2. $UU^* = I$,
3. the columns of U form an orthonormal basis.

Given an $m \times m$ unitary matrix U and a vector $v \in \mathbb{C}^m$, the coordinates of v relative t U are given by the vector U^*v. Eigenvectors, eigenvalues, and determinants of squar matrices over $\mathbb{C}$ are defined just as they are for matrices over $\mathbb{R}$. The absolute value c the determinant of a matrix A is denoted by $| A |$. Thus $| A |=| \det(A) |$.

Some important properties of determinants of matrices over $\mathbb{C}$ are the following. Le A and B be $m \times m$ matrices.

1. If B is obtained from A by multiplication of a row or column of A by a constan $c \in \mathbb{C}$, then $\det(B) = c \det(A)$.
2. If $\mathcal{U}$ is a subset of $\mathbb{C}^m$ and $\mathcal{V}$ is the image of $\mathcal{U}$ under the linear transformatio determined by A:

$$\mathcal{V} = \{Ax : x \in \mathcal{U}\},$$

then

$$(\text{the volume of } \mathcal{U}) =| A |^2 \times (\text{the volume of } \mathcal{V}).$$

3. $\det(AB) = \det(A)\det(B)$.
4. $\det^*(A) = \det(A^*)$.
5. $| U |= 1$ if U is unitary.
6. The columns of A span $\mathbb{C}^n$ if and only if $\det(A) \neq 0$.
7. The equation $p(\lambda) = \det(\lambda I - A)$ defines a polynomial p of degree m called th *characteristic polynomial* of A.
8. The zeros $\lambda_1, \lambda_2, \ldots, \lambda_m$ of the characteristic polynomial of A, repeated accordin to multiplicity, are the eigenvalues of A, and $\det(A) = \prod_{i=1}^n \lambda_i$. The eigenvalue can be complex valued with nonzero imaginary parts.

A matrix K is called *Hermitian symmetric* if $K = K^*$. If K is a Hermitian symmetric $m \times$ m matrix, then the eigenvalues $\lambda_1, \lambda_2, \ldots, \lambda_m$, are real valued (not necessarily distinct and there exists an orthonormal basis consisting of the corresponding eigenvector $\varphi_1, \varphi_2, \ldots, \varphi_m$. Let U be the unitary matrix with columns $\varphi_1, \ldots, \varphi_m$ and let Λ be th diagonal matrix with diagonal entries given by the eigenvalues

$$\Lambda = \begin{pmatrix} \lambda_1 & & & \\ & \lambda_2 & & \\ & & \ddots & \\ & & & \lambda_m \end{pmatrix}.$$

Then the relations among the eigenvalues and eigenvectors may be written as $KU =$ $U\Lambda$. Therefore $K = U\Lambda U^*$ and $\Lambda = U^*KU$. A Hermitian symmetric $m \times m$ matrix is *positive semidefinite* if $\alpha^* A\alpha \geq 0$ for all $\alpha \in \mathbb{C}^m$. A Hermitian symmetric matrix is positive semidefinite if and only if its eigenvalues are nonnegative.

Many questions about matrices over $\mathbb{C}$ can be addressed using matrices over $\mathbb{R}$. If Z is an $m \times m$ matrix over $\mathbb{C}$, then Z can be expressed as $Z = A+Bj$, for some $m \times m$ matrices A and B over $\mathbb{R}$. Similarly, if x is a vector in $\mathbb{C}^m$ then it can be written as $x = u+jv$ for vectors $u, v \in \mathbb{R}^m$. Then $Zx = (Au - Bv) + j(Bu + Av)$. There is a one-to-one and onto mapping from $\mathbb{C}^m$ to $\mathbb{R}^{2m}$ defined by $u + jv \to \binom{u}{v}$. Multiplication of x by the matrix Z is thus equivalent to multiplication of $\binom{u}{v}$ by $\widetilde{Z} = \begin{pmatrix} A & -B \\ B & A \end{pmatrix}$. We will show that

$$|Z|^2 = \det(\widetilde{Z}), \tag{11.6}$$

so that Property 2 for determinants of matrices over $\mathbb{C}$ follows from Property 2 for determinants of matrices over $\mathbb{R}$.

It remains to prove (11.6). Suppose that A^{-1} exists and examine the two $2m \times 2m$ matrices

$$\begin{pmatrix} A & -B \\ B & A \end{pmatrix} \text{ and } \begin{pmatrix} A & 0 \\ B & A + BA^{-1}B \end{pmatrix}. \tag{11.7}$$

The second matrix is obtained from the first by left multiplying each sub-block in the right column of the first matrix by $A^{-1}B$, and adding the result to the left column. Equivalently, the second matrix is obtained by right multiplying the first matrix by $\begin{pmatrix} I & A^{-1}B \\ 0 & I \end{pmatrix}$. But $\det \begin{pmatrix} I & A^{-1}B \\ 0 & I \end{pmatrix} = 1$, so that the two matrices in (11.7) have the same determinant. Equating the determinants of the two matrices in (11.7) yields $\det(\widetilde{Z}) = \det(A)\det(A + BA^{-1}B)$. Similarly, the following four matrices have the same determinant:

$$\begin{pmatrix} A+Bj & 0 \\ 0 & A-Bj \end{pmatrix}, \begin{pmatrix} A+Bj & A-Bj \\ 0 & A-Bj \end{pmatrix},$$
$$\begin{pmatrix} 2A & A-Bj \\ A-Bj & A-Bj \end{pmatrix}, \begin{pmatrix} 2A & 0 \\ A-Bj & \frac{A+BA^{-1}B}{2} \end{pmatrix}. \tag{11.8}$$

Equating the determinants of the first and last of the matrices in (11.8) yields that $|Z|^2 = \det(Z)\det^*(Z) = \det(A+Bj)\det(A-Bj) = \det(A)\det(A+BA^{-1}B)$. Combining these observations yields that (11.6) holds if A^{-1} exists. Since each side of (11.6) is a continuous function of A, (11.6) holds in general.

12 Solutions to even numbered problems

1.2 A ballot problem There are $\binom{6}{4} = 15$ possibilities for the positions of the winnin ballots, and the event in question can be written as $\{110110, 110101, 111001, 11101C$ $111100\}$, so the event has probability $\frac{5}{15} = \frac{1}{3}$. It can be shown in general that if k o the ballots are for the winning candidate and $n - k$ are for the losing candidate, then th winning candidate has a strict majority throughout the counting with probability $\frac{2k-n}{n}$ This remains true even if the cyclic order of the ballots counted is fixed, with only th identity of the first ballot counted being random and uniform over the n possibilities.

1.4 Independent vs. mutually exclusive (a) If E is an event independent of itself, the $P(E) = P(E \cap E) = P(E)P(E)$. This can happen if $P(E) = 0$. If $P(E) \neq 0$ the canceling a factor of $P(E)$ on each side yields $P(E) = 1$. In summary, either $P(E) =$ or $P(E) = 1$.

(b) In general, $P(A \cup B) = P(A) + P(B) - P(AB)$. On one hand, if A and B are independen then $P(A \cup B) = P(A) + P(B) - P(A)P(B) = 0.3 + 0.4 - (0.3)(0.4) = 0.58$. On th other hand, if the events A and B are mutually exclusive, then $P(AB) = 0$ and therefor $P(A \cup B) = 0.3 + 0.4 = 0.7$.

(c) If $P(A) = 0.6$ and $P(B) = 0.8$, then the two events could be independent. Howeve if A and B were mutually exclusive, then $P(A) + P(B) = P(A \cup B) \leq 1$, so it would no possible for A and B to be mutually exclusive if $P(A) = 0.6$ and $P(B) = 0.8$.

1.6 Frantic search Let D, T, B, and O denote the events that the glasses are in th drawer, on the table, in the briefcase, or in the office, respectively. These four event partition the probability space.

(a) Let E denote the event that the glasses were not found in the first drawer search: $P(T|E) = \frac{P(TE)}{P(E)} = \frac{P(E|T)P(T)}{P(E|D)P(D)+P(E|D^c)P(D^c)} = \frac{(1)(0.06)}{(0.1)(0.9)+(1)(0.1)} = \frac{0.06}{0.19} \approx 0.315$.

(b) Let F denote the event that the glasses were not found after the first drawer searc and first table search:

$$
\begin{aligned}
P(B|F) &= \frac{P(BF)}{P(F)} \\
&= \frac{P(F|B)P(B)}{P(F|D)P(D) + P(F|T)P(T) + P(F|B)P(B) + P(F|O)P(O)} \\
&= \frac{(1)(0.03)}{(0.1)(0.9) + (0.1)(0.06) + (1)(0.03) + (1)(0.01)} \approx 0.22.
\end{aligned}
$$

(c) Let G denote the event that the glasses were not found after the two drawer searches, two table searches, and one briefcase search:

$$\begin{aligned}
P(O|G) &= \frac{P(OG)}{P(G)} \\
&= \frac{P(G|O)P(O)}{P(G|D)P(D)+P(G|T)P(T)+P(G|B)P(B)+P(G|O)P(O)} \\
&= \frac{(1)(0.01)}{(0.1)^2(0.9)+(0.1)^2(0.06)+(0.1)(0.03)+(1)(0.01)} \approx 0.4225.
\end{aligned}$$

1.8 Conditional probabilities – basic computations of iterative decoding
(a) Here is one of several approaches. Note that the n pairs $(B_1, Y_1), \ldots, (B_n, Y_n)$ are mutually independent, and $\lambda_i(b_i) \stackrel{def}{=} P(B_i = b_i | Y_i = y_i) = \frac{q_i(y_i|b_i)}{q_i(y_i|0)+q_i(y_i|1)}$. Therefore

$$\begin{aligned}
&P(B = 1|Y_1 = y_1, \ldots, Y_n = y_n) \\
&= \sum_{b_1,\ldots,b_n: b_1 \oplus \cdots \oplus b_n = 1} P(B_1 = b_1, \ldots, B_n = b_n | Y_1 = y_1, \ldots, Y_n = y_n) \\
&= \sum_{b_1,\ldots,b_n: b_1 \oplus \cdots \oplus b_n = 1} \prod_{i=1}^{n} \lambda_i(b_i).
\end{aligned}$$

(b) Using the definitions,

$$\begin{aligned}
P(B = 1|Z_1 = z_1, \ldots, Z_k = z_k) &= \frac{p(1, z_1, \ldots, z_k)}{p(0, z_1, \ldots, z_k) + p(1, z_1, \ldots, z_k)} \\
&= \frac{\frac{1}{2}\prod_{j=1}^{k} r_j(1|z_j)}{\frac{1}{2}\prod_{j=1}^{k} r_j(0|z_j) + \frac{1}{2}\prod_{j=1}^{k} r_j(1|z_j)} \\
&= \frac{\eta}{1+\eta} \quad \text{where } \eta = \prod_{j=1}^{k} \frac{r_j(1|z_j)}{r_j(0|z_j)}.
\end{aligned}$$

1.10 Blue corners (a) There are 24 ways to color five corners so that at least one face has four blue corners (there are six choices of the face, and for each face there are four choices for which additional corner to color blue). Since there are $\binom{8}{5} = 56$ ways to select five out of eight corners, $P(B|\text{exactly five corners colored blue}) = 24/56 = 3/7$.
(b) By counting the number of ways that B can happen for different numbers of blue corners we find $P(B) = 6p^4(1-p)^4 + 24p^5(1-p)^3 + 24p^6(1-p)^2 + 8p^7(1-p) + p^8$.
1.12 Recognizing cumulative distribution functions (a) Valid (draw a sketch). $P\{X^2 \le 5\} = P\{X \le -\sqrt{5}\} + P\{X \ge \sqrt{5}\} = F_1(-\sqrt{5}) + 1 - F_1(\sqrt{5}) = \frac{e^{-5}}{2}$.
(b) Invalid. $F(0) > 1$. Another reason is that F is not nondecreasing.
(c) Invalid, not right continuous at 0.

1.14 CDF and characteristic function of a mixed type random variable

(a) Range of X is $[0, 0.5]$. For $0 \leq c \leq 0.5$, $P\{X \leq c]\} = P\{U \leq c+0.5\} = c+0.5$ Thus

$$F_X(c) = \begin{cases} 0 & c < 0 \\ c+0.5 & 0 \leq c \leq 0.5 \\ 1 & c \geq 0.5 \end{cases}.$$

(b) $\Phi_X(u) = 0.5 + \int_0^{0.5} e^{jux}dx = 0.5 + \frac{e^{ju/2}-1}{ju}$.

1.16 Conditional expectation for uniform density over a triangular region

(a) The triangle has base and height one, so the area of the triangle is 0.5. Thus the joint pdf is 2 inside the triangle.

(b)

$$f_X(x) = \int_{-\infty}^{\infty} f_{XY}(x,y)dy = \begin{cases} \int_0^{x/2} 2dy = x & \text{if } 0 < x < 1 \\ \int_{x-1}^{x/2} 2dy = 2-x & \text{if } 1 < x < 2 \\ 0 & \text{else} \end{cases}.$$

(c) In view of part (b), the conditional density $f_{Y|X}(y|x)$ is not well defined unless $0 < x < 2$. In general we have

$$f_{Y|X}(y|x) = \begin{cases} \frac{2}{x} & \text{if } 0 < x \leq 1 \text{ and } y \in [0, \frac{x}{2}] \\ 0 & \text{if } 0 < x \leq 1 \text{ and } y \notin [0, \frac{x}{2}] \\ \frac{2}{2-x} & \text{if } 1 < x < 2 \text{ and } y \in [x-1, \frac{x}{2}] \\ 0 & \text{if } 1 < x < 2 \text{ and } y \notin [x-1, \frac{x}{2}] \\ \text{not defined} & \text{if } x \leq 0 \text{ or } x \geq 2 \end{cases}.$$

Thus, for $0 < x \leq 1$, the conditional distribution of Y is uniform over the interval $[0,$ For $1 < x \leq 2$, the conditional distribution of Y is uniform over the interval $[x-1, \frac{x}{2}$

(d) Finding the midpoints of the intervals that Y is conditionally uniformly distributed over, or integrating x against the conditional density found in part (c), yields:

$$E[Y|X=x] = \begin{cases} \frac{x}{4} & \text{if } 0 < x \leq 1 \\ \frac{3x-2}{4} & \text{if } 1 < x < 2 \\ \text{not defined} & \text{if } x \leq 0 \text{ or } x \geq 2 \end{cases}.$$

1.18 Density of a function of a random variable

(a) $P(X \geq 0.4|X \leq 0.8) = P(0.4 \leq X \leq 0.8|X \leq 0.8) = (0.8^2 - 0.4^2)/0.8^2 = \frac{3}{4}$.

(b) The range of Y is the interval $[0, +\infty)$. For $c \geq 0$,
$P\{-\ln(X) \leq c\} = P\{\ln(X) \geq -c\} = P\{X \geq e^{-c}\} = \int_{e^{-c}}^1 2xdx = 1 - e^{-2c}$ so

$f_Y(c) = \begin{cases} 2\exp(-2c) & c \geq 0 \\ 0 & \text{else} \end{cases}$. That is, Y is an exponential random variable with parameter 2.

1.20 Functions of independent exponential random variables

(a) Z takes values in the positive real line. So let $z \geq 0$.

$$\begin{aligned} P\{Z \leq z)\} &= P\{\min\{X_1, X_2\} \leq z\} = P\{X_1 \leq z \text{ or } X_2 \leq z\} \\ &= 1 - P\{X_1 > z \text{ and } X_2 > z\} = 1 - \{P(X_1 > z]P\{X_2 > z\} \\ &= 1 - e^{-\lambda_1 z}e^{-\lambda_2 z} = 1 - e^{-(\lambda_1+\lambda_2)z}. \end{aligned}$$

Differentiating yields that

$$f_Z(z) = \begin{cases} (\lambda_1 + \lambda_2)e^{-(\lambda_1+\lambda_2)z} & z \geq 0 \\ 0 & z < 0 \end{cases}.$$

That is, Z has the exponential distribution with parameter $\lambda_1 + \lambda_2$.
(b) R takes values in the positive real line and by independence the joint pdf of X_1 and X_2 is the product of their individual densities. So for $r \geq 0$

$$\begin{aligned} P\{R \leq r\} &= P\left\{\frac{X_1}{X_2} \leq r\right\} = P\{X_1 \leq rX_2\} \\ &= \int_0^\infty \int_0^{rx_2} \lambda_1 e^{-\lambda_1 x_1} \lambda_2 e^{-\lambda_2 x_2} dx_1 dx_2 \\ &= \int_0^\infty (1 - e^{-r\lambda_1 x_2})\lambda_2 e^{-\lambda_2 x_2} dx_2 = 1 - \frac{\lambda_2}{r\lambda_1 + \lambda_2}. \end{aligned}$$

Differentiating yields that

$$f_R(r) = \begin{cases} \frac{\lambda_1\lambda_2}{(\lambda_1 r+\lambda_2)^2} & r \geq 0 \\ 0 & r < 0 \end{cases}.$$

1.22 Gaussians and the Q function (a) $\mathrm{Cov}(3X + 2Y, X + 5Y + 10) = 3\mathrm{Cov}(X,X) + 10\mathrm{Cov}(Y,Y) = 3\mathrm{Var}(X) + 10\mathrm{Var}(Y) = 13$.
(b) $X + 4Y$ is $N(0, 17)$, so $P\{X + 4Y \geq 2\} = P\{\frac{X+4Y}{\sqrt{17}} \geq \frac{2}{\sqrt{17}}\} = Q(\frac{2}{\sqrt{17}})$.
(c) $X - Y$ is $N(0, 2)$, so $P\{(X - Y)^2 > 9\} = P\{(X - Y) \geq 3 \text{ or } X - Y \leq -3\} = 2P\{\frac{X-Y}{\sqrt{2}} \geq \frac{3}{\sqrt{2}}\} = 2Q(\frac{3}{\sqrt{2}})$.
1.24 Working with a joint density (a) The density must integrate to one, so $c = 4/19$.
(b)

$$f_X(x) = \begin{cases} \frac{4}{19}\int_1^2 (1+xy)dy = \frac{4}{19}[1 + \frac{3x}{2}] & 2 \leq x \leq 3 \\ 0 & \text{else} \end{cases},$$

$$f_Y(y) = \begin{cases} \frac{4}{19}\int_2^3 (1+xy)dx = \frac{4}{19}[1 + \frac{5y}{2}] & 1 \leq y \leq 2 \\ 0 & \text{else} \end{cases}.$$

Therefore $f_{X|Y}(x|y)$ is well defined only if $1 \leq y \leq 2$. For $1 \leq y \leq 2$:

$$f_{X|Y}(x|y) = \begin{cases} \frac{1+xy}{1+\frac{5}{2}y} & 2 \leq x \leq 3 \\ 0 & \text{for other } x \end{cases}.$$

1.26 Density of a difference *Method 1* The joint density is the product of the marginals, and for any $c \geq 0$, the probability $P\{|X - Y| \leq c\}$ is the integral of the joint density over the region of the positive quadrant such that $\{|x - y| \leq c\}$, which by symmetry is one minus twice the integral of the density over the region $\{y \geq 0 \text{ and } y \leq y + c\}$. Thus, $P\{|X-Y| \leq c\} = 1 - 2\int_0^\infty \exp(-\lambda(y+c))\lambda\exp(-\lambda y)dy = 1 - \exp(-\lambda c)$. Thus, $f_Z(c) = \begin{cases} \lambda\exp(-\lambda c) & c \geq 0 \\ 0 & \text{else} \end{cases}$. That is, Z has the exponential distribution with parameter λ.

Method 2 The problem can be solved without calculation by the memoryless propert of the exponential distribution, as follows. Suppose X and Y are lifetimes of identica lightbulbs which are turned on at the same time. One of them will burn out first. At tha time, the other lightbulb will be the same as a new lightbulb, and $|X - Y|$ is equal to hov much longer that lightbulb will last.

1.28 Some characteristic functions (a) Differentiation yields $jE[X] = \Phi'(0) = 2j$ c $E[X] = 2$, and $j^2 E[X^2] = \Phi''(0) = -14$, so $\mathrm{Var}(x) = 14 - 2^2 = 10$. In fact, this is th characteristic function of an $N(10, 2^2)$ random variable.

(b) Evaluation of the derivatives at zero requires l'Hôpital's rule, and is a little tediou A simpler way is to use the Taylor series expansion $\exp(ju) = 1 + (ju) + (ju)^2/2!$ $(ju)^3/3!...$ The result is $E[X] = 0.5$ and $\mathrm{Var}(X) = 1/12$. In fact, this is the characteristi function of a $U(0, 1)$ random variable.

(c) Differentiation is straightforward, yielding $E[X] = \mathrm{Var}(X) = \lambda$. In fact, this is th characteristic function of a $Poi(\lambda)$ random variable.

1.30 A transformation of jointly continuous random variables (a) We are using th mapping, from the square region $\{(u, v) : 0 \leq u, v \leq 1\}$ in the u–v plane to th triangular region with corners (0,0), (3,0), and (3,1) in the x–y plane, given by

$$x = 3u,$$
$$y = uv.$$

The mapping is one-to-one, meaning that for any (x, y) in the range we can recove (u, v). Indeed, the inverse mapping is given by

$$u = \frac{x}{3},$$
$$v = \frac{3y}{x}.$$

The Jacobian determinant of the transformation is

$$J(u, v) = \det \begin{pmatrix} \frac{\partial x}{\partial u} & \frac{\partial x}{\partial v} \\ \frac{\partial y}{\partial u} & \frac{\partial y}{\partial v} \end{pmatrix} = \det \begin{pmatrix} 3 & 0 \\ v & u \end{pmatrix} = 3u \neq 0, \quad \text{for all } u, v \in (0, 1)^2.$$

Therefore the required pdf is

$$f_{X,Y}(x, y) = \frac{f_{U,V}(u, v)}{|J(u, v)|} = \frac{9u^2v^2}{|3u|} = 3uv^2 = \frac{9y^2}{x}$$

within the triangle with corners (0,0), (3,0), and (3,1), and $f_{X,Y}(x, y) = 0$ elsewhere.

(b) Integrating out y from the joint pdf yields

$$f_X(x) = \begin{cases} \int_0^{\frac{x}{3}} \frac{9y^2}{x} dy = \frac{x^2}{9} & \text{if } 0 \leq x \leq 3 \\ 0 & \text{else} \end{cases}.$$

Therefore the conditional density $f_{Y|X}(y|x)$ is well defined only if $0 \leq x \leq 3$. Fo $0 \leq x \leq 3$,

$$f_{Y|X}(y|x) = \frac{f_{X,Y}(x, y)}{f_X(x)} = \begin{cases} \frac{81y^2}{x^3} & \text{if } 0 \leq y \leq \frac{x}{3} \\ 0 & \text{else} \end{cases}.$$

1.32 Opening a bicycle combination lock The time required has possible values from 2 seconds to 20 000 seconds. It is well approximated (within 2 seconds) by a continuous type random variable T that is uniformly distributed on the interval $[0, 20\,000]$. In fact, if we were to round T up to the nearest multiple of 2 seconds we would get a random variable with the exact distribution of time required. Then $E[T] = \frac{2\times 10^4}{2}$ seconds $= 10\,000$ seconds $= 166.66$ minutes, and the standard deviation of T is $\frac{20\,000}{\sqrt{12}} = 5773.5$ seconds $= 96.22$ minutes.

1.34 Computing some covariances (a) $\text{Cov}(X+Y, X-Y) = \text{Cov}(X,X) - \text{Cov}(X,Y) + \text{Cov}(Y,X) - \text{Cov}(Y,Y) = \text{Var}(X) - \text{Var}(Y) = 0$.
(b) $\text{Cov}(3X+Z, 3X+Y) = 9\text{Var}(X) + 3\text{Cov}(X,Y) + 3\text{Cov}(Z,X) + \text{Cov}(Z,Y) = 9\times 20 + 3\times 10 + 3\times 10 + 5 = 245$.
(c) Since $E[X+Y] = 0$, $E[(X+Y)^2] = \text{Var}(X+Y) = \text{Var}(X) + 2\text{Cov}(X,Y) + \text{Var}(Y) = 20 + 2\times 10 + 20 = 60$.

1.36 Jointly distributed variables
(a) $E[\frac{V^2}{1+U}] = E[V^2]E[\frac{1}{1+U}] = \int_0^\infty v^2\lambda e^{-\lambda v}dv \int_0^1 \frac{1}{1+u}du = (\frac{2}{\lambda^2})(\ln(2)) = \frac{2\ln 2}{\lambda^2}$.
(b) $P\{U \le V\} = \int_0^1 \int_u^\infty \lambda e^{-\lambda v} dv du = \int_0^1 e^{-\lambda u} du = (1 - e^{-\lambda})/\lambda$.
(c) The support of both f_{UV} and f_{YZ} is the strip $[0,1] \times [0,\infty)$, and the mapping $(u,v) \to (y,z)$ defined by $y = u^2$ and $z = uv$ is one-to-one. Indeed, the inverse mapping is given by $u = y^{\frac{1}{2}}$ and $v = zy^{-\frac{1}{2}}$. The absolute value of the Jacobian determinant of the forward mapping is $|\frac{\partial(x,y)}{\partial(u,v)}| = \begin{vmatrix} 2u & 0 \\ v & u \end{vmatrix} = 2u^2 = 2y$. Thus,

$$f_{Y,Z}(y,z) = \begin{cases} \frac{\lambda}{2y} e^{-\lambda z y^{-\frac{1}{2}}} & (y,z) \in [0,1] \times [0,\infty) \\ 0 & \text{otherwise} \end{cases}.$$

2.2 The limit of the product is the product of the limits
(a) There exists n_1 so large that $|y_n - y| \le 1$ for $n \ge n_1$. Thus, $|y_n| \le L$ for all n, where $L = \max\{|y_1|, |y_2|, \ldots, |y_{n_1-1}|, |y|+1\}$.
(b) Given $\epsilon > 0$, there exists n_ϵ so large that $|x_n - x| \le \frac{\epsilon}{2L}$ and $|y_n - y| \le \frac{\epsilon}{2(|x|+1)}$. Thus, for $n \ge n_\epsilon$,

$$|x_n y_n - xy| \le |(x_n - x)y_n| + |x(y_n - y)| \le |x_n - x|L + |x||y_n - y| \le \frac{\epsilon}{2} + \frac{\epsilon}{2} \le \epsilon.$$

So $x_n y_n \to xy$ as $n \to \infty$.

2.4 Limits of some deterministic series (a) Convergent. This is the power series expansion for e^x, which is everywhere convergent, evaluated at $x = 3$. The value of the sum is thus e^3. Another way to show the series is convergent is to note that for $n \ge 3$ the nth term can be bounded above by $\frac{3^n}{n!} = \frac{3^3}{3!}\frac{3}{4}\frac{3}{5}\cdots\frac{3}{n} \le (4.5)(\frac{3}{4})^{n-3}$. Thus, the sum is bounded by a constant plus a geometric series, so it is convergent.
(b) Convergent. Let $0 < \eta < 1$. Then $\ln n < n^\eta$ for all large enough n. Also, $n+2 \le 2n$ for all large enough n, and $n+5 \ge n$ for all n. Therefore, the nth term in the series is bounded above, for all sufficiently large n, by $\frac{2n \cdot n^\eta}{n^3} = 2n^{\eta-2}$. Therefore, the sum in (b) is bounded above by finitely many terms of the sum, plus $2\sum_{n=1}^\infty n^{\eta-2}$, which is finite, because, for $\alpha > 1$, $\sum_{n=1}^\infty n^{-\alpha} < 1 + \int_1^\infty x^{-\alpha}dx = \frac{\alpha}{\alpha-1}$, as shown in Example 11.2.

(c) Not convergent. Let $0 < \eta < 0.2$. Then $\log(n+1) \leq n^{\eta}$ for all n large enough, s
for n large enough the nth term in the series is greater than or equal to $n^{-5\eta}$. The serie
is therefore divergent. We used the fact that $\sum_{n=1}^{\infty} n^{-\alpha}$ is infinite for any $0 \leq \alpha \leq 1$
because it is greater than or equal to the integral $\int_1^{\infty} x^{-\alpha} dx$, which is infinite fo
$0 \leq \alpha \leq 1$.

2.6 Convergence of alternating series (a) For $n \geq 0$, let I_n denote the interval wi
endpoints s_n and s_{n+1}. It suffices to show that $I_0 \supset I_1 \supset I_2 \supset \ldots$ If n is eve
then $I_n = [s_{n+1}, s_{n+1} + b_{n+1}] \supset [s_{n+1}, s_{n+1} + b_{n+2}] = I_{n+1}$. Similarly, if n is od
$I_n = [s_{n+1} - b_{n+1}, s_{n+1}] \supset [s_{n+1} - b_{n+2}, s_{n+1}] = I_{n+1}$. So in general, for any n, $I_n \supset I_{n+}$
(b) Given $\epsilon > 0$, let N_ϵ be so large that $b_{N_\epsilon} < \epsilon$. It remains to prove that $|s_n - s_m| \leq$
whenever $n \geq N_\epsilon$ and $m \geq N_\epsilon$. Without loss of generality, we can assume that $n \leq n$
Since $I_m \subset I_n$ it follows that $s_m \in I_n$ and therefore that $|s_m - s_n| \leq b_{n+1} \leq \epsilon$.

2.8 Convergence of sequences of random variables (a) The distribution of X_n is th
same for all n, so the sequence converges in distribution to any random variab
with the distribution of X_1. To check for mean square convergence, use the fa
$\cos(a)\cos(b) = (\cos(a+b) + \cos(a-b))/2$ to calculate that $E[X_n X_m] = \frac{1}{2}$ if $n = m$ a
$E[X_n X_m] = 0$ if $n \neq m$. Therefore, $\lim_{n,m\to\infty} E[X_n X_m]$ does not exist, so the sequen
(X_n) does not satisfy the Cauchy criteria for m.s. convergence, so it does not conver
in the m.s. sense. Since it is a bounded sequence, it therefore does not converge in th
p. sense either. (Because for bounded sequences, convergence p. implies convergen
m.s.) Therefore the sequence does not converge in the a.s. sense either. In summary, t
sequence converges in distribution but not in the other three senses. (Another approa
is to note that the distribution of $X_n - X_{2n}$ is the same for all n, so that the sequen
does not satisfy the Cauchy criteria for convergence in probability.)
(b) If ω is such that $0 < \Theta(\omega) < 2\pi$, then $|1 - \frac{\Theta(\omega)}{\pi}| < 1$ so that $\lim_{n\to\infty} Y_n(\omega) =$
for such ω. Since $P\{0 < \Theta(\omega) < 2\pi\} = 1$, it follows that (Y_n) converges to zero in t
a.s. sense, and hence also in the p. and d. senses. Since the sequence is bounded, it al
converges to zero in the m.s. sense.

2.10 Convergence of random variables on (0,1] (a) (d. only) The graphs of X_n and
CDF are shown in Figure 12.1, and the CDF is given by:

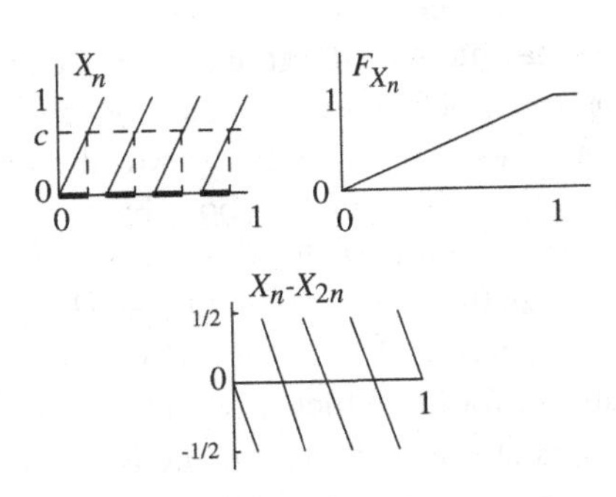

Figure 12.1 X_n, F_{X_n}, and $X_n - X_{2n}$

$$F_{X_n}(c) = \begin{cases} 0, \text{ if } c \leq 0 \\ P\{\omega : n\omega - \lceil n\omega \rceil \leq c\} = n\frac{c}{n} = c \\ 1 \text{ if } c \leq 1 \end{cases}.$$

Thus X_n is uniformly distributed over $[0, 1]$ for all n. So trivially X_n converges in distribution to U, where U is uniformly distributed on $[0, 1]$. A simple way to show that (X_n) does not converge in probability for this problem is to consider the distribution of $X_n - X_{2n}$. The graph of $X_n - X_{2n}$ is shown in Figure 12.1. Observe that for any $n \geq 1$, if $0 \leq \epsilon \leq 0.5$, then

$$P\{|X_n - X_{2n}| \geq \epsilon\} = 1 - 2\epsilon.$$

Therefore, $P\{|X_n - X_m| \geq \epsilon\}$ does not converge to zero as $n, m \to \infty$. By the Cauchy criteria for convergence in probability, (X_n) does not converge to any random variable in probability. It therefore does not converge in the m.s. sense or a.s. sense either.

(b) (a.s, p., d., not m.s.) For any $\omega \in (0, 1] = \Omega$, $X_n(\omega) = 0$ for $n > \frac{1}{\omega}$. Therefore $\lim_{n\to\infty} X_n(\omega) = 0$ for all $\omega \in \Omega$. Hence $\lim_{n\to\omega} X_n = 0$ a.s. (so $\lim_{n\to\infty} X_n = 0$ $d.$ and $p.$ also).

It remains to check whether (X_n) converges in the m.s. sense. If X_n converges in the m.s. sense, then it must converge to the same random variable in the p. sense. But as already shown, X_n converges to 0 in the p. sense. So if X_n converges in the m.s. sense, the limit must be the zero random variable. However, $E[X_n - 0|^2] = \int_0^{\frac{1}{n}} n^4x^2dx = \frac{n}{3} \to +\infty$ as $n \to \infty$. Therefore (X_n) does not converge in the m.s. sense.

(c) (a.s, p., d., not m.s.) For any $\omega \in \Omega$ fixed, the deterministic sequence $X_n(\omega)$ converges to zero. So $X_n \to 0$ a.s. The sequence thus also converges in p. and d. If the sequence converged in the m.s. sense, the limit would also have to be zero, but

$$E[|X_n - 0|^2] = E[|X_n|^2] = \frac{1}{n^2}\int_0^1 \frac{1}{\omega}d\omega = +\infty \not\to 0.$$

The sequence thus does not converge in the m.s. sense.

(d) (a.s, p., d., not m.s.) For any $\omega \in \Omega$ fixed, except the single point 1 which has zero probability, the deterministic sequence $X_n(\omega)$ converges to zero. So $X_n \to 0$ a.s. The sequence also converges in p. and d. If the sequence converged in the m.s. sense, the limit would also have to be zero, but

$$E[|X_n - 0|^2] = E[|X_n|^2] = n^2 \int_0^1 \omega^{2n}d\omega = \frac{n^2}{2n+1} \not\to 0.$$

The sequence thus does not converge in the m.s. sense.

(e) (d. only) For ω fixed and irrational, the sequence does not even come close to settling down, so intuitively we expect the sequence does not converge in any of the three strongest senses: a.s., m.s., or p. To prove this, it suffices to prove that the sequence does not converge in p. Since the sequence is bounded, convergence in probability is equivalent to convergence in the m.s. sense, so it also would suffice to prove the sequence does not converge in the m.s. sense. The Cauchy criteria for m.s. convergence would be violated if $E[(X_n - X_{2n})^2] \not\to 0$ as $n \to \infty$. By the double angle formula, $X_{2n}(\omega) = 2\omega \sin(2\pi n\omega)\cos(2\pi n\omega)$ so that

$$E[(X_n - X_{2n})^2] = \int_0^1 \omega^2(\sin(2\pi n\omega))^2(1 - 2\cos(2\pi n\omega))^2 d\omega,$$

and this integral clearly does not converge to zero as $n \to \infty$. In fact, following th heuristic reasoning below, the limit can be shown to equal $E[\sin^2(\Theta)(1-2\cos(\Theta))^2]/$: where Θ is uniformly distributed over the interval $[0, 2\pi]$. So the sequence (X_n) doe not converge in m.s., p., or a.s. senses.

The sequence does converge in the distribution sense. We shall give a heuristi derivation of the limiting CDF. Note that the CDF of X_n is given by

$$F_{X_n}(c) = \int_0^1 I_{\{f(\omega)\sin(2\pi n\omega) \le c\}} d\omega, \quad (12.1)$$

where f is the function defined by $f(\omega) = \omega$. As $n \to \infty$, the integrand in (12.1) jump between zero and one more and more frequently. For any small $\epsilon > 0$, we can imagin partitioning $[0, 1]$ into intervals of length ϵ. The number of oscillations of the integran within each interval converges to infinity, and the factor $f(\omega)$ is roughly constant ove each interval. The fraction of a small interval for which the integrand is one nearl converges to $P\{f(\omega)\sin(\Theta) \le c\}$, where Θ is a random variable that is uniforml distributed over the interval $[0, 2\pi]$, and ω is a fixed point in the small interval. So th CDF of X_n converges for all constants c to:

$$\int_0^1 P\{f(\omega)\sin(\Theta) \le c\}\, d\omega. \quad (12.2)$$

(Note: The following observations can be used to make the above argument rigorou: The integrals in (12.1) and (12.2) would be equal if f were constant within each interva of the form $(\frac{i}{n}, \frac{i+1}{n})$. If f is continuous on $[0, 1]$, it can be approximated by such ste functions with maximum approximation error converging to zero as $n \to \infty$. Detail are left to the reader.)

2.12 A Gaussian sequence We begin by considering convergence in distribution. B induction on k, X_k is a Gaussian random variable with mean zero for all k. The varianc of X_k, denoted by σ_k^2, is determined by the following recursion: $\sigma_0^2 = 0$ and $\sigma_{k+1}^2 =$ $\frac{\sigma_k^2+\sigma^2}{4}$. This can be solved to get $\sigma_k^2 = \sigma^2(\frac{1}{4} + \frac{1}{4^2} + \cdots + \frac{1}{4^k})$ so by the formula for th sum of a geometric series, $\lim_{k\to\infty} \sigma_k^2 = \sigma_\infty^2 \triangleq \frac{\sigma^2}{3}$. The CDF of X_k is given by $F_k(c) =$ $P\left\{\frac{X_k}{\sigma_k} \le \frac{c}{\sigma_k}\right\} = \Phi\left(\frac{c}{\sigma_k}\right)$, where Φ is the standard normal CDF. Since Φ is a continuou function, it follows that F_k converges pointwise to the CDF $F_\infty(c) = \Phi(\frac{c}{\sigma_\infty})$, so tha (X_k) converges in distribution with the limit having the $N(0, \sigma_\infty^2)$ distribution.

The sequence does *not* converge in p. Let $\epsilon > 0$. Consider $P\{|D_k| \ge \epsilon\}$ wher $D_k = X_{k+1} - X_k$. By the recursion, $D_k = \frac{X_k+W_k}{2} - X_k = \frac{W_k-X_k}{2}$. D_k is a Gaussian randon variable and $\mathrm{Var}(D_k) = \frac{\sigma^2+\sigma_k^2}{4} \ge \frac{\sigma^2}{4}$. Therefore, $P\{|D_k| \ge \epsilon\} = P\{\frac{2|D_k|}{\sigma} \ge \frac{2\epsilon}{\sigma}\} \ge$ $2Q(\frac{2\epsilon}{\sigma}) > 0$. So $P\{|D_k| \ge \epsilon\} \not\to 0$ as $k \to \infty$ so $P\{|X_n - X_m| \ge \epsilon\} \not\to 0$ as $m, n \to \infty$. That is, (X_n) is not a Cauchy sequence in probability, and hence does not converge i probability. The sequence thus also does not converge in the a.s. or m.s. sense.

2.14 Convergence of a sequence of discrete random variables (a) F_n is shown i Figure 12.2. Since $F_n(x) = F_X\left(x - \frac{1}{n}\right)$, $\lim_{n\to\infty} F_n(x) = F_X(x-)$ all x

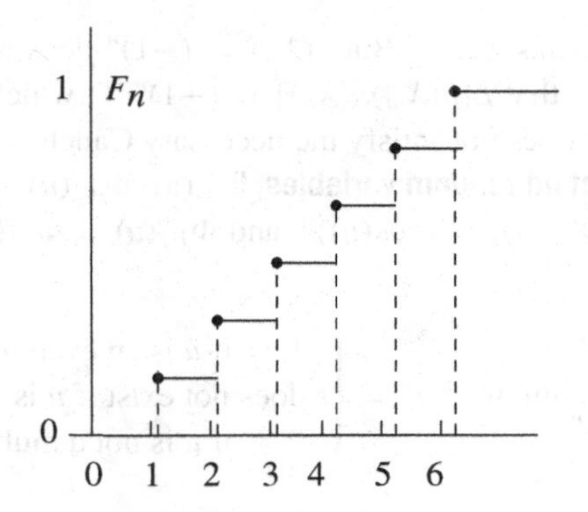

Figure 12.2 Graph of F_n.

So $\lim_{n\to\infty} F_n(x) = F_X(x)$ unless $F_X(x) \neq F_X(x-)$ i.e., unless $x = 1, 2, 3, 4, 5,$ or 6.
(b) F_X is continuous at x unless $x \in \{1, 2, 3, 4, 5, 6\}$.
(c) Yes, $\lim_{n\to\infty} X_n = X$ $d.$ by definition.
2.16 Convergence of a minimum (a) The sequence (X_n) converges to zero in all four senses. Here is one proof, and there are others. For any ϵ with $0 < \epsilon < 1$, $P\{|X_n - 0| \geq \epsilon\} = P\{U_1 \geq \epsilon, \ldots, U_n \geq \epsilon\} = (1-\epsilon)^n$, which converges to zero as $n \to \infty$. Thus, by definition, $X_n \to 0$ p. Thus, the sequence converges to zero in d. sense and, since it is bounded, in the m.s. sense. For each ω, as a function of n, the sequence of numbers $X_1(\omega), X_2(\omega), \ldots$ is a nonincreasing sequence of numbers bounded below by zero. Thus, the sequence X_n converges in the a.s. sense to some limit random variable. If a limit of random variables exists in different senses, the limit random variable has to be the same, so the sequence (X_n) converges a.s. to zero.
(b) For n fixed, the variable Y_n is distributed over the interval $[0, n^\theta]$, so let c be a number in that interval. Then $P\{Y_n \leq c\} = P\{X_n \leq cn^{-\theta}\} = 1 - P\{X_n > cn^{-\theta}\} = 1 - (1 - cn^{-\theta})^n$. Thus, if $\theta = 1$, $\lim_{n\to\infty} P\{Y_n \leq c\} = 1 - \lim_{n\to\infty}(1 - \frac{c}{n})^n = 1 - \exp(-c)$ for any $c \geq 0$. Therefore, if $\theta = 1$, the sequence (Y_n) converges in distribution, and the limit distribution is the exponential distribution with parameter one.
2.18 Limits of functions of random variables (a) Yes. Since g is a continuous function, if a sequence of numbers a_n converges to a limit a, then $g(a_n)$ converges to $g(a)$. Therefore, for any ω such that $\lim_{n\to\infty} X_n(\omega) = X(\omega)$, $\lim_{n\to\infty} g(X_n(\omega)) = g(X(\omega))$. If $X_n \to X$ a.s., then the set of all such ω has probability one, so $g(X_n) \to g(X)$ a.s.
(b) Yes. A direct proof is to first note that $|g(b) - g(a)| \leq |b - a|$ for any numbers a and b. So, if $X_n \to X$ m.s., then $E[|g(X_n) - g(X)|^2] \leq E[|X - X_n|^2] \to 0$ as $n \to \infty$. Therefore $g(X_n) \to g(X)$ m.s. A slightly more general proof would be to use the continuity of g (implying uniform continuity on bounded intervals) to show that $g(X_n) \to g(X)$ p., and then, since g is bounded, use the fact that convergence in probability for a bounded sequence implies convergence in the m.s. sense.
(c) No. For a counter example, let $X_n = (-1)^n/n$. Then $X_n \to 0$ deterministically, and hence in the a.s. sense. But $h(X_n) = (-1)^n$, which converges with probability zero, not with probability one.
(d) No. For a counter example, let $X_n = (-1)^n/n$. Then $X_n \to 0$ deterministically, and

hence in the m.s. sense. But $h(X_n) = (-1)^n$ does not converge in the m.s. sense. (Fo a proof, note that $E[h(X_m)h(X_n)] = (-1)^{m+n}$, which does not converge as $m, n \to \infty$ Thus, $h(X_n)$ does not satisfy the necessary Cauchy criteria for m.s. convergence.)

2.20 Sums of iid random variables, II (a) $\Phi_{X_1}(u) = \frac{1}{2}e^{ju} + \frac{1}{2}e^{-ju} = \cos(u)$, s $\Phi_{S_n}(u) = \Phi_{X_1}(u)^n = (\cos(u))^n$, and $\Phi_{V_n}(u) = \Phi_{S_n}(u/\sqrt{n}) = \cos(u/\sqrt{n})^n$.

(b)

$$\lim_{n\to\infty} \Phi_{S_n}(u) = \begin{cases} 1 & \text{if } u \text{ is an even multiple of } \pi \\ \text{does not exist} & \text{if } u \text{ is an odd multiple of } \pi \\ 0 & \text{if } u \text{ is not a multiple of } \pi \end{cases},$$

$$\lim_{n\to\infty} \Phi_{V_n}(u) = \lim_{n\to\infty} \left(1 - \frac{1}{2}\left(\frac{u}{\sqrt{n}}\right)^2 + o\left(\frac{u^2}{n}\right)\right)^n = e^{-\frac{u^2}{2}}.$$

(c) S_n does not converge in distribution, because, for example, $\lim_{n\to\infty} \Phi_{S_n}(\pi) = \lim_{n\to\infty}(-1)^n$ does not exist. So S_n does not converge in the m.s., a.s., or p. sens either. The limit of Φ_{V_n} is the characteristic function of the $N(0, 1)$ distribution, so tha (V_n) converges in distribution and the limit distribution is $N(0, 1)$. It will next be prove that V_n does not converge in probability. The intuitive idea is that if m is much large than n, then most of the random variables in the sum defining V_m are independent o the variables defining V_n. Hence, there is no reason for V_m to be close to V_n with hig probability. The proof below looks at the case $m = 2n$. Note that

$$\begin{aligned} V_{2n} - V_n &= \frac{X_1 + \ldots + X_{2n}}{\sqrt{2n}} - \frac{X_1 + \ldots + X_n}{\sqrt{n}} \\ &= \frac{\sqrt{2} - 2}{2}\left\{\frac{X_1 + \ldots + X_n}{\sqrt{n}}\right\} + \frac{1}{\sqrt{2}}\left\{\frac{X_{n+1} + \ldots + X_{2n}}{\sqrt{n}}\right\}. \end{aligned}$$

The two terms within the two pairs of braces are independent, and by the central limit theorem, each converges in distribution to the $N(0, 1)$ distribution. Thu $\lim_{n\to\infty} d.\ V_{2n} - V_n = W$, where W is a normal random variable with mean 0 an $\text{Var}(W) = \left(\frac{\sqrt{2}-2}{2}\right)^2 + \left(\frac{1}{\sqrt{2}}\right)^2 = 2 - \sqrt{2}$. Thus, $\lim_{n\to\infty} P(|V_{2n} - V_n| > \epsilon) \neq 0$ so by the Cauchy criteria for convergence in probability, V_n does not converge in probability Hence V_n does not converge in the a.s. sense or m.s. sense either.

2.22 Convergence and robustness of the sample median (a) We show that $Y_n \stackrel{a.s.}{\to} c^*$ It suffices to prove that for any c_0 and c_1 with $c_0 < c^* < c_1$,

$$P\{Y_n \le c_1 \text{ for all } n \text{ sufficiently large}\} = 1, \qquad (12.3)$$

$$P\{Y_n \ge c_0 \text{ for all } n \text{ sufficiently large}\} = 1. \qquad (12.4)$$

Since c^* is the unique solution to $F(c^*) = 0.5$, it follows that $F(c_1) > 0.5$. By the strong law of large numbers,

$$\frac{I_{\{X_1 \le c_1\}} + \ldots + I_{\{X_{2n+1} \le c_1\}}}{2n+1} \stackrel{a.s.}{\to} F_X(c_1).$$

In words, it means that the fraction of the variables $X_1, \ldots, X_{2n+1}$ that are less than or equal to c_1 converges to $F_X(c_1)$. Since $F_X(c_1) > 0.5$, it follows that

$$P\left\{\frac{I_{\{X_1\le c_1\}} + \ldots + I_{\{X_{2n+1}\le c_1\}}}{2n+1} > 0.5 \text{ for all n large enough}\right\} = 1,$$

which, in turn, implies (12.3). The proof of (12.4) is similar, and omitted.
(b) The event $\{|Y_n| > c\}$ is a subset of the union of the events $\{|X_i| \ge c \text{ for all } i \in A\}$ over all $A \subset \{1, \ldots, 2n+1\}$ with $|A| = n+1$. There are less than $\binom{2n+1}{n+1}$ such subsets A of $\{1, \ldots, 2n+1\}$, and for any one of them, $P\{|X_i| \ge c \text{ for all } i \in A\} = P\{|X_1| \ge c\}^{n+1}$. Now $\binom{2n+1}{n+1} \le 2^{2n+1}$ because the number of subsets of $\{1, \ldots, 2n+1\}$ of cardinality $n+1$ is less than or equal to the total number of subsets. Thus (b) follows by the union bound.
(c) Note that for $c > 0$, $P\{|X_1| \ge c\} = 2\int_c^\infty \frac{1}{\pi(1+u^2)}du \le \frac{2}{\pi}\int_c^\infty \frac{1}{u^2}du = \frac{2}{\pi c}$. By the result of part (b) with $n = 1$, $P\{|Y_1| \ge c\} \le 8\left(\frac{2}{\pi c}\right)^2 = \frac{32}{(\pi c)^2}$. Thus, $E[|Y_1|] = \int_0^\infty P\{|Y_1| \ge c\}dc \le 1 + \int_1^\infty P\{|Y_1| \ge c\}dc \le 1 + \int_1^\infty \frac{32}{(\pi c)^2}dc \le 1 + \frac{32}{\pi^2}$.

2.24 Normal approximation for quantization error The mean of each roundoff error is zero and the variance is $\int_{-0.5}^{0.5} u^2 du = \frac{1}{12}$. Thus, $E[S] = 0$ and $\mathrm{Var}(S) = \frac{100}{12} = 8.333$. Thus, $P\{|S| \ge 5\} = P\left\{\left|\frac{S}{\sqrt{8.333}}\right| \ge \frac{5}{\sqrt{8.333}}\right\} \approx 2Q(\frac{5}{\sqrt{8.333}}) = 2Q(1.73) = 2(1 - \Phi(1.732)) = 0.083$.

2.26 Applications of Jensen's inequality (a) The convex function is $\varphi(u) = \frac{1}{u}$ and the random variable is X.
(b) The convex function is $\varphi(u) = u^2$, and the random variable is X^2.
(c) The convex function is $\varphi(u) = u \ln u$, and the random variable is $L = f(Y)/g(Y)$, where Y has probability density g. Indeed, Jensen's inequality is $E[\varphi(L)] \ge \varphi(E[L])$. But $E[\varphi(L)] = \int_A \left(\frac{f(y)}{g(y)} \ln \frac{f(y)}{g(y)}\right) g(y)dy = D(f|g)$, and $E[Y] = \int_A \left(\frac{f(y)}{g(y)}\right) g(y)dy = \int_A f(y)dy = 1$ and $\varphi(1) = 0$, so that Jensen's inequality becomes $D(f|g) \ge 0$. Another solution is to use the function $\varphi(u) = -\ln u$ and the random variable $Z = g(X)/f(X)$, where X has density f. Indeed, Jensen's inequality is $E[\varphi(Z)] \ge \varphi(E[Z])$. But $E[\varphi(Z)] = \int_A -\ln \frac{g(x)}{f(x)} f(x)dx = D(f|g)$, and $E[Z] = \int_A \left(\frac{g(x)}{f(x)}\right) f(x)dx = \int_A g(x)dx = 1$ and $\varphi(1) = 0$, so that Jensen's inequality becomes $D(f|g) \ge 0$.

2.28 Understanding the Markov inequality
(a) $P\{|X| \ge 10\} = P\{X^4 \ge 10^4\} \le \frac{E[X^4]}{10^4} = 0.003$.
(b) Equality holds if $P\{X = 10\} = 0.003$ and $P\{X = 0\} = 0.997$. (We could have guessed this answer as follows. The inequality in part (a) is obtained by taking expectations on each side of the following inequality: $I_{\{|X|\ge 10\}} \le \frac{X^4}{10^4}$. In order for equality to hold, we need $I_{\{|X|\ge 10\}} = \frac{X^4}{10^4}$ with probability one. This requires $X \in \{-10, 0, 10\}$ with probability one.)

2.30 Portfolio allocation Let

$$Z_n = \begin{cases} 2 & \text{if you win on day } n \\ \frac{1}{2} & \text{if you lose on day } n \end{cases}.$$

Then $W_n = \prod_{k=1}^n (1 - \alpha + \alpha Z_n)$.
(a) For $\alpha = 0$, $W_n \equiv 1$ (so $W_n \to 1$ a.s., m.s., p., d.).

(b) For $\alpha = 1$, $W_n = \exp\left(\sum_{k=1}^{n} \ln(Z_k)\right)$. The exponent is a simple random walk, sam as S_n in *6. (Does not converge in any sense. It can be show that with probability on W_n is bounded neither below nor above.)
(c) $\ln W_n = \sum_{k=1}^{n} \ln(1 - \alpha + \alpha Z_k)$. By the strong law of large numbers, $\lim_{n\to\infty} \frac{\ln W_n}{n} = R(\alpha)$ a.s., where $R(\alpha) = E[\ln(1-\alpha+\alpha Z_n)] = \frac{1}{2}[\ln(1+\alpha)+\ln(1-\frac{\alpha}{2})$ Intuitively, this means that $W_n \approx e^{nR(\alpha)}$ as $n \to \infty$ in some sense. To be precise, means there is a set of ω with probability one, so that for any ω in the set and any $\epsilon >$ there is a finite number $n_\epsilon(\omega)$ such that $e^{n(R(\alpha)-\epsilon)} \le W_n \le e^{n(R(\alpha)+\epsilon)}$ for $n \ge n(\epsilon)$. Th number $R(\alpha)$ is the growth exponent in the a.s. sense. If $0 < \alpha < 1$, then $R(\alpha) > 0$ an $W_n \to \infty$ *a.s.* as $n \to \infty$. Therefore, if $0 < \alpha < 1$, $W_n \to +\infty$ *p.* and *d.* as well, but doesn't make sense to say $W_n \to +\infty$ *m.s.*
(d) $EW_n = \prod_{k=1}^{n} EZ_k = \left(1 + \frac{\alpha}{2}\right)^n$ which is maximized by $\alpha = 1$. Even so, most peop would prefer to use some α with $0 < \alpha < 1$, in order to enjoy a positive rate of growt
(e) The function $R(\alpha)$ achieves a maximum value of $\frac{1}{2}\ln\frac{9}{8} \approx .0589$ over $0 \le \alpha \le 1$ $\alpha = 0.5$. For $\alpha = \frac{1}{2}$, $\lim_{n\to\infty} \frac{1}{n} \ln W_n = 0.0589$ *a.s.*, or $W_n \sim e^{n(0.0589)}$. The fact $\alpha =$ maximizes the growth rate shows that a "diversified strategy" has a higher growth ra than either "pure" strategy.

2.32 Some large deviations (a) The required bound is provided by Chernoff's inequal ity for any $c > 0.5$ because the Us have mean 0.5. If $c = 0.5$ the probabilit is exactly 0.5 for all n and doesn't satisfy the required bound for any $b >$ Hence $c < 0.5$ doesn't work either. In summary, the bound holds precisely whe $c > 0.5$.
(b) The probability in question is equal to $P\{X_1 + \cdots + X_n > 0\}$, where $X_k = U_k - cU_{n+}$ for $1 \le k \le n$. The Xs are iid and $E[X_k] = \frac{1-c}{2}$. So if $c < 1$ the required boun is provided by Chernoff's inequality applied to the Xs. If $c = 1$ the probabilit is exactly 0.5 for all n and does not satisfy the required bound for any $b >$ Hence $c > 1$ does not work either. In summary, such $b > 1$ exists if and onl if $c < 1$.

2.34 A rapprochement between the CLT and large deviations (a) Different iating with respect to θ yields $M'(\theta) = (\frac{dE[\exp(\theta X)]}{d\theta})/E[\exp(\theta X)]$ and $M''(\theta$ $= \left(\frac{d^2E[X\exp(\theta X)]}{(d\theta)^2}E[\exp(\theta X)] - (\frac{dE[\exp(\theta X)]}{d\theta})^2\right)/E[\exp(\theta X)]^2$. Interchanging differ entiation and expectation yields $\frac{d^kE[\exp(\theta X)]}{(d\theta)^k} = E[X^k\exp(\theta X)]$. Thus, $M'(\theta) =$ $E[X\exp(\theta X)]/E[\exp(\theta X)]$, which is the mean for the tilted distribution f_θ, and $M''(\theta) =$ $\left(E[X^2\exp(\theta X)]E[\exp(\theta X)] - E[X\exp(\theta X)]^2\right)/E[\exp(\theta X)]^2$, which is the secon moment, minus the first moment squared, or simply the variance, for the tilted density f_θ
(b) In particular, $M'(0) = 0$ and $M''(0) = \mathrm{Var}(X) = \sigma^2$, so the second order Taylor' approximation for M near zero is $M(\theta) = \theta^2\sigma^2/2$. Therefore, $\ell(a)$ for small a satisfie $\ell(a) \approx \max_\theta(a\theta - \frac{\theta^2\sigma^2}{2}) = \frac{a^2}{2\sigma^2}$, so as $n \to \infty$, the large deviations upper boun behaves as $P\{S_n \ge b\sqrt{n}\} \le \exp(-n\ell(b/\sqrt{n})) \approx \exp(-n\frac{b^2}{2\sigma^2 n}) = \exp(-\frac{b^2}{2\sigma^2})$. Th exponent is the same as in the bound/approximation to the central limit approximatio described in the problem statement. Thus, for moderately large b, the central lim theorem approximation and large deviations bound/approximation are consistent wit each other.

2.36 Large deviations of a mixed sum Modifying the derivation for iid random variables, we find that for $\theta \geq 0$:

$$
\begin{aligned}
P\left\{\frac{S_n}{n} \geq a\right\} &\leq E[e^{\theta(S_n - an)}] \\
&= E[e^{\theta X_1}]^{nf} E[e^{\theta Y_1}]^{n(1-f)} e^{-n\theta a} \\
&= \exp(-n[\theta a - fM_X(\theta) - (1-f)M_Y(\theta)]),
\end{aligned}
$$

where M_X and M_Y are the log moment generating functions of X_1 and Y_1 respectively. Therefore,

$$
l(f,a) = \max_\theta \theta a - fM_X(\theta) - (1-f)M_Y(\theta),
$$

where

$$
M_X(\theta) = \begin{cases} -\ln(1-\theta) & \theta < 1 \\ +\infty & \theta \geq 1 \end{cases}, \quad M_Y(\theta) = \ln \sum_{k=0}^{\infty} \frac{e^{\theta k} e^{-1}}{k!} = \ln(e^{e^\theta - 1}) = e^\theta - 1.
$$

Note that $l(a,0) = a\ln a + 1 - a$ (large deviations exponent for the $Poi(1)$ distribution) and $l(a,1) = a - 1 - \ln(a)$ (large deviations exponent for the $Exp(1)$ distribution). For $0 < f < 1$ we compute $l(f,a)$ by numerical optimization. The result is

f	0	0+	1/3	2/3	1
$l(f,4)$	2.545	2.282	1.876	1.719	1.614

Note: $l(4,f)$ is discontinuous in f at $f = 0$. In fact, adding only one exponentially distributed random variable to a sum of Poisson random variables can change the large deviations exponent.

2.38 Bennett's inequality and Bernstein's inequality (a)

$$
\begin{aligned}
E[e^{\theta X_i}] &= E\left[1 + \theta X_i + \sum_{k=2}^{\infty} \frac{(\theta X_i)^k}{k!}\right] \\
&\leq E\left[1 + \sum_{k=2}^{\infty} \frac{|\theta X_i|^k}{k!}\right] \\
&\leq E\left[1 + \frac{X_i^2}{L^2} \sum_{k=2}^{\infty} \frac{(\theta L)^k}{k!}\right] \\
&\leq 1 + \frac{d_i^2}{L^2}(e^{\theta L} - 1 - \theta L) \\
&\leq \exp\left(\frac{d_i^2}{L^2}(e^{\theta L} - 1 - \theta L)\right).
\end{aligned}
$$

(b) The function to be maximized is a differentiable concave function of θ, so the maximizing θ is found by setting the derivative with respect to θ to zero, yielding

$$
\alpha - \frac{\sum_{i=1}^n d_i^2}{L}(e^{\theta L} - 1) = 0
$$

or $\theta = \frac{1}{L}\ln\left(1 + \frac{\alpha L}{\sum_{i=1}^n d_i^2}\right)$.

(c) This follows the proof of the Chernoff inequality. By the Markov inequality, for any $\theta > 0$,

$$P\left\{\sum_{i=1}^n X_i \geq \alpha\right\} \leq E\left[\exp\left(-\theta\alpha + \theta\sum_{i=1}^n X_i\right)\right]$$
$$\leq \exp\left(-\left[\theta\alpha - \frac{\sum_{i=1}^n d_i^2}{L^2}(e^{\theta L} - 1 - \theta L)\right]\right).$$

Plugging in the optimal value of θ found in part (b), which is positive as required, and rearranging yields Bennet's inequality.

(d) By complex analysis, the radius of convergence of the Taylor series of $\ln(1+u)$ about $u = 0$ is one. Thus, for $|u| < 1$, $\ln(1+u) = u - \frac{u^2}{2} + \frac{u^3}{3} - \ldots$ Hence

$$\frac{\varphi(u)}{u^2} = \frac{1}{2} + \frac{u}{2\cdot 3} - \frac{u^2}{3\cdot 4} + \frac{u^3}{4\cdot 5} - \ldots,$$

which implies, for $0 < u < 1$,

$$\left|\frac{\varphi(u)}{u^2} - \frac{1}{2}\right| \leq \frac{u}{6}.$$

(e) Straightforward substitution.

2.40 The sum of products of a sequence of uniform random variables

(a) Yes. $E[(B_k - 0)^2] = E[A_1^2]^k = (\frac{5}{8})^k \to 0$ as $k \to \infty$. Thus, $B_k \stackrel{m.s.}{\to} 0$.

(b) Yes. Each sample path of the sequence B_k is monotone nonincreasing and bounded below by zero, and is hence convergent. Thus, $\lim_{k\to\infty} B_k$ a.s. exists. (The limit has to be the same as the m.s. limit, so B_k converges to zero almost surely.)

(c) If $j \leq k$, then $E[B_jB_k] = E[A_1^2 \ldots A_j^2 A_{j+1} \ldots A_k] = (\frac{5}{8})^j(\frac{3}{4})^{k-j}$. Therefore,

$$E[S_nS_m] = E[\sum_{j=1}^n B_j \sum_{k=1}^m B_k] = \sum_{j=1}^n\sum_{k=1}^m E[B_jB_k] \to \sum_{j=1}^\infty\sum_{k=1}^\infty E[B_jB_k] \quad (12.5)$$
$$= 2\sum_{j=1}^\infty\sum_{k=j+1}^\infty \left(\frac{5}{8}\right)^j\left(\frac{3}{4}\right)^{k-j} + \sum_{j=1}^\infty\left(\frac{5}{8}\right)^j$$
$$= 2\sum_{j=1}^\infty\sum_{l=1}^\infty \left(\frac{5}{8}\right)^j\left(\frac{3}{4}\right)^l + \sum_{j=1}^\infty\left(\frac{5}{8}\right)^j$$
$$= \left(\sum_{j=1}^\infty\left(\frac{5}{8}\right)^j\right)\left(2\sum_{l=1}^\infty\left(\frac{3}{4}\right)^l + 1\right) \quad (12.6)$$
$$= \frac{5}{3}(2\cdot 3 + 1) = \frac{35}{3}.$$

A visual way to derive (12.6), is to note that (12.5) is the sum of all entries in the infinite 2-d array:

$$\begin{array}{cccc} \vdots & \vdots & \vdots & \iddots \\ \left(\frac{5}{8}\right)\left(\frac{3}{4}\right)^2 & \left(\frac{5}{8}\right)^2\left(\frac{3}{4}\right) & \left(\frac{5}{8}\right)^3 & \cdots \\ \left(\frac{5}{8}\right)\left(\frac{3}{4}\right) & \left(\frac{5}{8}\right)^2 & \left(\frac{5}{8}\right)^2\left(\frac{3}{4}\right) & \cdots \\ \left(\frac{5}{8}\right) & \left(\frac{5}{8}\right)\left(\frac{3}{4}\right) & \left(\frac{5}{8}\right)\left(\frac{3}{4}\right)^2 & \cdots \end{array}$$

Therefore, $\left(\frac{5}{8}\right)^j\left(2\sum_{l=1}^{\infty}\left(\frac{3}{4}\right)^l+1\right)$ is readily seen to be the sum of the jth term on the diagonal, plus all terms directly above or directly to the right of that term.

(d) Mean square convergence implies convergence of the mean. Thus, the mean of the limit is $\lim_{n\to\infty} E[S_n] = \lim_{n\to\infty}\sum_{k=1}^{n} E[B_k] = \sum_{k=1}^{\infty}\left(\frac{3}{4}\right)^k = 3$. The second moment of the limit is the limit of the second moments, namely $\frac{35}{3}$, so the variance of the limit is $\frac{35}{3}-3^2=\frac{8}{3}$.

(e) Yes. Each sample path of the sequence S_n is monotone nondecreasing and is hence convergent. Thus, $\lim_{n\to\infty} S_n$ a.s. exists. The limit has to be the same as the m.s. limit.

3.2 Linear approximation of the cosine function over an interval

$\widehat{E}[Y|\Theta] = E[Y] + \frac{\mathrm{Cov}(\Theta,Y)}{\mathrm{Var}(\Theta)}(\Theta - E[\Theta])$, where $E[Y] = \frac{1}{\pi}\int_0^{\pi}\cos(\theta)d\theta = 0$, $E[\Theta] = \frac{\pi}{2}$, $\mathrm{Var}(\Theta) = \frac{\pi^2}{12}$, $E[\Theta Y] = \int_0^{\pi}\frac{\theta\cos(\theta)}{\pi}d\theta = \frac{\theta\sin(\theta)}{\pi}\Big|_0^{\pi} - \int_0^{\pi}\frac{\sin(\theta)}{\pi}d\theta = -\frac{2}{\pi}$, and $\mathrm{Cov}(\Theta,Y) = E[\Theta Y] - E[\Theta]E[Y] = -\frac{2}{\pi}$. Therefore, $\widehat{E}[Y|\Theta] = -\frac{24}{\pi^3}(\Theta - \frac{\pi}{2})$, so the optimal choice is $a = \frac{12}{\pi^2}$ and $b = -\frac{24}{\pi^3}$.

3.4 Valid covariance matrix Set $a = 1$ to make K symmetric. Choose b so that the determinants of the following seven matrices are nonnegative:

$$(2) \quad (1) \quad (1) \quad \begin{pmatrix} 2 & 1 \\ 1 & 1 \end{pmatrix} \quad \begin{pmatrix} 2 & b \\ b & 1 \end{pmatrix} \quad \begin{pmatrix} 1 & 0 \\ 0 & 1 \end{pmatrix} \quad K \text{ itself.}$$

The fifth matrix has determinant $2-b^2$ and $\det(K) = 2-1-b^2 = 1-b^2$. Hence K is a valid covariance matrix (i.e. symmetric and positive semidefinite) if and only if $a=1$ and $-1 \le b \le 1$.

3.6 Conditional probabilities with joint Gaussians II

(a) $P\{|X-1| \ge 2\} = P\{X \le -1 \text{ or } X \ge 3\} = P\{\frac{X}{2} \le -\frac{1}{2}\} + P\{\frac{X}{2} \ge \frac{3}{2}\} = \Phi(-\frac{1}{2}) + 1 - \Phi(\frac{3}{2})$.

(b) Given $Y=3$, the conditional density of X is Gaussian, $E[X] + \frac{\mathrm{Cov}(X,Y)}{\mathrm{Var}(Y)}(3-E[Y]) = 1$ and $\mathrm{Var}(X) - \frac{\mathrm{Cov}(X,Y)^2}{\mathrm{Var}(Y)} = 4 - \frac{6^2}{18} = 2$.

(c) The estimation error $X - E[X|Y]$ is Gaussian, has mean zero and variance 2, and is independent of Y. (The variance of the error was calculated to be 2 in part (b).) Thus the probability is $\Phi(-\frac{1}{\sqrt{2}}) + 1 - \Phi(\frac{1}{\sqrt{2}})$, which can also be written as $2\Phi(-\frac{1}{\sqrt{2}})$ or $2(1-\Phi(\frac{1}{\sqrt{2}}))$.

3.8 An MMSE estimation problem (a) $E[XY] = 2\int_0^1\int_{2x}^{1+x} xy\,dx\,dy = \frac{5}{12}$. The other moments can be found in a similar way. Alternatively, note that the marginal densities are given by

$$f_X(x) = \begin{cases} 2(1-x) & 0 \le x \le 1 \\ 0 & \text{else} \end{cases}, \qquad f_Y(y) = \begin{cases} y & 0 \le y \le 1 \\ 2-y & 1 \le y \le 2 \\ 0 & \text{else} \end{cases},$$

so that $E[X] = \frac{1}{3}$, $\mathrm{Var}(X) = \frac{1}{18}$, $E[Y] = 1$, $\mathrm{Var}(Y) = \frac{1}{6}$, $\mathrm{Cov}(X, Y) = \frac{5}{12} - \frac{1}{3} = \frac{1}{12}$. So

$$\widehat{E}[X \mid Y] = \frac{1}{3} + \frac{1}{12}(\frac{1}{6})^{-1}(Y-1) = \frac{1}{3} + \frac{Y-1}{2},$$
$$E[e^2] = \frac{1}{18} - (\frac{1}{12})(\frac{1}{6})^{-1}(\frac{1}{12}) = \frac{1}{72} \quad = \text{the MMSE for } \widehat{E}[X|Y].$$

Inspection of Figure 12.3 shows that for $0 \le y \le 2$, the conditional distribution of X given $Y = y$ is the uniform distribution over the interval $[0, y/2]$ if $0 \le y \le 1$ and over the interval $[y-1, y/2]$ if $1 \le y \le 2$. The conditional mean of X given $Y = y$ is thus the midpoint of that interval, yielding:

$$E[X|Y] = \begin{cases} \frac{Y}{4} & 0 \le Y \le 1 \\ \frac{3Y-2}{4} & 1 \le Y \le 2 \end{cases}.$$

To find the corresponding MSE, note that given Y, the conditional distribution of X is uniform over some interval. Let $L(Y)$ denote the length of the interval. Then

$$E[e^2] = E[E[e^2|Y]] = E[\frac{1}{12}L(Y)^2]$$
$$= 2\left(\frac{1}{12}\int_0^1 y(\frac{y}{2})^2 dy\right) = \frac{1}{96}.$$

For this example, the MSE for the best estimator is 25% smaller than the MSE for the best linear estimator.

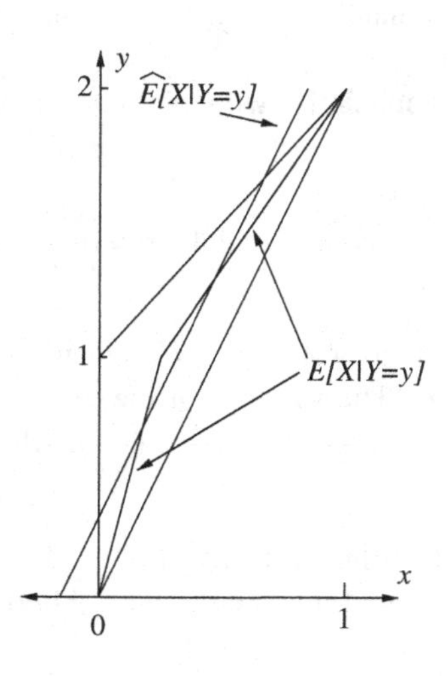

Figure 12.3 $E[X|Y = y]$ and $\widehat{E}[X|Y = y]$.

(b)

$$E[X] = \int_{-\infty}^{\infty} |y| \frac{1}{\sqrt{2\pi}} e^{-y^2/2} dy = 2\int_0^{\infty} \frac{y}{\sqrt{2\pi}} e^{-\frac{1}{2}y^2} dy = \sqrt{\frac{2}{\pi}} \text{ and } E[Y] = 0,$$

$$\text{Var}(Y) = 1, \text{Cov}(X, Y) = E[|Y|Y] = 0 \text{ so } \widehat{E}[X|Y] = \sqrt{\frac{2}{\pi}} + \frac{0}{1}Y \equiv \sqrt{\frac{2}{\pi}}.$$

That is, the best linear estimator is the constant $E[X]$. The corresponding MSE is $\text{Var}(X) = E[X^2] - E[X]^2 = E[Y^2] - \frac{2}{\pi} = 1 - \frac{2}{\pi}$. Note that $|Y|$ is a function of Y with mean square error $E[(X - |Y|)^2] = 0$. Nothing can beat that, so $|Y|$ is the MMSE estimator of X given Y. So $|Y| = E[X|Y]$. The corresponding MSE is 0, or 100% smaller than the MSE for the best linear estimator.

3.10 Conditional Gaussian comparison

(a) $p_a = P\{X \geq 2\} = P\{\frac{X}{\sqrt{10}} \geq \frac{2}{\sqrt{10}}\} = Q(\frac{2}{\sqrt{10}}) = Q(0.6324)$.

(b) By the theory of conditional distributions for jointly Gaussian random variables, the conditional distribution of X given $Y = y$ is Gaussian, with mean $\widehat{E}[X|Y = y]$ and variance σ_e^2, which is the MSE for estimation of X by $\widehat{E}[X|Y]$. Since X and Y are mean zero and $\frac{\text{Cov}(X,Y)}{\text{Var}(Y)} = 0.8$, we have $\widehat{E}[X|Y = y] = 0.8y$, and $\sigma_e^2 = \text{Var}(X) - \frac{\text{Cov}(X,Y)^2}{\text{Var}(Y)} = 3.6$. Hence, given $Y = y$, the conditional distribution of X is $N(0.8y, 3.6)$. Therefore, $P(X \geq 2|Y = y) = Q(\frac{2-(0.8)y}{\sqrt{3.6}})$. In particular, $p_b = P(X \geq 2|Y = 3) = Q(\frac{2-(0.8)3}{\sqrt{3.6}}) = Q(-0.2108)$.

(c) Given the event $\{Y \geq 3\}$, the conditional pdf of Y is obtained by setting the pdf of Y to zero on the interval $(-\infty, 3)$, and then renormalizing by $P\{Y \geq 3\} = Q(\frac{3}{\sqrt{10}})$ to make the density integrate to one. We can write this as

$$f_{Y|Y\geq 3}(y) = \begin{cases} \frac{f_Y(y)}{1-F_Y(3)} = \frac{e^{-y^2/20}}{Q(\frac{3}{\sqrt{10}})\sqrt{20\pi}} & y \geq 3 \\ 0 & \text{else} \end{cases}$$

Using this density, by considering the possible values of Y, we have

$$\begin{aligned} p_c = P(X \geq 2|Y \geq 3) &= \int_3^{\infty} P(X \geq 2, Y \in dy|Y \geq 3) \\ &= \int_3^{\infty} P(X \geq 2|Y = y)P(Y \in dy|Y \geq 3) \\ &= \int_3^{\infty} Q(\frac{2-(0.8)y}{\sqrt{3.6}}) f_{Y|Y\geq 3}(y)dy. \end{aligned}$$

(*Alternative*) The same expression can be derived in a more conventional fashion as follows:

$$\begin{aligned} p_e = P(X \geq 2|Y \geq 3) &= \frac{P\{X \geq 2, Y \geq 3\}}{P\{Y \geq 3\}} \\ &= \int_3^{\infty} \left[\int_2^{\infty} f_{X|Y}(x|y)dx\right] f_Y(y)dy/P\{Y \geq 3\} \\ &= \int_3^{\infty} Q\left(\frac{2-(0.8)y}{\sqrt{3.6}}\right) f_Y(y)dy/(1 - F_Y(3)) \\ &= \int_3^{\infty} Q(\frac{2-(0.8)y}{\sqrt{3.6}}) f_{Y|Y\geq 3}(y)dy. \end{aligned}$$

(d) We will show that $p_a < p_b < p_c$. The inequality $p_a < p_b$ follows from parts (a) an (b) and the fact the function Q is decreasing. By part (c), p_c is an average of $Q(\frac{2-(0.8)y}{\sqrt{3.6}})$ with respect to y over the region $y \in [3, \infty)$ (using the pdf $f_{Y|Y\geq 3}$). But everywhere i that region, $Q(\frac{2-(0.8)y}{\sqrt{3.6}}) > p_b$, showing that $p_c > p_b$.

3.12 An estimator of an estimator To show that $\widehat{E}[X|Y]$ is the LMMSE estimator o $E[X|Y]$, it suffices by the orthogonality principle to note that $\widehat{E}[X|Y]$ is linear in $(1, Y$ and to prove that $E[X|Y] - \widehat{E}[X|Y]$ is orthogonal to 1 and to Y. However $E[X|Y] - \widehat{E}[X|Y]$ can be written as the difference of two random variables $(X - \widehat{E}[X|Y])$ and $(X - E[X|Y])$, which are each orthogonal to 1 and to Y. Thus, $E[X|Y] - \widehat{E}[X|Y]$ is also orthogonal to and to Y, and the result follows.

Here is a generalization, which can be proved in the same way. Suppose $\mathcal{V}_0$ and $\mathcal{V}$ are two closed linear subspaces of random variables with finite second moments, suc that $\mathcal{V}_0 \supset \mathcal{V}_1$. Let X be a random variable with finite second moment, and let X_i^* be th variable in $\mathcal{V}_i$ with the minimum mean square distance to X, for $i = 0$ or $i = 1$. The X_1^* is the variable in $\mathcal{V}_1$ with the minimum mean square distance to X_0^*.

Another solution to the original problem can be obtained by using the formula fo $\widehat{E}[Z|Y]$ applied to $Z = E[X|Y]$:

$$\widehat{E}[E[X|Y]|Y] = E[E[X|Y]] + \mathrm{Cov}(Y, E[X|Y])\mathrm{Var}(Y)^{-1}(Y - E[Y]),$$

which can be simplified using $E[E[X|Y]] = E[X]$ and

$$\begin{aligned}\mathrm{Cov}(Y, E[X|Y]) &= E[Y(E[X|Y] - E[X])] \\ &= E[YE[X|Y]] - E[Y]E[X] \\ &= E[E[XY|Y]] - E[Y]E[X] \\ &= E[XY] - E[X]E[Y] = \mathrm{Cov}(X, Y)\end{aligned}$$

to yield the desired result.

3.14 Some identities for estimators (a) True. The random variable $E[X|Y]\cos(Y)$ ha the following two properties:

- It is a function of Y with finite second moments (because $E[X|Y]$ is a function o Y with finite second moment and $\cos(Y)$ is a bounded function of Y).
- $(X\cos(Y) - E[X|Y]\cos(Y)) \perp g(Y)$ for any g with $E[g(Y)^2] < \infty$ (because fo such g, $E[(X\cos(Y) - E[X|Y]\cos(Y))g(Y)] = E[(X - E[X|Y])\widetilde{g}(Y)] = 0$, wher $\widetilde{g}(Y) = g(Y)\cos(Y)$).

Thus, by the orthogonality principle, $E[X|Y]\cos(Y)$ is equal to $E[X\cos(Y)|Y]$.

(b) True. The lefthand side is the projection of X onto $\{g(Y) : E[g(Y)^2] < \infty\}$ an the righthand side is the projection of X onto the space $\{f(Y^3) : E[f(Y^3)^2] < \infty\}$ But these two spaces are the same, because for each function g there is the functio $f(u) = g(u^{1/3})$. The point is that the function y^3 is an invertible function, so any functio of Y can also be written as a function of Y^3.

(c) False. For example, let X be uniform on the interval [0, 1] and let Y be identicall zero. Then $E[X^3|Y] = E[X^3] = \frac{1}{4}$ and $E[X|Y]^3 = E[X]^3 = \frac{1}{8}$.

(d) False. For example, let $P\{X = Y = 1\} = P\{X = Y = -1\} = 0.5$. Then $E[X|Y] = Y$ while $E[X|Y^2] = 0$. The point is that the function y^2 is not invertible, so that no every function of Y can be written as a function of Y^2. Equivalently, Y^2 can give less information than Y.

(e) False. For example, let X be uniformly distributed on $[-1, 1]$, and let $Y = X$. Then $\widehat{E}[X|Y] = Y$ while $\widehat{E}[X|Y^3] = E[X] + \frac{\mathrm{Cov}(X,Y^3)}{\mathrm{Var}(Y^3)}(Y^3 - E[Y^3]) = \frac{E[X^4]}{E[X^6]}Y^3 = \frac{7}{5}Y^3$.

(f)) True. The given implies that the mean, $E[X]$, has the minimum MSE over all possible functions of Y (i.e. $E[X] = E[X|Y]$). Therefore, $E[X]$ also has the minimum MSE over all possible affine functions of Y, so $\widehat{E}[X|Y] = E[X]$. Thus, $E[X|Y] = E[X] = \widehat{E}[X|Y]$.

3.16 Some simple examples Of course there are many valid answers for this problem – we only give one.

(a) Let X denote the outcome of a roll of a fair die, and let $Y = 1$ if X is odd and $Y = 2$ if X is even. Then $E[X|Y]$ has to be linear. In fact, since Y has only two possible values, any function of Y can be written in the form $a+bY$. That is, any function of Y is linear. (There is no need to even calculate $E[X|Y]$ here, but we note that it is given by $E[Y|X] = X+2$.)

(b) Let X be an N(0,1) random variable, and let W be independent of X, with $P\{W = 1\} = P\{W = -1\} = \frac{1}{2}$. Finally, let $Y = XW$. The conditional distribution of Y given W is $N(0, 1)$, for either possible value of W, so the unconditional value of Y is also $N(0, 1)$. However, $P\{X - Y = 0\} = 0.5$, so that $X - Y$ is not a Gaussian random variable, so X and Y are not jointly Gaussian.

(c) Let (X, Y, Z) take on the four values $(0, 0, 0), (1, 1, 0), (1, 0, 1), (0, 1, 1)$ with equal probability. Then any pair of these variables takes the values $(0, 0), (0, 1), (1, 0), (1, 1)$ with equal probability, indicating pairwise independence. But $P\{(X, Y, Z) = (0, 0, 1) = 0 \neq P\{X = 0\}P\{Y = 0\}P\{Z = 1\} = \frac{1}{8}$. So the three random variables are not independent.

3.18 Estimating a quadratic (a) Recall the fact that $E[Z^2] = E[Z]^2 + \mathrm{Var}(Z)$ for any second order random variable Z. The idea is to apply the fact to the conditional distribution of X given Y. Given Y, the conditional distribution of X is Gaussian with mean ρY and variance $1 - \rho^2$. Thus, $E[X^2|Y] = (\rho Y)^2 + 1 - \rho^2$.

(b)

$$\begin{aligned}\mathrm{MSE} &= E[(X^2)^2] - E[(E[X^2|Y])^2] \\ &= E[X^4] - \rho^4 E[Y^4] - 2\rho^2 E[Y^2](1-\rho^2) - (1-\rho^2)^2 = 2(1-\rho^4).\end{aligned}$$

(c) Since $\mathrm{Cov}(X^2, Y) = E[X^2 Y] = 0$, it follows that $\widehat{E}[X^2|Y] = E[X^2] = 1$. That is, the best linear estimator in this case is just the constant estimator, 1.

3.20 An innovations sequence and its application (a) $\widetilde{Y}_1 = Y_1$, (Note: $E[\widetilde{Y}_1^2] = 1$), $\widetilde{Y}_2 = Y_2 - \frac{E[Y_2\widetilde{Y}_1]}{E[\widetilde{Y}_1^2]}\widetilde{Y}_1 = Y_2 - 0.5Y_1$ (Note: $E[\widetilde{Y}_2^2] = 0.75$), $\widetilde{Y}_3 = Y_3 - \frac{E[Y_3\widetilde{Y}_1]}{E[\widetilde{Y}_1^2]}\widetilde{Y}_1 - \frac{E[Y_3\widetilde{Y}_2]}{E[\widetilde{Y}_2^2]}\widetilde{Y}_2 = Y_3 - (0.5)\widetilde{Y}_1 - \frac{1}{3}\widetilde{Y}_2 = Y_3 - \frac{1}{3}Y_1 - \frac{1}{3}Y_2$. Summarizing,

$$\begin{pmatrix}\widetilde{Y}_1\\ \widetilde{Y}_2\\ \widetilde{Y}_3\end{pmatrix} = A\begin{pmatrix}Y_1\\ Y_2\\ Y_3\end{pmatrix} \quad \text{where } A = \begin{pmatrix}1 & 0 & 0\\ -\frac{1}{2} & 1 & 0\\ -\frac{1}{3} & -\frac{1}{3} & 1\end{pmatrix}\begin{pmatrix}Y_1\\ Y_2\\ Y_3\end{pmatrix}.$$

(b) $\mathrm{Cov}\begin{pmatrix}Y_1\\ Y_2\\ Y_3\end{pmatrix} = \begin{pmatrix}1 & 0.5 & 0.5\\ 0.5 & 1 & 0.5\\ 0.5 & 0.5 & 1\end{pmatrix}$ $\quad \mathrm{Cov}\left(X, \begin{pmatrix}Y_1\\ Y_2\\ Y_3\end{pmatrix}\right) = (0\ 0.25\ 0.25)$

$$\mathrm{Cov}\begin{pmatrix}\widetilde{Y}_1\\ \widetilde{Y}_2\\ \widetilde{Y}_3\end{pmatrix} = A\begin{pmatrix}1 & 0.5 & 0.5\\ 0.5 & 1 & 0.5\\ 0.5 & 0.5 & 1\end{pmatrix}A^T = \begin{pmatrix}1 & 0 & 0\\ 0 & \frac{3}{4} & 0\\ 0 & 0 & \frac{2}{3}\end{pmatrix},$$

$$\mathrm{Cov}\left(X, \begin{pmatrix}\widetilde{Y}_1\\ \widetilde{Y}_2\\ \widetilde{Y}_3\end{pmatrix}\right) = (0\ 0.25\ 0.25)A^T = (0\ \tfrac{1}{4}\ \tfrac{1}{6}).$$

(c) $a = \frac{\mathrm{Cov}(X,\widetilde{Y}_1)}{E[\widetilde{Y}_1^2]} = 0, \quad b = \frac{\mathrm{Cov}(X,\widetilde{Y}_2)}{E[\widetilde{Y}_2^2]} = \frac{1}{3}, \quad c = \frac{\mathrm{Cov}(X,\widetilde{Y}_3)}{E[\widetilde{Y}_3^2]} = \frac{1}{4}.$

3.22 A Kalman filtering example

(a)

$$\widehat{x}_{k+1|k} = f\widehat{x}_{k|k-1} + K_k(y_k - \widehat{x}_{k|k-1}),$$
$$\sigma_{k+1}^2 = f^2(\sigma_k^2 - \sigma_k^2(\sigma_k^2+1)^{-1}\sigma_k^2) + 1 = \frac{\sigma_k^2 f^2}{1+\sigma_k^2} + 1,$$

and $\quad K_k = f(\frac{\sigma_k^2}{1+\sigma_k^2}).$

(b) Since $\sigma_k^2 \le 1 + f^2$ for $k \ge 1$, the sequence (σ_k^2) is bounded for any value of f.

3.24 A variation of Kalman filtering Equations (3.20) and (3.21) hold as before, yielding

$$\widehat{x}_{k|k} = \widehat{x}_{k|k-1} + \frac{\sigma_{k|k-1}^2 \widetilde{y}_k}{1+\sigma_{k|k-1}^2},$$
$$\sigma_{k|k}^2 = \sigma_{k|k-1}^2 - \frac{\sigma_{k|k-1}^2}{1+\sigma_{k|k-1}^2} = \frac{1}{1+\sigma_{k|k-1}^2},$$

where we write σ instead of Σ and $\widetilde{y}_k = y_k - \widehat{x}_{k|k-1}$ as usual. Since $w_k = y_k - x_k$, it follows that $x_{k+1} = (f-1)x_k + y_k$, so (3.22) and (3.23) get replaced by

$$\begin{aligned}\widehat{x}_{k+1|k} &= \widehat{E}[(f-1)x_k + y_k|y^k]\\ &= (f-1)\widehat{x}_{k|k} + y_k,\\ \sigma_{k+1|k}^2 &= (f-1)^2\sigma_{k|k}^2.\end{aligned}$$

Combining the equations above yields

$$\widehat{x}_{k+1|k} = f\widehat{x}_k + K_k(y_k - \widehat{x}_{k|k-1}), \qquad K_k = \frac{1+f\sigma_{k|k-1}^2}{1+\sigma_{k|k-1}^2},$$
$$\sigma_{k+1|k}^2 = \frac{(f-1)^2\sigma_{k|k-1}^2}{1+\sigma_{k|k-1}^2}.$$

For $f = 1$ we find $\widehat{x}_{k+1|k} = y_k$ and $\sigma_{k+1|k}^2 = 0$ because $x_{k+1} = y_k$.

3.26 An innovations problem (a) $E[Y_n] = E[U_1 \dots U_n] = E[U_1]\dots E[U_n] = 2^{-n}$ and $E[Y_n^2] = E[U_1^2 \dots U_n^2] = E[U_1^2]\dots E[U_n^2] = 3^{-n}$, so $\mathrm{Var}(Y_n) = 3^{-n} - (2^{-n})^2 = 3^{-n} - 4^{-n}$.

(b) $E[Y_n|Y_0,\dots,Y_{n-1}] = E[Y_{n-1}U_n|Y_0,\dots,Y_{n-1}] = Y_{n-1}E[U_n|Y_0,\dots,Y_{n-1}] = Y_{n-1}E[U_n] = Y_{n-1}/2$.

(c) Since the conditional expectation found in (b) is linear, it follows that $\widehat{E}[Y_n|Y_0,\dots,Y_{n-1}] = E[Y_n|Y_0,\dots,Y_{n-1}] = Y_{n-1}/2$.

(d) $\widetilde{Y}_0 = Y_0 = 1$, and $\widetilde{Y}_n = Y_n - Y_{n-1}/2$ (also equal to $U_1 \ldots U_{n-1}(U_n - \frac{1}{2})$) for $n \geq 1$.
(e) For $n \geq 1$, $\mathrm{Var}(\widetilde{Y}_n) = E[(\widetilde{Y}_n)^2] = E[U_1^2 \ldots U_{n-1}^2 (U_n - \frac{1}{2})^2] = 3^{-(n-1)}/12$ and $\mathrm{Cov}(X_M, \widetilde{Y}_n) = E[(U_1 + \ldots + U_M)\widetilde{Y}_n] = E[(U_1 + \ldots + U_M)U_1 \ldots U_{n-1}(U_n - \frac{1}{2})]$ $= E[U_n(U_1 \cdots U_{n-1})(U_n - \frac{1}{2})] = 2^{-(n-1)}\mathrm{Var}(U_n) = 2^{-(n-1)}/12$. Since $\widetilde{Y}_0 = 1$ and all the other innovations variables are mean zero, we have

$$\begin{aligned}
\widehat{E}[X_M | Y_0, \ldots, Y_M] &= \frac{M}{2} + \sum_{n=1}^{M} \frac{\mathrm{Cov}(X_M, \widetilde{Y}_n)\widetilde{Y}_n}{\mathrm{Var}(\widetilde{Y}_n)} \\
&= \frac{M}{2} + \sum_{n=1}^{M} \frac{2^{-n+1}/12}{3^{-n+1}/12}\widetilde{Y}_n \\
&= \frac{M}{2} + \sum_{n=1}^{M} \left(\frac{3}{2}\right)^{n-1} \widetilde{Y}_n.
\end{aligned}$$

3.28 Linear innovations and orthogonal polynomials for the uniform distribution
(a)

$$E[U^n] = \int_{-1}^{1} \frac{u^n}{2} du = \frac{u^{n+1}}{2(n+1)}\bigg|_{-1}^{1} = \begin{cases} \frac{1}{n+1} & n \text{ even} \\ 0 & n \text{ odd} \end{cases}.$$

(b) The formula for the linear innovations sequence yields:
$\widetilde{Y}_1 = U$, $\widetilde{Y}_2 = U^2 - \frac{1}{3}$, $\widetilde{Y}_3 = U^3 - \frac{3U}{5}$, and
$\widetilde{Y}_4 = U^4 - \frac{E[U^4 \cdot 1]}{E[1^2]} \cdot 1 - \frac{E[U^4(U^2 - \frac{1}{3})]}{E[(U^2 - \frac{1}{3})^2]}(U^2 - \frac{1}{3}) = U^4 - \frac{1}{5} - \left(\frac{\frac{1}{7} - \frac{1}{15}}{\frac{1}{5} - \frac{2}{3} + 1}\right)(U^2 - 1) =$ $U^4 - \frac{6}{7}U^2 + \frac{3}{35}$. Note: These mutually orthogonal (with respect to the uniform distribution on [-1,1]) polynomials 1, U, $U^2 - \frac{1}{3}$, $U^3 - \frac{3}{5}U$, $U^4 - \frac{6}{7}U^2 + \frac{3}{35}$ are (up to constant multiples) known as the Legendre polynomials.

3.30 Example of extended Kalman filter
(a) Taking the derivative, we have $H_k = \cos(2\pi fk + \widehat{x}_{k|k-1})$. Writing σ_k^2 for $\Sigma_{k|k-1}$, the Kalman filter equation, $\widehat{x}_{k+1|k} = \widehat{x}_{k|k-1} + K_k\widetilde{y}_k$, becomes expanded to

$$\widehat{x}_{k+1|k} = \widehat{x}_{k|k-1} + \frac{\sigma_k^2 \cos(2\pi fk + \widehat{x}_{k|k-1})}{\cos^2(2\pi fk + \widehat{x}_{k|k-1})\sigma_k^2 + r}\left(y_k - \sin(\widehat{x}_{k|k-1} + 2\pi fk)\right).$$

(b) To check that the feedback is in the right direction, we consider two cases. First, if $\widehat{x}_{k|k-1}$ and x_k are such that the cos term is positive, that means the sin term is locally increasing in $\widehat{x}_{k|k-1}$. In that case if the actual phase x_k is slightly ahead of the estimate $\widehat{x}_{k|k-1}$, then the conditional expectation of $\widetilde{y}_k = y_k - \sin(2\pi fk + \widehat{x}_{k|k-1})$ is positive, and this difference gets multiplied by the positive cosine term, so the expected change in the phase estimate is positive. So the filter is changing the estimated phase in the right direction. Second, similarly, if $\widehat{x}_{k|k-1}$ and x_k are such that the cos term is negative, that means the sin term is locally decreasing in $\widehat{x}_{k|k-1}$. In that case if the actual phase x_k is slightly ahead of the estimate $\widehat{x}_{k|k-1}$, then the conditional expectation of $\widetilde{y}_k = y_k - \sin(2\pi fk + \widehat{x}_{k|k-1})$ is negative, and this difference gets multiplied by the negative cosine term, so the expected change in the phase estimate is positive. So, again, the filter is changing the estimated phase in the right direction.

4.2 Correlation function of a product
$R_X(s,t) = E[Y_sZ_sY_tZ_t] = E[Y_sY_tZ_sZ_t] = E[Y_sY_t]E[Z_sZ_t] = R_Y(s,t)R_Z(s,t)$.
4.4 Another sinusoidal random process
(a) Since $E[X_1] = E[X_2] = 0$, $E[Y_t] \equiv 0$. In addition,

$$\begin{aligned}R_Y(s,t) &= E[X_1^2]\cos(2\pi s)\cos(2\pi t) - 2E[X_1X_2]\cos(2\pi s)\sin(2\pi t)\\&\quad + E[X_2^2]\sin(2\pi s)\sin(2\pi t)\\&= \sigma^2(\cos(2\pi s)\cos(2\pi t) + \sin(2\pi s)\sin(2\pi t))\\&= \sigma^2\cos(2\pi(s-t)) \text{ (a function of } s-t \text{ only).}\end{aligned}$$

So $(Y_t : t \in \mathbb{R})$ is WSS.
(b) If X_1 and X_2 are independent Gaussian random variables, then $(Y_t : t \in \mathbb{R})$ is real-valued Gaussian WSS random process and is hence stationary.
(c) A simple solution to this problem is to take X_1 and X_2 to be independent, mea zero, variance σ^2 random variables with different distributions. For example, X_1 coul be $N(0,\sigma^2)$ and X_2 could be discrete with $P(X_1 = \sigma) = P(X_1 = -\sigma) = \frac{1}{2}$. The $Y_0 = X_1$ and $Y_{3/4} = X_2$, so Y_0 and $Y_{3/4}$ do not have the same distribution, so that Y i not stationary.
4.6 A random process corresponding to a random parabola (a) The mean function i $\mu_X(t) = 0 + 0t + t^2 = t^2$ and the covariance function is given by

$$\begin{aligned}C_X(s,t) &= \mathrm{Cov}(A + Bs + s^2, A + Bt + t^2)\\&= \mathrm{Cov}(A,A) + st\mathrm{Cov}(B,B) = 1 + st.\end{aligned}$$

Thus, $\widehat{E}[X_5|X_1] = \mu_X(5) + \frac{C_X(5,1)}{C_X(1,1)}(X_1 - \mu_X(1)) = 25 + \frac{6}{2}(X_1 - 1)$.
(b) A and B are jointly Gaussian and X_1 and X_5 are linear combinations of A and B, s X_1 and X_5 are jointly Gaussian. Thus, $E[X_5|X_1] = \widehat{E}[X_5|X_1]$.
(c) Since $X_0 = A$ and $X_1 = A + B + 1$, it follows that $B = X_1 - X_0 - 1$. Thu $X_t = X_0 + (X_1 - X_0 - 1)t + t^2$. So $X_0 + (X_1 - X_0 - 1)t + t^2$ is a linear estimator of X based on (X_0, X_1) with zero MSE, so it is the LMMSE estimator.
4.8 Brownian motion: Ascension and smoothing (a) Since the increments of W ove nonoverlapping intervals are independent, mean zero Gaussian random variables,

$$\begin{aligned}P\{W_r \le W_s \le W_t\} &= P\{W_s - W_r \ge 0,\ W_t - W_s \ge 0\}\\&= P\{W_s - W_r \ge 0\}P\{W_t - W_s \ge 0\} = \frac{1}{2}\cdot\frac{1}{2} = \frac{1}{4}.\end{aligned}$$

(b) Since W is a Gaussian process, the three random variables W_r, W_s, W_t are jointl Gaussian. They also all have mean zero, so that

$$\begin{aligned}&E[W_s|W_r, W_t]\\&= \widehat{E}[W_s|W_r, W_t]\\&= (\mathrm{Cov}(W_s,W_r), \mathrm{Cov}(W_s,W_t))\begin{pmatrix}\mathrm{Var}(X_r) & \mathrm{Cov}(X_r,X_t)\\ \mathrm{Cov}(X_t,X_r) & \mathrm{Var}(X_t)\end{pmatrix}^{-1}\begin{pmatrix}W_r\\W_t\end{pmatrix}\\&= (r,s)\begin{pmatrix}r & r\\ r & t\end{pmatrix}^{-1}\begin{pmatrix}W_r\\W_t\end{pmatrix}\\&= \frac{(t-s)W_r + (s-r)W_t}{t-r},\end{aligned}$$

where we use the fact $\begin{pmatrix} a & b \\ c & d \end{pmatrix}^{-1} = \frac{1}{ad-bc}\begin{pmatrix} d & -b \\ -c & a \end{pmatrix}$. As s varies from r to t, $E[W_s|W_r, W_t]$ is obtained by linearly interpolating between W_r and W_t.

4.10 Empirical distribution functions as random processes
(a) $E[\widehat{F}_n(t)] = \frac{1}{n}\sum_{k=1}^n E[I_{\{X_k\le t\}}] = \frac{1}{n}\sum_{k=1}^n F(t) = F(t)$.

$$\begin{aligned}
C(s,t) &= \mathrm{Cov}\left(\frac{1}{n}\sum_{k=1}^n I_{\{X_k\le s\}}, \frac{1}{n}\sum_{l=1}^n I_{\{X_l\le t\}}\right) \\
&= \frac{1}{n^2}\sum_{k=1}^n\sum_{l=1}^n \mathrm{Cov}\left(I_{\{X_k\le s\}}, I_{\{X_l\le t\}}\right) \\
&= \frac{1}{n^2}\sum_{k=1}^n \mathrm{Cov}\left(I_{\{X_k\le s\}}, I_{\{X_k\le t\}}\right) \\
&= \frac{1}{n}\mathrm{Cov}\left(I_{\{X_1\le s\}}, I_{\{X_1\le t\}}\right),
\end{aligned}$$

where we use the fact that for $k \neq l$, the random variables $I_{\{X_k\le s\}}$ and $I_{\{X_l\le t\}}$ are independent, and hence, uncorrelated, and the random variables X_k are identically distributed. If $s \le t$, then

$$\begin{aligned}
\mathrm{Cov}\left(I_{\{X_1\le s\}}, I_{\{X_1\le t\}}\right) &= E\left[I_{\{X_1\le s\}}I_{\{X_1\le t\}}\right] - E\left[I_{\{X_1\le s\}}\right]E\left[I_{\{X_1\le t\}}\right] \\
&= E\left[I_{\{X_1\le s\}}\right] - F(s)F(t) = F(s) - F(s)F(t).
\end{aligned}$$

Similarly, if $s \ge t$,

$$\mathrm{Cov}\left(I_{\{X_1\le s\}}, I_{\{X_1\le t\}}\right) = F(t) - F(s)F(t).$$

Thus, in general, $\mathrm{Cov}\left(I_{\{X_1\le s\}}, I_{\{X_1\le t\}}\right) = F(s\wedge t) - F(s)F(t)$, where $s\wedge t = \min\{s,t\}$, and so $C(s,t) = \frac{F(s\wedge t)-F(s)F(t)}{n}$.
(b) The convergence follows by the strong law of large numbers applied to the iid random variables $I_{\{X_k\le t\}}, k \ge 1$.
(c) Let $U_k = F(X_k)$ for all $k \ge 1$ and suppose that F is a continuous CDF. Fix $v \in (0,1)$. Then, since F is a continuous CDF, there exists a value t such that $F(t) = v$. Then $P\{F(X_k) \le v\} = P\{X_k \le t\} = F(t) = v$. Therefore, as suggested in the hint, the Us are uniformly distributed over $[0,1]$. For any k, under the assumptions on F, $\{X_k \le t\}$ and $\{F(X_k) \le v\}$ are the same events. Summing the indicator functions over k and dividing by n yields that $\widehat{F}_n(t) = \widehat{G}_n(v)$, and therefore that $|\widehat{F}_n(t) - F(t)| = |\widehat{G}_n(v) - v|$.

Taking the supremum over all $t \in \mathbb{R}$, or over all $v \in (0,1)$, while keeping $F(t) = v$, shows that $D_n = \sup_{0<v<1}|\widehat{G}_n(v) - v|$, and, since $G(v) = v$, the LHS of this equation is just D_n for the case of the uniformly distributed random variables $U_k, k \ge 1$.
(d) Observe, for t fixed, that $X_n(t) = \frac{\sum_{k=1}^n (I_{\{X_n\le t\}} - F(t))}{\sqrt{n}}$ and the random variables $I_{\{X_n\le t\}} - F(t)$ have mean zero and variance $F(t)(1-F(t))$. Therefore, by the central limit theorem, for each t fixed, $X_n(t)$ converges in distribution and the limit is Gaussian with mean zero and variance $C(t,t) = F(t)(1-F(t))$.

(e) The covariance is n times the covariance function found in part (a), with $F(t) = t$. The result is $s \wedge t - st$, as claimed in the problem statement.

(Note: The distance D_n is known as the Kolmogorov–Smirnov statistic, and by pursuing the method of this problem further, the limiting distribution of $\sqrt{n}D_n$ can be found and it is equal to the distribution of the maximum of a Brownian bridge, a result due to J.L. Doob.)

4.12 MMSE prediction for a Gaussian process based on two observations

(a) Since $R_X(0) = 5, R_X(1) = 0$, and $R_X(2) = -\frac{5}{9}$, the covariance matrix is $\begin{pmatrix} 5 & 0 & -\frac{5}{9} \\ 0 & 5 & 0 \\ -\frac{5}{9} & 0 & 5 \end{pmatrix}$.

(b) The variables are mean zero; $E[X(4)|X(2)] = \frac{\mathrm{Cov}(X(4),X(2))}{\mathrm{Var}(X(2))}X(2) = -\frac{X(2)}{9}$.

(c) $X(3)$ is uncorrelated with $(X(2), X(4))^T$; the variables are jointly Gaussian; $X(3)$ is independent of $(X(2), X(4))^T$. So $E[X(4)|X(2)] = E[X(4)|X(2), X(3)] = -\frac{X(2)}{9}$.

4.14 Poisson process probabilities (a) The numbers of arrivals in the disjoint intervals are independent, Poisson random variables with mean λ. Thus, the probability is $(\lambda e^{-\lambda})^3 = \lambda^3 e^{-3\lambda}$.

(b) The event is the same as the event that the numbers of counts in the intervals [0,1], [1,2], and [2,3] are 020, 111, or 202. The probability is thus
$e^{-\lambda}(\frac{\lambda^2}{2}e^{-\lambda})e^{-\lambda} + (\lambda e^{-\lambda})^3 + (\frac{\lambda^2}{2}e^{-\lambda})e^{-\lambda}(\frac{\lambda^2}{2}e^{-\lambda}) = (\frac{\lambda^2}{2} + \lambda^3 + \frac{\lambda^4}{4})e^{-3\lambda}$.

(c) This is the same as the probability the counts are 020, divided by the answer to part (b), or $\frac{\lambda^2}{2}/(\frac{\lambda^2}{2} + \lambda^3 + \frac{\lambda^4}{4}) = 2\lambda^2/(2 + 4\lambda + \lambda^2)$.

4.16 Adding jointly stationary Gaussian processes

(a) $R_Z(s,t) = E\left[\left(\frac{X(s)+Y(s)}{2}\right)\left(\frac{X(t)+Y(t)}{2}\right)\right] = \frac{1}{4}[R_X(s-t) + R_Y(s-t) + R_{XY}(s-t) + R_{YX}(s-t)]$. So $R_Z(s,t)$ is a function of $s-t$. Also, $R_{YX}(s,t) = R_{XY}(t,s)$. Thus, $R_Z(\tau) = \frac{1}{4}[2e^{-|\tau|} + \frac{e^{-|\tau-3|}}{2} + \frac{e^{-|\tau+3|}}{2}]$.

(b) Yes, the mean function of Z is constant ($\mu_Z \equiv 0$) and $R_Z(s,t)$ is a function of $s-t$ only, so Z is WSS. However, Z is obtained from the jointly Gaussian processes X and Y by linear operations, so Z is a Gaussian process. Since Z is Gaussian and WSS, it is stationary.

(c) $P\{X(1) < 5Y(2) + 1\} = P\left\{\frac{X(1)-5Y(2)}{\sigma} \le \frac{1}{\sigma}\right\} = \Phi\left(\frac{1}{\sigma}\right)$, where $\sigma^2 = \mathrm{Var}(X(1) - 5Y(2)) = R_X(0) - 10R_{XY}(1-2) + 25R_Y(0) = 1 - \frac{10e^{-4}}{2} + 25 = 26 - 5e^{-4}$.

4.18 A linear evolution equation with random coefficients

(a) $P_{k+1} = E[(A_kX_k + B_k)^2] = E[A_k^2X_k^2] + 2E[A_kX_k]E[B_k] + E[B_k^2] = \sigma_A^2P_k + \sigma_B^2$.

(b) Yes. Think of n as the present time. The future values $X_{n+1}, X_{n+2}, \ldots$ are all functions of X_n and $(A_k, B_k : k \ge n)$. But the variables $(A_k, B_k : k \ge n)$ are independent of $X_0, X_1, \ldots X_n$. Thus, the future is conditionally independent of the past, given the present.

(c) No. For example, $X_1 - X_0 = X_1 = B_1$, and $X_2 - X_1 = A_2B_1 + B_2$, and clearly B_1 and $A_2B_1 + B_2$ are not independent. (Given $B_1 = b$, the conditional distribution of $A_2B_1 + B_2$ is $N(0, \sigma_A^2b^2 + \sigma_B^2)$, which depends on b.)

(d) Suppose $s, t \in \mathbb{Z}$ with $s < t$. Then $R_Y(s,t) = E[Y_s(A_{t-1}Y_{t-1} + B_{t-1})] = E[A_{t-1}]$ $E[Y_sY_{t-1}] + E[Y_s]E[B_{t-1}] = 0$. Thus, $R_Y(s,t) = \begin{cases} P_k & \text{if } s = t = k \\ 0 & \text{else} \end{cases}$.
(e) The variables $Y_1, Y_2, \ldots$ are already orthogonal by part (d) (and the fact the variables have mean zero). Thus, $\widetilde{Y}_k = Y_k$ for all $k \geq 1$.

4.20 A Poisson spacing probability (a) $x(t)$ is simply the probability that either zero or one arrivals happens in an interval of length t. So $x(t) = (1 + \lambda t)e^{-\lambda t}$.
(b) Consider $t \geq 1$ and a small $h > 0$. For an interval of length $t+h$, if there is no arrival in the first h time units, then the conditional probability of success is $x(t)$. If there is an arrival in the first h time units, then the conditional probability of success is the product of the probability of no arrivals for the next unit of time, times the probability of success for an interval of length $t - 1$. Thus,

$$x(t+h) = (1 - \lambda h)x(t) + \lambda h e^{-\lambda} x(t-1) + o(h),$$

where the $o(h)$ term accounts for the possibility of two or more arrivals in an interval of length h and the exact time of arrival given there is one arrival in the first h time units. Thus, $\frac{x(t+h)-x(t)}{h} = -\lambda x(t) + \lambda e^{-\lambda} x(t-1) + \frac{o(h)}{h}$. Taking $h \to 0$ yields $x'(t) = -\lambda x(t) + \lambda e^{-\lambda} x(t-1)$.
(c) The function $y(t) = e^{-\theta t}$ satisfies the equation $y'(t) = -\lambda y(t) + \lambda e^{-\lambda} y(t-1)$ for all $t \in \mathbb{R}$ if $\theta = -\lambda + \lambda e^{\theta - \lambda}$, which has a unique positive solution θ^*. By the ordering property mentioned in the statement of part (b), the inequalities to be proved in part (c) are true for all $t \geq 0$ if they are true for $0 \leq t \leq 1$, so the tightest choices of c_0 and c_1 are given by $c_0 = \min_{0 \leq t \leq 1} x(t)e^{\theta * t}$ and $c_1 = \max_{0 \leq t \leq t} x(t)e^{\theta * t}$.
(d) Given there are k arrivals during $[0, t]$, we can view the times as uniformly distributed over the region $[0, t]^k$, which has volume t^k. By shrinking times between arrivals by exactly one, we see there is a one-to-one correspondence between vectors of k arrival times in $[0, t]$ such that A_t is true, and vectors of k arrival times in $[0, t - k + 1]$. So the volume of the set of vectors of k arrival times in $[0, t]$ such that A_t is true is $(t - k + 1)^k$. This explains the fact given at the beginning of part (d). The total number of arrivals during $[0, t]$ has the Poisson distribution with mean λt. Therefore, using the law of total probability,

$$x(t) = \sum_{k=0}^{\lceil t \rceil} \frac{e^{-\lambda t}(\lambda t)^k}{k!} \left(\frac{t - k + 1}{t} \right)^k = \sum_{k=0}^{\lceil t \rceil} \frac{e^{-\lambda t}(\lambda(t - k + 1))^k}{k!}.$$

4.22 A fly on a cube (a), (b) See the figures. For part (a), each two-headed line represents a directed edge in each direction and all directed edges have probability 1/3.
(c) Let a_i be the mean time for Y to first reach state zero starting in state i. Conditioning on the first time step yields $a_1 = 1 + \frac{2}{3}a_2$, $a_2 = 1 + \frac{2}{3}a_1 + \frac{1}{3}a_3$, $a_3 = 1 + a_2$. Using the first and third of these equations to eliminate a_1 and a_3 from the second equation yields a_2, and then a_1 and a_3. The solution is $(a_1, a_2, a_3) = (7, 9, 10)$. Therefore, $E[\tau] = 1 + a_1 = 8$.

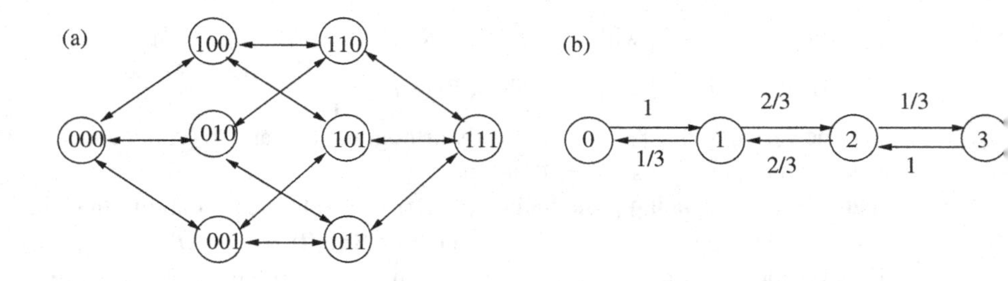

4.24 A random process created by interpolation
(a)

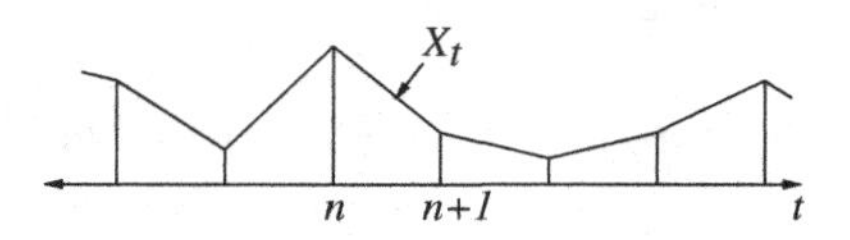

(b) X_t is the sum of two random variables, $(1-a)U_t$, which is uniformly distributed o the interval $[0, 1-a]$, and aU_{n+1}, which is uniformly distributed on the interval $[0, a$ Thus, the density of X_t is the convolution of the densities of these two variables:

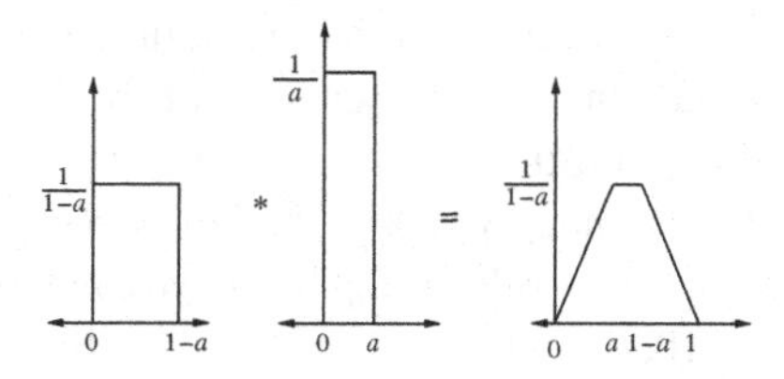

(c) $C_X(t,t) = \frac{a^2+(1-a)^2}{12}$ for $t = n + a$. Since this depends on t, X is not WSS.
(d) $P\{\max_{0\leq t\leq 10} X_t \leq 0.5\} = P\{U_k \leq 0.5 \text{ for } 0 \leq k \leq 10\} = (0.5)^{11}$.
4.26 Restoring samples (a) Yes. The possible values of X_k are $\{1, \ldots, k-1\}$. Give X_k, X_{k+1} is equal to X_k with probability $\frac{X_k}{k}$ and is equal to $X_k + 1$ with probability $1 - \frac{X_k}{k}$ Another way to say this is that the one-step transition probabilities for the transitio from X_k to X_{k+1} are given by

$$p_{ij} = \begin{cases} \frac{i}{k} & \text{for } j = i \\ 1 - \frac{i}{k} & \text{for } j = i+1 \\ 0 & \text{else} \end{cases} .$$

(b) $E[X_{k+1}|X_k] = X_k(\frac{X_k}{k}) + (X_k + 1)\left(1 - \frac{X_k}{k}\right) = X_k + 1 - \frac{X_k}{k}$.
(c) The Markov property of X, the information equivalence of X_k and M_k, and part (b) imply that $E[M_{k+1}|M_2, \ldots, M_k] = E[M_{k+1}|M_k] = \frac{1}{k+1}(X_k + 1 - \frac{X_k}{k}) \neq M_k$, so that $(M_k$ does not form a martingale sequence.
(d) Using the transition probabilities mentioned in part (a) again, yields (with som tedious algebra steps not shown)

$$E[D_{k+1}^2|X_k] = \left(\frac{X_k}{k+1} - \frac{1}{2}\right)^2 \left(\frac{X_k}{k}\right) + \left(\frac{X_k+1}{k+1} - \frac{1}{2}\right)^2 \left(\frac{k-X_k}{k}\right)$$
$$= \frac{1}{4k(k+1)^2}\left((4k-8)X_k^2 - (4k-8)kX_k + k(k-1)^2\right)$$
$$= \frac{1}{(k+1)^2}\left\{k(k-2)D_k^2 + \frac{1}{4}\right\}.$$

(e) Since, by the tower property of conditional expectations, $v_{k+1} = E[D_{k+1}^2] = E[E[D_{k+1}^2|X_k]]$, taking the expectation on each side of the equation found in part (d) yields

$$v_{k+1} = \frac{1}{(k+1)^2}\left\{k(k-2)v_k + \frac{1}{4}\right\},$$

and the initial condition $v_2 = 0$ holds. The desired inequality, $v_k \le \frac{1}{4k}$, is thus true for $k = 2$. For the purpose of proof by induction, suppose that $v_k \le \frac{1}{4k}$ for some $k \ge 2$. Then,

$$v_{k+1} \le \frac{1}{(k+1)^2}\left\{k(k-2)\frac{1}{4k} + \frac{1}{4}\right\}$$
$$= \frac{1}{4(k+1)^2}\{k-2+1\} \le \frac{1}{4(k+1)}.$$

So the desired inequality is true for $k+1$. Therefore, by proof by induction, $v_k \le \frac{1}{4k}$ for all k. Hence, $v_k \to 0$ as $k \to \infty$. By definition, this means that $M_k \stackrel{m.s.}{\to} \frac{1}{2}$ as $k \to \infty$. (We could also note that, since M_k is bounded, the convergence also holds in probability, and also it holds in distribution.)

4.28 An M/M/1/B queueing system

(a) $Q = \begin{pmatrix} -\lambda & \lambda & 0 & 0 & 0 \\ 1 & -(1+\lambda) & \lambda & 0 & 0 \\ 0 & 1 & -(1+\lambda) & \lambda & 0 \\ 0 & 0 & 1 & -(1+\lambda) & \lambda \\ 0 & 0 & 0 & 1 & -1 \end{pmatrix}$.

(b) The equilibrium vector $\pi = (\pi_0, \pi_1, \ldots, \pi_B)$ solves $\pi Q = 0$. Thus, $\lambda\pi_0 = \pi_1$. Also, $\lambda\pi_0 - (1+\lambda)\pi_1 + \pi_2 = 0$, which with the first equation yields $\lambda\pi_1 = \pi_2$. Continuing this way yields that $\pi_n = \lambda\pi_{n-1}$ for $1 \le n \le B$. Thus, $\pi_n = \lambda^n\pi_0$. Since the probabilities must sum to one, $\pi_n = \lambda^n/(1+\lambda+\ldots+\lambda^B)$.

4.30 Identification of special properties of two discrete-time processes (II) (a) (yes, yes, no). The process is Markov by its description. Think of a time k as the present time. Given the number of cells alive at the present time k (i.e. given X_k) the future evolution does not depend on the past. To check for the martingale property in discrete time, it suffices to check that $E[X_{k+1}|X_1, \ldots, X_k] = X_k$. But this equality is true because for each cell alive at time k, the expected number of cells alive at time $k+1$ is one $(=0.5 \times 0 + 0.5 \times 2)$. The process does not have independent increments, because, for example, $P(X_2 - X_1 = 0|X_1 - X_0 = -1) = 1$ and $P(X_2 - X_1 = 0|X_1 - X_0 = 1) = 1/2$. So $X_2 - X_1$ is not independent of $X_1 - X_0$.

(b) (yes, yes, no). Let k be the present time. Given Y_k, the future values are all determined by $Y_k, U_{k+1}, U_{k+2}, \ldots$ Since $U_{k+1}, U_{k+2}, \ldots$ is independent of $Y_0, \ldots, Y_k$, the future of

Y is conditionally independent of the past, given the present value Y_k. So Y is Marko
The process Y is a martingale because

$$E[Y_{k+1}|Y_1,\ldots,Y_k] = E[U_{k+1}Y_k|Y_1,\ldots,Y_k]$$
$$= Y_kE[U_{k+1}|Y_1,\ldots,Y_k] = Y_kE[U_{k+1}] = Y_k.$$

The process Y does not have independent increments; for example $Y_1 - Y_0 = U_1 - 1$ clearly not independent of $Y_2 - Y_1 = U_1(U_2 - 1)$. (To argue this further we could no that the conditional density of $Y_2 - Y_1$ given $Y_1 - Y_0 = y - 1$ is the uniform distributic over the interval $[-y, y]$, which depends on y.)

4.32 Identification of special properties of two continuous-time processes (II) (a) (ye no,no) Z is Markov because W is Markov and the mapping from W_t to Z_t is invertibl So W_t and Z_t have the same information. To see if W^3 is a martingale we suppose $s \leq$ and use the independent increment property of W to get:

$$E[W_t^3|W_u, 0 \leq u \leq s] = E[W_t^3|W_s] = E[(W_t - W_s + W_s)^3|W_s] =$$
$$3E[(W_t - W_s)^2]W_s + W_s^3 = 3(t-s)W_s + W_s^3 \neq W_s^3.$$

Therefore, W^3 is not a martingale. If the increments were independent, then since W is the increment $W_s - W_0$, it would have to be that $E[(W_t - W_s + W_s)^3|W_s]$ does nc depend on W_s. But it does. So the increments are not independent.

(b) (no, no, no) R is not Markov because knowing R_t for a fixed t does not quit determine Θ to be one of two values. But for one of these values R has a positiv derivative at t, and for the other R has a negative derivative at t. If the past c R just before t were also known, then θ could be completely determined, whic would give more information about the future of R. So R is not Markov. R is nc a martingale. For example, observing R on a finite interval totally determines R. S $E[R_t|(R_u, 0 \leq u \leq s] = R_t$, and if $s - t$ is not an integer, $R_s \neq R_t$. R does not hav independent increments. For example the increments $R(0.5) - R(0)$ and $R(1.5) - R(1$ are identical random variables, not independent random variables.

4.34 Moving balls (a) The states of the "relative-position process" can be taken to b 111, 12, and 21. The state 111 means that the balls occupy three consecutive position the state 12 means that one ball is in the left most occupied position and the other tw balls are one position to the right of it, and the state 21 means there are two balls i the leftmost occupied position and one ball one position to the right of them. With th states in the order 111, 12, 21, the one-step transition probability matrix is given b

$$P = \begin{pmatrix} 0.5 & 0.5 & 0 \\ 0 & 0 & 1 \\ 0.5 & 0.5 & 0 \end{pmatrix}.$$

(b) The equilibrium distribution π of the process is the probability vector satisfyin $\pi = \pi P$, from which we find $\pi = (\frac{1}{3}, \frac{1}{3}, \frac{1}{3})$. That is, all three states are equally likel in equilibrium.

(c) Over a long period of time, we expect the process to be in each of the states abou a third of the time. After each visit to states 111 or 12, the leftmost position of th configuration advances one position to the right. After a visit to state 21, the next stat will be 12, and the leftmost position of the configuration does not advance. Thus, after 2/3 of the slots there will be an advance. So the long-term speed of the balls is 2/3. Another approach is to compute the mean distance the moved ball travels in each slot, and divide by three.

(d) The same states can be used to track the relative positions of the balls as in discrete time. The generator matrix is given by $Q = \begin{pmatrix} -0.5 & 0.5 & 0 \\ 0 & -1 & 1 \\ 0.5 & 0.5 & -1 \end{pmatrix}$. (Note that if the state is 111 and if the leftmost ball is moved to the rightmost position, the state of the relative-position process is 111 the entire time. That is, the relative-position process misses such jumps in the actual configuration process.) The equilibrium distribution can be determined by solving the equation $\pi Q = 0$, and the solution is found to be $\pi = (\frac{1}{3}, \frac{1}{3}, \frac{1}{3})$ as before. When the relative-position process is in states 111 or 12, the leftmost position of the actual configuration advances one position to the right at rate one, while when the relative-position process is in state is 21, the rightmost position of the actual configuration cannot directly move right. The long-term average speed is thus 2/3, as in the discrete-time case.

4.36 Mean hitting time for a continuous-time Markov process

$$Q = \begin{pmatrix} -1 & 1 & 0 \\ 10 & -11 & 1 \\ 0 & 5 & -5 \end{pmatrix}, \qquad \pi = \left(\frac{50}{56}, \frac{5}{56}, \frac{1}{56}\right).$$

Consider X_h to get

$$\begin{aligned} a_1 &= h + (1-h)a_1 + ha_2 + o(h), \\ a_2 &= h + 10a_1 + (1-11h)a_2 + o(h), \end{aligned}$$

or equivalently $1 - a_1 + a_2 + \frac{o(h)}{h} = 0$ and $1 + 10a_1 - 11a_2 + \frac{o(h)}{h} = 0$. Let $h \to 0$ to get $1 - a_1 + a_2 = 0$ and $1 + 10a_1 - 11a_2 = 0$, or $a_1 = 12$ and $a_2 = 11$.

4.38 Poisson splitting This is basically the previous problem in reverse. This solution is based directly on the definition of a Poisson process, but there are other valid approaches. Let X be a Poisson random variable, and let each of X individuals be independently assigned a type, with type i having probability p_i, for some probability distribution $p_1, \ldots, p_K$. Let X_i denote the number assigned type i. Then,

$$\begin{aligned} &P(X_1 = i_1, X_2 = i_2, \ldots, X_K = i_K) \\ &= P(X = i_1 + \cdots + i_K)\frac{(i_1 + \ldots + i_K)!}{i_1!\, i_2! \ldots i_K!} p_1^{k_1} \cdots p_K^{i_K} \\ &= \prod_{j=1}^{K} \frac{e^{-\lambda_j}\lambda_j^{i_j}}{i_j!}, \end{aligned}$$

where $\lambda_i = \lambda p_i$. Thus, independent splitting of a Poisson number of individuals yields that the number of each type i is Poisson, with mean $\lambda_i = \lambda p_i$ and they are independent of each other.

Now suppose that N is a rate λ Poisson process, and that N_i is the process of typ
i points, given independent splitting of N with split distribution $p_1, \ldots, p_K$. By th
definition of a Poisson process, the following random variables are independent, wi
the ith having the $Poi(\lambda(t_{i+1} - t_i))$ distribution:

$$N(t_1) - N(t_0) \quad N(t_2) - N(t_1) \quad \ldots \quad N(t_p) - N(t_{p-1}). \qquad (12.7)$$

Suppose each column of the following array is obtained by independent splitting of th
corresponding variable in (12.7):

$$\begin{array}{cccc} N_1(t_1) - N_1(t_0) & N_1(t_2) - N_1(t_1) & \ldots & N_1(t_p) - N_1(t_{p-1}) \\ N_2(t_1) - N_2(t_0) & N_2(t_2) - N_2(t_1) & \ldots & N_2(t_p) - N_2(t_{p-1}) \\ \vdots & \vdots & \ldots & \vdots \\ N_K(t_1) - N_K(t_0) & N_K(t_2) - N_K(t_1) & \ldots & N_K(t_p) - N_K(t_{p-1}) \end{array} . \qquad (12.8)$$

Then by the splitting property of Poisson random variables described above, we g
that all elements of the array (12.8) are independent, with the appropriate means. B
definition, the ith process N_i is a rate λp_i random process for each i, and because
the independence of the rows of the array, the K processes $N_1, \ldots, N_K$ are mutual
independent.

4.40 Some orthogonal martingales based on Brownian motion Throughout
the solution of this problem, let $0 < s < t$, and let $Y = W_t - W_s$. Note that Y
independent of $\mathcal{W}_s$ and it has the $N(0, t-s)$ distribution.
(a) $E[M_t|\mathcal{W}_s] = M_s E[\frac{M_t}{M_s}|\mathcal{W}_s]$. Now $\frac{M_t}{M_s} = \exp(\theta Y - \frac{\theta^2(t-s)}{2})$. Therefore,
$E[\frac{M_t}{M_s}|\mathcal{W}_s] = E[\frac{M_t}{M_s}] = 1$. Thus $E[M_t|\mathcal{W}_s] = M_s$. By the hint, M is a martingale.
(b) $W_t^2 - t = (W_s + Y)^2 - s - (t-s) = W_s^2 - s + 2W_sY + Y^2 - (t-s)$, but
$E[2W_sY|\mathcal{W}_s] = 2W_sE[Y|\mathcal{W}_s] = 2W_sE[Y] = 0$, and
$E[Y^2 - (t-s)|\mathcal{W}_s] = E[Y^2 - (t-s)] = 0$. So $E[2W_sY + Y^2 - (t-s)|\mathcal{W}_s] = 0$; the martinga
property follows from the hint. Similarly, $W_t^3 - 3tW_t = (Y + W_s)^3 - 3(s + t - s)(Y + W_s$
$= W_s^3 - 3sW_s + 3W_s^2Y + 3W_s(Y^2 - (t-s)) + Y^3 - 3tY$. Since Y is independent of $\mathcal{W}_s$ an
$E[Y] = E[Y^2 - (t-s)] = E[Y^3] = 0$, $E[3W_s^2Y + 3W_s(Y^2 - (t-s)) + Y^3 - 3tY|\mathcal{W}_s] = ($
so the martingale property follows from the hint.
(c) Fix distinct nonnegative integers m and n. Then

$$\begin{aligned} E[M_n(s)M_m(t)] &= E[E[M_n(s)M_m(t)|\mathcal{W}_s]] \quad \text{property of cond. expectation} \\ &= E[M_n(s)E[M_m(t)|\mathcal{W}_s]] \quad \text{property of cond. expectation} \\ &= E[M_n(s)M_m(s)] \quad \text{martingale property} \\ &= 0 \quad \text{orthogonality of variables at a fixed time.} \end{aligned}$$

5.2 A variance estimation problem with Poisson observation
(a)

$$P\{N = n\} = E[P(N = n|X)] = E[\frac{(X^2)^n e^{-X^2}}{n!}]$$

$$= \int_{-\infty}^{\infty} \frac{x^{2n} e^{-x^2}}{n!} \frac{e^{-\frac{x^2}{2\sigma^2}}}{\sqrt{2\pi\sigma^2}} dx.$$

(b) To arrive at a simple answer, we could set the derivative of $P\{N=n\}$ with respect to σ^2 equal to zero either before or after simplifying. Here we simplify first, using the fact that if X is an $N(0,\widetilde{\sigma}^2)$ random variable, then $E[X^{2n}] = \frac{\widetilde{\sigma}^{2n}(2n)!}{n!2^n}$. Let $\widetilde{\sigma}^2$ be such that $\frac{1}{2\widetilde{\sigma}^2} = 1 + \frac{1}{2\sigma^2}$, or equivalently, $\widetilde{\sigma}^2 = \frac{\sigma^2}{1+2\sigma^2}$. Then the above integral can be written as follows:

$$P\{N=n\} = \frac{\widetilde{\sigma}}{\sigma}\int_{-\infty}^{\infty} \frac{x^{2n}}{n!} \frac{e^{\frac{-x^2}{2\widetilde{\sigma}^2}}}{\sqrt{2\pi\widetilde{\sigma}^2}} dx$$
$$= \frac{c_1\widetilde{\sigma}^{2n+1}}{\sigma} = \frac{c_1\sigma^{2n}}{(1+2\sigma^2)^{\frac{2n+1}{2}}},$$

where the constant c_1 depends on n but not on σ^2. Taking the logarithm of $P\{N=n\}$ and calculating the derivative with respect to σ^2, we find that $P\{N=n\}$ is maximized at $\sigma^2 = n$. That is, $\widehat{\sigma}^2_{ML}(n) = n$.

5.4 Estimation of Bernoulli parameter in Gaussian noise by EM algorithm

(a)

$$P(Z_1 = 1|Y_1 = u, \theta) = \frac{P(Z_1 = 1, Y_1 = u|\theta)}{P(Y_1 = u|\theta)}$$
$$= \frac{\theta \exp(-\frac{(u-1)^2}{2})}{\theta \exp(-\frac{(u-1)^2}{2}) + (1-\theta)\exp(\frac{(u+1)^2}{2})}$$
$$= \frac{\theta e^u}{\theta e^u + (1-\theta)e^{-u}}.$$

So $\varphi(u|\theta) = P(Z_1 = 1|Y_1 = u, \theta) - P(Z_1 = -1|Y_1 = u, \theta) = \frac{\theta e^u - (1-\theta)e^{-u}}{\theta e^u + (1-\theta)e^{-u}}$.

(b)

$$p_{cd}(y, z|\theta) = \prod_{t=1}^{T}\left\{\theta^{\frac{1+z_t}{2}}(1-\theta)^{\frac{1-z_t}{2}} \frac{1}{\sqrt{2\pi}} \exp(-\frac{(y_t - z_t)^2}{2})\right\}$$
$$= \theta^{\frac{T+\sum_t z_t}{2}}(1-\theta)^{\frac{T-\sum_t z_t}{2}} e^{\sum_t y_t z_t} R(y),$$

where $R(y)$ depends on y only.

$$Q(\theta|\theta^{(k)}) = \frac{T+\sum_t \varphi(y_t, \theta^{(k)})}{2}\ln(\theta) + \frac{T-\sum_t \varphi(y_t, \theta^{(k)})}{2}\ln(1-\theta)$$
$$+ \sum_{t=1}^{T} \varphi(y_t, \theta^{(k)}) y_t + R_1(y, \theta^{(k)}),$$

where $R_1(y, \theta^{(k)})$ depends on y and $\theta^{(k)}$ only. Maximizing over θ yields

$$\theta^{(k+1)}(y) = \frac{T + \sum_{t=1}^{T} \varphi(y_t, \theta^{(k)})}{2T}.$$

5.6 Transformation of estimators and estimators of transformations (a) Yes, because the transformation is invertible.
(b) Yes, because the transformation is invertible.
(c) Yes, because the transformation is linear, the pdf of $3 + 5\Theta$ is a scaled version of the pdf of Θ.
(d) No, because the transformation is not linear.
(e) Yes, because the MMSE estimator is given by the conditional expectation, which is linear. That is, $3 + 5E[\Theta|Y] = E[3 + 5\Theta|Y]$.
(f) No. Typically $E[\Theta^3|Y] \neq E[\Theta|Y]^3$.
5.8 Finding a most likely path Finding the path z to maximize the posterior probability given the sequence 021201 is the same as maximizing $p_{cd}(y, z|\theta)$. Due to the form of the parameter $\theta = (\pi, A, B)$, for any path $z = (z_1, \ldots, z_6)$, $p_{cd}(y, z|\theta)$ has the form $c^6 a^i$ for some $i \geq 0$. Similarly, the variable $\delta_j(t)$ has the form $c^t a^i$ for some $i \geq 0$. Since $a < 1$, larger values for $p_{cd}(y, z|\theta)$ and $\delta_j(t)$ correspond to smaller values of i. Rather than keeping track of products, such as $a^i a^j$, we keep track of the exponents of the products, which for $a^i a^j$ would be $i + j$. Thus, the problem at hand is equivalent to finding a path from left to right in the trellis indicated in Figure 12.4(a) with minimum weight, where the weight of a path is the sum of all the numbers indicated on the vertices and edges of the graph. Figure 12.4(b) shows the result of running the Viterbi algorithm. The value of $\delta_j(t)$ has the form $c^t a^i$, where the values of i are indicated by the numbers in boxes. Of the two paths reaching the final states of the trellis, the upper one, namely the

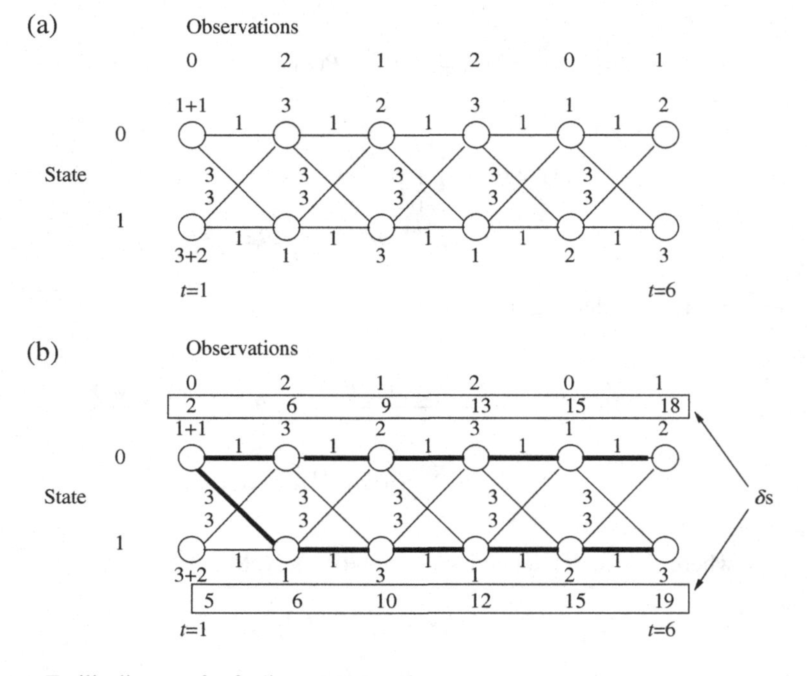

Figure 12.4 Trellis diagram for finding a MAP path.

path 000000, has the smaller exponent, 18, and therefore, the larger probability, namely c^6a^{18}. Therefore, 000000 is the MAP path.

5.10 Estimation of the parameter of an exponential in exponential noise

(a) By assumption, Z has the exponential distribution with parameter θ, and given $Z = z$, the conditional distribution of $Y - z$ is the exponential distribution with parameter one (for any θ). So $f_{cd}(y, z|\theta) = f(z|\theta)f(y|z, \theta)$ where

$$f(z|\theta) = \begin{cases} \theta e^{-\theta z} & z \geq 0 \\ 0 & \text{else} \end{cases} \quad \text{and for } z \geq 0: \ f(y|z,\theta) = \begin{cases} e^{-(y-z)} & y \geq z \\ 0 & \text{else} \end{cases}.$$

(b)

$$f(y|\theta) = \int_0^y f_{cd}(y,z|\theta)dz = \begin{cases} \frac{\theta e^{-y}(e^{(1-\theta)y}-1)}{1-\theta} & \theta \neq 1 \\ ye^{-y} & \theta = 1 \end{cases}.$$

(c)

$$\begin{aligned} Q(\theta|\theta^{(k)}) &= E[\ \ln f_{cd}(Y,Z|\theta)\ |y, \theta^{(k)}] \\ &= \ln\theta + (1-\theta)E[Z|y,\theta^{(k)}] - y, \end{aligned}$$

which is a concave function of θ. The maximum over θ can be identified by setting the derivative with respect to θ equal to zero, yielding:
$\theta^{(k+1)} = \arg\max_\theta Q(\theta|\theta^{(k)}) = \frac{1}{E[Z|y,\theta^{(k)}]} = \frac{1}{\varphi(y,\theta^{(k)})}$.

(d)

$$\begin{aligned} Q(\theta|\theta^{(k)}) &= E[\ \ln f_{cd}(Y,Z|\theta)\ |y, \theta^{(k)}] \\ &= \sum_{t=1}^{T} E[\ \ln f(y_t, Z_t|\theta)\ |y_t, \theta^{(k)}] \\ &= T\ln\theta + (1-\theta)\sum_{t=1}^{T}\varphi\left(y_t,\theta^{(k)}\right) - \sum_{t=1}^{T} y_t, \end{aligned}$$

which is a concave function of θ. The maximum over θ can be identified by setting the derivative with respect to θ equal to zero, yielding:

$$\theta^{(k+1)} = \arg\max_\theta Q(\theta|\theta^{(k)}) = \frac{T}{\sum_{t=1}^{T}\varphi\left(y_t,\theta^{(k)}\right)}.$$

5.12 Maximum likelihood estimation for HMMs (a) *Approach one* Note that $p(y|z) = \prod_{t=1}^{T} b_{z_t,y_t}$. Thus, for fixed y, $p(y|z)$ is maximized with respect to z by selecting z_t to maximize b_{z_t,y_t} for each t. Thus, $(\widehat{Z}_{\mathrm{ML}}(y))_t = \arg\max_i b_{i,y_t}$ for $1 \leq t \leq T$.
Approach two Let $\widetilde{\pi}_i = \frac{1}{N_s}$ and $\widetilde{A}_{i,j} = \frac{1}{N_s}$ for all states i,j of the hidden Markov process Z. The HMM for parameter $\widetilde{\theta} \triangleq (\widetilde{\pi}, \widetilde{A}, B)$ is such that all N_s^T possible values for Z are equally likely, and the conditional distribution of Y given Z is the same as for the HMM with parameter θ. Use the Viterbi algorithm with parameter $\widetilde{\theta}$ to compute $\widehat{Z}_{\mathrm{MAP}}$, and that is equal to $\widehat{Z}_{\mathrm{ML}}$ for the HMM with parameter θ.
(b) Let $\widetilde{\pi}_i = 1$ if $\pi_i > 0$ and $\widetilde{\pi}_i = 0$ if $\pi_i = 0$ for $1 \leq i \leq N_s$. Similarly, let $\widetilde{A}_{i,j} = 1$ if $a_{i,j} > 0$ and $\widetilde{A}_{i,j} = 0$ if $a_{i,j} = 0$ for $1 \leq i,j \leq N_s$. While $\widetilde{\pi}$ and the rows of $\widetilde{A}$ are not

normalized to sum to one, they can still be used in the Viterbi algorithm. Under param eter $\widetilde{\theta} \stackrel{\triangle}{=} (\widetilde{\pi}, \widetilde{A}, B)$, every choice of possible trajectory for Z has weight one, every othe trajectory has weight zero, and the conditional distribution of Y given Z is the same a for the HMM with parameter θ. Use the Viterbi algorithm with parameter $\widetilde{\theta}$ to comput $\widehat{Z}_{\text{MAP}}$, and that is equal to the constrained estimator $\widehat{Z}_{\text{ML}}$ for the HMM with parameter θ
(c) Note that $P(Y = y|Z_1 = i) = \beta_i(1)b_{i,y_1}$, where $\beta_i(1)$ can be computed for all i usin the backward algorithm. Therefore, $\widehat{Z}_{1,\text{ ML}}(y) = \arg\max_i \beta_i(1)b_{i,y_1}$.
(d) Note that $P(Y = y|Z_{t_o} = i) = \frac{\gamma_i(t_o)P\{Y=y\}}{P\{Z_{t_o}=i\}}$, where $\gamma_i(t_o)$ can be computed b the forward backward algorithm, and $P\{Z_{t_o} = i\} = (\pi A^{t_o-1})_i$. Then $\widehat{Z}_{t_o,\text{ ML}}(y) = \arg\max_i \frac{\gamma_i(t_o)}{P\{Z_{t_o}=i\}}$.

5.14 Specialization of Baum–Welch algorithm for no hidden data

(a) Suppose the sequence $y = (y_1, \ldots, y_T)$ is observed. If $\theta^{(0)} = \theta = (\pi, A, B)$ is suc that B is the identity matrix, and all entries of π and A are nonzero, then directly by th definitions (without using the αs and βs):

$$\gamma_i(t) \stackrel{\triangle}{=} P(Z_t = i|Y_1 = y_1, \ldots, Y_T = y_T, \theta) = I_{\{y_t=i\}},$$

$$\xi_{ij}(t) \stackrel{\triangle}{=} P(Z_t = i, Z_{t+1} = j|Y_1 = y_1, \ldots, Y_T = y_T, \theta) = I_{\{(y_t,y_{t+1})=(i,j)\}}.$$

Thus, (5.27)–(5.29) for the first iteration, $t = 0$, become

$$\pi_i^{(1)} = I_{\{y_1=i\}} \quad \text{i.e. the probability vector for } \mathcal{S} \text{ with all mass on } y_1,$$

$$a_{i,j}^{(1)} = \frac{\text{number of } (i,j) \text{ transitions observed}}{\text{number of visits to } i \text{ up to time } T-1},$$

$$b_{il}^{(1)} = \frac{\text{number of times the state is } i \text{ and the observation is } l}{\text{number of times the state is } i}.$$

It is assumed that B is the identity matrix, so that each time the state is i the observatio should also be i. Thus, $b_{il}^{(1)} = I_{\{i=l\}}$ for any state i that is visited. That is consistent wit the assumption that B is the identity matrix. (Alternatively, since B is fixed to be th identity matrix, we could just work with estimating π and A, and simply not conside B as part of the parameter to be estimated.) The next iteration will give the same value of π and A. Thus, the Baum–Welch algorithm converges in one iteration to the fina value $\theta^{(1)} = (\pi^{(1)}, A^{(1)}, B^{(1)})$ already described. Note that, by Lemma 5.3, $\theta^{(1)}$ is th ML estimate.

(b) In view of part (a), the ML estimates are $\pi = (1, 0)$ and $A = \begin{pmatrix} \frac{2}{3} & \frac{1}{3} \\ \frac{1}{3} & \frac{2}{3} \end{pmatrix}$. Thi estimator of A results from the fact that, of the first 21 times, the state was zero 1 times, and 8 of those 12 times the next state was a zero. So $a_{00} = 8/12 = 2/3$ is th ML estimate. Similarly, the ML estimate of a_{11} is $6/9$, which simplifies to $2/3$.

5.16 Extending the forward-backward algorithm

(a) Forward equations: $\mu_j(t, t+1) = \sum_{i \in \mathcal{S}} \mu_i(t-1, t) b_{iy_t} a_{ij} \quad \mu_i(-1, 0) = 1,$
Backward equations: $\mu_j(t, t-1) = \sum_{i \in \mathcal{S}} \mu_i(t+1, t) b_{iy_t} a_{ji} \quad \mu_i(T+1, T) = 1,$

$$\gamma_i(t) = \frac{\mu_i(t-1, t)\mu_i(t+1, t)b_{iy_t}}{\sum_j \mu_j(t-1, t)\mu_j(t+1, t)b_{jy_t}}. \qquad (12.9$$

(b)

$$\mu_i(t-1,t) = \sum_{z_1,\cdots,z_{t-1}} \left(a_{z_1z_2}a_{z_2z_3}\dots a_{z_{t-1}i} \prod_{s=1}^{t-1} b_{z_s,y_s} \right), \tag{12.10}$$

$$\mu_i(t+1,t) = \sum_{z_{t+1},\cdots,z_T} \left(a_{iz_{t+1}}a_{z_{t+1}z_{t+2}}\dots a_{z_{T-1}z_T} \prod_{s=t+1}^{T} b_{z_s,y_s} \right). \tag{12.11}$$

(To bring out the symmetry more, we could let $\widetilde{A}_{ij} = a_{ji}$ (corresponds to A^T) and rewrite (12.11) as

$$\mu_i(t+1,t) = \sum_{z_T,\cdots,\, z_{t+1}} \left(\widetilde{A}_{z_Tz_{T-1}}\widetilde{A}_{z_{T-1}z_{T-2}}\dots \widetilde{A}_{z_{t+2}z_{t+1}}\widetilde{A}_{z_{t+1}i} \prod_{s=t+1}^{T} b_{z_s,y_s} \right). \tag{12.12}$$

Observe that (12.10) and (12.12) are the same up to time reversal.)

A partially probabilistic interpretation can be given to the messages as follows. First, consider how to find the marginal distribution of Z_t for some t. It is obtained by summing out all values of the other variables in the complete data probability function, with Z_t fixed at i. For $Z_t = i$ fixed, the numerator in the joint probability function factors into three terms involving disjoint sets of variables:

$$\begin{aligned}\left(b_{z_1y_1}a_{z_1z_2}b_{z_2y_2}\dots a_{z_{t-2}z_{t-1}}b_{z_{t-1}y_{t-1}}a_{z_{t-1}i}\right) &\times \left(b_{iy_t}\right)\\ &\times \left(a_{iz_{t+1}}b_{z_{t+1}y_{t+1}}a_{z_{t+1},z_{t+2}}\dots a_{z_{T-1}z_T}b_{z_Ty_T}\right).\end{aligned}$$

Let $G_i(t-1,t)_i$ denote the sum of the first factor over all $(z_1, y_1, z_2, \dots, z_{t-1}, y_{t-1})$, let $G_i^o(t)$ denote the sum of the second factor over all y_t and let $G_i(t+1,t)$ denote the sum of the third factor over all $(z_{t+1}, y_{t+1}, \dots, z_T, y_T)$. Then the marginal distribution of Z_t can be represented as

$$P\{Z_t = i\} = \frac{G_i(t-1,t)G_i^o(t)G_i(t+1,t)}{G},$$

and the constant G can be expressed as $G = \sum_j G_j(t-1,t)G_j^o(t)G_j(t+1,t)$. Note that the Gs depend on the joint distribution but do not depend on specific values of the observation. They are simply factors in the prior (i.e. before observations are incorporated) distribution of Z_t.

For fixed $y_1, \dots, y_{t-1}$, using the definition of conditional probability yields that

$$P(Y_1 = y_1, \dots, Y_{t-1} = y_{t-1} | Z_t = i) = \frac{\mu_i(t-1,t)}{G_i(t-1,t)},$$

or equivalently,

$$\mu_i(t-1,t) = P(Y_1 = y_1, \dots, Y_{t-1} = y_{t-1} | Z_t = i)G_i(t-1,t). \tag{12.13}$$

Equation (12.13) gives perhaps the closest we can get to a probabilistic interpretation of $\mu_i(t-1,t)$. In words, $\mu_i(t-1,t)$ is the product of the likelihood of the observations $(y_1, \dots, y_{t-1})$ and a factor $G_t(t-1,t)$, not depending on the observations, that contributes to the unconditional prior distribution of Z_t. A similar interpretation holds for $\mu_i(t+1,t)$. Also, $b_{y_t i}$ can be thought of as a message from the observation node c the graph at time t to the node for z_t, and $b_{y_t i} = P(Y_t = y_t|Z_t = i)G_i^o(t)$. Combinin these observations shows that the numerator in (12.9) is given by:

$$\begin{aligned}&\mu_i(t-1,t)\mu_i(t+1,t)b_{iy_t}\\&\quad = P(Y_1 = y_1,\dots,Y_T = Y_T|Z_t = i)G_i(t-1,t)G_i^o(t)G_i(t+1,t)\\&\quad = P(Y_1 = y_1,\dots,Y_T = Y_T|Z_t = i)P(Z_t = i)G\\&\quad = P(Y_1 = y_1,\dots,Y_T = Y_T, Z_t = i)G.\end{aligned}$$

(c) Comparison of the numerator in (12.9) to the definition of $p_{cd}(y,z|\theta)$ given in th problem statement shows that the numerator in (12.9) is the sum of $p_{cd}(y,z|\theta)G$ ove all values of $\{z : z_t = i\}$ for y fixed, so it is $P(Y = y, Z_t = i|\theta)G$. Thus,

$$\text{RHS of (12.9)} = \frac{P(Y = y, Z_t = i|\theta)G}{\sum_j P(Y = y, Z_t = j|\theta)G} = \frac{P(Y = y, Z_t = i|\theta)}{P(Y = y|\theta)} = \gamma_i(t).$$

5.18 Baum–Welch saddlepoint It turns out that $\pi^{(k)} = \pi^{(0)}$ and $A^{(k)} = A^{(0)}$, for eac $k \geq 0$. Also, $B^{(k)} = B^{(1)}$ for each $k \geq 1$, where $B^{(1)}$ is the matrix with identical rows such that each row of $B^{(1)}$ is the empirical distribution of the observation sequence. Fo example, if the observations are binary valued, and if there are $T = 100$ observations, o which 37 observations are zero and 63 are 1, then each row of $B^{(1)}$ would be $(0.37, 0.63$ Thus, the EM algorithm converges in one iteration, and unless $\theta^{(0)}$ happens to be a loca maximum or local minimum, the EM algorithm converges to an inflection point of th likelihood function.

One intuitive explanation for this assertion is that since all the rows of $B^{(0)}$ ar the same, then the observation sequence is initially believed to be independent o the state sequence, and the state process is initially believed to be stationary. Hence even if there is, for example, notable time variation in the observed data sequence there is no way to change beliefs in a particular direction in order to increase th likelihood. In real computer experiments, the algorithm may still eventually reach near maximum likelihood estimate, due to round-off errors in the computations whicl allow the algorithm to break away from the inflection point.

The assertion can be proved by use of the update equations for the Baum–Welcl algorithm. It is enough to prove the assertion for the first iteration only, for then i follows for all iterations by induction.

Since the rows of $B^{(0)}$ are all the same, we write b_l to denote $b_{il}^{(0)}$ for an arbitrar value of i. By induction on t, we find $\alpha_i(t) = b_{y_1}\dots b_{y_t}\pi_i^{(0)}$ and $\beta_j(t) = b_{y_{t+1}}\dots b_{y_T}$. I particular, $\beta_j(t)$ does not depend on j. So the vector $(\alpha_i\beta_i : i \in \mathcal{S})$ is proportional to $\pi^{(0)}$ and therefore $\gamma_i(t) = \pi_i^{(0)}$. Similarly, $\xi_{i,j}(t) = P(Z_t = i, Z_{t+1} = j|y,\theta^{(0)}) = \pi_i^{(0)}a_{i,j}^{(0)}$. B (5.27), $\pi^{(1)} = \pi^{(0)}$, and by (5.28), $A^{(1)} = A^{(0)}$. Finally, (5.29) gives

$$b_{i,l}^{(1)} = \frac{\sum_{t=1}^{T}\pi_i I_{\{y_t=l\}}}{T\pi_i} = \frac{\text{number of times } l \text{ is observed}}{T}.$$

5.20 Constraining the Baum–Welch algorithm A quite simple way to deal with this problem is to take the initial parameter $\theta^{(0)} = (\pi, A, B)$ in the Baum–Welch algorithm to be such that $a_{ij} > 0$ if and only if $\overline{a}_{ij} = 1$ and $b_{il} > 0$ if and only if $\overline{b}_{il} = 1$. (These constraints are added in addition to the usual constraints that π, A, and B have the appropriate dimensions, with π and each row of A and b being probability vectors.) After all, it makes sense for the initial parameter value to respect the constraint. And if it does, then the same constraint will be satisfied after each iteration, and no changes are needed to the algorithm itself.

6.2 A two station pipeline in continuous time (a) $\mathcal{S} = \{00, 01, 10, 11\}$.
(b)

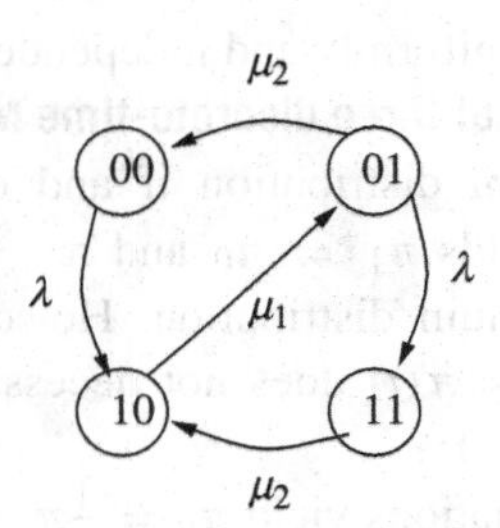

(c)

$$Q = \begin{pmatrix} -\lambda & 0 & \lambda & 0 \\ \mu_2 & -\mu_2 - \lambda & 0 & \lambda \\ 0 & \mu_1 & -\mu_1 & 0 \\ 0 & 0 & \mu_2 & -\mu_2 \end{pmatrix}.$$

(d) $\eta = (\pi_{00} + \pi_{01})\lambda = (\pi_{01} + \pi_{11})\mu_2 = \pi_{10}\mu_1$. If $\lambda = \mu_1 = \mu_2 = 1.0$ then $\pi = (0.2, 0.2, 0.4, 0.2)$ and $\eta = 0.4$.
(e) Let $\tau = \min\{t \geq 0 : X(t) = 00\}$, and define $h_s = E[\tau | X(0) = s]$, for $s \in \mathcal{S}$. We wish to find h_{11}. Taking into account the probabilities for the first jump of the process yields the following system of equations for the mean hitting times:

$$\begin{aligned} h_{00} &= 0, \\ h_{01} &= \tfrac{1}{\mu_2+\lambda} + \tfrac{\mu_2 h_{00}}{\mu_2+\lambda} + \tfrac{\lambda h_{11}}{\mu_2+\lambda}, \\ h_{10} &= \tfrac{1}{\mu_1} + h_{01}, \\ h_{11} &= \tfrac{1}{\mu_2} + h_{10}. \end{aligned}$$

For if $\lambda = \mu_1 = \mu_2 = 1.0$, this yields $\begin{pmatrix} h_{00} \\ h_{01} \\ h_{10} \\ h_{11} \end{pmatrix} = \begin{pmatrix} 0 \\ 3 \\ 4 \\ 5 \end{pmatrix}$. Thus, $h_{11} = 5$ is the required answer.

6.4 A simple Poisson process calculation Suppose $0 < s < t$ and $0 \le i \le k$.

$$P(N(s) = i | N(t) = k) = \frac{P\{N(s) = i, N(t) = k\}}{P\{N(t) = k\}}$$

$$= \left(\frac{e^{-\lambda s}(\lambda s)^i}{i!}\right)\left(\frac{e^{-\lambda(t-s)}(\lambda(t-s))^{k-i}}{(k-i)!}\right)\left(\frac{e^{-\lambda t}(\lambda t)^k}{k!}\right)^{-1}$$

$$= \binom{k}{i}\left(\frac{s}{t}\right)^i\left(\frac{t-s}{t}\right)^{k-1}.$$

That is, given $N(t) = k$, the conditional distribution of $N(s)$ is binomial. This coul have been deduced with no calculation, using the fact that given $N(t) = k$, the locatio of the k points are uniformly and independently distributed on the interval $[0, t]$.

6.6 On distributions of three discrete-time Markov processes (a) A probability vect π is an equilibrium distribution if and only if π satisfies the balance equation $\pi = \pi P$. This yields $\pi_1 = \pi_0$ and $\pi_2 = \pi_3 = \pi_1/2$. Thus, $\pi = \left(\frac{1}{3}, \frac{1}{3}, \frac{1}{6}, \frac{1}{6}\right)$ i the unique equilibrium distribution. However, this Markov process is periodic wi period 2, so $\lim_{t\to\infty} \pi(t)$ does not necessarily exist. (The limit exists if and only $\pi_0(0) + \pi_2(0) = 0.5$.)

(b) The balance equations yield $\pi_n = \frac{1}{n}\pi_{n-1}$ for all $n \ge 1$, so that $\pi_n = \frac{\pi_0}{n!}$. Thus, th Poisson distribution with mean one, $\pi_n = \frac{e^{-1}}{n!}$, is the unique equilibrium distributio Since there is an equilibrium distribution and the process is irreducible and aperiodi all states are positive recurrent and $\lim_{t\to\infty} \pi(t)$ exists and is equal to the equilibriu distribution for any choice of initial distribution.

(c) The balance equations yield $\pi_n = \frac{n-1}{n}\pi_{n-1}$ for all $n \ge 1$, so that $\pi_n = \frac{\pi_0}{n}$. But sinc $\sum_{n=1}^{\infty} \frac{1}{n} = \infty$, there is no way to normalize this distribution to make it a probabilit distribution. Thus, there does not exist an equilibrium distribution. The process is thu transient or null recurrent: $\lim_{t\to\infty} \pi_n(t) = 0$ for each state n. (It can be shown that th process is recurrent. Indeed,

$$P(\text{not return to } 0 | X(0) = 0) = \lim_{n\to\infty} P(\text{hit } n \text{ before return to } 0 | X(0) = 0)$$

$$= \lim_{n\to\infty} 1 \cdot \frac{1}{2} \cdot \frac{2}{3} \cdot \cdots \cdot \frac{n-1}{n} = 0.)$$

6.8 A Markov process on a ring $Q = \begin{pmatrix} -a-1 & a & 1 \\ 1 & -b-1 & b \\ c & 1 & -c-1 \end{pmatrix}$ and simpl algebra yields $(1 + c + cb, 1 + a + ac, 1 + b + ba)Q = (0, 0, 0)$. (Since the row sums c Q are zero it suffices to check two of the equations. By symmetry in fact it suffices t check just the first equation.)

(b) The long term rate of jumps from state 1 to state 2 is $\pi_1 a$ and the long term rat of jumps from state 2 to 1 is π_2. The difference is the mean cycle rate: $\theta = \pi_1 a - \pi_2$ Similarly, $\theta = \pi_2 b - \pi_3$ and $\theta = \pi_3 c - \pi_1$.

Alternatively, the average rate of clockwise jumps per unit time is $\pi_1 a + \pi_2 b + \pi_3$ and the average rate of counterclockwise jumps is one. So the net rate of jumps in th

clockwise direction is $\pi_1 a+\pi_2 b+\pi_3 c-1$. Since there are three jumps to a cycle, divide by three to get $\theta=(\pi_1 a+\pi_2 b+\pi_3 c-1)/3$.
(c) By part (a), $\pi=(1+c+cb, 1+a+ac, 1+b+ba)/Z$ where $Z=3+a+b+c+ab+ac+bc$. So then using part (b), $\theta=\frac{(1+c+bc)a-1-a-ac}{Z}=\frac{abc-1}{3+a+b+c+ab+ac+bc}$. The mean net cycle rate is zero if and only if $abc=1$. (Note: The nice form of the equilibrium for this problem, which generalizes to rings of any integer circumference, is a special case of the tree based formula for equilibrium distributions that can be found, for example, in the book of Freidlin and Wentzell, *Random perturbations of dynamical systems.*)

6.10 A mean hitting time problem

(a)

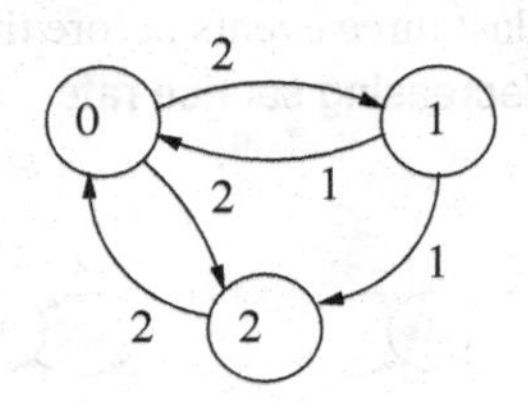

$\pi Q=0$ implies $\pi=(\frac{2}{7},\frac{2}{7},\frac{3}{7})$.
(b) Clearly $a_1=0$. Condition on the first step. The initial holding time in state i has mean $-\frac{1}{q_{ii}}$ and the next state is j with probability $p^J_{ij}=\frac{-q_{ij}}{q_{ii}}$. Thus $\begin{pmatrix} a_0 \\ a_2 \end{pmatrix}=\begin{pmatrix} -\frac{1}{q_{00}} \\ -\frac{1}{q_{22}} \end{pmatrix}+\begin{pmatrix} 0 & p^J_{02} \\ p^J_{20} & 0 \end{pmatrix}\begin{pmatrix} a_0 \\ a_2 \end{pmatrix}$. Solving: $\begin{pmatrix} a_0 \\ a_2 \end{pmatrix}=\begin{pmatrix} 1 \\ 1.5 \end{pmatrix}$.
(c) Clearly $\alpha_2(t)=0$ for all t:

$$\alpha_0(t+h)=\alpha_0(t)(1+q_{00}h)+\alpha_1(t)q_{10}h+o(h),$$
$$\alpha_1(t+h)=\alpha_0(t)q_{01}h+\alpha_1(t)(1+q_{11}h)+o(h).$$

Subtract $\alpha_i(t)$ from each side and let $h\to 0$; $(\frac{\partial\alpha_0}{\partial t},\frac{\partial\alpha_1}{\partial t})=(\alpha_0,\alpha_1)\begin{pmatrix} q_{00} & q_{01} \\ q_{10} & q_{11} \end{pmatrix}$ with the initial condition $(\alpha_0(0),\alpha_1(0))=(1,0)$. (Note: the matrix involved here is the Q matrix with the row and column for state 2 removed.)
(d) Similarly,

$$\beta_0(t-h)=(1+q_{00}h)\beta_0(t)+q_{01}h\beta_1(t)+o(h),$$
$$\beta_1(t-h)=q_{10}h\beta_0(t)+(1+q_{11}h)\beta_1(t)+o(h).$$

Subtract $\beta_i(t)$s, divide by h and let $h\to 0$ to get:
$\begin{pmatrix} -\frac{\partial\beta_0}{\partial t} \\ -\frac{\partial\beta_1}{\partial t} \end{pmatrix}=\begin{pmatrix} q_{00} & q_{01} \\ q_{10} & q_{11} \end{pmatrix}\begin{pmatrix} \beta_0 \\ \beta_1 \end{pmatrix}$ with $\begin{pmatrix} \beta_0(t_f) \\ \beta_1(t_f) \end{pmatrix}=\begin{pmatrix} 1 \\ 1 \end{pmatrix}$.

6.12 Markov model for a link with resets (a) Let $\mathcal{S}=\{0,1,2,3\}$, where the state is the number of packets passed since the last reset.
(b) By the PASTA property, the dropping probability is π_3. We can find the equilibrium distribution π by solving the equation $\pi Q=0$. The balance equation for state 0 is $\lambda\pi_0=\mu(1-\pi_0)$ so that $\pi_0=\frac{\mu}{\lambda+\mu}$. The balance equation for state $i\in\{1,2\}$ is

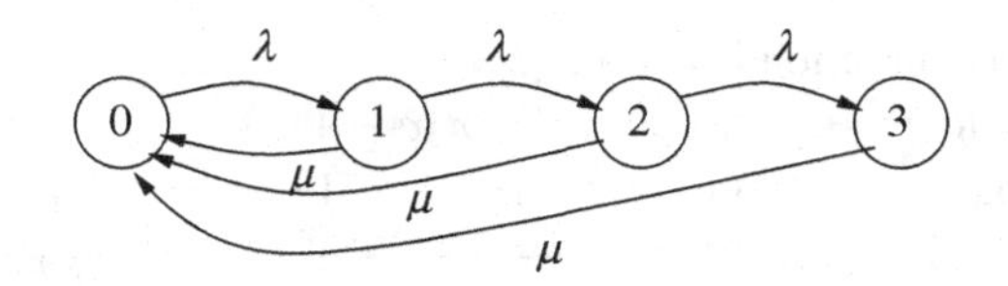

$\lambda\pi_{i-1} = (\lambda + \mu)\pi_i$, so that $\pi_1 = \pi_0(\frac{\lambda}{\lambda+\mu})$ and $\pi_2 = \pi_0(\frac{\lambda}{\lambda+\mu})^2$. Finally, $\lambda\pi_2 = \mu\pi_3$ so that $\pi_3 = \pi_0(\frac{\lambda}{\lambda+\mu})^2\frac{\lambda}{\mu} = \frac{\lambda^3}{(\lambda+\mu)^3}$. The dropping probability is $\pi_3 = \frac{\lambda^3}{(\lambda+\mu)^3}$. (This formula for π_3 can be deduced with virtually no calculation from the properties of merged Poisson processes. Fix a time t. Each event is a packet arrival with probability $\frac{\lambda}{\lambda+\mu}$ and is a reset otherwise. The types of different events are independent. Finally, $\pi_3(t)$ is the probability that the last three events before time t were arrivals. The formula follows.)

6.14 A queue with decreasing service rate

(a)

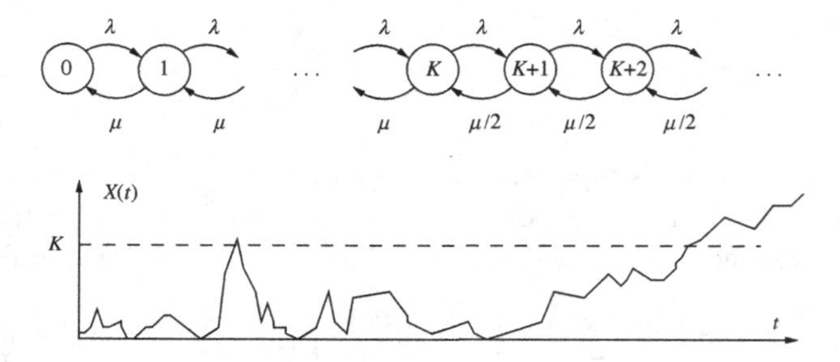

(b) $S_2 = \sum_{k=0}^{\infty}(\frac{\mu}{2\lambda})^k 2^{k\wedge K}$, where $k \wedge K = \min\{k, K\}$. Thus, if $\lambda < \frac{\mu}{2}$ then $S_2 < +\infty$ and the process is recurrent. $S_1 = \sum_{k=0}^{\infty}(\frac{2\lambda}{\mu})^k 2^{-k\wedge K}$, so if $\lambda < \frac{\mu}{2}$ then $S_1 < +\infty$ and the process is positive recurrent. In this case, $\pi_k = (\frac{2\lambda}{\mu})2^{-k\wedge K}\pi_0$, where

$$\pi_0 = \frac{1}{S_1} = \left[\frac{1-(\lambda/\mu)^K}{1-(\lambda/\mu)} + \frac{(\lambda/\mu)^K}{1-(2\lambda/\mu)}\right]^{-1}.$$

(c) If $\lambda = \frac{2\mu}{3}$, the queue appears to be stable until it fluctuates above K. Eventually the queue length will grow to infinity at rate $\lambda - \frac{\mu}{2} = \frac{\mu}{6}$. See figure above.

6.16 An M/M/1 queue with impatient customers

(a)

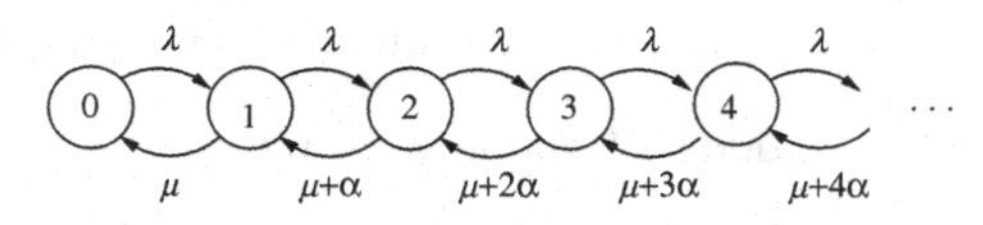

(b) The process is positive recurrent for all λ, μ if $\alpha > 0$, and $p_k = \frac{c\lambda^k}{\mu(\mu+\alpha)\cdots(\mu+(k-1)\alpha)}$ where c is chosen so that the p_ks sum to one.

(c) If $\alpha = \mu$, $p_k = \frac{c\lambda^k}{k!\mu^k} = \frac{c\rho^k}{k!}$. Therefore, $(p_k : k \geq 0)$ is the Poisson distribution with mean ρ. Furthermore, p_D is the mean departure rate by defecting customers, divided by the mean arrival rate λ. Thus,

$$p_D = \frac{1}{\lambda}\sum_{k=1}^{\infty} p_k(k-1)\alpha = \frac{\rho - 1 + e^{-\rho}}{\rho} \to \begin{cases} 1 & \text{as } \rho \to \infty \\ 0 & \text{as } \rho \to 0 \end{cases},$$

where l'Hôpital's rule can be used to find the limit as $\rho \to 0$.

6.18 A queue with blocking

(a)

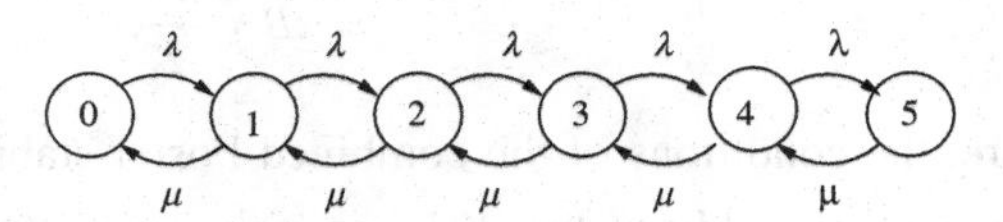

$\pi_k = \frac{\rho^k}{1+\rho+\rho^2+\rho^3+\rho^4+\rho^5} = \frac{\rho^k(1-\rho)}{1-\rho^6}$ for $0 \le k \le 5$.

(b) $p_B = \pi_5$ by the PASTA property.

(c) $\overline{W} = \overline{N_W}/(\lambda(1-p_B))$ where $\overline{N_W} = \sum_{k=1}^{5}(k-1)\pi_k$. Or $\overline{W} = \overline{N}/(\lambda(1-p_B)) - \frac{1}{\mu}$ (i.e. $\overline{W}$ is equal to the mean time in system minus the mean time in service).

(d)

$$\pi_0 = \frac{1}{\lambda(\text{mean cycle time for visits to state zero})} = \frac{1}{\lambda(1/\lambda + \text{mean busy period duration})}.$$

Therefore, the mean busy period duration is $\frac{1}{\lambda}[\frac{1}{\pi_0} - 1] = \frac{\rho-\rho^6}{\lambda(1-\rho)} = \frac{1-\rho^5}{\mu(1-\rho)}$.

6.20 On two distributions seen by customers (a) As can be seen in the picture,

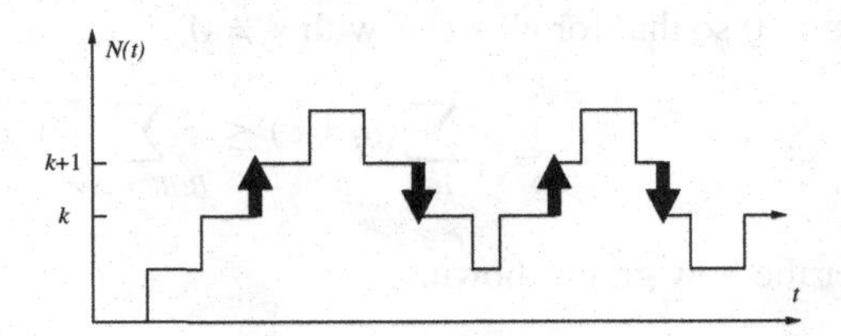

between any two transitions from state k to $k+1$ there is a transition from state $k+1$ to k, and vice versa. Thus, the number of transitions of one type is within one of the number of transitions of the other type. This establishes that $|D(k,t) - R(k,t)| \le 1$ for all k.

(b) Observe that

$$\left|\frac{D(k,t)}{\alpha_t} - \frac{R(k,t)}{\delta_t}\right| \le \left|\frac{D(k,t)}{\alpha_t} - \frac{R(k,t)}{\alpha_t}\right| + \left|\frac{R(k,t)}{\alpha_t} - \frac{R(k,t)}{\delta_t}\right|$$
$$\le \frac{1}{\alpha_t} + \frac{R(k,t)}{\alpha_t}\left|1 - \frac{\alpha_t}{\delta_t}\right|$$
$$\le \frac{1}{\alpha_t} + \left|1 - \frac{\alpha_t}{\delta_t}\right| \to 0 \text{ as } t \to \infty.$$

Thus, $\frac{D(k,t)}{\alpha_t}$ and $\frac{R(k,t)}{\delta_t}$ have the same limits, if the limits of either exist.

6.22 Positive recurrence of reflected random walk with negative drift
Let $V(x) = \frac{1}{2}x^2$. Then

$$\begin{aligned} PV(x) - V(x) &= E[\frac{(x+B_n+L_n)^2}{2}] - \frac{x^2}{2} \\ &\leq E[\frac{(x+B_n)^2}{2}] - \frac{x^2}{2} \\ &= x\overline{B} + \frac{\overline{B^2}}{2}. \end{aligned}$$

Therefore, the conditions of the combined Foster stability criteria and moment bound corollary apply, yielding that X is positive recurrent, and $\overline{X} \leq \frac{\overline{B^2}}{-2\overline{B}}$. (This bound is somewhat weaker than Kingman's moment bound, discussed later in the book: $\overline{X} \leq \frac{\mathrm{Var}(B)}{-2\overline{B}}$.)

6.24 An inadequacy of a linear potential function Suppose x is on the positive x_2 axis (i.e. $x_1 = 0$ and $x_2 > 0$). Then, given $X(t) = x$, during the slot, queue 1 will increase to 1 with probability $a(1-d_1) = 0.42$, and otherwise stay at zero. Queue 2 will decrease by one with probability 0.4, and otherwise stay the same. Thus, the drift of V, $E[V(X(t+1)) - V(x)|X(t)) = x]$ is equal to 0.02. Therefore, the drift is strictly positive for infinitely many states, whereas the Foster–Lyapunov condition requires that the drift be negative off a finite set C. So, the linear choice for V does not work for this example.

6.26 Opportunistic scheduling (a) The lefthand side of (6.35) is the arrival rate to the set of queues in s, and the righthand side is the probability that some queue in s is eligible for service in a given time slot. The condition is necessary for the stability of the set of queues in s.
(b) Fix $\epsilon > 0$ so that for all $s \in E$ with $s \neq \emptyset$,

$$\sum_{i \in s}(a_i + \epsilon) \leq \sum_{B: B \cap s \neq \emptyset} w(B).$$

Consider the flow graph shown.

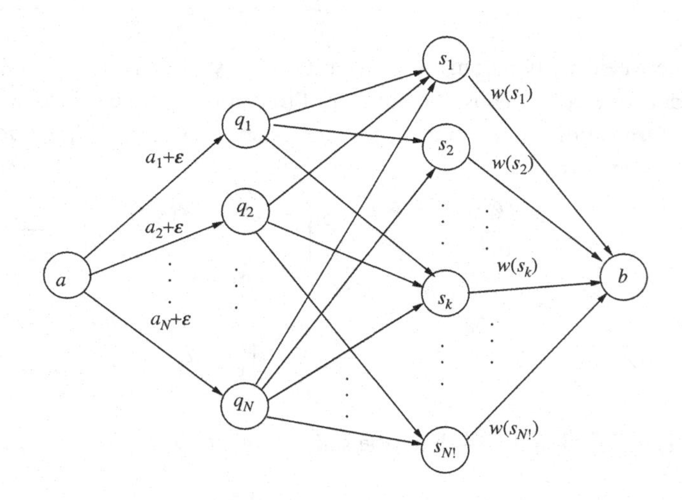

In addition to the source node a and sink node b, there are two columns of nodes in the graph. The first column of nodes corresponds to the N queues, and the second column of nodes corresponds to the 2^N subsets of E. There are three stages of links in the graph. The capacity of a link (a, q_i) in the first stage is $a_i + \epsilon$, there is a link (q_i, s_j) in the second stage if and only if $q_i \in s_j$, and each such link has capacity greater than the sum of the capacities of all the links in the first stage, and the weight of a link (s_k, t) in the third stage is $w(s_k)$.

We claim that the minimum of the capacities of all $a - b$ cuts is $v* = \sum_{i=1}^N (a_i + \epsilon)$. Here is a proof. The $a - b$ cut $(\{a\} : V - \{a\})$ (here V is the set of nodes in the flow network) has capacity v^*, so to prove the claim, it suffices to show that any other $a - b$ cut has capacity greater than or equal to v^*. Fix any $a - b$ cut $(A : B)$. Let $\widetilde{A} = A \cap \{q_1, \ldots, q_N\}$, or in words, $\widetilde{A}$ is the set of nodes in the first column of the graph (i.e. set of queues) that are in A. If $q_i \in \widetilde{A}$ and $s_j \in B$ such that (q_i, s_j) is a link in the flow graph, then the capacity of $(A : B)$ is greater than or equal to the capacity of link (q_i, s_j), which is greater than v^*, so the required inequality is proved in that case. Thus, we can suppose that A contains all the nodes s_j in the second column such that $s_j \cap \widetilde{A} \neq \emptyset$. Therefore,

$$
\begin{aligned}
C(A : B) &\geq \sum_{i \in \{q_1, \ldots, q_N\} - \widetilde{A}} (a_i + \epsilon) + \sum_{s \subset E : s \cap \widetilde{A} \neq \emptyset} w(s) \\
&\geq \sum_{i \in \{q_1, \ldots, q_N\} - \widetilde{A}} (a_i + \epsilon) + \sum_{i \in \widetilde{A}} (a_i + \epsilon) = v^*,
\end{aligned}
\tag{12.14}
$$

where the inequality in (12.14) follows from the choice of ϵ. The claim is proved.

Therefore there is an $a - b$ flow f which saturates all the links of the first stage of the flow graph. Let $u(i, s) = f(q_i, s)/f(s, b)$ for all i, s such that $f(s, b) > 0$. That is, $u(i, s)$ is the fraction of flow on link (s, b) which comes from link (q_i, s). For those s such that $f(s, b) = 0$, define $u(i, s)$ in some arbitrary way, respecting the requirements $u(i, s) \geq 0$, $u(i, s) = 0$ if $i \notin s$, and $\sum_{i \in E} u(i, s) = I_{\{s \neq \emptyset\}}$. Then $a_i + \epsilon = f(a, q_i) = \sum_s f(q_i, s) = \sum_s f(s, b) u(i, s) \leq \sum_s w(s) u(i, s) = \mu_i(u)$, as required.
(c) Let $V(x) = \frac{1}{2} \sum_{i \in E} x_i^2$. Let $\delta(t)$ denote the identity of the queue given a potential service at time t, with $\delta(t) = 0$ if no queue is given potential service. Then $P(\delta(t) = i | S(t) = s) = u(i, s)$. The dynamics of queue i are given by $X_i(t+1) = X_i(t) + A_i(t) - R_i(\delta(t)) + L_i(t)$, where $R_i(\delta) = I_{\{\delta = i\}}$. Since $\sum_{i \in E} (A_i(t) - R_i(\delta_i(t)))^2 \leq \sum_{i \in E} (A_i(t))^2 + (R_i(\delta_i(t)))^2 \leq N + \sum_{i \in E} A_i(t)^2$ we have

$$PV(x) - V(x) \leq \left(\sum_{i \in E} x_i (a_i - \mu_i(u)) \right) + K \tag{12.15}$$

$$\leq -\epsilon \left(\sum_{i \in E} x_i \right) + K, \tag{12.16}$$

where $K = \frac{N}{2} + \sum_{i=1}^N K_i$. Thus, under the necessary stability conditions we have that under the vector of scheduling probabilities u, the system is positive recurrent, and

$$\sum_{i \in E} \overline{X}_i \leq \frac{K}{\epsilon}. \tag{12.17}$$

(d) If u could be selected as a function of the state, x, then the righthand side of (12.1
would be minimized by taking $u(i,s) = 1$ if i is the smallest index in s such th
$x_i = \max_{j\in s} x_j$. This suggests using the *longest connected first* (LCF) policy, in whic
the longest connected queue is served in each time slot. If P^{LCF} denotes the one-ste
transition probability matrix for the LCF policy, then (12.15) holds for any u, if P
replaced by P^{LCF}. Therefore, under the necessary condition and ϵ as in part (b), (12.1
also holds with P replaced by P^{LCF}, and (12.17) holds for the LCF policy.

6.28 Stability of two queues with transfers (a) System is positive recurrent for some
if and only if $\lambda_1 < \mu_1 + \nu$, $\lambda_2 < \mu_2$, and $\lambda_1 + \lambda_2 < \mu_1 + \mu_2$.

(b)

$$\begin{aligned}QV(x) &= \sum_{y:y\neq x} q_{xy}\,(V(y) - V(x)) \\ &= \frac{\lambda_1}{2}[(x_1+1)^2 - x_1^2] + \frac{\lambda_2}{2}[(x_2+1)^2 - x_2^2] + \frac{\mu_1}{2}[(x_1-1)_+^2 - x_1^2] \\ &\quad + \frac{\mu_2}{2}[(x_2-1)_+^2 - x_2^2] + \frac{u\nu I_{\{x_1\geq 1\}}}{2}[(x_1-1)^2 \\ &\quad - x_1^2 + (x_2+1)^2 - x_2^2].\end{aligned} \tag{12.18}$$

(c) If the righthand side of ((12.18)) is changed by dropping the positive part symbo
and dropping the factor $I_{\{x_1\geq 1\}}$, then it is not increased, so that

$$\begin{aligned}QV(x) &\leq x_1(\lambda_1 - \mu_1 - u\nu) + x_2(\lambda_2 + u\nu - \mu_2) + K \\ &\leq -(x_1+x_2)\min\{\mu_1 + u\nu - \lambda_1, \mu_2 - \lambda_2 - u\nu\} + K,\end{aligned} \tag{12.19}$$

where $K = \frac{\lambda_1+\lambda_2+\mu_1+\mu_2+2\nu}{2}$. To get the best bound on $\overline{X_1} + \overline{X_2}$, we select u
maximize the min term in (12.19), or $u = u^*$, where u^* is the point in $[0,1]$ nearest t
$\frac{\mu_1+\mu_2-\lambda_1-\lambda_2}{2\nu}$. For $u = u^*$, we find $QV(x) \leq -\epsilon(x_1+x_2) + K$ where $\epsilon = \min\{\mu_1 + \nu -$
$\lambda_1, \mu_2 - \lambda_2, \frac{\mu_1+\mu_2-\lambda_1-\lambda_2}{2}\}$. Which of the three terms is smallest in the expression for
corresponds to the three cases $u^* = 1$, $u^* = 0$, and $0 < u^* < 1$, respectively. It is easy t
check that this same ϵ is the largest constant such that the stability conditions (with stric
inequality relaxed to less than or equal) hold with (λ_1, λ_2) replaced by $(\lambda_1+\epsilon, \lambda_2+\epsilon)$.

7.2 Lack of sample path continuity of a Poisson process (a) The sample path of N
continuous over $[0,T]$ if and only if it has no jumps in the interval, equivalently,
and only if $N(T) = 0$. So $P(N$ is continuous over the interval [0,T] $) = \exp(-\lambda T)$. B
continuity of probability (Lemma 1.1),

$$\begin{aligned}P(N \text{ is continuous over } [0,+\infty)) &= \lim_{n\to\infty} P(N \text{ is continuous over } [0,n]) \\ &= \lim_{n\to\infty} e^{-\lambda n} = 0.\end{aligned}$$

(b) Since $P(N$ is continuous over $[0,+\infty)) \neq 1$, N is not a.s. sample continuou
However N is m.s. continuous. One proof is to simply note that the correlation functio
given by $R_N(s,t) = \lambda(s\wedge t) + \lambda^2 st$, is continuous. A more direct proof is to note that fo
fixed t, $E[|N_s - N_t|^2] = \lambda|s-t| + \lambda^2|s-t|^2 \to 0$ as $s \to t$.

7.4 Some statements related to the basic calculus of random processes

(a) False. $\lim_{t\to\infty} \frac{1}{t}\int_0^t X_s ds = Z \neq E[Z]$ (except in the degenerate case that Z ha
variance zero).

(b) False. One reason is that the function is continuous at zero, but not everywhere. For another, we would have $\mathrm{Var}(X_1 - X_0 - X_2) = 3R_X(0) - 4R_X(1) + 2R_X(2) = 3 - 4 + 0 = -1$.
(c) True. In general, $R_{X'X}(\tau) = R'_X(\tau)$. Since R_X is an even function, $R'_X(0) = 0$. Thus, for any t, $E[X'_t X_t] = R_{X'X}(0) = R'_X(0) = 0$. Since the process X has mean zero, it follows that $\mathrm{Cov}(X'_t, X_t) = 0$ as well. Since X is a Gaussian process, and differentiation is a linear operation, X_t and X'_t are jointly Gaussian. Summarizing, for t fixed, X'_t and X_t are jointly Gaussian and uncorrelated, so they are independent. (Note: X'_s is not necessarily independent of X_t if $s \neq t$.)

7.6 Continuity of a process passing through a nonlinearity (a) Fix $t \in \mathbb{R}$ and let (s_n) be a sequence converging to t. Let $\epsilon > 0$ be arbitrary. Let $[a, b]$ be an interval so large that $P\{X_t \in [a, b]\} \geq \epsilon$. Let δ with $0 < \delta < 1$ be so small that $|G(x) - G(x')| \leq \epsilon$ whenever $x, x' \in [a - 1, b + 1]$ with $|x - x'| \leq \delta$. Since $X_{s_n} \to X_t$ m.s. it follows that $X_{s_n} \to X_t$ in probability, so there exists N so large that $P\{|X_{s_n} - X_t| > \delta\} \leq \epsilon$ whenever $n \geq N$. Then for $n \geq N$,

$$\begin{aligned} P\{|Y_{s_n} - Y_t| > \epsilon\} &\leq P\{|Y_{s_n} - Y_t| > \epsilon, X_t \in [a, b]\} + P\{X_t \notin [a, b]\} \\ &\leq P\{|X_{s_n} - X_t| > \delta\} + \epsilon \\ &\leq 2\epsilon. \end{aligned}$$

Therefore, $Y_{s_n} \to Y_t$ in probability as $n \to \infty$. Since the Ys are bounded, the convergence also holds in the m.s. sense. Thus, Y is m.s. continuous at an arbitrary t, so Y is a m.s. continuous process.
(b) Let $X_t = t$ (a deterministic process) and $G(x) = I_{\{x \geq 0\}}$. Then $Y_t = I_{\{t \geq 0\}}$, which is not continuous at $t = 0$, and so is not a m.s. continuous process.
(c) Let $X_t \equiv U$, (a process content in time) where U has the exponential distribution with parameter one. Let $G(x) = e^x$. Then $E[Y_t^2] = \int_0^\infty (e^u)^2 e^{-u} du = \infty$, so that Y is not even a second order random process, so Y is not a m.s. continuous random process.

7.8 Integral of OU process (a) The process Y has mean zero because X has mean zero. For $s \leq t$,

$$\begin{aligned} R_Y(s, t) &= \int_0^s \int_0^t e^{-|u-v|} dv du \\ &= \int_0^s \left[\int_0^u d^{v-u} dv + \int_u^t e^{u-v} dv \right] du \\ &= \int_0^s 1 - e^{-u} + 1 - e^{u-t} du \\ &= 2s - 1 + e^{-s} + e^{-t} - e^{s-t}, \end{aligned}$$

so in general, $R_Y(s, t) = 2(s \wedge t) - 1 + e^{-s} + e^{-t} - e^{-|s-t|}$.
(b) For $t > 0$, Y_t has the $N(0, \sigma_t^2)$ distribution with $\sigma_t^2 = R_Y(t, t) = 2(t - 1 + e^{-t})$. Therefore, $P\{|Y_t| \geq g(t)\} = 2Q\left(\frac{g(t)}{\sigma_t}\right)$, which, since $Q(0.81) \approx 0.25$, means we want $g(t) = Q^{-1}(0.25)\sigma_t \approx (0.81)\sqrt{2(t - 1 + e^{-t})} \approx (1.15)\sqrt{t - 1 + e^{-t}}$.

(c) Since

$$\begin{aligned} R_Z(s,t) &= f(\alpha)^2 R_Y(\alpha s, \alpha t) \\ &= f(\alpha)^2 \left[2\alpha(s \wedge t) - 1 + e^{-\alpha s} + e^{-\alpha t} - e^{-\alpha|s-t|}\right] \\ &\sim f(\alpha)^2 2\alpha(s \wedge t) \quad \text{as } \alpha \to \infty, \end{aligned}$$

the choice $f(\alpha) = \frac{1}{\sqrt{2\alpha}}$ works. Intuitively, speeding up the process X causes the duratio of the memory in X to decrease.

7.10 Cross correlation between a process and its m.s. derivative
Fix $t, u \in T$. By assumption, $\lim_{s\to t} \frac{X_s - }{s-t} = X'_t$ *m.s.* Therefore, by Corollary 2.1 $E\left[\left(\frac{X_s - X_t}{s-t}\right) X_u\right] \to E[X'_t X_u]$ as $s \to t$. Equivalently,

$$\frac{R_X(s,u) - R_X(t,u)}{s-t} \to R_{X'X}(t,u) \quad \text{as } s \to t.$$

Hence $\partial_1 R_X(s,u)$ exists, and $\partial_1 R_X(t,u) = R_{X'X}(t,u)$.

7.12 A windowed Poisson process (a) The sample paths of X are piecewise constan integer valued with initial value zero. They jump by +1 at each jump of N, and jump b −1 one time unit after each jump of N.
(b) Method 1: If $|s-t| \geq 1$ then X_s and X_t are increments of N over disjoint interval and are therefore independent, so $C_X(s,t) = 0$. If $|s-t| < 1$, then there are three disjoir intervals, I_0, I_1, and I_2, with $I_0 = [s, s+1] \cup [t, t+1]$, such that $[s, s+1] = I_0 \cup I_1$ an $[t, t+1] = I_0 \cup I_2$. Thus, $X_s = D_0 + D_1$ and $X_t = D_0 + D_2$, where D_i is the increment c N over the interval I_i. The three increments D_1, D_2, and D_3 are independent, and D_0 is Poisson random variable with mean and variance equal to λ times the length of I_0, whic is $1-|s-t|$. Therefore, $C_X(s,t) = \mathrm{Cov}(D_0+D_1, D_0+D_2) = \mathrm{Cov}(D_0, D_0) = \lambda(1-|s-t|$. Summarizing, $C_X(s,t) = \begin{cases} \lambda(1-|s-t|) & \text{if } |s-t| < 1 \\ 0 & \text{else} \end{cases}$.
Method 2: $C_X(s,t) = \mathrm{Cov}(N_{s+1} - N_s, N_{t+1} - N_t) = \lambda[\min(s+1, t+1) - \min(s+1, t)$ - $\min(s, t+1) - \min(s,t)]$. This answer can be simplified to the one found by Method by considering the cases $|s-t| > 1$, $t < s < t+1$, and $s < t < s+1$ separately.
(c) No. X has a −1 jump one time unit after each +1 jump, so the value X_t for a "present time t tells less about the future, $(X_s : s \geq t)$, than the past, $(X_s : 0 \leq s \leq t)$, tells abou the future.
(d) Yes, recall that $R_X(s,t) = C_X(s,t) - \mu_X(s)\mu_X(t)$. Since C_X and μ_X are continuou functions, so is R_X, so that X is m.s. continuous.
(e) Yes. Using the facts $C_X(s,t)$ is a function of $s-t$ alone, and $C_X(s) \to 0$ as $s \to \infty$ we find, as in the section on ergodicity, $\mathrm{Var}(\frac{1}{t}\int_0^t X_s ds) = \frac{2}{t}\int_0^t (1 - \frac{s}{t}) C_X(s) ds \to 0$ a $t \to \infty$.

7.14 A singular integral with a Brownian motion (a) The integral $\int_\epsilon^1 \frac{w_t}{t} dt$ exists in th m.s. sense for any $\epsilon > 0$ because w_t/t is m.s. continuous over $[\epsilon, 1]$. To see if the limi exists we apply the correlation form of the Cauchy criteria (Proposition 2.11). Usin different letters as variables of integration and the fact $R_w(s,t) = s \wedge t$ (the minimur of s and t), yields that as $\epsilon, \epsilon' \to 0$,

$$
\begin{aligned}
E\left[\int_\epsilon^1 \frac{w_s}{s}ds \int_{\epsilon'}^1 \frac{w_t}{t}dt\right] &= \int_\epsilon^1\int_{\epsilon'}^1 \frac{s\wedge t}{st}dsdt \\
&\to \int_0^1\int_0^1 \frac{s\wedge t}{st}dsdt \\
&= 2\int_0^1\int_0^t \frac{s\wedge t}{st}dsdt = 2\int_0^1\int_0^t \frac{s}{st}dsdt \\
&= 2\int_0^1\int_0^t \frac{1}{t}dsdt = 2\int_0^1 1dt = 2.
\end{aligned}
$$

Thus the m.s. limit defining the integral exists. The integral has the $N(0,2)$ distribution.
(b) As $a, b \to \infty$,

$$
\begin{aligned}
E\left[\int_1^a \frac{w_s}{s}ds \int_1^b \frac{w_t}{t}dt\right] &= \int_1^a\int_1^b \frac{s\wedge t}{st}dsdt \\
&\to \int_1^\infty\int_1^\infty \frac{s\wedge t}{st}dsdt \\
&= 2\int_1^\infty\int_1^t \frac{s\wedge t}{st}dsdt = 2\int_1^\infty\int_1^t \frac{s}{st}dsdt \\
&= 2\int_1^\infty\int_1^t \frac{1}{t}dsdt = 2\int_1^\infty \frac{t-1}{t}dt = \infty,
\end{aligned}
$$

so that the m.s. limit does not exist, and the integral is not well defined.

7.16 Recognizing m.s. properties (a) Yes, m.s. continuous since R_X is continuous. No, not m.s. differentiable since $R'_X(0)$ doesn't exist. Yes, m.s. integrable over finite intervals since m.s. continuous. Yes, mean ergodic in m.s. since $R_X(T) \to 0$ as $|T| \to \infty$.
(b) Yes, no, yes, for the same reasons as in part (a). Since X is mean zero, $R_X(T) = C_X(T)$ for all T. Thus

$$
\lim_{|T|\to\infty} C_X(T) = \lim_{|T|\to\infty} R_X(T) = 1.
$$

Since the limit of C_X exists and is not zero, X is not mean ergodic in the m.s. sense.
(c) Yes, no, yes, yes, for the same reasons as in (a).
(d) No, not m.s. continuous since R_X is not continuous. No, not m.s. differentiable since X is not even m.s. continuous. Yes, m.s. integrable over finite intervals, because the Riemann integral $\int_a^b\int_a^b R_X(s,t)dsdt$ exists and is finite, for the region of integration is a simple bounded region and the integrand is piece-wise constant.
(e) Yes, m.s. continuous since R_X is continuous. No, not m.s. differentiable. For example,

$$
\begin{aligned}
E\left[\left(\frac{X_t - X_0}{t}\right)^2\right] &= \frac{1}{t^2}\left[R_X(t,t) - R_X(t,0) - R_X(0,t) + R_X(0,0)\right] \\
&= \frac{1}{t^2}\left[\sqrt{t} - 0 - 0 + 0\right] \to +\infty \text{ as } t \to 0.
\end{aligned}
$$

Yes, m.s. integrable over finite intervals since m.s. continuous.

7.18 A stationary Gaussian process (a) No. All mean zero stationary, Gaussian Mark processes have autocorrelation functions of the form $R_X(t) = A\rho^{|t|}$, where $A \geq 0$ an $0 \leq \rho \leq 1$ for continuous time (or $|\rho| \leq 1$ for discrete time).
(b) $E[X_3|X_0] = \widehat{E}[X_3|X_0] = \frac{R_X(3)}{R_X(0)}X_0 = \frac{X_0}{10}$. The error is Gaussian with mean zero an variance MSE = $\mathrm{Var}(X_3) - \mathrm{Var}(\frac{X_0}{10}) = 1 - 0.01 = 0.99$. So $P\{|X_3 - E[X_3|X_0]| \geq 10\}$ = $2Q(\frac{10}{\sqrt{0.99}})$.
(c) $R_{X'}(\tau) = -R_X''(\tau) = \frac{2-6\tau^2}{(1+\tau^2)^3}$. In particular, since $-R_X''$ exists and is continuous, X i continuously differentiable in the m.s. sense.
(d) The vector has a joint Gaussian distribution because X is a Gaussian process an differentiation is a linear operation. $\mathrm{Cov}(X_\tau, X_0') = R_{XX'}(\tau) = -R_X'(\tau) = \frac{2\tau}{(1+\tau^2)^2}$. I particular, $\mathrm{Cov}(X_0, X_0') = 0$ and $\mathrm{Cov}(X_1, X_0') = \frac{2}{4} = 0.5$. Also, $\mathrm{Var}(X_0') = R_{X'}(0) =$ 2

So $(X_0, X_0', X_1)^T$ is $N\left(\begin{pmatrix} 0 \\ 0 \\ 0 \end{pmatrix}, \begin{pmatrix} 1 & 0 & 0.5 \\ 0 & 2 & 0.5 \\ 0.5 & 0.5 & 1 \end{pmatrix}\right)$.

7.20 Correlation ergodicity of Gaussian processes (a) Fix h and let $Y_t = X_{t+h}X$ Clearly Y is stationary with mean $\mu_Y = R_X(h)$. Observe that

$$\begin{aligned} C_Y(\tau) &= E[Y_\tau Y_0] - \mu_Y^2 \\ &= E[X_{\tau+h}X_\tau X_h X_0] - R_X(h)^2 \\ &= R_X(h)^2 + R_X(\tau)R_X(\tau) + R_X(\tau+h)R_X(\tau-h) - R_X(h)^2. \end{aligned}$$

Therefore, $C_Y(\tau) \to 0$ as $|\tau| \to \infty$. Hence Y is mean ergodic, so X is correlatio ergodic.
(b) $X_t = A\cos(t+\Theta)$, where A is a random variable with positive variance, Θ i uniformly distributed on the interval $[0, 2\pi]$, and A is independent of Θ. Note tha $\mu_X = 0$ because $E[\cos(t+\Theta)] = 0$. Also, $|\int_0^T X_t dt| = |A\int_0^T \cos(t+\Theta)dt| \leq 2|A$ so $\left|\frac{\int_0^T X_t dt}{T}\right| \leq \frac{2|A|}{T} \to 0$ in the m.s. sense. So X is m.s. ergodic. Similarly, we hav $\frac{\int_0^T X_t^2 dt}{T} \to \frac{A^2}{2}$ in the m.s. sense. The limit is random, so X_t^2 is not mean ergodic, so X i not correlation ergodic. (The definition is violated for $h = 0$.)
Alternatively $X_t = \cos(Vt + \Theta)$ where V is a positive random variable with nonzer variance, Θ is uniformly distributed on the interval $[0, 2\pi]$, and V is independent of Θ In this case, X is correlation ergodic as before. But $\int_0^T X_t X_{t+h} dt \to \frac{\cos(Vh)}{2}$ in the m.s sense. This limit is random, at least for some values of h, so Y is not mean ergodic so is not correlation ergodic.

7.22 Gaussian review question (a) Since X is Markovian, the best estimator of X given (X_0, X_1) is a function of X_1 alone. Since X is Gaussian, such estimator is linea in X_1. Since X is mean zero, the estimator is given by $\mathrm{Cov}(X_2, X_1)\mathrm{Var}(X_1)^{-1}X_1$ = $e^{-1}X_1$. Thus $E[X_2|X_0, X_1] = e^{-1}X_1$. No function of (X_0, X_1) is a better estimator But $e^{-1}X_1$ is equal to $p(X_0, X_1)$ for the polynomial $p(x_0, x_1) = x_1/e$. This is th optimal polynomial. The resulting mean square error is given by MMSE = $\mathrm{Var}(X_2)$ - $(\mathrm{Cov}(X_1X_2)^2)/\mathrm{Var}(X_1) = 9(1 - e^{-2})$.

(b) Given $(X_0 = \pi, X_1 = 3)$, X_2 is $N(3e^{-1}, 9(1-e^{-2}))$ so

$$P(X_2 \geq 4 | X_0 = \pi, X_1 = 3) = P\left\{ \frac{X_2 - 3e^{-1}}{\sqrt{9(1-e^{-2})}} \geq \frac{4 - 3e^{-1}}{\sqrt{9(1-e^{-2})}} \right\}$$
$$= Q\left(\frac{4 - 3e^{-1}}{\sqrt{9(1-e^{-2})}} \right).$$

7.24 KL expansion of a simple random process (a) Yes, because $R_X(\tau)$ is twice continuously differentiable.
(b) No. $\lim_{t\to\infty} \frac{2}{t}\int_0^t (\frac{t-\tau}{t}) C_X(\tau) d\tau = 50 + \lim_{t\to\infty} \frac{100}{t}\int_0^t (\frac{t-\tau}{t}) \cos(20\pi\tau) d\tau = 50 \neq 0$. Thus, the necessary and sufficient condition for mean ergodicity in the m.s. sense does not hold.
(c) *Approach one* Since $R_X(0) = R_X(1)$, the process X is periodic with period one (actually, with period 0.1). Thus, by the theory of WSS periodic processes, the eigenfunctions can be taken to be $\varphi_n(t) = e^{2\pi jnt}$ for $n \in \mathbb{Z}$. (Still have to identify the eigenvalues.)
Approach two The identity $\cos(\theta) = \frac{1}{2}(e^{j\theta} + e^{-j\theta})$, yields

$$\begin{aligned} R_X(s-t) &= 50 + 25e^{20\pi j(s-t)} + 25e^{-20\pi j(s-t)} \\ &= 50 + 25e^{20\pi js}e^{-20\pi jt} + 25e^{-20\pi js}e^{20\pi jt} \\ &= 50\varphi_0(s)\varphi_0^*(t) + 25\varphi_1(s)\varphi_1^*(t) + 25\varphi_2(s)\varphi_2^*(t), \end{aligned}$$

for the choice $\varphi_0(t) \equiv 1$, $\varphi_1(t) = e^{20\pi jt}$ and $\varphi_2 = e^{-20\pi jt}$. The eigenvalues are thus 50, 25, and 25. The other eigenfunctions can be selected to fill out an orthonormal basis, and the other eigenvalues are zero.
Approach three For $s, t \in [0, 1]$ we have

$$\begin{aligned} R_X(s,t) &= 50 + 50\cos(20\pi(s-t)) \\ &= 50 + 50\cos(20\pi s)\cos(20\pi t) + 50\sin(20\pi s)\sin(20\pi t) \\ &= 50\varphi_0(s)\varphi_0^*(t) + 25\varphi_1(s)\varphi_1^*(t) + 25\varphi_2(s)\varphi_2^*(t), \end{aligned}$$

for the choice $\varphi_0(t) \equiv 1$, $\varphi_1(t) = \sqrt{2}\cos(20\pi t)$ and $\varphi_2 = \sqrt{2}\sin(20\pi t)$. The eigenvalues are thus 50, 25, and 25. The other eigenfunctions can be selected to fill out an orthonormal basis, and the other eigenvalues are zero.
(Note: the eigenspace for eigenvalue 25 is two dimensional, so the choice of eigenfunctions spanning that space is not unique.)

7.26 KL expansion for derivative process (a) Since $\varphi_n'(t) = (2\pi jn)\varphi_n(t)$, the equation given in the problem statement leads to: $X'(t) = \sum_n \langle X, \varphi_n\rangle \varphi_n'(t) = \sum_n [(2\pi jn)\langle X, \varphi_n\rangle]\varphi_n(t)$, which is a KL expansion, because the functions φ_n are orthonormal in $L^2[0,1]$ and the coordinates are orthogonal random variables. Thus,

$$\psi_n(t) = \varphi_n(t), \quad \langle X', \psi_n\rangle = (2\pi jn)\langle X_n, \varphi_n\rangle, \text{ and } \mu_n = (2\pi n)^2\lambda_n \text{ for } n \in \mathbb{Z}.$$

(Recall that the eigenvalues are equal to the means of the squared magnitudes of the coordinates.)
(b) Note that $\varphi_1' = 0$, $\varphi_{2k}'(t) = -(2\pi k)\varphi_{2k+1}(t)$ and $\varphi_{2k+1}'(t) = (2\pi k)\varphi_{2k}(t)$. This is similar to part (a). The same basis functions can be used for X' as for X, but the $(2k)$th

and $(2k+1)$th coordinates of X' come from the $(2k+1)$th and $(2k)$th coordinates of X
respectively, for all $k \geq 1$. Specifically, we can take

$$\begin{aligned} &\psi_n(t) = \varphi_n(t) \text{ for } n \geq 0, && \mu_0 = 0 \text{ (because } \langle X', \psi_0 \rangle = 0), \\ &\langle X', \psi_{2k} \rangle = (2\pi k)\langle X, \varphi_{2k+1} \rangle, && \mu_{2k} = (2\pi k)^2 \lambda_{2k+1}, \\ &\langle X', \psi_{2k+1} \rangle = -(2\pi k)\langle X, \varphi_{2k} \rangle, && \mu_{2k+1} = (2\pi k)^2 \lambda_{2k}, \qquad \text{for } k \geq 1. \end{aligned}$$

Defining ψ_0 was optional because the corresponding eigenvalue is zero.
(c) Note that $\varphi_n'(t) = \frac{(2n+1)\pi}{2}\psi_n(t)$, where $\psi_n(t) = \sqrt{2}\cos\left(\frac{(2n+1)\pi t}{2}\right)$, $n \geq 0$. That is, ψ_n is the same as φ_n, but with sin replaced by cos . Or equivalently, by the hint, we discover that ψ_n is obtained from φ_n by time-reversal: $\psi_n(t) = \varphi_n(1-t)(-1)^n$. Thus, the functions ψ_n are orthonormal. As in part (a), we also have $\langle X', \psi_n \rangle = \frac{(2n+1)\pi}{2}\langle X, \varphi_n \rangle$, and therefore, $\mu_n = (\frac{(2n+1)\pi}{2})^2 \lambda_n$. (The set of eigenfunctions is not unique – for example, some could be multiplied by -1 to yield another valid set.)
(d) Differentiating the KL expansion of X yields

$$X_t' = \langle X, \varphi_1 \rangle \varphi_1'(t) + \langle X, \varphi_2 \rangle \varphi_2'(t) = \langle X, \varphi_1 \rangle c_1\sqrt{3} - \langle X, \varphi_2 \rangle c_2\sqrt{3}.$$

That is, the random process X' is constant in time. So its KL expansion involves only one nonzero term, with the eigenfunction $\psi_1(t) = 1$ for $0 \leq t \leq 1$. Then $\langle X', \psi_1 \rangle = \langle X, \varphi_1 \rangle c_1\sqrt{3} - \langle X, \varphi_2 \rangle c_2\sqrt{3}$, and therefore $\mu_1 = 3c_1^2\lambda_1 + 3c_2^2\lambda_2$.

7.28 KL expansion of a Brownian bridge The (eigenfunction, eigenvalue) pairs satisfy $\int_0^1 R_B(t,s)\varphi(s)ds = \lambda\varphi(t)$. Since $R_B(t,s) \to 0$ as $t \to 0$ or $t \to 1$ and the function φ is continuous (and hence bounded) on $[0,1]$ by Mercer's theorem, it follows that $\varphi(0) = \varphi(1) = 0$. Inserting the expression for R_B, into the eigen relation yields

$$\int_0^1 ((s \wedge t) - st)\varphi(s)ds = \lambda\varphi(t),$$

or

$$\int_0^t (1-t)s\varphi(s)ds + \int_t^1 t(1-s)\varphi(s)ds = \lambda\varphi(t).$$

Differentiating both sides with respect to t, yields

$$-\int_0^t s\varphi(s)ds + \int_t^1 (1-s)\varphi(s)ds = \lambda\varphi'(t),$$

where we use the fact that the terms coming from differentiating the limit of integration t cancel out. Differentiating a second time with respect to t yields $-t\varphi(t) - (1-t)\varphi(t) = \lambda\varphi''(t)$, or $\varphi''(t) = \frac{1}{\lambda}\varphi(t)$. The solutions to this second order equation have the form $A\sin\left(\frac{t}{\sqrt{\lambda}}\right) + B\cos\left(\frac{t}{\sqrt{\lambda}}\right)$. Since $\varphi = 0$ at the endpoints 0 and 1, $B = 0$ and $\sin\left(\frac{1}{\sqrt{\lambda}}\right) = 0$. Thus, $\frac{1}{\sqrt{\lambda}} = n\pi$ for some integer $n \geq 1$, so that $\varphi(t) = A\sin(n\pi t)$ for some $n \geq 1$. Normalizing φ to have energy one yields $\varphi_n(t) = \sqrt{2}\sin(n\pi t)$ with the corresponding eigenvalue $\lambda_n = \frac{1}{(n\pi)^2}$. Thus, the Brownian bridge has the KL representation

$$B(t) = \sum_{n=1}^{\infty} B_n \sqrt{2}\sin(n\pi t),$$

where the (B_n) are independent with B_n having the $N\left(0, \frac{1}{(n\pi)^2}\right)$ distribution.

7.30 Mean ergodicity of a periodic WSS random process

$$\frac{1}{t}\int_0^t X_u du = \frac{1}{t}\int_0^t \sum_n \widehat{X}_n e^{2\pi jnu/T} du = \sum_{n\in\mathbb{Z}} a_{n,t}\widehat{X}_n,$$

where $a_0 = 1$, and for $n \neq 0$, $|a_{n,t}| = |\frac{1}{t}\int_0^t e^{2\pi jnu/T}du| = |\frac{e^{2\pi jnt/T}-1}{2\pi jnt/T}| \leq \frac{T}{\pi nt}$. The $n \neq 0$ terms are not important as $t \to \infty$. Indeed,

$$E\left[\left|\sum_{n\in\mathbb{Z},n\neq 0} a_{n,t}\widehat{X}_n\right|^2\right] = \sum_{n\in\mathbb{Z},n\neq 0} |a_{n,t}|^2 p_X(n) \leq \frac{T^2}{\pi^2 t^2}\sum_{n\in\mathbb{Z},n\neq 0} p_X(n) \to 0$$

as $t \to \infty$. Therefore, $\frac{1}{t}\int_0^t X_u du \to \widehat{X}_0$ *m.s.* The limit has mean zero and variance $p_X(0)$. For mean ergodicity (in the m.s. sense), the limit should be zero with probability one, which is true if and only if $p_X(0) = 0$. That is, the process should have no zero frequency, or DC, component. (Note: More generally, if X were not assumed to be mean zero, then X would be mean ergodic if and only if $\mathrm{Var}(\widehat{X}_0) = 0$, or equivalently, $p_X(0) = \mu_X^2$, or equivalently, $\widehat{X}_0$ is a constant a.s.)

8.2 A second order stochastic differential equation (a) For deterministic, finite energy signals x and y, the given relationship in the frequency domain becomes $((j\omega)^2 + j\omega + 1)\widehat{y}(\omega) = \widehat{x}(\omega)$, so the transfer function is $H(\omega) = \frac{\widehat{y}(\omega)}{\widehat{x}(\omega)} = \frac{1}{(j\omega)^2+j\omega+1} = \frac{1}{1-\omega^2+j\omega}$. Note that $|H(\omega)|^2 = \frac{1}{(1-\omega^2)^2+\omega^2} = \frac{1}{1-\omega^2+\omega^4}$. Therefore, $S_Y(\omega) = \frac{1}{1-\omega^2+\omega^4}S_X(\omega)$.

(b) Letting $\eta = \omega^2$, the denominator in H is $1 - \eta + \eta^2$, which takes its minimum value $\frac{3}{4}$ when $\eta = 1/2$. Thus, $\max_\omega |H(\omega)|^2 = \frac{4}{3}$, and the maximum is achieved at $\omega = \pm\sqrt{0.5}$. If the power of X is one then the power of Y is less than or equal to $\frac{4}{3}$, with equality if and only if all the power in X is at $\pm\sqrt{0.5}$. For example, X could take the form $X_t = \sqrt{2}\cos(2\pi\sqrt{0.5}t + \Theta)$, where Θ is uniformly distributed over $[0, 2\pi]$.

(c) Similarly, for the power of Y to be small for an X with power one, the power spectral density of X should be concentrated on high frequencies, where $H(\omega) \approx 0$. This can make the power of Y arbitrarily close to zero.

8.4 On the cross spectral density Follow the hint. Let U be the output if X is filtered by H^ϵ and V be the output if Y is filtered by H^ϵ. The Schwarz inequality applied to random variables U_t and V_t for t fixed yields $|R_{UV}(0)|^2 \leq R_U(0)R_V(0)$, or equivalently,

$$\left|\int_{J^\epsilon} S_{XY}(\omega)\frac{d\omega}{2\pi}\right|^2 \leq \int_{J^\epsilon} S_X(\omega)\frac{d\omega}{2\pi}\int_{J^\epsilon} S_Y(\omega)\frac{d\omega}{2\pi},$$

which implies that

$$|\epsilon S_{XY}(\omega_o) + o(\epsilon)|^2 \leq (\epsilon S_X(\omega_o) + o(\epsilon))(\epsilon S_Y(\omega_o) + o(\epsilon)).$$

Letting $\epsilon \to 0$ yields the desired conclusion.

8.6 Filtering a Gauss Markov process (a) The process Y is the output when X is passed through the linear time-invariant system with impulse response function

$h(\tau) = e^{-\tau}I_{\{\tau\geq 0\}}$. Thus, X and Y are jointly WSS, and $R_{XY}(\tau) =$ $R_X * \widetilde{h}(\tau) = \int_{t=-\infty}^{\infty} R_X(t)\widetilde{h}(\tau - t)dt = \int_{-\infty}^{\infty} R_X(t)h(t - \tau)d$
$= \begin{cases} \frac{1}{2}e^{-\tau} & \tau \geq 0 \\ (\frac{1}{2} - \tau)e^{\tau} & \tau \leq 0 \end{cases}.$

(b) X_5 and Y_5 are jointly Gaussian, mean zero, with $\mathrm{Var}(X_5) = R_X(0) = 1$, an $\mathrm{Cov}(Y_5, X_5) = R_{XY}(0) = \frac{1}{2}$, so $E[Y_5|X_5 = 3] = (\mathrm{Cov}(Y_5, X_5)/\mathrm{Var}(X_5))3 = 3/2$.

(c) Yes, Y is Gaussian, because X is a Gaussian process and Y is obtained from X b linear operations.

(d) No, Y is not Markov. For example, we see that $S_Y(\omega) = \frac{2}{(1+\omega^2)^2}$, which does not hav the form required for a stationary mean zero Gaussian process to be Markov (namel $\frac{2A}{\alpha^2+\omega^2}$). Another explanation is that, if t is the present time, given Y_t, the future of Y i determined by Y_t and $(X_s : s \geq t)$. The future could be better predicted by knowin something more about X_t than Y_t gives alone, which is provided by knowing the pas of Y.

(Note: the $\mathbb{R}^2$-valued process $((X_t, Y_t) : t \in \mathbb{R})$ is Markov.)

8.8 A stationary two-state Markov process $\pi P = \pi$ implies $\pi = (\frac{1}{2}, \frac{1}{2})$ is the equilit rium distribution so $P\{X_n = 1\} = P\{X_n = -1\} = \frac{1}{2}$ for all n. Thus $\mu_X = 0$. For $n \geq 1$

$$\begin{aligned}
R_X(n) &= P(X_n = 1, X_0 = 1) + P(X_n = -1, X_0 = -1) \\
&\quad - P(X_n = -1, X_0 = 1) - P(X_n = 1, X_0 = -1) \\
&= \frac{1}{2}\left[\frac{1}{2} + \frac{1}{2}(1-2p)^n\right] + \frac{1}{2}\left[\frac{1}{2} + \frac{1}{2}(1-2p)^n\right] - \\
&\quad \frac{1}{2}\left[\frac{1}{2} - \frac{1}{2}(1-2p)^n\right] - \frac{1}{2}\left[\frac{1}{2} - \frac{1}{2}(1-2p)^n\right] \\
&= (1-2p)^n.
\end{aligned}$$

So in general, $R_X(n) = (1-2p)^{|n|}$. The corresponding power spectral density is give by:

$$\begin{aligned}
S_X(\omega) &= \sum_{n=-\infty}^{\infty} (1-2p)^n e^{-j\omega n} \\
&= \sum_{n=0}^{\infty}((1-2p)e^{-j\omega})^n + \sum_{n=0}^{\infty}((1-2p)e^{j\omega})^n - 1 \\
&= \frac{1}{1-(1-2p)e^{-j\omega}} + \frac{1}{1-(1-2p)e^{j\omega}} - 1 \\
&= \frac{1-(1-2p)^2}{1-2(1-2p)\cos(\omega) + (1-2p)^2}.
\end{aligned}$$

8.10 A linear estimation problem

$$\begin{aligned}
E[|X_t - Z_t|^2] &= E[(X_t - Z_t)(X_t - Z_t)^*] \\
&= R_X(0) + R_Z(0) - R_{XZ}(0) - R_{ZX}(0) \\
&= R_X(0) + h * \widetilde{h} * R_Y(0) - 2Re(\widetilde{h} * R_{XY}(0)) \\
&= \int_{-\infty}^{\infty} S_X(\omega) + |H(\omega)|^2 S_Y(\omega) - 2Re(H^*(\omega)S_{XY}(\omega))\frac{d\omega}{2\pi}.
\end{aligned}$$

The hint with $\sigma^2 = S_Y(\omega)$, $z_o = S_{XY}(\omega)$, and $z = H(\omega)$ implies $H_{\text{opt}}(\omega) = \frac{S_{XY}(\omega)}{S_Y(\omega)}$.

8.12 The accuracy of approximate differentiation

(a) $S_{X'}(\omega) = S_X(\omega)|H(\omega)|^2 = \omega^2 S_X(\omega)$.

(b) $k(\tau) = \frac{1}{2a}(\delta(\tau+a) - \delta(\tau-a))$ and $K(\omega) = \int_{-\infty}^{\infty} k(\tau)e^{-j\omega t}d\tau = \frac{1}{2a}(e^{j\omega a} - e^{-j\omega a}) = \frac{j\sin(a\omega)}{a}$. By l'Hôpital's rule, $\lim_{a\to 0} K(\omega) = \lim_{a\to 0} \frac{j\omega\cos(a\omega)}{1} = j\omega$.

(c) D is the output of the linear system with input X and transfer function $H(\omega) - K(\omega)$. The output thus has power spectral density $S_D(\omega) = S_X(\omega)|H(\omega) - K(\omega)|^2 = S_X(\omega)|\omega - \frac{\sin(a\omega)}{a}|^2$.

(d) Or, $S_D(\omega) = S_{X'}(\omega)|1 - \frac{\sin(a\omega)}{a\omega}|^2$. Suppose $0 < a \le \frac{\sqrt{0.6}}{\omega_o}$ ($\approx \frac{0.77}{\omega_o}$). Then by the bound given in the problem statement, if $|\omega| \le \omega_o$ then $0 \le 1 - \frac{\sin(a\omega)}{a\omega} \le \frac{(a\omega)^2}{6} \le \frac{(a\omega_o)^2}{6} \le 0.1$, so that $S_D(\omega) \le (0.01)S_{X'}(\omega)$ for ω in the baseband. Integrating this inequality over the band yields that $E[|D_t|^2] \le (0.01)E[|X'_t|^2]$.

8.14 Filtering Poisson white noise (a) Since $\mu_{N'} = \lambda$, $\mu_X = \lambda\int_{-\infty}^{\infty} h(t)dt$. Also, $C_X = h * \widetilde{h} * C_{N'} = \lambda h * \widetilde{h}$. (In particular, if $h(t) = I_{\{0\le t<1\}}$, then $C_X(\tau) = \lambda(1 - |\tau|)_+$, as already found in Problem 4.19.)

(b) In the special case, in between arrival times of N, X decreases exponentially, following the equation $X' = -X$. At each arrival time of N, X has an upward jump of size one. Formally, we can write, $X' = -X + N'$. For a fixed time t_o, which we think of as the present time, the process after time t_o is the solution of the above differential equation, where the future of N' is independent of X up to time t_o. Thus, the future evolution of X depends only on the current value, and random variables independent of the past. Hence, X is Markov.

8.16 Linear and nonlinear reconstruction from samples

(a) $E[X_t] = \sum_n E[g(t-n-U)]E[B_n] \equiv 0$ because $E[B_n] = 0$ for all n.

$$
\begin{aligned}
R_X(s,t) &= E\left[\sum_{n=-\infty}^{\infty} g(s-n-U)B_n \sum_{m=-\infty}^{\infty} g(t-m-U)B_m\right] \\
&= \sigma^2 \sum_{n=-\infty}^{\infty} E[g(s-n-U)g(t-n-U)] \\
&= \sigma^2 \sum_{n=-\infty}^{\infty} \int_0^1 g(s-n-u)g(t-n-u)du \\
&= \sigma^2 \sum_{n=-\infty}^{\infty} \int_n^{n+1} g(s-v)g(t-v)dv = \sigma^2 \int_{-\infty}^{\infty} g(s-v)g(t-v)dv \\
&= \sigma^2 \int_{-\infty}^{\infty} g(s-v)\widetilde{g}(v-t)dv = \sigma^2(g*\widetilde{g})(s-t).
\end{aligned}
$$

So X is WSS with mean zero and $R_X = \sigma^2 g * \widetilde{g}$.

(b) By part (a), the power spectral density of X is $\sigma^2|G(\omega)|^2$. If g is a baseband signal, so that $|G(\omega)^2| = 0$ for $\omega \ge \omega_o$, then by the sampling theorem for WSS baseband random processes, X can be recovered from $(X(nT) : n \in \mathbb{Z})$ as long as $T \le \frac{\pi}{\omega_o}$.

(c) For this case, $G(2\pi f) = \text{sinc}^2(f)$, which is not supported in any finite interval. So part (a) does not apply. The sample paths of X are continuous and piecewise linear,

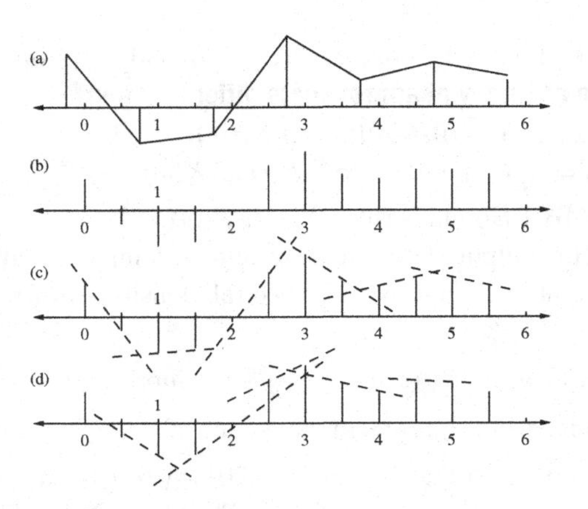

Figure 12.5 Nonlinear reconstruction of a signal from samples

and at least two sample points fall within each linear portion of X. Either all pairs of samples of the form $(X_n, X_{n+0.5})$ fall within linear regions (happens when $0.5 \le U \le 1$), or all pairs of samples of the form $(X_{n+0.5}, X_{n+1})$ fall within linear regions (happens when $0 \le U \le 0.5$). We can try reconstructing X using both cases. With probability one, only one of the cases will yield a reconstruction with change points having spacing one. That must be the correct reconstruction of X. The algorithm is illustrated in Figure 12.5. Figure 12.5(a) shows a sample path of B and a corresponding sample path of X, for $U = 0.75$. Thus, the breakpoints of X are at times of the form $n + 0.75$ for integers n. Figure 12.5(b) shows the corresponding samples, taken at integer multiples of $T = 0.5$. Figure 12.5(c) shows the result of connecting pairs of the form $(X_n, X_{n+0.5})$, and Figure 12.5(d) shows the result of connecting pairs of the form $(X_{n+0.5}, X_{n+1})$. Of these two, only Figure 12.5(c) yields breakpoints with unit spacing. Thus, the dashed lines in Figure 12.5(c) are connected to reconstruct X.

8.18 An approximation of white noise (a) Since $E[B_k B_l^*] = I_{\{k=l\}}$,

$$E\left[\left|\int_0^1 N_t dt\right|^2\right] = E\left[\left|A_T T \sum_{k=1}^K B_k\right|^2\right] = (A_T T)^2 E\left[\sum_{k=1}^K B_k \sum_{l=1}^K B_l^*\right]$$
$$= (A_T T)^2 \sigma^2 K = A_T^2 T \sigma^2.$$

(b) The choice of scaling constant A_T such that $A_T^2 T \equiv 1$ is $A_T = \frac{1}{\sqrt{T}}$. Under this scaling the process N approximates white noise with power spectral density σ^2 as $T \to 0$.
(c) If the constant scaling $A_T = 1$ is used, then $E[|\int_0^1 N_t dt|^2] = T\sigma^2 \to 0$ as $T \to 0$.

8.20 Synthesizing a random process with specified spectral density Recall from Example 8.6, a Gaussian random process Z with a rectangular spectral density $S_Z(2\pi f) = I_{\{-f_o \le f \le f_o\}}$ can be represented as (note, if $\frac{1}{T} = 2f_o$, then $\frac{t-nT}{T} = 2f_o t - n$):

$$Z_t = \sum_{n=-\infty}^{\infty} A_n \left(\sqrt{2f_o}\right) \operatorname{sinc}(2f_o t - n),$$

where the A_ns are independent, $N(0, 1)$ random variables. (To double check that Z is scaled correctly, note that the total power of Z is equal to both the integral of the psd and to $E[Z_0^2]$.) The desired psd S_X can be represented as the sum of two rectangular psds: $S_X(2\pi f) = I_{\{-20\leq f\leq 20\}} + I_{\{-10\leq f\leq 10\}}$, and the psd of the sum of two independent WSS processes is the sum of the psds, so X could be represented as:

$$X_t = \sum_{n=-\infty}^{\infty} A_n\left(\sqrt{40}\right)\mathrm{sinc}(40t-n) + \sum_{n=-\infty}^{\infty} B_n\left(\sqrt{20}\right)\mathrm{sinc}(20t-n),$$

where the As and Bs are independent $N(0, 1)$ random variables. This requires 60 samples per unit simulation time.

An approach using fewer samples is to generate a random process Y with psd $S_Y(\omega) = I_{\{-20\leq f\leq 20\}}$ and then filter Y using a filter with impulse response H with $|H|^2 = S_X$. For example, we could simply take $H(2\pi f) = \sqrt{S_X(2\pi f)} = I_{\{-20\leq f\leq 20\}} + \left(\sqrt{2}-1\right)I_{\{-10\leq f\leq 10\}}$, so X could be represented as:

$$X = \left(\sum_{n=-\infty}^{\infty} A_n\left(\sqrt{40}\right)\mathrm{sinc}(40t-n)\right) * h,$$

where $h(t) = (40)\mathrm{sinc}(40t) + \left(\sqrt{2}-1\right)(20)\mathrm{sinc}(20t)$. This approach requires 40 samples per unit simulation time.

8.22 Finding the envelope of a deterministic signal (a) $\widehat{z}(2\pi f) = 2[\widehat{x}(2\pi(f+f_c))]_{LP} = \delta(f+f_c-1000) + \delta(f+f_c-1001)$. If $f_c = 1000.5$ then $\widehat{z}(2\pi f) = \delta(f+0.5) + \delta(f-0.5)$. Therefore $z(t) = 2\cos(\pi t)$ and $|z(t)| = 2|\cos(\pi t)|$.
(b) If $f_c = 995$ then $\widehat{z}(2\pi f) = \delta(f-5) + \delta(f-6)$. Therefore $z(t) = e^{j2\pi(5.5)t}2\cos(\pi t)$ and $|z(t)| = 2|\cos(\pi t)|$.
(c) In general, the complex envelope in the frequency domain is given by $\widehat{z}(2\pi f) = 2[\widehat{x}(2\pi(f+f_c))]_{LP}$. If a somewhat different carrier frequency $\tilde{f}_c = f_c + \Delta f_c$ is used, the complex envelope of x using $\tilde{f}_c$ is the original complex envelope, shifted to the left in the frequency domain by Δf. This is equivalent to multiplication by $e^{-j2\pi(\Delta f)t}$ in the time domain. Since $|e^{-j2\pi(\Delta f)t}| \equiv 1$, the real envelope is unchanged.

8.24 A narrowband Gaussian process (a) The power spectral density S_X, which is the Fourier transform of R_X, can be found graphically as shown in Figure 12.6.
(b) A sample path of X generated by computer simulation is pictured in Figure 12.7. Several features of the sample path are apparent. The carrier frequency is 30 Hz, so for a period of time of the order of a tenth of a second, the signal resembles a pure sinusoidal signal with frequency near 30 Hz. On the other hand, the one sided root mean squared bandwidth of the baseband signals U and V is 2.7 Hz, so that the envelope of X varies significantly over intervals of length 1/3 of a second or more. The mean square value of the real envelope is given by $E[|Z_t|^2] = 2$, so the amplitude of the real envelope process $|Z_t|$ fluctuates about $\sqrt{2} \approx 1.41$.
(c) The power spectral densities $S_U(2\pi f)$ and $S_V(2\pi f)$ are equal, and are equal to the Fourier transform of $\mathrm{sinc}(6\tau)^2$, shown in Figure 12.6. The cross spectral density S_{UV} is zero since the upper lobe of S_X is symmetric about 30Hz.

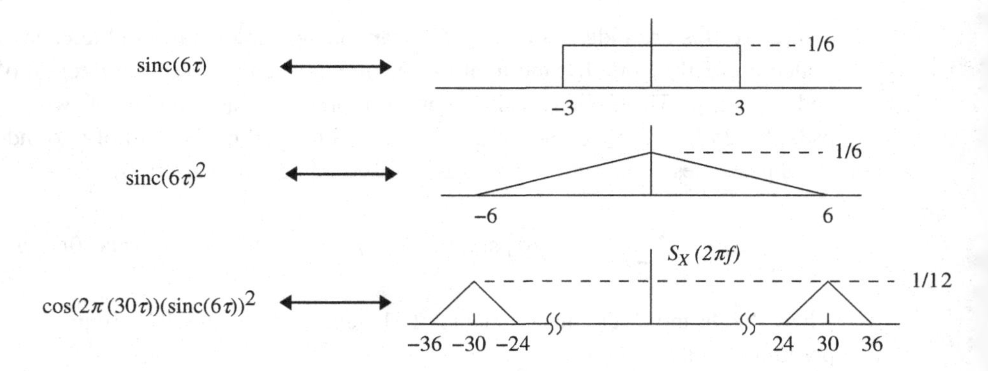

Figure 12.6 Taking the Fourier transform.

(d) The real envelope process is given by $|Z_t| = \sqrt{U_t^2 + V_t^2}$. For t fixed, U_t and V_t are independent $N(0, 1)$ random variables. The processes U and V have unit power since their power spectral densities integrate to one. The variables U_t and V_t for t fixed are uncorrelated even if $S_{XY} \neq 0$, since R_{XY} is an odd function. Thus $|Z_t|$ has the Rayleigh density with $\sigma^2 = 1$. Hence

$$P(|Z_{33}| \geq 5) = \int_5^\infty \frac{r}{\sigma^2} e^{-r\frac{2}{2\sigma^2}} dr = e^{-\frac{5^2}{2\sigma^2}} = e^{-\frac{25}{2}} = 3.7 \times 10^{-6}.$$

8.26 Another narrowband Gaussian process (version 2) (a) Since the white noise has mean zero, so does X, and

$$S_X(2\pi f) = \frac{N_o}{2}|H(2\pi f)|^2 = \begin{cases} \frac{N_o}{2} & 19.10 \leq |f| \leq 19.11 \\ \frac{N_o}{2}\frac{19.12-|f|}{0.01} & 19.11 \leq |f| \leq 19.12 \\ 0 & \text{else} \end{cases}.$$

(b) For any t, X_t is a real valued $N(0, \sigma^2)$ random variable with σ^2 = (the power of X) $= \int_{-\infty}^{\infty} S_X(2\pi f) df = \frac{3N_o}{2} \times 10^7$. So $P\{X_{25} > 2\} = Q(2/\sigma) = Q\left(2/\sqrt{\frac{3N_o}{2} \times 10^7}\right)$.
(c) See Figure 12.8.
(d) For t fixed, the real and imaginary parts of Z_t are independent, $N(0, \sigma^2)$ random variables. So by definition, Z_t is a proper complex normal random variable with variance $2\sigma^2$. It follows that the real envelope $|Z_t|$ has the Rayleigh(σ^2) distribution, with density $f(r) = \frac{r}{\sigma^2} \exp(-\frac{r^2}{2\sigma^2})$ for $r \geq 0$.
9.2 A smoothing problem Write $\widehat{X}_5 = \int_0^3 g(s) Y_s ds + \int_7^{10} g(s) y_s ds$. The mean square error is minimized over all linear estimators if and only if $(X_5 - \widehat{X}_5) \perp Y_u$ for $u \in [0, 3] \cup [7, 10]$, or equivalently

$$R_{XY}(5, u) = \int_0^3 g(s) R_Y(s, u) ds + \int_7^{10} g(s) R_Y(s, u) ds \quad \text{for } u \in [0, 3] \cup [7, 10].$$

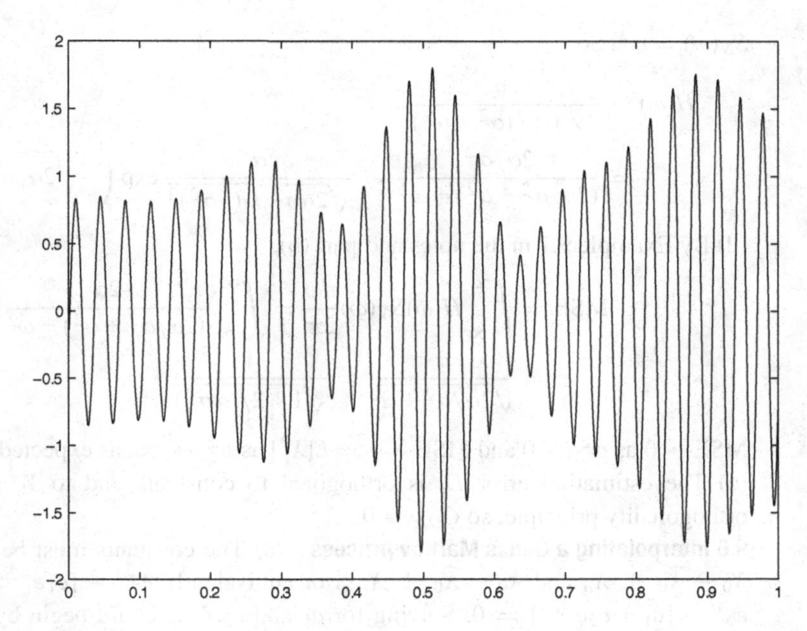

Figure 12.7 A sample path of X.

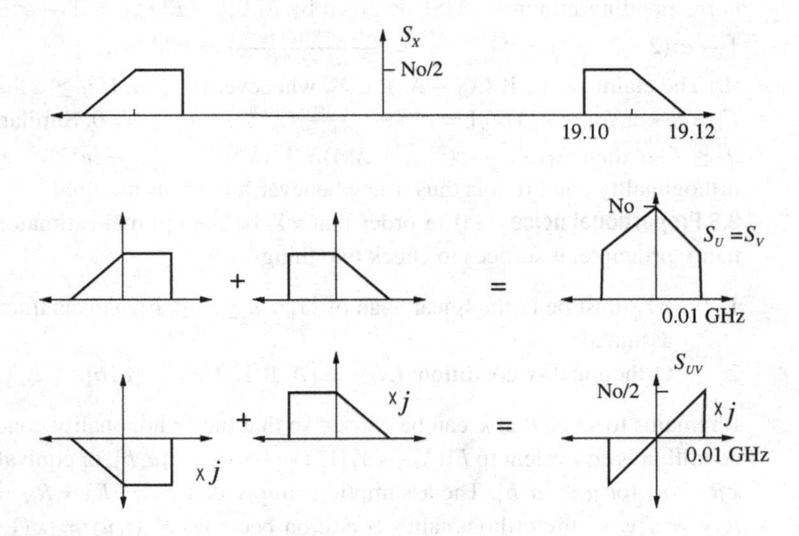

Figure 12.8 Problem 8.26c.

9.4 A standard noncausal estimation problem (a) $\widehat{g}(\omega) = \int_0^\infty g(t)e^{-j\omega t}dt + \int_{-\infty}^0 g(t)e^{-j\omega t}dt = \frac{1}{\alpha+j\omega} + \frac{1}{\alpha-j\omega} = \frac{2\alpha}{\omega^2+\alpha^2}$.
(So $\int_{-\infty}^\infty \frac{1}{\omega^2+\alpha^2}\frac{d\omega}{2\pi} = \frac{1}{2\alpha}$.)
(b) $\int_{-\infty}^\infty \frac{1}{a+b\omega^2}\frac{d\omega}{2\pi} = \int_{-\infty}^\infty \frac{1/b}{a/b+\omega^2}\frac{d\omega}{2\pi} = \frac{1/b}{2\sqrt{a/b}} = \frac{1}{2\sqrt{ab}}$.
(c) By Example 9.1 $H(\omega) = \frac{S_X(\omega)}{S_X(\omega)+S_N(\omega)}$. By the given and part (a), $S_X(\omega) = \frac{2\alpha}{\omega^2+\alpha^2}$ and

$S_N(\omega) = \sigma^2$. So

$$H(\omega) = \frac{2\alpha}{2\alpha + \sigma^2(\alpha^2+\omega^2)}$$
$$= \frac{2\alpha/\sigma^2}{(2\alpha/\sigma^2+\alpha^2)+\omega^2} \leftrightarrow \frac{\alpha}{\sqrt{2\alpha\sigma^2+(\alpha\sigma^2)^2}} \exp\left(-\sqrt{2\alpha/\sigma^2+\alpha^2}|t|\right).$$

(d) By Example 9.1 in the notes and part (b),

$$\text{MSE} = \int_{-\infty}^{\infty} H(\omega)S_N(\omega)\frac{d\omega}{2\pi} = \int_{-\infty}^{\infty} \frac{2\alpha}{(2\alpha/\sigma^2+\alpha^2)+\omega^2}\frac{d\omega}{2\pi}$$
$$= \frac{\alpha}{\sqrt{2\alpha/\sigma^2+\alpha^2}} = \frac{1}{\sqrt{1+2/(\alpha\sigma^2)}}.$$

MSE$\to 0$ as $\sigma^2 \to 0$ and MSE$\to 1 = E[X_t^2]$ as $\sigma^2 \to \infty$, as expected.
(e) The estimation error D_t is orthogonal to constants and to Y_s for all s by the orthogonality principle, so $C_{D,Y} \equiv 0$.
9.6 Interpolating a Gauss Markov process (a) The constants must be selected so that $X_0 - \widehat{X}_0 \perp X_a$ and $X_0 - \widehat{X}_0 \perp X_{-a}$, or equivalently $e^{-a} - [c_1e^{-2a} + c_2] = 0$ and $e^{-a} - [c_1 + c_2e^{-2a}] = 0$. Solving for c_1 and c_2 (one could begin by subtracting the two equations) yields $c_1 = c_2 = c$ where $c = \frac{e^{-a}}{1+e^{-2a}} = \frac{1}{e^a+e^{-a}} = \frac{1}{2\cosh(a)}$. The corresponding minimum MSE is given by $E[X_0^2] - E[\widehat{X}_0^2] = 1 - c^2E[(X_{-a}+X_a)^2] = 1 - c^2(2+2e^{-2a}) = \frac{e^{2a}-e^{-2a}}{(e^a+e^{-a})^2} = \frac{(e^a-e^{-a})(e^a+e^{-a})}{(e^a+e^{-a})^2} = \tanh(a)$.
(b) The claim is true if $(X_0 - \widehat{X}_0) \perp X_u$ whenever $|u| \geq a$. If $u \geq a$ then $E[(X_0 - c(X_{-a}+X_a))X_u] = e^{-u} - \frac{1}{e^a+e^{-a}}(e^{-a-u}+e^{a+u}) = 0$. Similarly if $u \leq -a$ then $E[(X_0 - c(X_{-a}+X_a))X_u] = e^u - \frac{1}{e^a+e^{-a}}(e^{a+u}+e^{-a+u}) = 0$. The orthogonality condition is thus true whenever $|u| \geq a$ as required.
9.8 Proportional noise (a) In order that κY_t be the optimal estimator, by the orthogonality principle, it suffices to check two things:

1. κY_t must be in the linear span of $(Y_u : a \leq u \leq b)$. This is true since $t \in [a,b]$ is assumed.
2. Orthogonality condition: $(X_t - \kappa Y_t) \perp Y_u$ for $u \in [a,b]$.

It remains to show that κ can be chosen so that the orthogonality condition is true. The condition is equivalent to $E[(X_t - \kappa Y_t)Y_u^*] = 0$ for $u \in [a,b]$, or equivalently $R_{XY}(t,u) = \kappa R_Y(t,u)$ for $u \in [a,b]$. The assumptions imply that $R_Y = R_X + R_N = (1+\gamma^2)R_X$ and $R_{XY} = R_X$, so the orthogonality condition becomes $R_X(t,u) = \kappa(1+\gamma^2)R_X(t,u)$ for $u \in [a,b]$, which is true for $\kappa = 1/(1+\gamma^2)$. The form of the estimator is proved. The MSE is given by $E[|X_t - \widehat{X}_t|^2] = E[|X_t|^2] - E[|\widehat{X}_t|]^2 = \frac{\gamma^2}{1+\gamma^2}R_X(t,t)$.
(b) Since S_Y is proportional to S_X, the factors in the spectral factorization of S_Y are proportional to the factors in the spectral factorization of X:

$$S_Y = (1+\gamma^2)S_X = \underbrace{\left(\sqrt{1+\gamma^2}S_X^+\right)}_{S_Y^+}\underbrace{\left(\sqrt{1+\gamma^2}S_X^-\right)}_{S_Y^-}.$$

That and the fact $S_{XY} = S_X$ imply that

$$H(\omega) = \frac{1}{S_Y^+}\left[\frac{e^{j\omega T}S_{XY}}{S_Y^-}\right]_+ = \frac{1}{\sqrt{1+\gamma^2}S_X^+}\left[\frac{e^{j\omega T}S_X^+}{\sqrt{1+\gamma^2}}\right]_+$$

$$= \frac{\kappa}{S_X^+(\omega)}\left[e^{j\omega T}S_X^+(\omega)\right]_+.$$

Therefore H is κ times the optimal filter for predicting X_{t+T} from $(X_s : s \le t)$. In particular, if $T < 0$ then $H(\omega) = \kappa e^{j\omega T}$, and the estimator of X_{t+T} is simply $\widehat{X}_{t+T|t} = \kappa Y_{t+T}$, which agrees with part (a).
(c) As already observed, if $T > 0$ then the optimal filter is κ times the prediction filter for X_{t+T} given $(X_s : s \le t)$.

9.10 Short answer filtering questions (a) The convolution of a causal function h with itself is causal, and H^2 has transform $h * h$. So if H is a positive type function then H^2 is positive type.
(b) Since the intervals of support of S_X and S_Y do not intersect, $S_X(2\pi f)S_Y(2\pi f) \equiv 0$. Since $|S_{XY}(2\pi f)|^2 \le S_X(2\pi f)S_Y(2\pi f)$ (by the first problem in Chapter 6) it follows that $S_{XY} \equiv 0$. Hence the assertion is true.
(c) Since sinc(f) is the Fourier transform of $I_{[-\frac{1}{2},\frac{1}{2}]}$, it follows that

$$[H]_+(2\pi f) = \int_0^{\frac{1}{2}} e^{-2\pi f j t}dt = \frac{1}{2}e^{-\pi j f/2}\text{sinc}\left(\frac{f}{2}\right).$$

9.12 A singular estimation problem (a) $E[X_t] = E[A]e^{j2\pi f_o t} = 0$, which does not depend on t. $R_X(s,t) = E[Ae^{j2\pi f_o s}(Ae^{j2\pi f_o t})^*] = \sigma_A^2 e^{j2\pi f_o(s-t)}$ is a function of $s-t$. Thus, X is WSS with $\mu_X = 0$ and $R_X(\tau) = \sigma_A^2 e^{j2\pi f_o \tau}$. Therefore, $S_X(2\pi f) = \sigma_A^2\delta(f - f_0)$, or equivalently, $S_X(\omega) = 2\pi\sigma_A^2\delta(\omega - \omega_0)$. (This makes $R_X(\tau) = \int_{-\infty}^{\infty} S_X(2\pi f)e^{j2\pi f\tau}df = \int_{-\infty}^{\infty} S_X(\omega)e^{j\omega\tau}\frac{d\omega}{2\pi}$.)
(b)

$$(h * X)_t = \int_{-\infty}^{\infty} h(\tau)X_{t-\tau}d\tau = \int_0^{\infty} \alpha e^{-(\alpha - j2\pi f_o)\tau}Ae^{j2\pi f_o(t-\tau)}d\tau$$

$$= \int_0^{\infty} \alpha e^{-(\alpha\tau)}d\tau Ae^{j2\pi f_o t} = X_t.$$

Another way to see this is to note that X is a pure tone sinusoid at frequency f_o, and $H(2\pi f_0) = 1$.
(c) In view of part (b), the mean square error is the power of the output due to the noise, or MSE $= (h * \widetilde{h} * R_N)(0) = \int_{-\infty}^{\infty}(h * \widetilde{h})(t)R_N(0 - t)dt = \sigma_N^2 h * \widetilde{h}(0) = \sigma_N^2||h||^2 = \sigma_N^2\int_0^{\infty}\alpha^2 e^{-2\alpha t}dt = \frac{\sigma_N^2\alpha}{2}$. The MSE can be made arbitrarily small by taking α small enough. That is, the minimum mean square error for estimation of X_t from $(Y_s : s \le t)$ is zero. Intuitively, the power of the signal X is concentrated at a single frequency, while the noise power in a small interval around that frequency is small, so that perfect estimation is possible.

9.14 A prediction problem The optimal prediction filter is given by $\frac{1}{S_X^+}\left[e^{j\omega T}S_X^+\right]$. Sinc $R_X(\tau) = e^{-|\tau|}$, the spectral factorization of S_X is given by

$$S_X(\omega) = \underbrace{\left(\frac{\sqrt{2}}{j\omega+1}\right)}_{S_X^+}\underbrace{\left(\frac{\sqrt{2}}{-j\omega+1}\right)}_{S_X^-},$$

so $[e^{j\omega T}S_X^+]_+ = e^{-T}S_X^+$ (see Figure 12.8). Thus the optimal prediction filter i $H(\omega) \equiv e^{-T}$, or in the time domain it is $h(t) = e^{-T}\delta(t)$, so that $\widehat{X}_{T+t|t} = e^{-T}X_t$. Thi simple form can be explained and derived another way. Since linear estimation is bein considered, only the means (assumed zero) and correlation functions of the processe matter. We can therefore assume without loss of generality that X is a real-value Gaussian process. By the form of R_X we recognize that X is Markov so the best estimat of X_{T+t} given $(X_s : s \le t)$ is a function of X_t alone. Since X is Gaussian with mean zerc the optimal estimator of X_{t+T} given X_t is $E[X_{t+T}|X_t] = \frac{\mathrm{Cov}(X_{t+T},X_t)X_t}{\mathrm{Var}(X_t)} = e^{-T}X_t$.

9.16 Spectral decomposition and factorization (a) Building up transform pairs b steps yields:

$$\begin{aligned}
\mathrm{sinc}(f) &\leftrightarrow I_{\{-\frac{1}{2}\le t\le\frac{1}{2}\}} \\
\mathrm{sinc}(100f) &\leftrightarrow 10^{-2}I_{\{-\frac{1}{2}\le\frac{t}{100}\le\frac{1}{2}\}} \\
\mathrm{sinc}(100f)e^{2\pi jfT} &\leftrightarrow 10^{-2}I_{\{-\frac{1}{2}\le\frac{t+T}{100}\le\frac{1}{2}\}} \\
\left[\mathrm{sinc}(100f)e^{j2\pi fT}\right]_+ &\leftrightarrow 10^{-2}I_{\{-50-T\le t\le 50-T\}\cap\{t\ge 0\}},
\end{aligned}$$

so

$$\begin{aligned}
||x||^2 &= 10^{-4}\text{length of } ([-50-T, 50-T]\cap[0,+\infty)) \\
&= \begin{cases} 10^{-2} & T\le -50 \\ 10^{-4}(50-T) & -50\le T\le 50 \\ 0 & T\ge 50 \end{cases}.
\end{aligned}$$

(b) By the hint, $1+3j$ is a pole of S. (Without the hint, the poles can be found by firs solving for values of ω^2 for which the denominator of S is zero.) Since S is real valued $1-3j$ must also be a pole of S. Since S is an even function, i.e. $S(\omega) = S(-\omega)$, $-(1+3j$

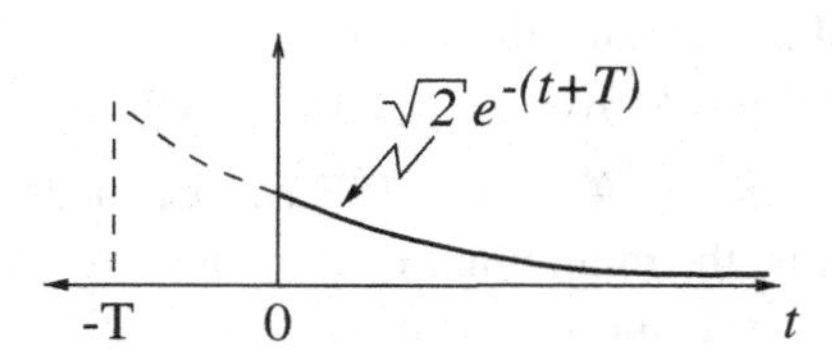

Figure 12.8 $\sqrt{2}e^{j\omega T}S_X^+$ in the time domain.

and $-(1-3j)$ must also be poles. Indeed, we find

$$S(\omega) = \frac{1}{(\omega-(1+3j))(\omega-(1-3j))(\omega+1+3j)(\omega+1-3j)},$$

or, multiplying each term by j (and using $j^4 = 1$) and rearranging terms:

$$S(\omega) = \underbrace{\frac{1}{(j\omega+3+j)(j\omega+3-j)}}_{S^+(\omega)} \underbrace{\frac{1}{(-j\omega+3+j)(-j\omega+3-j)}}_{S^-(\omega)},$$

or $S^+(\omega) = \frac{1}{(j\omega^2)+6j\omega+10}$. The choice of S^+ is unique up to a multiplication by a unit magnitude constant.

9.18 Estimation of a random signal, using the KL expansion (a) Note that $(Y,\varphi_j) = (X,\varphi_j) + (N,\varphi_j)$ for all j, where the variables $(X,\varphi_j), j \geq 1$ and $(N,\varphi_j), j \geq 1$ are all mutually orthogonal, with $E[|(X,\varphi_j)|^2] = \lambda_j$ and $E[|(N,\varphi_j)|^2] = \sigma^2$. Observation of the process Y is linearly equivalent to observation of $((Y,\varphi_j) : j \geq 1)$. Since these random variables are orthogonal and all random variables are mean zero, the MMSE estimator is the sum of the projections onto the individual observations, (Y,φ_j). But for fixed i, only the ith observation, $(Y,\varphi_i) = (X,\varphi_i) + (N,\varphi_i)$, is not orthogonal to (X,φ_i). Thus, the optimal linear estimator of (X,φ_i) given Y is $\frac{\mathrm{Cov}((X,\varphi_i),(Y,\varphi_i))}{\mathrm{Var}((Y,\varphi_i))}(Y,\varphi_i) = \frac{\lambda_i(Y,\varphi_i)}{\lambda_i+\sigma^2}$. The mean square error is (using the orthogonality principle): $E[|(X,\varphi_i)|^2] - E[|\frac{\lambda_i(Y,\varphi_i)}{\lambda_i+\sigma^2}|^2] = \lambda_i - \frac{\lambda_i^2(\lambda_i+\sigma^2)}{(\lambda_i+\sigma^2)^2} = \frac{\lambda_i\sigma^2}{\lambda_i+\sigma^2}$.
(b) Since $f(t) = \sum_j (f,\varphi_j)\varphi_j(t)$, we have $(X,f) = \sum_j (f,\varphi_j)(X,\varphi_j)$. That is, the random variable to be estimated is the sum of the random variables of the form treated in part (a). Thus, the best linear estimator of (X,f) given Y can be written as the corresponding weighted sum of linear estimators:

$$(\text{MMSE estimator of } (X,f) \text{ given } Y) = \sum_i \frac{\lambda_i(Y,\varphi_i)(f,\varphi_i)}{\lambda_i+\sigma^2}.$$

The error in estimating (X,f) is the sum of the errors for estimating the terms $(f,\varphi_j)(X,\varphi_j)$, and those errors are orthogonal. Thus, the mean square error for (X,f) is the sum of the mean square errors of the individual terms:

$$(\text{MSE}) = \sum_i \frac{\lambda_i\sigma^2|(f,\varphi_i)|^2}{\lambda_i+\sigma^2}.$$

9.20 Linear innovations and spectral factorization *First approach*: The first approach is motivated by the fact that $\frac{1}{S_Y^+}$ is a whitening filter. Let $\mathcal{H}(z) = \frac{\beta}{S_X^+(z)}$ and let Y be the output when X is passed through a linear time-invariant system with z-transform $\mathcal{H}(z)$. We prove that Y is the innovations process for X. Since $\mathcal{H}$ is positive type and $\lim_{|z|\to\infty}\mathcal{H}(z) = 1$, it follows that $Y_k = X_k + h(1)X_{k-1} + h(2)X_{k-2} + \ldots$ Since $\mathcal{S}_Y(z) = \mathcal{H}(z)H^*(1/z^*)\mathcal{S}_X(z) \equiv \beta^2$, it follows that $R_Y(k) = \beta^2 I_{\{k=0\}}$. In particular,

$$Y_k \perp \text{linear span of } \{Y_{k-1}, Y_{k-2}, \ldots\}.$$

Since $\mathcal{H}$ and $1/\mathcal{H}$ both correspond to causal filters, the linear span of $\{Y_{k-1}, Y_{k-2}, \ldots\}$ is the same as the linear span of $\{X_{k-1}, X_{k-2}, \ldots\}$. Thus, the above orthogonality condition becomes,

$$X_k - (-h(1)X_{k-1} - h(2)X_{k-2} - \ldots) \perp \text{linear span of } \{X_{k-1}, X_{k-2}, \ldots\}.$$

Therefore $-h(1)X_{k-1} - h(2)X_{k-2} - \ldots$ must equal $\widehat{X}_{k|k-1}$, the one-step predictor for X_k. Thus, (Y_k) is the innovations sequence for (X_k). The one-step prediction error is $E[|Y_k|^2] = R_Y(0) = \beta^2$.

Second approach: The filter $\mathcal{K}$ for the optimal one-step linear predictor $(\widehat{X}_{k+1|k})$ is given by (take $T = 1$ in the general formula):

$$\mathcal{K} = \frac{1}{\mathcal{S}_X^+}\left[z\mathcal{S}_X^+\right]_+.$$

The z-transform $z\mathcal{S}_X^+$ corresponds to a function in the time domain with value β at time -1, and value zero at all other negative times, so $[z\mathcal{S}_X^+]_+ = z\mathcal{S}_X^+ - z\beta$. Hence $\mathcal{K}(z) = z - \frac{z\beta}{\mathcal{S}_X^+(z)}$. If X is filtered using $\mathcal{K}$, the output at time k is $\widehat{X}_{k+1|k}$. So if X is filtered using $1 - \frac{\beta}{\mathcal{S}_X^+(z)}$, the output at time k is $\widehat{X}_{k|k-1}$. So if X is filtered using $\mathcal{H}(z) = 1 - (1 - \frac{\beta}{\mathcal{S}_X^+(z)}) = \frac{\beta}{\mathcal{S}_X^+(z)}$ then the output at time k is $X_k - \widehat{X}_{k|k-1} = \widetilde{X}_k$, the innovations sequence. The output $\widetilde{X}$ has $\mathcal{S}_{\widetilde{X}}(z) \equiv \beta^2$, so the prediction error is $R_{\widetilde{X}}(0) = \beta^2$.

9.22 A discrete-time Wiener filtering problem To begin,

$$\frac{z^T S_{XY}(z)}{S_Y^-(z)} = \frac{z^T}{\beta(1-\rho/z)(1-z_o\rho)} + \frac{z^{T+1}}{\beta(\frac{1}{z_o}-\rho)(1-z_o z)}.$$

The righthand side corresponds in the time domain to the sum of an exponential function supported on $-T, -T+1, -T+2, \ldots$ and an exponential function supported on $-T-1, -T-2, \ldots$ If $T \geq 0$ then only the first term contributes to the positive part, yielding

$$\left[\frac{z^T S_{XY}}{S_Y^-}\right]_+ = \frac{z_o^T}{\beta(1-\rho/z)(1-z_o\rho)},$$

$$\mathcal{H}(z) = \frac{z_o^T}{\beta^2(1-z_o\rho)(1-z_o/z)} \quad \text{and} \quad h(n) = \frac{z_o^T}{\beta^2(1-z_o\rho)} z_o^n I_{\{n\geq 0\}}.$$

On the other hand if $T \leq 0$ then

$$\left[\frac{z^T S_{XY}}{S_Y^-}\right]_+ = \frac{z^T}{\beta(1-\rho/z)(1-z_o\rho)} + \frac{z(z^T - z_o^T)}{\beta(\frac{1}{z_o}-\rho)(1-z_o z)},$$

so

$$\mathcal{H}(z) = \frac{z^T}{\beta^2(1-z_o\rho)(1-z_o/z)} + \frac{z(z^T-z_o^T)(1-\rho/z)}{\beta^2(\frac{1}{z_o}-\rho)(1-z_o z)(1-z_o/z)}.$$

Inverting the z-transforms and arranging terms yields that the impulse response function for the optimal filter is given by

$$h(n) = \frac{1}{\beta^2(1-z_o^2)}\left\{ z_o^{|n+T|} - \left(\frac{z_o - \rho}{\frac{1}{z_o} - \rho}\right) z_o^{n+T} \right\} I_{\{n \geq 0\}}. \qquad (12.20)$$

Graphically, h is the sum of a two-sided symmetric exponential function, slid to the right by $-T$ and set to zero for negative times, minus a one-sided exponential function on the nonnegative integers. (This structure can be deduced by considering that the optimal causal estimator of X_{t+T} is the optimal causal estimator of the optimal noncausal estimator of X_{t+T}.) Going back to the z-transform domain, we find that $\mathcal{H}$ can be written as

$$\mathcal{H}(z) = \left[\frac{z^T}{\beta^2(1-z_o/z)(1-z_o z)}\right]_+ - \frac{z_o^T(z_o - \rho)}{\beta^2(1-z_o^2)(\frac{1}{z_o} - \rho)(1 - z_o/z)}. \qquad (12.21)$$

Although it is helpful to think of the cases $T \geq 0$ and $T \leq 0$ separately, interestingly enough, the expressions (12.20) and (12.21) for the optimal h and H hold for any integer value of T.

9.24 Estimation given a strongly correlated process

(a)

$$R_X = g * \widetilde{g} \leftrightarrow \mathcal{S}_X(z) = \mathcal{G}(z)\mathcal{G}^*(1/z^*),$$
$$R_Y = k * \widetilde{k} \leftrightarrow \mathcal{S}_Y(z) = \mathcal{K}(z)\mathcal{K}^*(1/z^*),$$
$$R_{XY} = g * \widetilde{k} \leftrightarrow \mathcal{S}_{XY}(z) = \mathcal{G}(z)\mathcal{K}^*(1/z^*).$$

(b) Note that $\mathcal{S}_Y^+(z) = \mathcal{K}(z)$ and $\mathcal{S}_Y^-(z) = \mathcal{K}^*(1/z^*)$. By the formula for the optimal causal estimator,

$$\mathcal{H}(z) = \frac{1}{\mathcal{S}_Y^+}\left[\frac{\mathcal{S}_{XY}}{\mathcal{S}_Y^-}\right]_+ = \frac{1}{\mathcal{K}(z)}\left[\frac{\mathcal{G}(z)\mathcal{K}^*(1/z^*)}{\mathcal{K}^*(1/z^*)}\right]_+ = \frac{[\mathcal{G}]_+}{\mathcal{K}} = \frac{\mathcal{G}(z)}{\mathcal{K}(z)}.$$

(c) The power spectral density of the estimator process $\widehat{X}$ is given by $\mathcal{H}(z)\mathcal{H}^*(1/z^*)\mathcal{S}_Y(z) = \mathcal{S}_X(z)$. Therefore, MSE $= R_X(0) - R_{\widehat{X}}(0) = \int_{-\pi}^{\pi} \mathcal{S}_X(e^{j\omega})\frac{d\omega}{2\pi} - \int_{-\pi}^{\pi} \mathcal{S}_{\widehat{X}}(e^{j\omega})\frac{d\omega}{2\pi} = 0$. A simple explanation for why the MSE is zero is the following. Using $\frac{1}{\mathcal{K}}$ inverts $\mathcal{K}$, so that filtering Y with $\frac{1}{\mathcal{K}}$ produces the process W. Filtering that with $\mathcal{G}$ then yields X. That is, filtering Y with $\mathcal{H}$ produces X, so the estimation error is zero.

9.26 Linear and nonlinear filtering (a) The equilibrium distribution π is the solution to $\pi Q = 0$; $\pi = (0.25, 0.25, 0.25, 0.25)$. Thus, for each t, Z_t takes each of its possible values with probability 0.25. In particular, $\mu_Z = \sum_{i \in \mathcal{S}}(0.25)i = (0.25)(3+1-1-3) = 0$. The Kolmogorov forward equation $\pi'(t) = \pi(t)Q$ and the fact $\sum_i \pi_i(t) = 1$ for all t yield $\pi_i'(t) = -3\lambda\pi_i(t) + \lambda(1 - \pi_i(t)) = -4\lambda\pi_i(t) + \lambda$ for each state i. Thus, $\pi(t) = \pi + (\pi(0) - \pi)e^{-4\lambda t}$. Considering the process starting in state i yields

$p_{i,j}(\tau) = 0.25 + (\delta_{i,j} - 0.25)e^{-4\lambda\tau}$. Therefore, for $\tau \geq 0$,

$$\begin{aligned} R_Z(\tau) = E[Z(\tau)Z(0)] &= \sum_{i\in\mathcal{S}}\sum_{j\in\mathcal{S}} ij\pi_i p_{i,j}(\tau) \\ &= (0.25)\sum_{i\in\mathcal{S}}\sum_{j\in\mathcal{S}} ij\delta_{i,j}e^{-4\lambda\tau} + (0.25)^2(1-e^{-4\lambda\tau})\underbrace{\sum_{i\in\mathcal{S}}\sum_{j\in\mathcal{S}} ij}_{=0} \\ &= (0.25)((-3)^2 + (-1)^2 + 1^2 + 3^2)e^{-4\lambda\tau} = 5e^{-4\lambda\tau}. \end{aligned}$$

So $R_Z(\tau) = 5e^{-4\lambda|\tau|}$.
(b) Thus, $S_Z(\omega) = \frac{40\lambda}{16\lambda^2+\omega^2}$. Also, $S_{YZ} = S_Z$. Thus, the optimal filter is given by

$$H_{\text{opt}}(\omega) = \frac{S_Z(\omega)}{S_Z(\omega) + S_N(\omega)} = \frac{40\lambda}{40\lambda + \sigma^2(16\lambda^2 + \omega^2)}.$$

The MSE is given by MSE$= \int_{-\infty}^{\infty} H_{\text{opt}}(\omega)S_N(\omega)\frac{d\omega}{2\pi} = \frac{5}{\sqrt{\frac{5}{2\lambda\sigma^2}+1}}$.
(c) It is known that $P\{|Z_t| \leq 3\} = 1$, so by hard limiting the estimator found in (b) to the interval $[-3, 3]$, a smaller MSE results. That is, let

$$\widehat{Z}_t^{(NL)} = \begin{cases} 3 & \text{if } \widehat{Z}_t \geq 3 \\ \widehat{Z}_t & \text{if } |\widehat{Z}_t| \leq 3 \\ -3 & \text{if } \widehat{Z}_t \leq -3 \end{cases}.$$

Then $(Z_t - \widehat{Z}_t^{(NL)})^2 \leq (Z_t - \widehat{Z}_t)^2$, and the inequality is strict on the positive probability event $\{|\widehat{Z}_t| \geq 3\}$.
(d) The initial distribution π for the hidden Markov model should be the equilibrium distribution, $\pi = (0.25, 0.25, 0.25, 0.25)$. By the definition of the generator matrix Q, the one-step transition probabilities for a length Δ time step are given by $p_{i,j}(\Delta) = \delta_{i,j} + q_{i,j}\Delta + o(\Delta)$. So we ignore the $o(\Delta)$ term and let $a_{i,j} = \lambda\Delta$ if $i \neq j$ and $a_{i,i} = 1 - 3\lambda\Delta$ for $i,j \in \mathcal{S}$. (*Alternatively*, we could let $a_{i,j} = p_{i,j}(\Delta)$, that is, use the exact transition probability matrix for time duration Δ.) If Δ is small enough, then Z will be constant over most of the intervals of length Δ. Given $Z = i$ over the time interval $[(k-1)\Delta, k\Delta]$, $\widetilde{Y}_k = \Delta i + \int_{(k-1)\Delta}^{k\Delta} N_t dt$ which has the $N(\Delta i, \Delta\sigma^2)$ distribution. Thus, we set $b_{i,y} = \frac{1}{\sqrt{2\pi\Delta\sigma^2}}\exp\left(-\frac{(y-i\Delta)^2}{2\Delta\sigma^2}\right)$.
10.2 A covering problem (a) Let X_i denote the location of the ith base station. Then $F = f(X_1, \ldots, X_m)$, where f satisfies the Lipschitz condition with constant $(2r-1)$. Thus, by the method of bounded differences based on the Azuma–Hoeffding inequality, $P\{|F - E[F]| \geq \gamma\} \leq 2\exp(-\frac{\gamma^2}{m(2r-1)^2})$.
(b) Using the Poisson method and associated bound technique, we compare to the case that the number of stations has a Poisson distribution with mean m. Note that the mean number of stations that cover cell i is $\frac{m(2r-1)}{n}$, unless cell i is near one of the boundaries. If cells 1 and n are covered, then all the other cells within distance r of either boundary are covered. Thus,

$$\begin{aligned} P\{X \geq m\} &\leq 2P\{\text{Poi}(m) \text{ stations is not enough}\} \\ &\leq 2ne^{-m(2r-1)/n} + P\{\text{cell 1 or cell } n \text{ is not covered}\} \\ &\to 0 \text{ as } n \to \infty \quad \text{if } m = \frac{(1+\epsilon)n \ln n}{2r-1}. \end{aligned}$$

For a bound going the other direction, note that if cells differ by $2r - 1$ or more then the events that they are covered are independent. Hence,

$$\begin{aligned} P\{X \leq m\} &\leq 2P\{\text{Poi}(m) \text{ stations cover all cells}\} \\ &\leq 2P\left\{\text{Poi}(m) \text{ stations cover cells } 1 + (2r-1)j, 1 \leq j \leq \frac{n-1}{2r-1}\right\} \\ &\leq 2\left(1 - e^{-\frac{m(2r-1)}{n}}\right)^{\frac{n-1}{2r-1}} \\ &\leq 2\exp\left(-e^{-\frac{m(2r-1)}{n}} \cdot \frac{n-1}{2r-1}\right) \\ &\to 0 \text{ as } n \to \infty \quad \text{if } m = \frac{(1-\epsilon)n \ln n}{2r-1}. \end{aligned}$$

Thus, in conclusion, we can take $g_1(r) = g_2(r) = \frac{1}{2r-1}$.

10.4 Stopping time properties (a) Suppose S and T are optional stopping times for some filtration $\mathcal{F}$. Then it suffices to note that:
$\{S \wedge T \leq n\} = \{S \leq n\} \cup \{T \leq n\} \in \mathcal{F}_n$, and
$\{S \vee T \leq n\} = \{S \leq n\} \cap \{T \leq n\} \in \mathcal{F}_n$.
$\{S + T \leq n\} = \cup_{0 \leq k \leq n} \{S \leq k\} \cap \{T \leq n - k\} \in \mathcal{F}_n$.
(b) Since X takes on values 0 and 1 only, events of the form $\{X_n \leq c\}$ are either empty or the whole probability space if $c < 0$ or if $c \geq 1$, so we can ignore such values of c. If $0 \leq c < 1$ and $n \geq 0$, then $\{X_n \leq c\} = \{T > n\}$. Thus, for each n $\{X_n \leq c\} \in \mathcal{F}_n$ if and only if $\{T \leq n\} \in \mathcal{F}_n$. Therefore, T is a stopping time if and only if X is adapted.
(c) (i)

A.1 $\emptyset \cap \{T \leq n\} = \emptyset \in \mathcal{F}_n$ for all n, so that $\emptyset \in \mathcal{F}_T$.
A.2 If $A \in \mathcal{F}^T$ then $A \cap \{T \leq n\} \in \mathcal{F}_n$ for all n. Also, $\{T \leq n\} \in \mathcal{F}_n$.
So $[A \cap \{T \leq n\}]^c \cap \{T \leq n\} = A^c \cap \{T \leq n\} \in \mathcal{F}_n$ for all n. Therefore, $A^c \in \mathcal{F}^T$.
A.3 If $A_i \in \mathcal{F}^T$ for all $i \geq 1$, then $A_i \cap \{T \leq n\} \in \mathcal{F}^n$ for all i, n. Therefore
$\cap_i (A_i \cap \{T \leq n\}) = (\cap_i A_i) \cap \{T \leq n\} \in \mathcal{F}_n$ for all n. Therefore, $\cap_i A_i \in \mathcal{F}_n$.

Thus $\mathcal{F}_T$ satisfies all three axioms of a σ-algebra so it is a σ-algebra.
(ii) To show that T is measurable with respect to a σ-algebra, we need events of the form $\{T \leq m\}$ to be in the σ-algebra, for any $m \geq 0$. For this event to be in $\mathcal{F}_T$, we need $\{T \leq m\} \cap \{T \leq n\} \in \mathcal{F}_n$ for any $n \geq 0$. But $\{T \leq m\} \cap \{T \leq n\} = \{T \leq m \wedge n\} \in \mathcal{F}_{m \wedge n} \in \mathcal{F}_n$, as desired.
(iii) Fix a constant c. Then for $n \geq 0$, $\{X_T \leq c\} \cap \{T = n\} = \{X_n \leq c\} \cap \{T = n\} \in \mathcal{F}_n$. Therefore, the event $\{X_T \leq c\}$ is in $\mathcal{F}_T$. Since c is arbitrary, X_T is $\mathcal{F}_T$ measurable.

10.6 Bounding the value of a game Let $X_t = (R_t, B_t)$, where R_t denotes the number of red marbles in the jar after t turns and B_t denotes the number of blue marbles in the jar after t turns, let u_t denote the decision taken by the player at the beginning of turn $t + 1$, and let $\mathcal{F}_t = \sigma(X_0, \ldots, X_t, u_0, \ldots, u_t)$. Then X is a controlled Markov process relative to the filtration $\mathcal{F}$.

(a) Suppose an initial state (r_o, b_o) and strategy $(u_t : t \geq 0)$ are fixed. Let $N_t = R_t + B_t$ (or equivalently, $N = V(X_t)$ for the potential function $V(r, b) = r + b$). Note that $E[N_{t+1} - N_t|\mathcal{F}_t] \geq -\frac{1}{2}$ for all t. Therefore the process M defined by $M_t = N_t + \frac{t}{2}$ is a submartingale relative to $\mathcal{F}$. Observe that $|M_{t+1} - M_t| \leq 2$, so that $E[|M_{t+1} - M_t||\mathcal{F}_t] \leq 2$. If $E[\tau] = +\infty$ then any lower bound on $E[\tau]$ is valid, so we can and do assume without loss of generality that $E[\tau] < \infty$. Therefore, by a version of the optional stopping theorem, $E[M_\tau] \geq E[M_0]$. But $M_\tau = \frac{\tau}{2}$ and $M_0 = a_o + b_o$. Thus, we find that $E[\tau] \geq 2(a_o + b_o)$ for any strategy of the player.

(b) Consider the strategy that selects two balls of the same color whenever possible. Let $V(X_t) = f(N_t)$ where $f(0) = 0$, $f(1) = 3$, and $f(n) = n + 3$ for $n \geq 2$. The function V was selected so that $E[V(X_{t+1}) - V(X_t)|\mathcal{F}_t] \leq -\frac{1}{2}$ whenever $X_t \neq (0, 0)$. Therefore, M is a supermartingale, where $M_t = V(X_{t\wedge\tau}) + \frac{t\wedge\tau}{2}$. Consequently, $E[M_t] \leq E[M_0]$ for all $t \geq 0$. That is, $E[V(X_{t\wedge\tau})] + E[\frac{t\wedge\tau}{2}] \leq 2f(a_o + b_o)$. Using this and the facts $E[V(X_{t\wedge\tau})] \geq 0$ and $f(a_o + b_o) \leq 3 + a_o + b_o$ yields that $E[t \wedge \tau] \leq 2(a_o + b_o) + 6$. Finally, $E[t \wedge \tau] \to E[\tau]$ as $t \to \infty$ by the monotone convergence theorem, so that $E[\tau] \leq 2(a_o + b_o) + 6$ for the specified strategy of the player.

References

Asmussen, S. (2003), *Applied Probability and Queues*, second edn, Springer, New York.

Baum, L., Petrie, T., Soules, G., & Weiss, N. (1970), "A maximization technique occurring in the statistical analysis of probabilisitic functions of Markov chains," *Ann. Math. Statist.* **41**, 164–171.

Dempster, A., Laird, N., & Rubin, B. (1977), "Maximum likelihood from incomplete data via the EM algorithm," *J. Royal Statistical Society* **39**(1), 1–38.

Doob, J. (1953), *Stochastic Processes*, Wiley, New York.

Foster, F. (1953), "On the stochastic matrices associated with certain queueing processes," *Ann. Math. Statist* **24**, 355–360.

Hajek, B. (2006), "Notes for ECE 567: Communication network analysis." Unpublished.

Kingman, J. (1962), "Some inequalities for the queue GI/G/1," *Biometrika* **49**(3/4), 315–324.

Kumar, P. & Meyn, S. (1995), "Stability of queueing networks and scheduling policies," *IEEE Trans. on Automatic Control* **40**(2), 251–260.

McDiarmid, C. (1989), "On the method of bounded differences," *Surveys in Combinatorics* **141**, 148–188.

McKeown, N., Mekkittikul, A., Anantharam, V., & Walrand, J. (1999), "Achieving 100% throughput in an input-queued switch," *IEEE Trans. Communications* **47**(8), 1260–1267.

Meyn, S. & Tweedie, R. (1993), *Markov Chains and Stochastic Stability*, Springer-Verlag, London.

Norris, J. (1997), *Markov Chains*, Cambridge University Press, Cambridge.

Rabiner, L. (1989), "A tutorial on hidden Markov models and selected applications in speech recognition," *Proceedings of the IEEE* **77**(2), 257–286.

Royden, H. (1968), *Real Analysis*, New York, Macmillan.

Tassiulas, L. (1997), "Scheduling and performance limits of networks with constantly-changing topology," *IEEE Trans. Information Theory* **43**(3), 1067–1073.

Tassiulas, L. & Ephremides, A. (1992), "Stability properties of constrained queueing systems and scheduling policies for maximum throughput in multihop radio networks," *IEEE Trans. on Automatic Control* **37**(12), 1936–1948.

Tassiulas, L. & Ephremides, A. (1993), "Dynamic server allocation to parallel queues with randomly varying connectivity," *IEEE Trans. Information Theory* **39**(2), 466–478.

Tweedie, R. (1983), "Existence of moments for stationary Markov chains," *Journal of Applied Probability* **20**(1), 191–196.

Varadhan, S. (2001), *Probability Theory Lecture Notes*, American Mathematical Society.

Wu, C. (1983), "On the convergence property of the EM algorithm," *The Annals of Statistics* **11**, 95–103.

Index

Printed in the United States
by Baker & Taylor Publisher Services

Printed in the United States
by Baker & Taylor Publisher Services